MATRIX AND FINITE ELEMENT ANALYSES OF STRUCTURES

(With Computer Programs in both FORTRAN and C)

MATRIX AND FINITE ELEMENT ANALYSES OF STRUCTURES

(With Computer Programs in both FORTRAN and C)

Madhujit Mukhopadhyay
B.E. (Civil), Ph. D. (Structures), D. Sc.
Former Dean and Professor
Indian Institute of Technology, Kharagpur

Abdul Hamid Sheikh
B.E. (Civil), M.E. (Structures), Ph. D. (Structures)
Associate Professor
Indian Institute of Technology, Kharagpur

Ane Books Pvt. Ltd.
New Delhi ♦ Chennai

Matrix and Finite Element Analyses of Structures

Madhujit Mukhopadhyay and Abdul Hamid Sheikh

Published by

Ane Books Pvt. Ltd.

4821, Parwana Bhawan, 1st Floor, 24 Ansari Road, Darya Ganj,
New Delhi - 110 002, Tel.: +91(011) 23276843-44, Fax: +91(011) 23276863
e-mail: kapoor@anebooks.com, Website: www.anebooks.com

Branch Office:

- Avantika Niwas, 1st Floor, 19 Doraiswamy Road, T. Nagar,
  Chennai - 600 017, Tel.: +91(044) 28141554, 28141209
  e-mail: anebookschennai@gmail.com, rathinam@anebooks.com

ISBN : 978-81-8052-075-4

Printed at : Sagar Colorscan, New Delhi.

To

Our Families

Ratna
Kuttus
Chuchu
Shaon
Arko

Daisy
Aniq

Preface

A revolutionary change has been brought about in the outlook of many fields of engineering and science with the advent of the electronic digital computer. Amongst them structural engineering forms an important discipline. With the innovations, improvements and rapid progress of the development of sophisticated computing systems, both in terms of hardware and software and with the fading away of the boundaries of different disciplines and the enlargement of the interdisciplinary work, never before has so much emphasis been laid in the computer aided analysis and design as it is today. The main objective of the book is to acquaint the engineers about the modern computer based techniques used in structural analysis.

This textbook is primarily written as a basic learning tool for senior undergraduate and postgraduate students in the departments dealing with the mechanics of solid and structural systems. The material of the book has been developed over a number of years of teaching senior undergraduate and postgraduate students of different disciplines of IIT, Kharagpur. The materials of the book have been presented in sufficiently simple manner. Basing on the knowledge of classical methods of structural analysis, the reader is introduced to 'Matrix and the Finite Element Methods' and then exposed to advanced topics in a more gradual manner. The book is self-contained in that most of the background materials required for the analysis of structures have been dealt with.

The presentation of the matter of the book is made as simple as possible so that engineers who are interested may use this for his own purpose. A large number of diagrams and illustrations make the text easy to understand. The book presents 'matrix', 'finite element', 'computer' and 'structural analysis' in a unified and integrated manner. The role of computers is of paramount importance to the present day structural analysis and design. Computer software and structural analysis have progressed side by side throughout the text as the matter presented in the text are primarily meant for large-sized structures. All the computer codes presented are in two languages — 'FORTRAN' and 'C'. The computer programs in 'C' are becoming quite popular now-a-days due to their versatility.

The senior author is grateful to his students – Dr Manoranjan Barik of National Institute of Technology, Rourkela and Dr K.S.R.K. Murthy of Indian Institute of Technology, Guwahati for the assistance rendered by them. Most of the computer programs in 'C' language have been developed by Mr Dhirendra Kumar Singh, an undergraduate student of the Department for which the authors are grateful to him.

The authors are very much thankful to Continuing Education Cell, IIT Kharagpur for providing financial support for the preparation of the manuscript.

Sincere thanks are due to Mr Parimal Kumar Roy for typing the manuscript and preparing a large number drawings. The authors are thankful to Chinmoy Mukherjee for preparing the necessary of drawings given to him. They also sincerely thank their wives Ratna and Daisy for the patience and forbearance shown by them during the preparation of the manuscript.

Dated, 25th June, 2004

Madhujit Mukhopadhyay
Abdul Hamid Sheikh

Contents

Preface *vii*

1. BASIC CONCEPTS OF STRUCTURAL ANALYSIS **1–12**

1.1 Types of structures 1
1.2 Objective of structural analysis 2
1.3 Materials and basic assumptions 3
1.4 Loads 3
1.5 General methods of analysis 4
1.5.1 Equilibrium conditions 4
1.5.2 Compatibility conditions 4
1.6 Force-displacement relationship 5
1.7 Statical indeterminacy 6
1.7.1 Plane structure 6
1.7.2 Space structures 8
1.8 Kinematic indeterminacy 9
1.9 Two approaches of structural analysis 10
Exercise 1 11

2. ENERGY PRINCIPLES **13–18**

2.1 Introduction 13
2.2 Principle of virtual work 13
2.3 Principle of complementary virtual work 14
2.4 Principle of minimum potential energy 15
2.5 Principle of minimum complementary energy 15
2.6 Castigliano's theorems 16
2.7 Determination of displacements 18
References and suggested readings 18

3. INTRODUCTION TO THE FLEXIBILITY AND STIFFNESS MATRIX METHODS 19–69

3.1 Introduction 19
3.2 The flexibility matrix method 19
3.3 The stiffness method 30
3.4 Incorporation of different loading conditions 48
3.5 Other types of loadings 49
3.5.1 Treatment by the flexibility matrix method 50
3.5.2 Treatment by the stiffness method 52
3.6 Incorporation of shear deformation 56
3.7 Relation between flexibility and stiffness matrices 57
3.8 Equivalent joint loads 59
3.9 Choice of the method of analysis 59
References and suggested readings 60
Exercise 3 61

4. DIRECT STIFFNESS METHOD 71–105

4.1 Introduction 71
4.2 Local and global coordinate system 71
4.3 Transformation of variables 72
4.3.1 Transformation of member coordinate axes 72
4.3.2 Transformation of member displacement matrix 73
4.3.3 Transformation of the member force matrix 74
4.3.4 Transformation of the member stiffness matrix 75
4.4 Transformation of the stiffness matrix of the member of a truss 75
4.5 Transformation of the stiffness matrix of the member of a rigid frame 77
4.6 Transformation of the stiffness matrix of the member of a grillage 80
4.7 Transformation of the stiffness matrix of the member of a space frame 81
4.8 Horizontally circular curved beam element 84
4.9 Overall stiffness matrix 85
4.10 Boundary conditions 90
4.10.1 Boundary conditions corresponding to skewed supports 90
4.11 Computation of internal forces 92
4.12 Computer program for the truss analysis by the direct stiffness method 92
4.13 Computer program for the frame analysis by direct stiffness method 98
4.14 Computer program for the grillage analysis by the direct stiffness method 100
References and suggested readings 103
Exercise 4 103

5. SUBSTRUCTURE TECHNIQUE FOR THE ANALYSIS OF STRUCTURAL SYSTEMS 107–124

5.1 Introduction 107
5.2 Basic concepts 107
5.3 Direct stiffness method restated 109
5.4 The substructure technique 110
5.5 An illustrative example 112
5.6 Computer program for the truss analysis by the substructure technique 117
References and suggested readings 122
Exercise 5 122

6. THE FLEXIBILITY MATRIX METHOD 125–138

6.1 Introduction 125
6.2 Element flexibility matrix 125
6.3 Principle of contragredience 126
6.4 The equilibrium matrix 127
6.5 Construction of the flexibility matrix of the structure 128
6.6 Matrix determination of the displacement vector 130
6.7 Determination of member forces 131
6.8 Procedure of the analysis of statically indeterminate structures 131
6.9 Illustrated examples 132
6.10 Choice of the released structure 136
References and suggested readings 137
Exercise 6 137

7. ELEMENTS OF ELASTICITY 139–153

7.1 Introduction 139
7.2 Some notations and relations in the theory of elasticity 139
7.2.1 Surface and body forces 139
7.2.2 Components of stresses 139
7.2.3 Components of strain 140
7.2.4 Stress-strain relationship 141
7.3 Two-dimensional problems 141
7.3.1 Plane stress 141
7.3.2 Plane strain 142
7.3.3 Differential equations of equilibrium 144
7.4 Bending of thin plates 145
7.4.1 Basic Assumptions 145

7.4.2 Deformation of the plate 145
7.4.3 Strain-displacement relationship 146
7.4.4 Stress-strain relationship 146
7.4.5 Equilibrium equations 148
7.4.6 Differential equation for deflection 149
7.4.7 Shearing forces 149

7.5 Boundary conditions 150
7.5.1 Simply supported edge 150
7.5.2 Clamped edge 150
7.5.3 Free edge 150
7.5.4 Elastically supported edge 152
7.5.5 Edge having elastic rotational restraint 152

7.6 Concluding remarks 153
References and suggested readings 153

8. INTRODUCTION TO THE FINITE ELEMENT METHOD 155–164

8.1 Introduction 155
8.2 The finite element method 155
8.3 Brief history of the development of the finite element method 156
8.4 Basic steps in the finite element method for the solution of static problems 158
8.5 Advantages and disadvantages of the finite element method 160
References and suggested readings 161

9. FINITE ELEMENT ANALYSIS OF PLANE ELASTICITY PROBLEMS 165–182

9.1 Introduction 165
9.2 Three-noded triangular element 165
9.2.1 Displacement function 166
9.2.2 Displacement function expressed in terms of nodal displacements 167
9.2.3 Strain–nodal parameter relationship 168
9.2.4 Stress-strain relationship 169
9.2.5 Derivation of the element stiffness matrix 169
9.2.6 Determination of element stresses 170

9.3 Criteria for the choice of the displacement function 171
9.4 Polynomial displacement functions 172
9.5 Verification of the convergence criteria of the displacement function of 3-noded triangular element 172

9.6 Number of terms in a polynomial 173
9.7 Four-noded rectangular element 173
9.7.1 Displacement function 174
9.7.2 Displacement function in terms of nodal displacements 175
9.7.3 Strain-nodal displacement relationship 176
9.7.4 Stress-strain relationship 176
9.7.5 Derivation of the element stiffness matrix 176
9.7.6 Evaluation of element stresses 177
9.8 A note on the rectangular element 177
9.9 A note on element stresses 177
9.10 Computer program for the plane stress analysis using three–noded triangular element 178
References and suggested reading 181
Exercise 9 181

10. ISOPARAMETRIC AND OTHER ELEMENT REPRESENTATIONS AND NUMERICAL INTEGRATIONS 183–216

10.1 Introduction 183
10.2 Shape function or interpolation function 183
10.3 Determination of shape functions 184
10.3.1 Linear 2-D element 185
10.3.2 Quadratic 2-D element 186
10.4 Plane stress isoparametric linear element 186
10.4.1 Displacement function in terms of nodal parameters 187
10.4.2 Strain-nodal parameter relationship 187
10.4.3 Evaluation of [B] matrix 188
10.4.4 Element stiffness matrix 189
10.4.5 Convergence of isoparametric elements 189
10.4.6 Concept of isoparametric element 190
10.5 Numerical integration 191
10.5.1 Gaussian quadrature formula 192
10.5.2 Gaussian integration of two variables 193
10.6 Lagrangian interpolation 195
10.7 Natural coordinates and higher order triangular elements 195
10.7.1 One-dimensional element 195
10.7.2 Higher order triangular elements 196
10.8 The quadratic triangle for the plane stress problem 197
10.9 Numerical integration of area coordinates 199

10.10 Triangular isoparametric elements for the analysis of plane stress problems 200

10.11 Allman's triangular plane stress element 201

10.12 Computer program for the solution of plane stress problem using isoparametric element 204

References and suggested readings 215

Exercise 10 216

11. FINITE ELEMENT ANALYSIS OF PLATE BENDING PROBLEMS 217–260

11.1 Introduction 217

11.2 Beam element 217

11.2.1 Displacement function 218

11.2.2 Displacement function in terms of nodal displacements 218

11.2.3 Strain (curvature) : nodal parameter relationship 219

11.2.4 Stress (moment) : strain (curvature) relationship 219

11.2.5 Derivation of the element stiffness matrix 220

11.2.6 Determination of equivalent loading on the beam 221

11.3 Rectangular plate bending element 222

11.3.1 Displacement function 222

11.3.1.1 Checking the continuity requirements of the displacement function 223

11.3.2 Displacement function expressed in terms of nodal Displacements 224

11.3.3 Strain-nodal parameter relationship 224

11.3.4 Stress (moment) – strain (curvature) relationship 225

11.3.5 Derivation of the element stiffness matrix 226

11.4 Parallelogram element of plate bending 228

11.4.1 Displacement function 228

11.4.2 Displacement function in terms of nodal displacements 229

11.4.3 Curvature-nodal parameter relationship 229

11.4.4 NMoment-curvature relationship 230

11.4.5 Element stiffness matrix 230

11.5 Hermitian polynomial interpolation 231

11.6 A conforming plate bending element 233

11.7 Isoparametric plate bending element 234

11.7.1 Displacement function 235

11.7.2 Strain-nodal displacement relationship 236

11.7.3 Stress-strain relationship 236

11.7.4 Element stiffness matrix 237
11.7.5 Reduced integration technique 237
11.7.5.1 Isoparametric quadratic element 237
11.7.5.2 Isoparametric linear element 238
11.8 Smoothed stresses 239
11.9 Triangular plate bending elements 239
11.10 DKT element 240
11.10.1 Constraint equations 242
11.10.2 Transformation matrix 242
11.10.3 Element stiffness matrix 245
11.11 The patch test 247
11.11.1 The patch test for the plane stress element 248
11.11.2 The patch test for plate bending elements 249
11.11.2.1 Prescribed displacement approach 249
11.11.2.2 Prescribed boundary load approach 250
11.12 Horizontally curved isoparametric beam 251
11.12.1 Displacement function in terms of nodal parameters 251
11.12.2 Stress-strain relations 252
11.12.3 Strain-displacement relationship 253
11.12.4 Element stiffness matrix 254
11.13 Non-uniform straight beam element 254
11.14 Computer program for isoparametric quadratic bending element 255
References and suggested readings 258
Exercise 11 259

12. FINITE ELEMENT ANALYSIS OF SHELLS 261–289
12.1 Introduction 261
12.2 Flat shell element 261
12.2.1 Transformation of the stiffness matrix and assembly 263
12.3 Shell of revolution 265
12.4 General shell finite element of triangular shape 267
12.4.1 Derivation of the stiffness matrix 268
12.4.2 Consistent load vector 275
12.4.3 Condensation of stiffness matrix 276
12.5 Isoparametric general shell element 276
12.5.1 Geometry of the shell element 277
12.5.2 Displacement field 278

12.5.3 Strains inside the element 280
12.5.4 Stress - strain relationship 282
12.5.5 Stiffness matrix of the shell element 283
12.6 Vertically curved beam element 283
12.7 Computer program for the finite element analysis of shallow shells of general shape using triangular element 285
References and suggested readings 288
Exercise 12 288

13. SEMI-ANALYTICAL AND SPLINE FINITE STRIP METHOD OF ANALYSES OF PLATE BENDING 291–312
13.1 Introduction 291
13.2 Beam function 292
13.3 Model of the plate 296
13.4 The displacement function 296
13.5 Curvature-nodal parameter relationship 297
13.6 Moment – curvature relationship 298
13.7 Strip stiffness matrix 298
13.8 Loading matrix 299
13.9 Force displacement relationship 300
13.10 Spline finite strip method of analysis of plate bending 302
13.10.1 The spline function 302
13.10.2 Displacement functions 304
13.10.3 Strain-displacement relationship 305
13.10.4 Stiffness matrix 305
13.10.5 The loading matrix 309
13.11 Computer program for the spline finite strip method of analysis of plates in bending 309
References and suggested readings 311
Exercise 13 311

14. DYNAMIC AND INSTABILITY ANALYSES BY THE FINITE ELEMENT METHOD 313–327
14.1 Introduction 313
14.2 Dynamic analysis 313
14.2.1 Torsional vibration of shafts 313
14.2.1.1 An example 315
14.2.2 Flexural vibration of beams 316

14.2.3 Inplane vibration of plates 317
14.2.4 Flexural vibration of plates 317
14.2.4.1 Vibration of isoparametric bending element 318
14.3 Elastic instability analysis 319
14.3.1 Column instability analysis 319
14.3.1.1 An illustrated example of a column 322
14.3.2 Plate instability analysis 323
References and suggested readings 325
Exercise 14 326

15. THE FINITE DIFFERENCE METHOD FOR THE ANALYSIS OF BEAMS AND PLATES 329–344

15.1 Introduction 329
15.2 Finite difference representation of derivatives 329
15.2.1 First derivative 330
15.2.2 Second derivative 330
15.2.3 Third derivative 330
15.2.4 Fourth derivative 331
15.3 Errors in the finite difference expressions 331
15.4 Equivalent concentrated load 332
15.5 Boundary conditions for beam bending 332
15.5.1 Simple support 332
15.5.2 Fixed end 333
15.5.3 Free end 333
15.6 A statically determinate static problem 334
15.7 A statically indeterminate nstatic problem 336
15.8 Free vibration of beams 337
15.9 Buckling of columns 339
15.10 Finite difference representation of the plate equation 340
15.10.1 A plate example 342
References and suggested readings 342
Exercise 15 343

16. ADAPTIVE FINITE ELEMENT ANALYSIS 345–368

16.1 Introduction 345
16.2 The adaptive finite element technique 345
16.3 Superconvergent patch recovery technique 346
16.4 Example of verification of SPR 349

16.5 Error estimation 351
16.6 ZZ-error estimator 354
16.7 ZZ-refinement framework 355
16.8 Adptive mesh generation 356
16.8.1 Mesh generation based on mapping 357
16.8.2 Delaunay triangulation method 358
16.8.3 Domain decomposition method (quadtree) 358
16.8.4 Advancing front technique 358
16.8.4.1 Structured and unstructured mesh 359
16.8.4.2 The background mesh 359
16.8.4.3 Geometrical representation of the domain 360
16.8.4.4 Generation of boundary nodes 361
16.8.4.5 Generation of triangular elements 363
16.8.4.6 Example of *h*-adaptive finite element analysis 365
References 367

17. GEOMETRICAL NON-LINEAR FINITE ELEMENT ANALYSIS 369–390
17.1 Introduction 369
17.2 Non-linear equation solving procedures 369
17.2.1 Direct iteration method 370
17.2.2 Newton - raphson method 371
17.2.3 Modified newton - raphson method 372
17.2.4 Incremental techniques 372
17.3 Formulation of the geometric non-linear problem 373
17.3.1 Equilibrium equations 373
17.3.2 Incremental equilibrium equation 374
17.4 Large deflection analysis of plates in *B*-notation 375
17.5 Large deflection analysis of plates in *N*-notation 380
17.6 Example of a pin-jointed bar 383
17.7 Computer program for geometrically non-linear analysis of plates 386
References and suggested readings 390

18. FINITE ELEMENT METHOD OF ANALYSIS OF STIFFENED PLATES 391–401
18.1 Introduction 391
18.2 Modeling the plate and the stiffener 392
18.3 Rectangular stiffened plate bending element 392
18.3.1 Stiffness matrix of the stiffener element 393

18.4 Isoparametric stiffened plate bending element 396
18.4.1 Stiffness matrix of arbitrarily oriented eccentric stiffener 397
References and suggested readings 401

19. SELECTED TOPICS **403–422**
19.1 Rayleigh-ritz method 403
19.1.1 An example 404
19.2 Rayleigh-ritz finite element method 405
19.3 Weighted residual methods 407
19.4 Galerkin method 408
19.4.1 An example of galerkin method 409
19.5 Galerkin finite element method 411
19.6 Torsional stiffness of prismatic beam element 413
19.7 Torsion of non-circular sections 415
19.8 Axi-symmetrical element 418
19.9 Three-dimensional elements 420
19.9.1 Linear element (8 nodes) 420
19.9.2 Quadratic element (20 nodes) 420
19.9.3 Cubic element (32 nodes) 420
19.9.4 16 Noded solid 421
19.9.5 24 Noded solid 422
References and suggested readings 422

Appendix I : Fixed end forces 423
Appendix II : List of computer programs in FORTRAN and C 425
Subject index 427

1

BASIC CONCEPTS OF STRUCTURAL ANALYSIS

1.1 TYPES OF STRUCTURES

A simply supported beam may be treated as a one-dimensional structure, and necessary quantities such as strength, rigidity or stability can be determined from the knowledge of strength of materials. The same beam may also be analysed as a two-dimensional or a three-dimensional problem and its solution based on the theory of elasticity is more rigorous and refined. Thus, the same structure can be idealised in different ways and the result of the analysis depends to a great extent on the idealisation of the structure. For the convenience of classification, structures are organised into the following three types:

1. Framed structures such as a truss, a frame, a beam or a grid are treated as a combination of linear or uniaxial members. All the members are considered long enough in comparison to their cross-sectional dimensions. The analysis of these structures follows a particular pattern. Some structures of this category are shown in Fig. 1.1.
2. Next category of structures are plates, shells and planar elements which are referred to as two-dimensional structures (Fig. 1.2). Just as the members of the framed structures are represented by their lengths, plates and shells are represented by their surfaces. The analysis of these structures requires the knowledge of the theory of elasticity.
3. Soil mass, rock foundations, complex machine parts *etc.* are three-dimensional bodies. For any detailed analysis, their actual behaviour in space is to be taken into account.

1.2 OBJECTIVE OF STRUCTURAL ANALYSIS

The analysis of all the above structures involves the determination of:

(*a*) displacements at salient points of the structure;

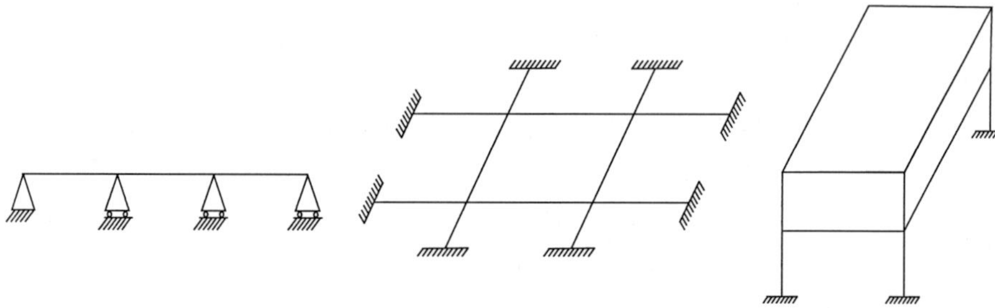

(a) Continuous beam (b) Grid (c) Frame

Fig. 1.1. Framed structures

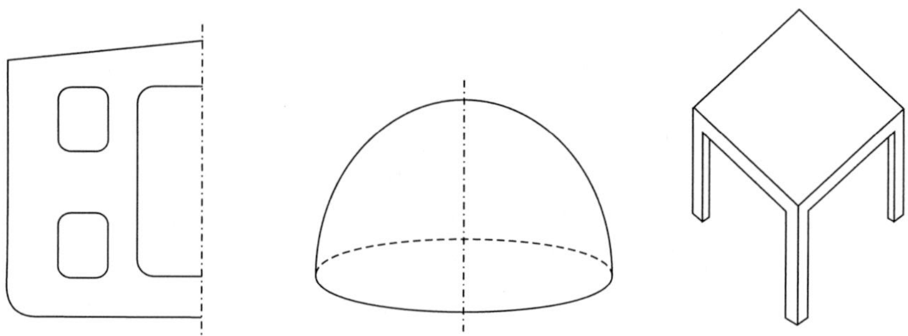

(a) Web frame of a tanker (b) Shell structure (c) Floor slab

Fig. 1.2. Two-dimensional structures

(*b*) internal forces ("force" is used here in general term. By force, we may mean a translatory force, a bending moment, a twisting moment *etc.*) in the members of the structure on the basis of:

(*i*) the given shape of the structure,

(*ii*) the sizes of the members,

(*iii*) the material properties of the members

and (*iv*) the applied loads.

In statically determinate structures, all the reactions can be determined from the equations of equilibrium and then all the necessary quantities such as bending moment, deflection *etc.* can be obtained from the knowledge of elementary mechanics of materials. Structures which are termed as statically indeterminate contain more unknown reactions and/or internal forces than can be solved from equations of statics. Solution of statically indeterminate structure therefore becomes more complex and additional equations over and above those obtained from statics are to be obtained on the basis of the deformation conditions of the structure.

The analysis of statically indeterminate structures finally reduces to the solution of a series of linear algebraic equations. The number of such equations will essentially depend on the method of analysis employed. The book is devoted primarily to the statically indeterminate framed structures, plates and shells.

The classical method of analysis of structures has its own limitations in that only simple structures can be dealt with. Modern structures are becoming complex in shape, and uncertainties associated with the design of structures are getting, reduced. An awareness has grown to analyse the structure in truer perspective. In so doing, many of the simplifying assumptions have to be done away with. With the advent of electronic digital computers, mammoth calculations no longer pose any problem. Its ultimate result is the development of matrix methods of structural analysis for one-dimensional structure and the finite element method[1] for plates, shells and more complicated three-dimensional problems. These recent computer-oriented methods are discussed in the following chapters.

1.3 MATERIALS AND BASIC ASSUMPTIONS

The analysis presented throughout unless otherwise specified, is based on the following assumptions:

1. The material is isotropic and homogeneous, *i.e.,* the property of the material is not changing from point to point.
2. The material with which the structure is made of is assumed to be linearly elastic, *i.e.,* it obeys Hooke's Law. For most cases, permissible stresses under working loads are sufficiently low so that the linear behaviour of the material may be justified.
3. Loads applied to the structure will not induce any effect of geometry changes.
4. Buckling aspects of the members in the general statical and dynamically analysis are neglected.

1.4 LOADS

A structure has to sustain the load coming on it. Before an analysis is made, load coming on the structure must be ascertained. Some of the important types of loadings are discussed here.

1. **Dead Load:** It is the self-weight of the structure.
2. **Live Load:** It is the useful load carried by the structure. Furniture and human weight in buildings, cranes in factory sheds, vehicles in bridges are examples of live load. This load is determined from code specifications.

---

1. Finite element method can be applied to one-dimensional structure as well. Whereas matrix methods of structural analysis have been developed based on intuitive reasoning, the finite element method follows a more formal mathematical procedure.

3. **Impact Load:** It is due to sudden application of the load on a structure. The jumping of a vehicle due to undulations on a surface causes impact which results in stresses much more than the value it would attain had the load been applied gradually.
4. **Wind, Earthquake and Blast Loads:** These loads cause accelerations of the mass of the structure and so must be dynamically considered. In fact that is the trend for "all highrise, tall and slender structures". For conventional structures, these loads are reduced to static loads, the values of which could be obtained by consulting the codes of practice or from recorded data.
5. **Self-straining Load:** It may occur due to the settlement of support, changes of temperature, creep of the material of the structure and due to prestressing effects.

For structures floating on the sea, there will be loads on the structure due to buoyancy and dynamic effects of the waves. If the floating structure is mobile, there will be forces due to its own motion. For fixed ocean structures, there will be horizontal forces due to water current. Loads to be encountered therefore, depend on the type of the structure.

1.5 GENERAL METHODS OF ANALYSIS

Any method of structural analysis must satisfy the following two conditions:

1. Equilibrium Conditions; 2. Compatibility Conditions.

1.5.1 Equilibrium Conditions

Not only the entire structure, but also any part of the structure considered as a free body should be in equilibrium under the action of the forces acting on it. The equations of equilibrium are obtained from statics. For a planar structure, there are three equations of equilibrium. For three-dimensional structures, number of equations of equilibrium is six. In the analysis to be followed the equilibrium conditions in terms of forces acting at the joints or nodes of the structure are formulated.

Similar consideration can be made for vibrating structures. The difference to be made in this case is that in addition to all the static forces, the inertia forces and the damping forces (if present) are to be taken into account using D' Alembert's principle.

1.5.2 Compatibility Conditions

Deformation of the structure, obtained should be compatible with the continuity of the structure and the support conditions. Compatibility conditions refer to the continuity of displacements throughout the structure *i.e.,* different parts of the structure must fit together without discontinuity at all stages of loading. To illustrate this, a few examples may be mentioned. There can be no rotation of the axis of the member at a fixed support. At the unyielding support of a continuous beam, there can be no deflection, and the rotation is same at both sides of the support. Well-known three moment theorem is based on this compatibility condition and such similar equations are frequently encountered in structural analysis.

Displacements at a joint must be identical to that of the individual members meeting at that joint. For a plane problem, there can be three possible placements at a joint-two translations and one rotation. For a three-dimensional problem, the number of possible displacements are six — three translations and three rotations.

1.6 FORCE-DISPLACEMENT RELATIONSHIP

When subjected to a force, a structural member undergoes deformation. A. relationship exists between the applied force and the displacement of the structure. This is usually done in one of the following two ways:

A uniform cantilever beam of Fig. 1.3 is subjected to a load P at the free end and the corresponding vertical deflection is X. The relationship between P and X may be given as:

$$X = FP \tag{1.1}$$

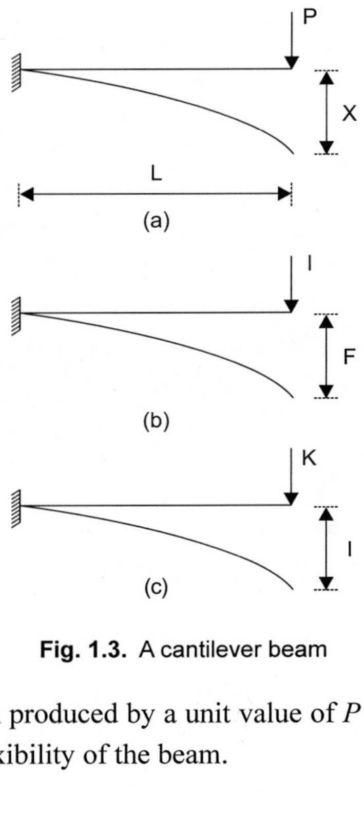

Fig. 1.3. A cantilever beam

where F is the deflection produced by a unit value of P at the point of deflection of P and is called the flexibility of the beam.

Another way of relating P and X can be done as follows:

$$P = KX \tag{1.2}$$

where K is the force corresponding to P required to produce unit displacement at the point of application of P and is called the stiffness of the beam. For the uniform beam of Fig. 1.3, F and K assume the following values:

$$\left.\begin{aligned} F &= \frac{L^3}{3EI} \\ K &= \frac{3EI}{L^3} \end{aligned}\right\} \tag{1.3}$$

where L is the length of the beam and EI is the flexural rigidity. Equation (1.3) reveals that the flexibility F and the stiffness K are inverse to one another.

The above concept has been extended to the solution of more general problems,which involve the relation; between the forces acting at a number of points in the structure and the corresponding displacements.

1.7 STATICAL INDETERMINACY

A structure remains in equilibrium under the action of external forces. A structure whether considered as a whole, or members or joints considered as freebody should be in equilibrium under the action of external forces. The analysis of the structure comprises the determination of reactions at supports and the internal forces (such as axial force in the member of a truss). The types of structures for which the equations of statics alone are not sufficient for the complete solution of the problem are referred to as statically indeterminate structures. For such structures, the number of unknown forces are more than the number of equations available from statics and the number of forces in excess of static equations are termed as degrees of statical indeterminacy or degrees of redundancy of the structure. Where the indeterminacy is due to the reactions, it is stated as externally indeterminate and where it is due to internal forces it is referred to as internally indeterminate. The redundancy of a structure may either be external or internal or both.

1.7.1 Plane Structure

For a truss-like structure having hinges at the joints (a joint is a point where two or more members intersect), following relationship holds good if the structure is statically determinate:

$$2j = m + 3 \tag{1.4}$$

where j is the number of joints including supports and m is the number of members of the truss. Equation (1.4) can be derived easily. For a statically determinate truss, number of unknown reactions to be determined from the equilibrium conditions of the overall structure is three. Other unknowns in the problem are the member forces which are m in number. Total number of equations at each joint is 2. By equating the number of unknowns to the number of equations of statics, eqn. (1.4) results.

For a statically indeterminate plane truss, the degree of indeterminacy is

$$i = (m + r) - 2j \tag{1.5}$$

where r is the number of unknown reactions. For the truss of Fig. 1.4, $m = 15$, $r = 4$ and $j = 8$. Therefore, the degree of statical indeterminacy is 3. It may be noted that the external indeterminacy is 1 and the internal indeterminacy is 2 for the truss.

Following similar reasoning it can be shown that a rigid plane frame is statically determinate if

$$3j = 3m + r - \alpha \tag{1.6}$$

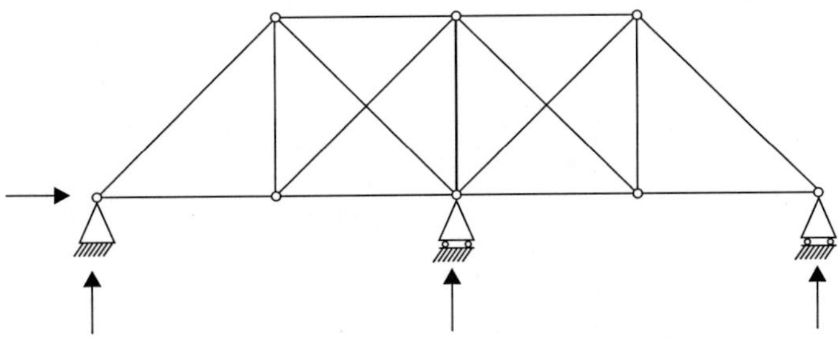

Fig. 1.4. Statically indeterminate plane truss

where α is the number of internal releases. If at any particular point in the structure, if the internal force remains zero for all possible types of loading, it is then said that a release has been introduced at that point in the structure. Introduction of a hinge at a point of a bending member means that bending moment at that point is zero irrespective of the load applied.

For the beam of Fig. 1.5 (*a*), $m = 2, j = 3, r = 4$ and $\alpha = 1$ (note that the hinge inside a member is not a joint). For this beam $3j = 9$ and $3m + r - \alpha = 9$. Therefore, eqn. (1.6) is satisfied and the beam is statically determinate. The frame of Fig. 1.5 (*b*) is also statically determinate where $3j = 9$, $m = 3$, $r = 4$ and $\alpha = 1$ eqn. (1.6) is satisfied.

The degree of indeterminacy of a rigid jointed plane structure is given by

$$i = 3m + r - 3j - \alpha \tag{1.7}$$

For the frame of Fig. 1.6, $j = 9$, $m = 10$, $r = 5$ and $\alpha = 0$. Therefore, the degree of indeterminacy of the frame is 8. For the grid of Fig. 1.7, the reactions at both supports are two couples and a vertical force. In this case $m = 2$, $j = 3$, $r = 6$ and $\alpha = 0$. Therefore, the degree of indeterminacy of grid is 3.

1.7.2 Space Structures

The degree of statical indeterminacy of a space truss is given by:

$$i = m + r - 3j \quad (1.8)$$

The degree of indeterminacy of a space frame is given by:

$$i = 6m + r - 6j - \alpha \quad (1.9)$$

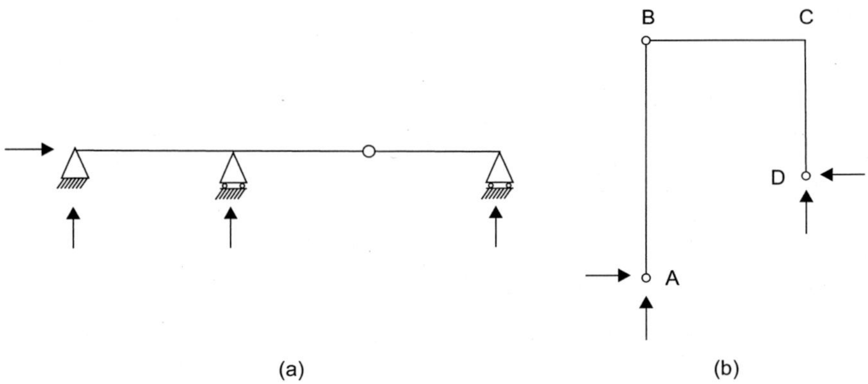

Fig. 1.5. Statically determinate structures

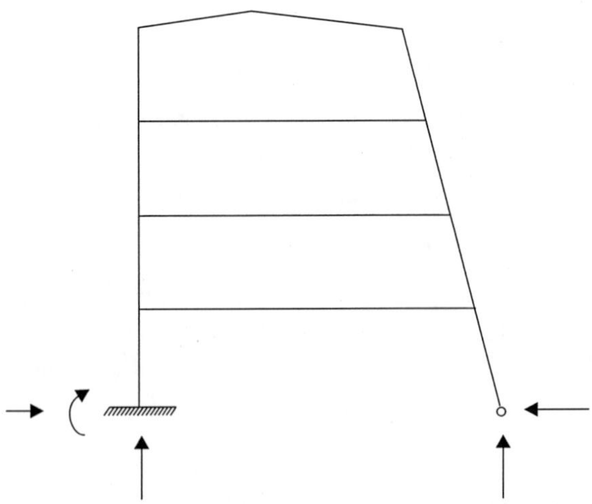

Fig. 1.6. A frame

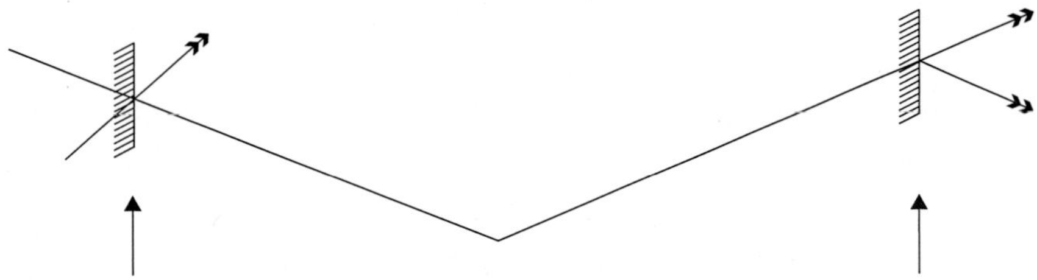

Fig. 1.7. A grid

For the space truss of Fig. 1.8, $m = 4$, $j = 5$, $r = 12$ which on substitution into eqn. (1.8) yields $i = 1$ which means that the given space truss is statically indeterminate to first degree. For the space frame of Fig. 1.9, $m = 8$, $j = 8$ and $r = 24$ and which on substitution into eqn. (1.9) gives $i = 24$.

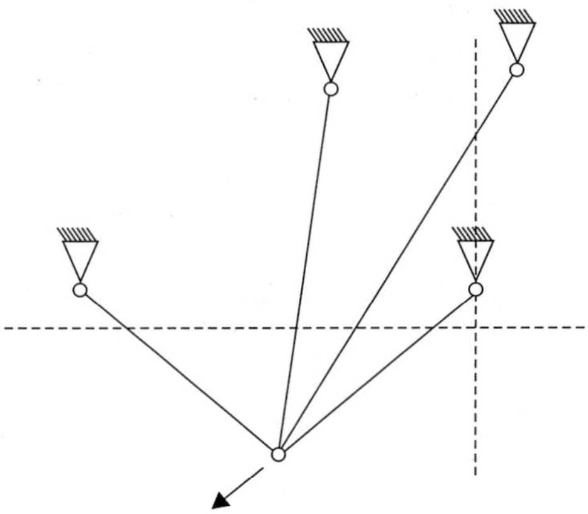

Fig. 1.8. A space truss

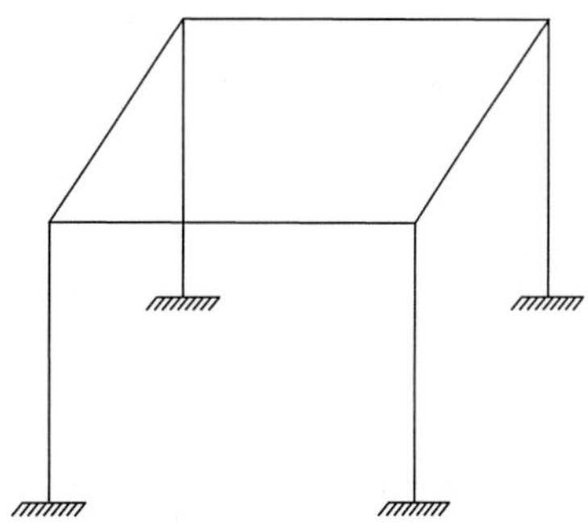

Fig. 1.9. A space frame

1.8 KINEMATIC INDETERMINACY

The joints of a structure subjected to load undergo rotations and translations. At each joint possible joint displacements can be labelled such that they can be treated as independent, *i.e.*, each displacement can be arbitrarily changed without affecting other displacements at that joint. The number of independent joint displacements of a structure is called the degree of kinematic indeterminacy or the number of degrees of freedom.

The propped cantilever AB of Fig. 1.10 has two joints A and B. Joint A is fixed; so there cannot be any displacement there. The vertical movement of joint B is restrained, but the horizontal translation and rotation of joint B are permitted. Hence, the

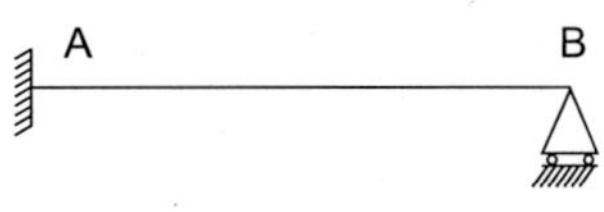

Fig. 1.10. A propped cantilever

degrees of freedom of the beam are 2. In the analysis of such beams, the axial deformation is neglected which means that the length of the beam remains same and the displacement corresponding to axial movement need be neglected. In that case the degree of kinematic indeterminacy becomes one.

Each of the three joints *A*, *B* and *C* of the rigid frame of Fig. 1.11 possesses three degrees of freedom - two translations and one rotation. Joint *E* has one degree of freedom - that of rotation. Total degrees of freedom then are 10. However, if the axial deformation of the members is neglected, the horizontal translation at joints *A*, *B* and *C* labelled as 2, 5 and 8 in the figure, would be equal. The independent joint displacements then would be the rotations at joints *A*, *B*, *C* and *E* and one horizontal translation. Total degrees of freedom of the frame then become 5.

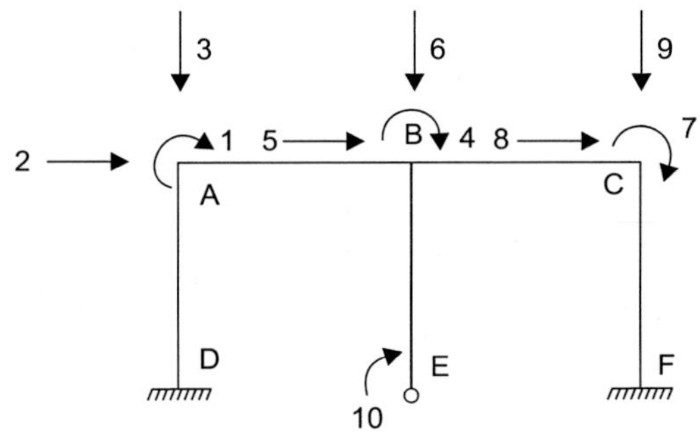

Fig 1.11. A rigid frame

1.9 TWO APPROACHES OF STRUCTURAL ANALYSIS

All methods of structural analysis for statically indeterminate structures fall under any of the following two categories: the force or the flexibility method and the displacement or the stiffness method.

In the force method, the structure is first made statically determinate by releasing certain redundants. The deformation of the released structure is thus inconsistent. For example, if a roller support is removed, the force at the support is eliminated and at the point of support, continuity is affected. To restore continuity in the displacement, additional forces corresponding to the "released" redundants are applied. This results in a set of simultaneous equations, the solution of which gives the values of the unknown forces. In this method, the forces are the unknowns.

In the displacement method, the structure is first made kinematically determinate *i.e.* inhibiting all possible joint displacements. In the restrained structure, joint equilibrium is violated. To restore the equilibrium, displacements at the restrained joints are allowed which results in a set of simultaneous equations of unknown displacements. In this method, the displacements are the unknowns.

There are thus two approaches of analysis of structure depending on the choice of modification of, the original structure. The philosophy of both methods is again the same and both methods finally result in a series of equations. In the subsequent chapters, these two methods are discussed in more details for the case of linear structures.

EXERCISE 1

Find out the degrees of statical and kinematic indeterminacies for the structures shown in Prob. 1.1.

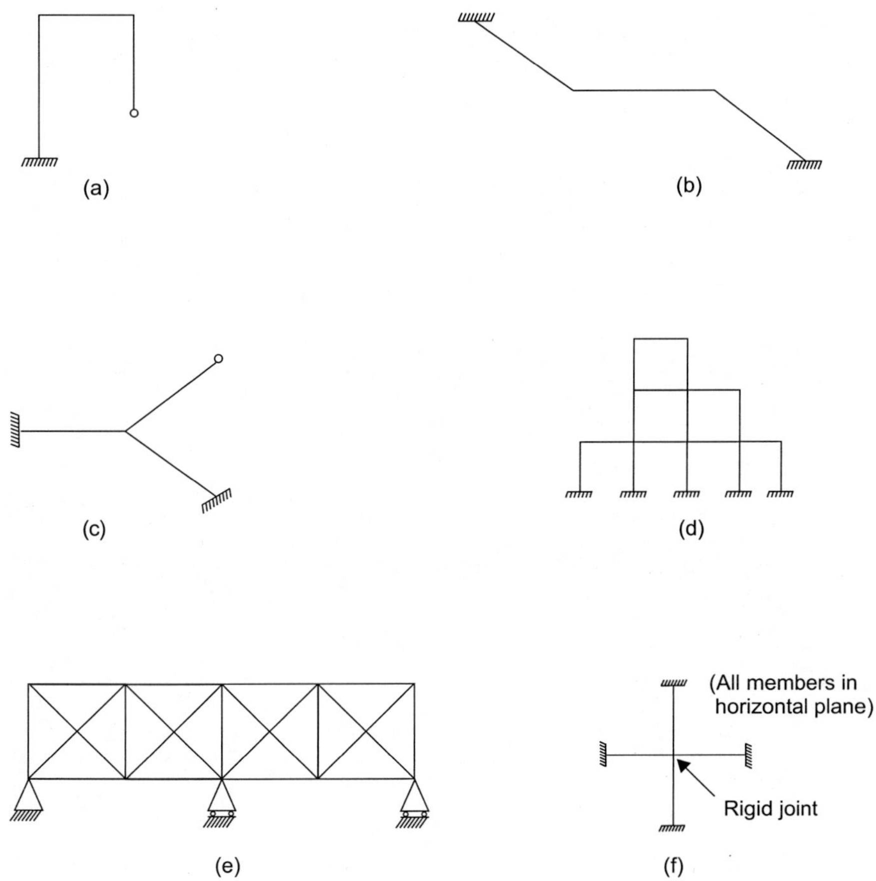

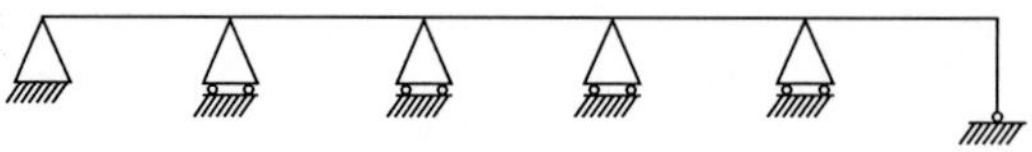

(g)

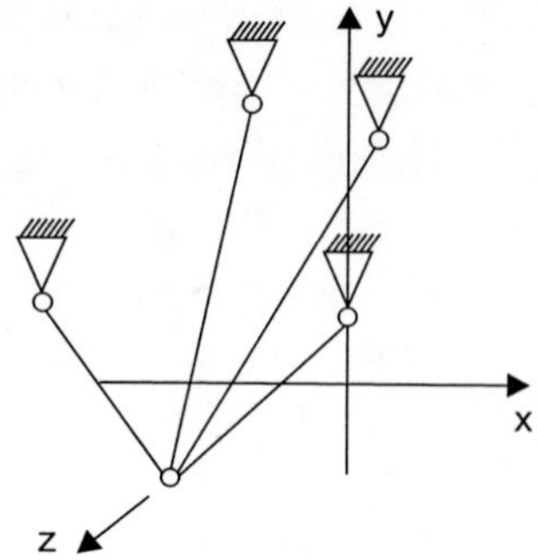

(h)

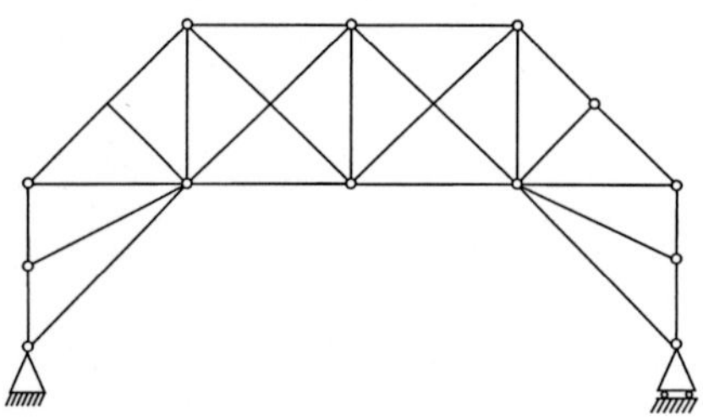

(I)

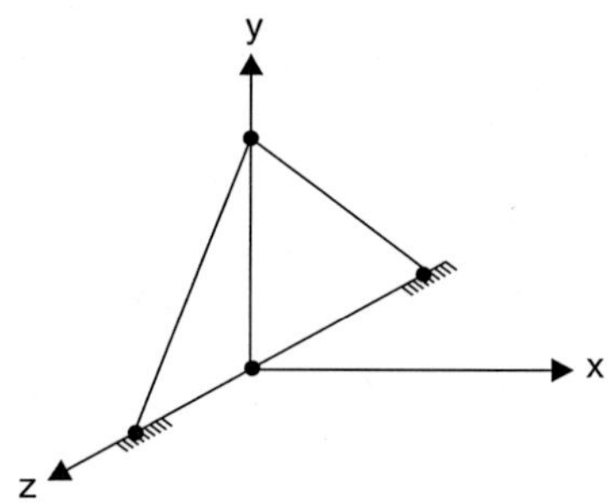

(j)

Prob 1.1. Various structures

❑❑❑

2

ENERGY PRINCIPLES

2.1 INTRODUCTION

The variational or energy methods are of primary importance in the finite element analysis [2.1–2.5]. These methods have been in use in the structural analysis for a considerable length of time. But the development of the finite element method has brought into being some of the sophisticated forms of these methods along with their scope and limitations. In this chapter, important energy theorems which have been applied in the subsequent chapters will be discussed (2.1, 2.5, 2.6).

2.2 PRINCIPLE OF VIRTUAL WORK

The principle of virtual work is a very useful approach to the solution of a variety of problems of structural mechanics. This principle forms the basis of the subsequent methods described in this chapter.

A body or a structure is considered to be in equilibrium under the action of body forces $\{P_B\}$ whose components along three reference axes x, y and z are P_{Bx}, P_{By} and P_{Bz} and surface forces $\{P_s\}$ whose components along x, y and z axes are P_{Sx}, P_{Sy} and P_{Sz} respectively. Entire structure is assumed to undergo fictitious displacement (termed as virtual displacement) such that prescribed boundary conditions of the structure are not violated. Components of the virtual displacement at each point are δu, δv, and δw. Total virtual work in the system consists of the work done by the different components—the external force as well as the internal force.

The external virtual work δW_E is the work done by the real loads acting on the structure due to the virtual displacement. These loads, include those distributed both over the entire surface and the entire volume of the structure. Therefore, the virtual work done by the external force is

$$\delta W_E = \int_s \{\delta u \quad \delta v \quad \delta w\} \begin{Bmatrix} P_{Sx} \\ P_{Sy} \\ P_{Sz} \end{Bmatrix} ds + \int_v \{\delta u \quad \delta v \quad \delta w\} \begin{Bmatrix} P_{Bx} \\ P_{By} \\ P_{Bz} \end{Bmatrix} dv \tag{2.1}$$

In equation (2.1), the integration is carried over the entire surface in the first term and over the entire volume in the second term. Equation (2.1) can be written as:

$$\delta W_E = \int_s \delta\{f\}^T \{P_S\} ds + \int_v \delta\{f\}^T \{P_B\} dv \tag{2.2}$$

where $$\{f\} = \begin{Bmatrix} u \\ v \\ w \end{Bmatrix}$$

For a general stress condition, there are, six components of stresses $\sigma_x, \sigma_y, \sigma_z, \tau_{xy}, \tau_{yz}$ and τ_{zx}. The six components of strain in the virtual displacement fields are $\delta\varepsilon_x, \delta\varepsilon_y, \delta\varepsilon_z, \delta\gamma_{xy}, \delta\gamma_{yz}$ and $\delta\gamma_{zx}$. The virtual internal work is, therefore

$$\delta U = \int_v \{\delta\varepsilon_x \quad \delta\varepsilon_y \quad \delta\varepsilon_z \quad \delta\gamma_{xy} \quad \delta\gamma_{yz} \quad \delta\gamma_{zx}\} \begin{Bmatrix} \sigma_x \\ \sigma_y \\ \sigma_z \\ \tau_{xy} \\ \tau_{yz} \\ \tau_{zx} \end{Bmatrix} dv$$

or, $$\delta U = \int_v \delta\{\varepsilon\}^T \{\sigma\} dv \tag{2.3}$$

The principle of virtual work states that the work done by the external forces due to the virtual displacement for a structure in equilibrium is equal to the work done by the internal forces for the virtual internal displacement. Mathematically,

$$\delta W_E = \delta U$$

or, $$-\delta U + \delta W_E = 0 \tag{2.4}$$

It may be noted that in the derivation of eqn. (2.4), the stress-strain relationship has not been assumed. As such the principle of virtual work holds good for linearly elastic as well as non-linear materials.

2.3 PRINCIPLE OF COMPLEMENTARY VIRTUAL WORK

Unlike the previous section, the work done by a system of virtual forces during actual deformation of the structure will now be investigated. The work done by the virtual forces during actual displacements is referred to as complementary virtual work. Imaginary system of body and surface forces are introduced into the actual displacement field. The external complementary virtual work done is therefore, given by

$$\delta W_E^* = \int_s \{f\}^T \delta\{P_S\} ds + \int_v \{f\}^T \delta\{P_B\} dv \tag{2.5}$$

A virtual state of stress $\delta\{\sigma\}$ results in the actual strain field $\{\varepsilon\}$ which results in the internal virtual work. It is expressed as:

$$\delta U^* = \int_v \{\varepsilon\}^T \delta\{\sigma\}\,dv \tag{2.6}$$

The principle of complementary virtual work states that the complementary virtual work done by the external virtual forces under the actual deformation of a structure in equilibrium is equal to the complementary work done by the virtual stresses due to internal strains. Mathematically,

$$-\delta U^* + \delta W_E^* = 0 \tag{2.7}$$

2.4 PRINCIPLE OF MINIMUM POTENTIAL ENERGY

Instead of conceiving the necessary forces and displacements in terms of virtual quantities, both the principle of virtual work and the principle of complementary virtual work may be interpreted in terms of mathematical variations.

The displacement field for a structure in equilibrium be $\{f\}$. Let us consider arbitrary displacements which are consistent with the constraints imposed, on the body. These arbitrary displacements differ from the actual displacements by $\delta\{f\}$. At the constraints, the variation $\delta\{f\}$ must vanish which introduces a concept equivalent to that of virtual displacement introduced in the earlier section.

The external virtual work may be interpreted in terms of potential energy of the system and may be considered as the work done during a variation $\delta\{f\}$ in the displacement field. The potential of the external forces is given by:

$$\delta W_E = -\,\delta V_E \tag{2.8}$$

Substituting δW_E from eqn. (2.8) into eqn. (2.4) yields

$$\delta U + \delta V_E = 0$$

or, $$\delta(U + V_E) = 0 \tag{2.9}$$

or, $$\delta\varnothing = 0$$

where $\varnothing = U + V_E$ is defined as the total potential energy of the system.

The principle of minimum potential energy states that among all displacement fields that satisfy the prescribed constrained conditions of the structure, those that satisfy equilibrium conditions, make the total potential energy a minimum.

Thus,

$$\delta^2\varnothing = \delta^2 (U + V_E) > 0 \tag{2.10}$$

2.5 PRINCIPLE OF MINIMUM COMPLEMENTARY ENERGY

Similarly, the external complementary virtual work δW_E^* may be conceived as the work done by a variation of external forces. A complementary potential function V_E^* may be introduced as

$$-\delta V_E^* = -\delta W_E^* \tag{2.11}$$

Substituting δW_E^* from eqn. (2.11) into eqn. (2.7) yields

$$\delta U^* + \delta V_E^* = 0$$

or, $$\delta(U^* + V_E^*) = 0$$

or, $$\delta\varnothing^* = 0 \tag{2.12}$$

The principle of minimum complementary energy states that among all states of stress that satisfy equilibrium conditions, those that satisfy the prescribed displacement boundary conditions make the total complementary energy a minimum.

It may be noted that in applying the principle of minimum complementary energy, it is the force which is to be varied, while in the principle of minimum potential energy, the displacement is to be varied.

2.6 CASTIGLIANO'S THEOREMS

Castigliano's theorems can be derived from the principle of minimum potential energy and the minimum complementary energy.

The principle of minimum potential energy states that

$$\delta\varnothing = \delta(U + V_E) = 0 \tag{2.13}$$

If a structure is in equilibrium under a set of forces $P_1, P_2, P_3, \ldots, P_n$ inducing displacements $X_1, X_2, X_3, \ldots, X_n$, then by the application of minimum potential energy, yields

$$\frac{\partial U}{\partial X_1}\delta X_1 + \frac{\partial U}{\partial X_2}\delta X_2 + \ldots + \frac{\partial U}{\partial X_n} - P_1\,\delta X_1 - P_2\,\delta X_2 \cdots - P_n\,\delta X_n = 0 \tag{2.14}$$

Rearranging the terms, gives

$$\left(\frac{\partial U}{\partial X_1} - P_1\right)\partial X_1 + \left(\frac{\partial U}{\partial X_2} - P_2\right)\partial X_2 + \cdots + \left(\frac{\partial U}{\partial X_n} - P_n\right)\partial X_n = 0 \tag{2.15}$$

Since, the variations are imaginary, the terms within the parentheses must vanish, therefore,

$$\left.\begin{aligned} \frac{\partial U}{\partial X_1} &= P_1 \\ \frac{\partial U}{\partial X_2} &= P_2 \\ &\vdots \\ &\vdots \\ \frac{\partial U}{\partial X_n} &= P_n \end{aligned}\right\} \tag{2.16}$$

Equation (2.16) is nothing but the mathematical form of what is known as Castigliano's first theorem. It states that if the strain energy U of an elastic structure is expressed in terms of generalised displacements, then the first partial derivative of U with respect to one of the generalised displacements gives the generalised force acting at that point.

Similarly, the principle of minimum complementary energy states that

$$\delta \varnothing^* = \delta(U^* + V_E^*) = 0 \tag{2.17}$$

For a structure in equilibrium under a set of forces P_1, P_2, ..., P_n inducing displacements X_1, X_2, ..., X_n then by the application of the principle of minimum complementary energy, yields

$$\frac{\partial U^*}{\partial P_1}\delta P_1 + \frac{\partial U^*}{\partial P_2}\delta P_2 + \cdots + \frac{\partial U^*}{\partial P_n}\delta P_n - X_1\,\delta P_1 - X_2\,\delta P_2 \cdots - X_n\,\delta P_n = 0$$

or,

$$\left(\frac{\partial U^*}{\partial P_1} - X_1\right)\delta P_1 + \left(\frac{\partial U^*}{\partial P_2} - X_2\right)\delta P_2 + \left(\frac{\partial U^*}{\partial P_n} - X_n\right)\delta P_n = 0 \tag{2.18}$$

Variations of P_1, P_2, ..., P_n are arbitrary and hence must vanish, therefore

$$\left.\begin{aligned} \frac{\partial U^*}{\partial P_1} &= X_1 \\ \frac{\partial U^*}{\partial P_2} &= X_2 \\ &\vdots \\ &\vdots \\ \frac{\partial U^*}{\partial P_n} &= X_n \end{aligned}\right\} \tag{2.19}$$

Equation (2.19) is known as Engesser's theorem, and is valid for both linear as well as non-linear elastic structures.

However, for linear elastic structure

$$U = U^* \tag{2.20}$$

So, eqn. (2.20) for elastic structure can be written as

$$\left.\begin{aligned} \frac{\partial U}{\partial P_1} &= X_1 \\ \frac{\partial U}{\partial P_2} &= X_2 \\ &\vdots \\ &\vdots \\ \frac{\partial U}{\partial P_n} &= X_n \end{aligned}\right\} \tag{2.21}$$

Equation (2.21) is the mathematical statement of Castigliano's second theorem. It states that if the strain energy U of a linearly elastic structure is expressed in terms of generalised forces, then the first partial derivative of U with respect to one of the generalised forces gives the corresponding displacements at those points.

2.7 DETERMINATION OF DISPLACEMENTS

If any arbitrary section of a one-dimensional structure of length L is subjected to the simultaneous action of axial force N, bending moment M, shear force V and twisting moment T, the total strain energy stored in the system is

$$U = \int_0^L \frac{N^2\,dx}{2EA} + \int_0^L \frac{M^2\,dx}{2EI} + \eta\int_0^L \frac{V^2}{2GA}dx + \int_0^L \frac{T^2\,dx}{2GJ} \tag{2.22}$$

where A is the cross-sectional area of the member, I is the second moment of area of the section, J is the polar moment of inertia of the section, E and G are the Young's modulus and shear modulus of elasticity of the member and η is a factor dependent on the cross-sectional properties of the member.

In order to determine a displacement X_i of the structure, Castigliano's second theorem yields *i.e.*, from eqn. (2.21) and eqn. (2.22)

$$X_i = \frac{\partial U}{\partial P_i} = \int_0^L \frac{N}{EA}\frac{\partial N}{\partial P_i}dx + \int_0^L \frac{M}{EI}\frac{\partial M}{\partial P_i}dx + \eta\int_0^L \frac{V}{GA}\frac{\partial V}{\partial P_i}dx + \int_0^L \frac{T}{GJ}\frac{\partial T}{\partial P_i}dx \tag{2.23}$$

The quantity $\partial N/\partial P_i$ is the value of the axial force at the arbitrary section due to a force corresponding to a unit value of P_i. Similarly, $\delta M/\delta P_i$ is the value of the bending moment at that section due to a force corresponding to a unit value of P_i and so on.

For determining the value of P_i, eqn. (2.23) indicates that two quantities N and $\partial N/\partial P_i$, M and $\partial M/\partial P_i$ and so on are to be multiplied over the entire length of the structure. This can be achieved by integrating the products. However, it becomes simple in many cases to determine displacements by multiplying the area diagrams given in Table 3.1 (next chapter).

No separate problem has been solved here either to demonstrate the application of energy principles or to calculate the displacement. These will be indicated in subsequent Chapters.

REFERENCES AND SUGGESTED READINGS

2.1. J.H. Argyris and S. Kesley, Energy Theorems and Structural Analysis, Butterworths, London, 1960.

2.2. E.B. Becker, G.F. Carey and J.T. Oden, Finite Element Method, v. I-V, Prentice Hall, 1981.

2.3. B.A. Finlayson, The Method of Weighted Residuals and Variational Principles, Academic Press, 1972.

2.4. R. Glowinsky, E.Y. Rodin and O.C. Zienkiewicz, Energy Methods in Finite Element Analysis, John Wiley & Sons, 1979.

2.5. O.C. Zienkiewicz and R.L Taylor, The Finite Element Method, v.I, 4th Edition, McGraw -Hill, 1989.

2.6. I.H. Shames and C.L. Dym, Energy and Finite Element Methods in Structural Mechanics, McGraw-Hill, 1985.

❑❑❑

3

INTRODUCTION TO THE FLEXIBILITY AND STIFFNESS MATRIX METHODS

3.1 INTRODUCTION

It has been mentioned in Chapter 1 that there are two basic methods of structural analysis the flexibility method and the stiffness method. The purpose of this chapter is to introduce these methods.

3.2 THE FLEXIBILITY MATRIX METHOD

The steps followed in the flexibility matrix method are described first. Then the method is illustrated with a number of examples.

First, the degree of statical indeterminacy of the structure is determined. The structure is then made statically determinate by removing the internal and/or external forces whose total number is equal to the degree of statical indeterminacy. This structure is referred to as the primary structure. Each force removed to convert the structure into a statically determinate one is called a released force or a redundant force. For example, if the end moment is released at a fixed support, then that support is converted into a simple support. The analyst is free to make a suitable choice of the releases. These released forces are the ones which are to be determined from the analysis.

Secondly, displacements corresponding to the releases are to be determined for the external load acting on the released structure. For example, if a roller support is removed in a continuous beam, the beam will deflect downward at that support point under the action of vertical loads acting downwards. The determination of this vertical deflection and likewise displacements at all similar release points is the second step to be followed.

Thirdly, the values of the displacements in the released structure corresponding to a unit value of the released force acting in the same direction one at a time at that point of release are to be determined.

In the actual structure there are either no or some specified displacements at the released points. So the displacements that are occurring at the points where the forces are released, are to be eliminated. For this purpose instead of a unit value of the released forces, the redundant forces are given some unknown values which are determined from the final equation after applying the principle of superposition of displacements. This becomes the fourth step.

Lastly, once the released forces are known, the deformation and other necessary quantities of the original indeterminate structure can be determined.

The sign convention to be adopted for the force or the flexibility method is first discussed. For beams, hogging moment is negative and sagging moment is positive and for frames, tension inside is positive and *vice versa*. For axial forces, tension is positive and compression negative.

The method is first explained in details with an example. Then a few more examples follow. The continuous beam of Fig. 3.1 is to be analysed by the flexibility method.

Step 1: The beam of Fig. 3.1 is statically indeterminate to the second degree. So two reactions are to be removed. There are various possible ways of making the choice. Three possible released structures are shown in Fig. 3.2 and the released forces are shown by dotted lines and are numbered 1 and 2. It may be noted that other possibilities also exist for making the structure of Fig. 3.1 statically determinate. In Fig. 3.2(*a*), the fixed end *A* is converted to a hinge thus releasing the end moment there and then at support *B*, a hinge is inserted thus removing the bending moment at *B*.

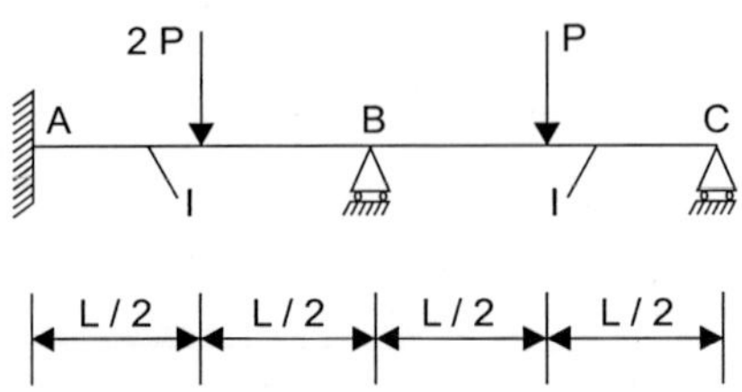

Fig. 3.1. A continuous beam

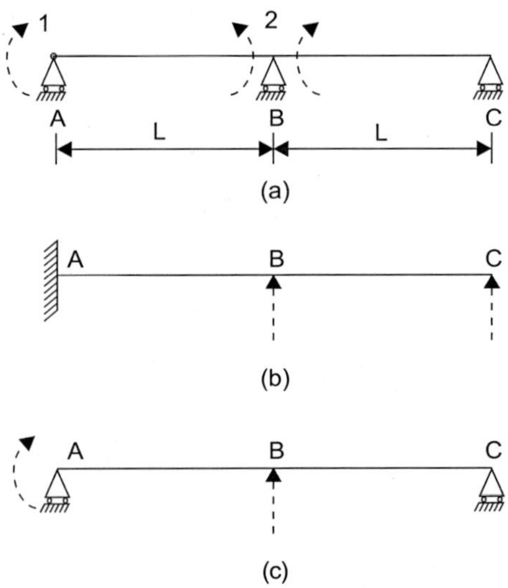

Fig. 3.2. Possible released structure

Thus, the redundant forces considered in the analysis in this case are the end moments at A and B. Another possible released structure is the cantilever beam of Fig. 3.2(b). In this case, the redundant forces are two support reactions at B and C. For the analysis presented here, the choice of the primary structure corresponding to Fig. 3.2(a) is made and the decision in this case is arbitrary. The redundant forces are R_1 and R_2 labelled as 1 and 2 in Fig. 3.2(a).

Step 2: The displacements produced at the released points due to the external loads for the structure of Fig. 3.2(a) are then determined.

The displacements in this case are X_1 and X_2 of Fig. 3.3(b). The bending moment diagram of Fig. 3.3(b) due to the external load is shown in Fig. 3.4(a) and is referred to as m_0 diagram. In Fig. 3.4(b), the bending moment diagram for the same beam has been shown for a unit load corresponding to R_1 applied at release point 1. This diagram is referred to as m_1 diagram. Similarly, m_2 diagram of Fig. 3.4(c) refers to the bending moment diagram of the same beam due to a unit load corresponding to R_2 applied at point 2. Considering only the bending part of the strain energy and applying eqn. (2.21), yields

$$X_1 = \frac{\partial U}{\partial R_1} = \int_0^L \frac{M}{EI}\frac{\partial M}{\partial R_1}\,dx \tag{3.1}$$

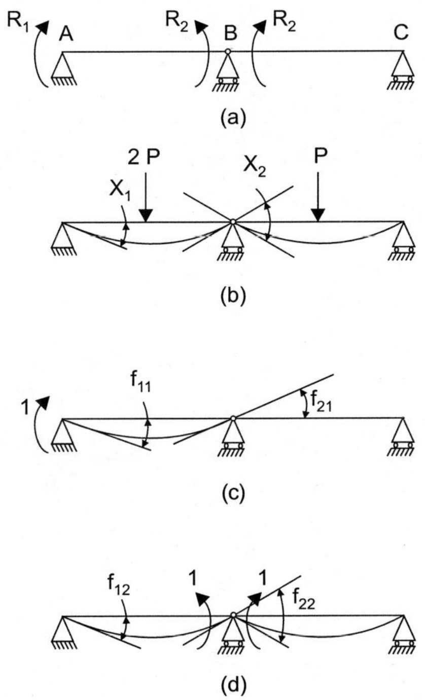

Fig. 3.3. The released structure and the steps followed in sequence

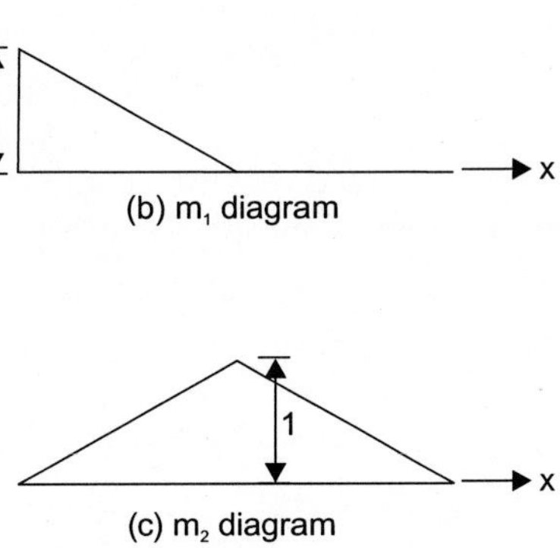

Fig. 3.4. Bending moment diagrams of the released structure

Equation (3.1) suggests that the value of X_1 can be obtained by multiplying the bending moment diagram due to the applied load which corresponds to m_0 diagram of Fig. 3.4(a) to the bending moment diagram due to a unit load corresponding to R_1 which corresponds to m_1 diagram of Fig. 3.4(b) divided by EI. From Table 3.1, by multiplying m_0 diagram with m_1 diagram, eqn. (3.1) yields

$$X_1 = \frac{1}{6}(1+\beta)\frac{KiL}{EI}$$

$$= \frac{1}{6}(1+1/2)\frac{PL}{2}\frac{L}{EI}$$

$$= \frac{PL^2}{8EI}$$

Similarly,

$$X_2 = \frac{1}{6}\left(1+\frac{1}{2}\right)\frac{PL}{2}\frac{L}{EI}+\frac{1}{6}\left(1+\frac{1}{2}\right)\frac{PL}{4}\frac{L}{EI}$$

$$= \frac{3}{16}\frac{PL^2}{EI}$$

Step 3: The displacements at the release points corresponding to a unit value of the released forces one at a time are next to be determined. They correspond to f_{11} and f_{21} of Fig. 3.3(c) for a unit value of R_1 and f_{12} and f_{22} of Fig. 3.3(d) for a unit value of R_2.

Looking at eqn. (3.1) it may be noticed that in order to determine f_{11}, m_1 diagram is to be multiplied by the same diagram and for f_{12}, m_1 diagram is to be multiplied by the m_2 diagram over the entire length of the beam. The process is repeated for other displacements. The necessary multiplication of the area diagrams can be carried out with the help of Table 3.1.

$$f_{11} = \frac{1}{3}\frac{1\times 1\times L}{EI} = \frac{L}{3EI}$$

$$f_{12} = f_{21} = \frac{1}{6}\frac{1\times 1\times L}{EI} = \frac{L}{6EI}$$

$$f_{22} = 2\times\frac{1}{3}\frac{1\times 1\times L}{EI} = \frac{2L}{3EI}$$

It may be noted that the quantities f_{11}, f_{12}, f_{21} and f_{22} are independent of the loading and as such are properties of the structure.

Step 4: If R_1 and R_2 are the redundant forces, the displacement at A due to R_1 and R_2, is given by

$$X_{R1} = f_{11}R_1 + f_{12}R_2$$

Similarly,

$$X_{R2} = f_{21}R_1 + f_{22}R_2$$

Now, as the total slope at A and relative slope at B are equal to zero, following equations result by applying the principle of superposition of displacements

$$\left.\begin{aligned} f_{11}R_1 + f_{12}R_2 + X_1 = 0 \\ f_{21}R_1 + f_{22}R_2 + X_2 = 0 \end{aligned}\right\} \tag{3.2}$$

Equation 3.2 written in matrix form becomes

$$\begin{bmatrix} f_{11} & f_{12} \\ f_{21} & f_{22} \end{bmatrix} \begin{Bmatrix} R_1 \\ R_2 \end{Bmatrix} + \begin{Bmatrix} X_1 \\ X_2 \end{Bmatrix} = \begin{Bmatrix} 0 \\ 0 \end{Bmatrix} \tag{3.3}$$

which in compact form is

$$[F]\{R\} + \{X\} = \{0\} \tag{3.4}$$

In order to determine $\{R\}$, eqn. (3.4) is rewritten as follows:

$$\{R\} = -[F]^{-1}\{X\} \tag{3.5}$$

The matrix $[F]$ is known as the flexibility matrix and its elements f_{11}, f_{12} *etc.* are called flexibility coefficients or influence coefficients. f_{ij} is defined as the displacement at point i in the direction of unknown "force" at i due to a unit value of the "force" corresponding to unknown force at j and applied at j. It may be noted that the flexibility matrix is symmetrical. This follows from Maxwell's reciprocal theorem and also from a careful study of eqn. (3.1). Substituting f_{11}, f_{12}, *etc.* and X_1 and X_2 for this problem of Fig. 3.1, eqn. (3.3) becomes

$$\frac{L}{EI}\begin{bmatrix} 2 & 1 \\ 1 & 4 \end{bmatrix} \begin{Bmatrix} R_1 \\ R_2 \end{Bmatrix} = -\frac{PL^2}{16EI}\begin{Bmatrix} 2 \\ 3 \end{Bmatrix}$$

$$[F]^{-1} = \frac{6EI}{7L}\begin{bmatrix} 4 & -1 \\ -1 & 2 \end{bmatrix}$$

Therefore,

$$\begin{Bmatrix} R_1 \\ R_2 \end{Bmatrix} = -\frac{6EI}{7L} \times \frac{PL^2}{16EI}\begin{bmatrix} 4 & -1 \\ -1 & 2 \end{bmatrix}\begin{Bmatrix} 2 \\ 3 \end{Bmatrix}$$

$$= -\frac{3PL}{56}\begin{Bmatrix} 5 \\ 4 \end{Bmatrix}$$

$$R_1 = -\frac{15PL}{56} \text{ and } R_2 = -\frac{3PL}{14}$$

Once R_1 and R_2 are determined, any other quantity of interest such as bending moment, shear force at any point, any reaction, rotation or deflection at any section may be obtained. A few more examples of the flexibility method are given below.

Example 3.1:

Analyse the rigid frame of Fig. 3.5

Two support reactions at A are considered here as redundant forces. The released structure with redundant forces denoted as R_1 and R_2 respectively are shown in Fig. 3.6 (*a*). m_0 diagram of Fig. 3.6(*b*) is the bending moment diagram of the released structure due to the external load, m_1 diagram is that corresponding to redundant force R_1. Similarly, m_2 diagram represents the variation of the bending moment of the released structure due to a unit value of the force corresponding to R_2, X_1 and X_2 are the displacements produced at the released points corresponding to the released forces R_1 and R_2 at A.

Following the same argument as discussed in the previous illustration, X_1 can be obtained by multiplying the area diagrams m_0 and m_1. This can be done with the help of Table 3.1. Thus,

$$X_1 = -\frac{1}{6}\times\frac{60\times6\times6}{EI}+\frac{1}{6}\times\frac{6\times(60+2\times120)\times6}{EI}$$

$$= \frac{1440}{EI}$$

$$X_2 = \frac{1}{2}\times\frac{120\times12\times12}{EI}$$

$$= \frac{8640}{EI}$$

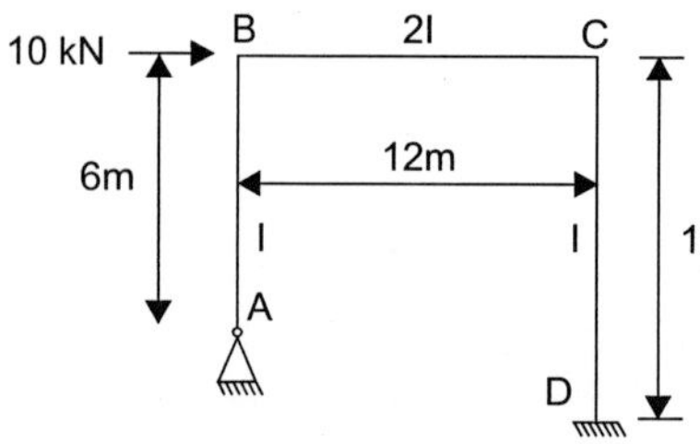

Fig. 3.5. A rigid plane frame

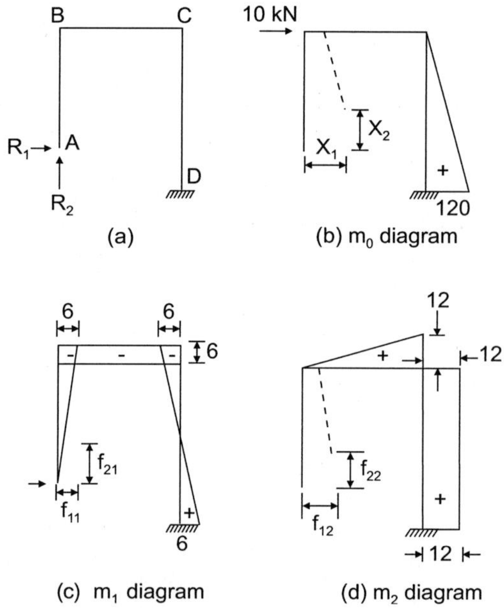

Fig. 3.6. Bending moment diagrams for the released frame in kNm

Next the flexibility coefficients are to be obtained in the same way as indicated in the previous illustration. They are

$$f_{11} = \frac{1}{3}\times3\times\frac{6\times6\times6}{EI}+\frac{6\times6\times12}{2\times EI}=\frac{432}{EI}$$

$$f_{22} = \frac{1}{3}\times\frac{12\times12\times12}{2\times EI}+\frac{12\times12\times12}{EI}=\frac{2016}{EI}$$

$$f_{12} = f_{21} = -\frac{216}{EI}$$

Similarly, assuming that there are no support displacements, the redundants R_1 and R_2 can be found by applying the principle of superposition as given in eqn. (3.3),

$$\frac{1}{EI} = \begin{bmatrix} 432 & -216 \\ -216 & 2016 \end{bmatrix} \begin{Bmatrix} R_1 \\ R_2 \end{Bmatrix} = \frac{1}{EI} \begin{Bmatrix} 1440 \\ 8640 \end{Bmatrix}$$

Solution of the above equation yields

$$\begin{Bmatrix} R_1 \\ R_2 \end{Bmatrix} = \begin{Bmatrix} -4.9 \\ -5.78 \end{Bmatrix}$$

Example 3.2:

For the frame with a tie-rod shown in Fig. 3.7, find out the force in the tie-rod. The tie-rod is capable of taking up axial tensile force only. The following are the particulars for the frame:

$$EA = 10 \times 10^3 \text{ kN m}^2 \text{ for the tie rod}$$

$$EI = 5 \times 10^3 \text{ kN for the inclined members}$$

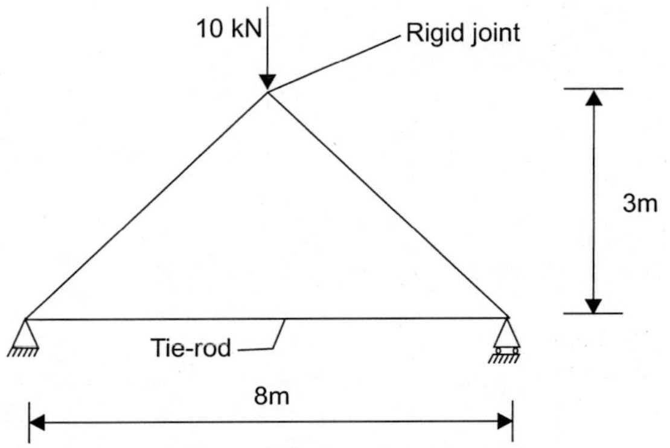

Fig. 3.7. A frame with a tie-rod

The force in the tie-rod is considered here as the redundant force. The tie-rod will take up axial force only and for the inclined members the energy due to the bending moment only is considered. X_1 can be determined exactly in the same way as has been done in the previous example.

$$X_1 = \frac{1}{3} \times \frac{5 \times 20 \times 3 \times 2}{5 \times 10^3} = 4 \times 10^{-2} \text{m}$$

For calculating the flexibility coefficient f_{11}, the contribution due to axial force in the tie-rod is to be taken into account. Therefore, if R_1 is the axial force in the tie-rod, then (from Fig. 3.8)

Table 3.1. Multiplication Table for Area Diagram

m_j / m_i	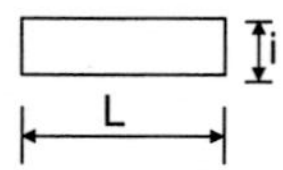	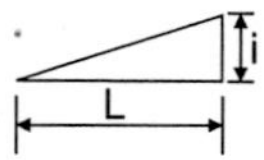	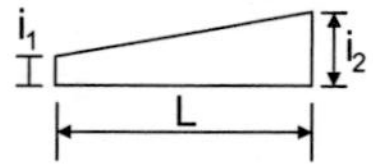	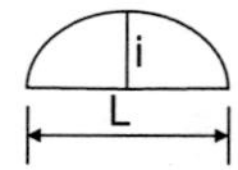	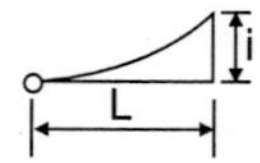	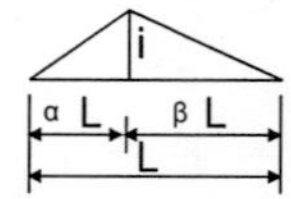
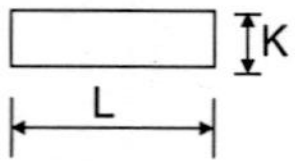	KiL	$\frac{1}{2} KiL$	$\frac{1}{2} K(i_1 + i_2)L$	$\frac{2}{3} KiL$	$\frac{1}{3} KiL$	$\frac{1}{2} KiL$
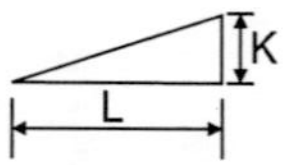	$\frac{1}{2} KiL$	$\frac{1}{3} KiL$	$\frac{1}{6} K(1i_1 + 2i_2)L$	$\frac{1}{3} KiL$	$\frac{1}{4} KiL$	$\frac{1}{6} (1 + \alpha) KiL$
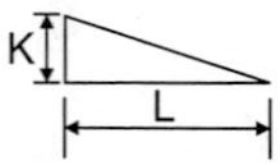	$\frac{1}{2} KiL$	$\frac{1}{6} KiL$	$\frac{1}{6} K(2i_1 + i_2)L$	$\frac{1}{3} KiL$	$\frac{1}{12} KiL$	$\frac{1}{6} (1 + \beta) KiL$
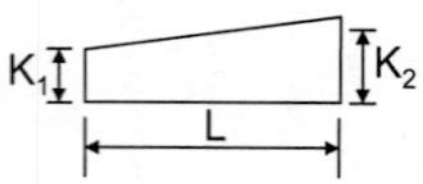	$\frac{1}{2} (K_1 + K_2)iL$	$\frac{1}{6} (K_1 + 2K_2)iL$	$\frac{1}{6} (2K_1i_1 + K_1i_2 + K_2i_1 + 2K_2i_2)L$	$\frac{1}{3} (K_1 + K_2)iL$	$\frac{1}{12} (K_1 + 3K_2)iL$	$\frac{1}{6} (1 + \beta)K + (1 + \beta)K_2iL$
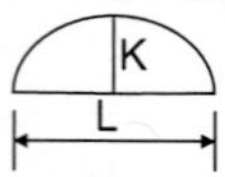	$\frac{2}{3} KiL$	$\frac{1}{3} KiL$	$\frac{1}{3} K(i_1 + i_2)L$	$\frac{8}{15} KiL$	$\frac{1}{5} KiL$	$\frac{1}{3} (1 + \alpha\beta)KiL$
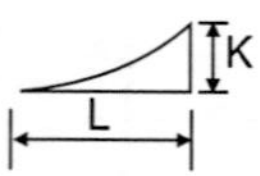	$\frac{1}{3} KiL$	$\frac{1}{4} KiL$	$\frac{1}{12} K(i_1 + 3i_2)L$	$\frac{1}{5} KiL$	$\frac{1}{5} KiL$	$\frac{1}{12} (1 + \alpha + \alpha^2) KiL$
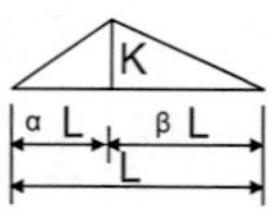	$\frac{1}{2} KiL$	$\frac{1}{6} (1 + \alpha)KiL$	$\frac{1}{6} KL[(1 + \beta)i_1 + (1 + \alpha)i_2]$	$\frac{1}{3} (1 + \alpha\beta) KiL$	$\frac{1}{2} (1 + \alpha + \alpha^2) KiL$	$\frac{1}{3} KiL$

$$f_{11} = \int_{\substack{\text{inclined}\\ \text{members}}} \frac{M}{EI}\frac{\partial M}{\partial R_1}dx + \int_{\text{tie-rod}} \frac{N}{EA}\frac{\partial N}{\partial R_1}dx$$

$$= \frac{1}{3} \times \frac{5\times3\times3\times2}{5\times10^3} + \frac{8\times1\times1}{10\times10^3}$$

$$= 68 \times 10^{-4}\text{m}$$

Applying the principle of superposition of displacements, yields

$$68 \times 10^{-4} R_1 + 4 \times 10^{-2} = 0$$

or

$$R_1 = -\frac{4}{68}\times10^2 = -5.889 \text{ kN}$$

It may be noted that the force R_1 in the tie-rod has been assumed as compressive and the negative sign of the answer indicates that the force is in fact, tensile.

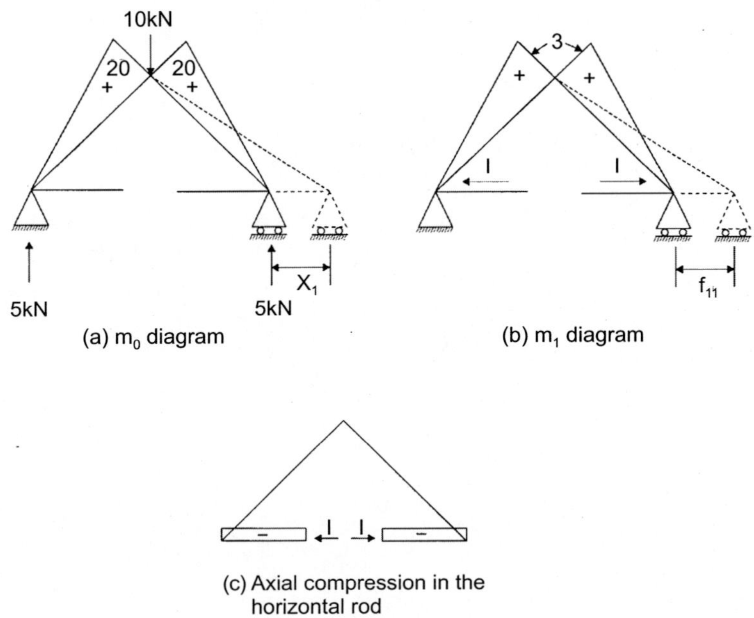

Fig. 3.8. Bending moment (in KNm) and axial force diagrams(in KN) of the released frame

Example 3.3:

For the plane truss of Fig. 3.9, the forces in the members are to be determined. Select suitable redundants and determine the forces in them. Axial rigidity of the horizontal and vertical members is EA and for diagonal members it is $2EA$.

The plane truss of Fig. 3.9 has two degrees of statical indeterminacy. The diagonals *BF* and *BD* are selected as redundants. These two bars are cut to make the structure statically determinate. It must be noticed at this stage that the cut bars remain a part of the released structure and its deformations are to be included in the calculation of displacements. For convenience, the members are numbered and the results are done in a tabular manner. Consider only the axial force components in Fig. 3.9. The displacement obtained by the application of Castigliano's second theorem then would be:

$$X_i = \sum N \frac{\partial N}{\partial P_i} \frac{L}{AE}$$

Therefore, displacements corresponding to the release due to external load and flexibility coefficients are given below. (referring to Fig. 3.10) (calculation of member forces corresponding to the releases are given in Table 3.2):

$$X_i = (-Q)\left(-\frac{3}{5}\right)\frac{3L}{AE} + \left(-\frac{Q}{2}\right)\left(-\frac{4}{5}\right)\frac{4L}{EA} + \left(-\frac{11}{12}Q\right)$$

$$\left(\frac{4}{5}\right)\frac{4L}{EA} + \left(-\frac{25}{48}Q \times \frac{5L}{2EA}\right) = -\,0.835\frac{QL}{AE}$$

$$X_2 = \left(\frac{11}{12}Q\right)\left(-\frac{4}{5}\right)\frac{4L}{AE} + \left(-\frac{55}{48}Q\right) \times 1 \times \frac{5L}{2AE} = -5.79\frac{QL}{AE}$$

$$f_{11} = \left[\left(-\frac{3}{5}\right)\left(-\frac{3}{5}\right)\frac{3L}{AE} + \left(-\frac{4}{5}\right)\left(-\frac{4}{5}\right)\frac{4L}{AE} + \frac{1 \times 1 \times 5L}{2AE}\right]2$$

$$= \frac{12.28L}{AE}$$

$$f_{12} = f_{21} = \left(-\frac{3}{5}\right)\left(-\frac{3}{5}\right)\frac{3L}{AE} = 1.08\frac{L}{AE}$$

$$f_{22} = \frac{12.28L}{AE}$$

Applying the principle of superposition for displacements, it yields

$$\frac{L}{AE}\begin{bmatrix} 12.28 & 1.08 \\ 1.08 & 12.28 \end{bmatrix}\begin{Bmatrix} R_1 \\ R_2 \end{Bmatrix} = \frac{QL}{AE}\begin{Bmatrix} 0.835 \\ 5.790 \end{Bmatrix}$$

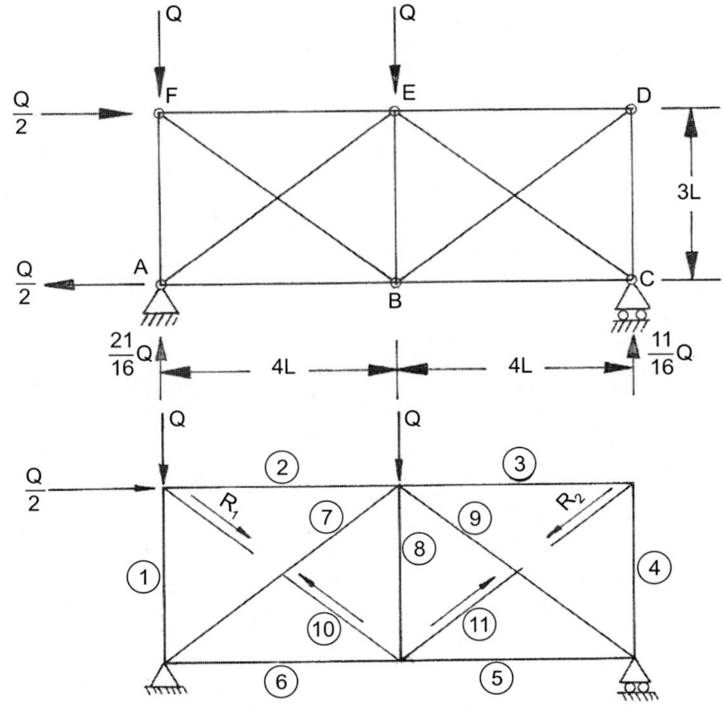

Fig. 3.9. A redundant truss

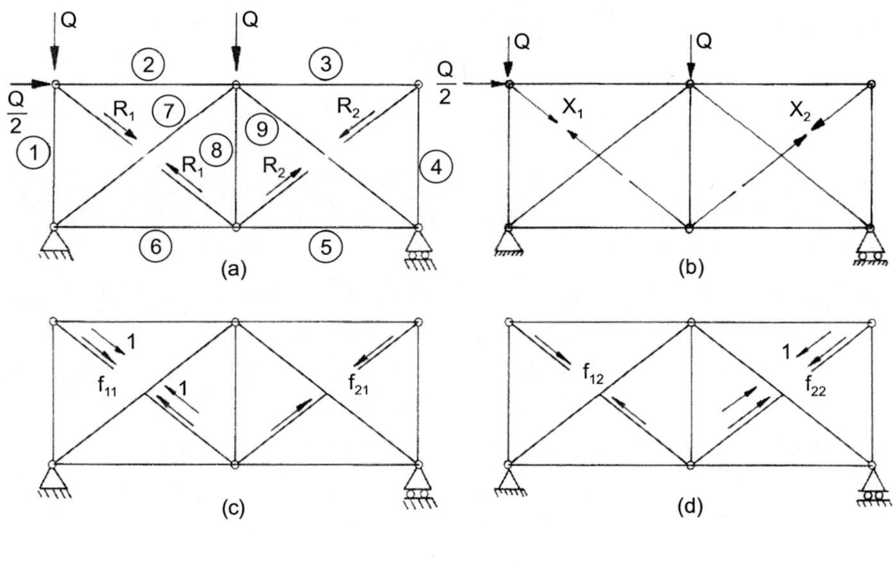

Fig. 3.10. Forces in the released truss

Table 3.2. Calculation of Forces in the Members of the Truss

Member	Forces Fig. 3.10(*b*)	Forces Fig. 3.10(*c*)	Forces Fig. 3.10(*d*)	Final forces
1	$-Q$	–3/5	0	$-1.016Q$
2	$-Q/2$	–4/5	0	$-0.5214Q$
3	0	0	–4/5	$-0.3753Q$
4	0	0	–3/5	$-0.2815Q$
5	$\frac{11}{12}Q$	0	–4/5	$0.5414Q$
6	$\frac{11}{12}Q$	~	0	$0.8953Q$
7	$-\frac{25}{48}Q$	1	0	$-0.4941Q$
8	0	–3/5	–3/5	$-0.2975Q$
9	$-\frac{55}{48}Q$	0	1	$-0.6767Q$
10	0	1	0	$-0.02674Q$
11	0	0	1	$0.4691Q$

The following are the values of R_1 and R_2:

$$R_1 = 0.023736Q$$

$$R_2 = 0.4691Q$$

3.3 THE STIFFNESS METHOD

The steps followed in the stiffness method are described first. Then the method is explained with examples. Finally, a number of examples demonstrating the applicability of the method to a variety of structural analysis problems is presented.

First step consists of the determination of the degrees of freedom of the structure. This involves the identification of separate elements of the structure. The intersections or interconnections between these elements are called joints or nodes. Displacements corresponding to each degree of freedom referred to as nodal displacements are numbered and positive directions are assumed.

The sign convention for the load is that if they are in the same direction as the nodal displacements, they are assumed to be positive and *vice versa* (Fig. 3.11).

Secondly, all the nodal displacements are prevented and fixed end forces for all the members at desired location under external loading are determined.

Restraining forces are obtained as a sum of the fixed end forces. Fixed end forces for uniform members for some typical loading cases are given in Appendix A.

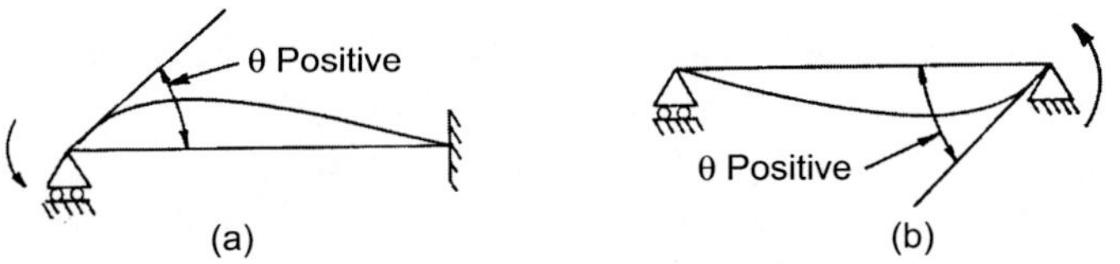

Fig. 3.11. Sign convention

Thirdly, the structure is assumed to deform corresponding to a unit value of one of the nodal displacements while all other displacements are restrained. The forces at all the required points are evaluated. The process is repeated for a unit value of each of the nodal displacements.

Fourthly, the displacements required to eliminate the restraining forces in step 2 are determined. This requires the application of the principle of superposition of forces. Final values of displacements are obtained from the solution of a set of simultaneous equations.

Finally, the end forces of the different members are obtained on the basis of superposing the fixed end forces on the restrained structure to those caused by nodal displacements.

In this method, the displacements corresponding to the degrees of freedom are determined and then, the forces in the members are evaluated.

The method is illustrated with an example. Let us consider the same continuous beam of Fig. 3.1 and analyse it by the displacement method.

Step 1: Referring to Fig. 3.12, the continuous beam has two degrees of freedom - the joint rotations at *B* and *C*. They are denoted as X_1 and X_2.

Step 2: All joints where displacements can take place *i.e.,* joints *B* and *C* in this case are clamped in Fig. 3.12(*b*) and the fixed end moments of the members *AB* and *BC* determined. This follows a standard procedure. The unbalanced moments at the joints are termed as R_1 and R_2.

$$R_1 = \text{fixed end moment at } B \text{ from the member } AB$$
$$+ \text{ fixed end moment at } B \text{ from the member } BC$$

The signs of the moments are to be properly taken into account. Therefore, referring to Table in Appendix I for the particular loading case,

$$R_1 = -\frac{2PL}{8} + \frac{PL}{8} = -\frac{PL}{8}$$

Similarly, R_2 can be determined. But joint *C* is the end of the member *BC*. The unbalanced moment is simply the fixed end moment at *C* for member *BC* under the given loading.

Therefore,

$$R_2 = -\frac{PL}{8}$$

Step 3: Joints are now allowed to deform by a unit value of the nodal displacement one at a time, while all other displacements are restrained. Referring to Fig. 3.12(*c*), the joint *B* is allowed to rotate by unity in the counterclockwise direction (*i.e.* positive) while joint *C* is kept fixed. The moment developed at *B* due to unit rotation at *B* is termed K_{11} and the moment induced at *C* due to this displacement is K_{21}. In order to determine K_{11}, contribution from both the members *AB* and *BC* is to be taken into account (referring to. Fig. 3.13),

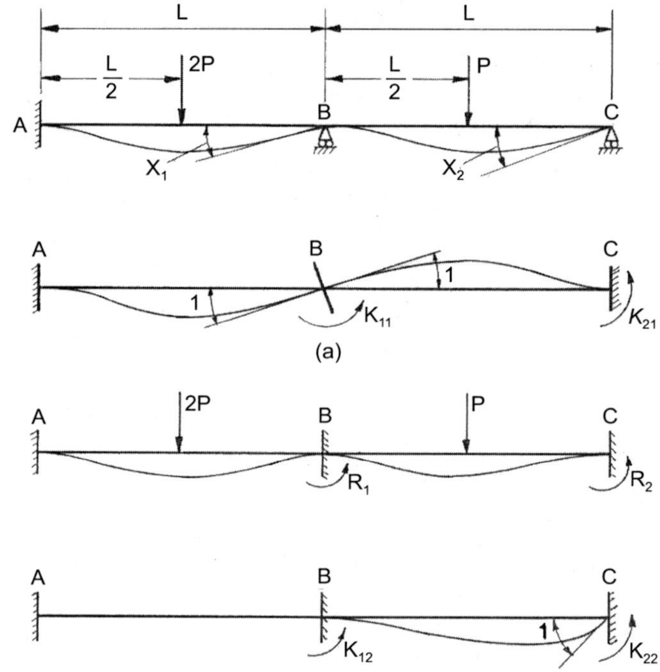

Fig. 3.12. Different steps in the displacement method

$$K_{11} = \frac{4EI}{L} + \frac{4EI}{L} + \frac{8EI}{L}$$

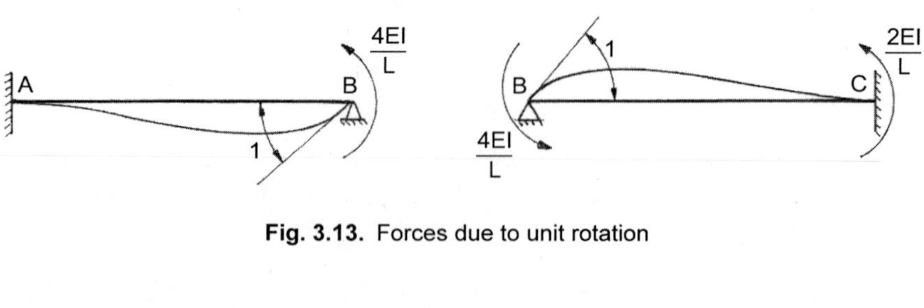

Fig. 3.13. Forces due to unit rotation

$$K_{21} = \frac{2EI}{L}$$

Similarly, $$K_{12} = \frac{2EI}{L} \text{ and } K_{22} = \frac{4EI}{L}$$

The total moment at B consisting of the unbalanced moment acting at that joint and the member end moments due to all possible rotations of the joints, is equal to zero. Similarly, end C being roller support, total moment at C will be equal to zero. This follows from equilibrium conditions of the joints. By the application of the principle of superposition of forces, following equations result.

$$\left.\begin{aligned} K_{11}X_1 + K_{12}X_2 + R_1 = 0 \\ K_{21}X_1 + K_{22}X_2 + R_2 = 0 \end{aligned}\right\} \tag{3.6}$$

Equation (3.6) written in matrix form becomes

$$\begin{bmatrix} K_{11} & K_{12} \\ K_{21} & K_{22} \end{bmatrix} \begin{Bmatrix} X_1 \\ X_2 \end{Bmatrix} + \begin{Bmatrix} R_1 \\ R_2 \end{Bmatrix} = \begin{Bmatrix} 0 \\ 0 \end{Bmatrix} \tag{3.7}$$

which in compact form becomes

$$[K]\{X\} + \{R\} = \{0\} \tag{3.8}$$

The matrix $[K]$ is called the stiffness matrix and its elements K_{11} and K_{12} *etc*. are called stiffness coefficients. K_{ij} is defined as the force developed at point i corresponding to the unknown displacement at i due to unit value of the displacement at j corresponding to the unknown displacement at j. It may be noted that the stiffness matrix is symmetric.

Substituting the value of K_{11}, K_{12} ..., R_2 into eqn. (3.7),

$$\frac{2EI}{L} = \begin{bmatrix} 4 & 1 \\ 1 & 2 \end{bmatrix} \begin{Bmatrix} X_1 \\ X_2 \end{Bmatrix} - \frac{PL}{8} \begin{Bmatrix} 1 \\ 1 \end{Bmatrix} = \begin{Bmatrix} 0 \\ 0 \end{Bmatrix}$$

Therefore,

$$\begin{Bmatrix} X_1 \\ X_2 \end{Bmatrix} = \frac{L}{14EI} \begin{bmatrix} 2 & -1 \\ -1 & 4 \end{bmatrix} \times \left(\frac{PL}{8}\right) \begin{Bmatrix} 1 \\ 1 \end{Bmatrix}$$

$$\begin{Bmatrix} X_1 \\ X_2 \end{Bmatrix} = \frac{PL^2}{112EI} \begin{Bmatrix} 1 \\ 3 \end{Bmatrix}$$

Usually, one is more interested in knowing the end forces rather than displacements for most of the problems. So in the stiffness method, the forces in the members are evaluated after the values of unknown displacements at joints are obtained. This can be done in a systematic manner.

If a member of a continuous beam is isolated as shown in Fig. 3.14, the end moments of the member are supposed to consist of the fixed end moment and bending moments developed at that end due to rotations at the two ends of the member.

In Fig. 3.14, the end moment at end 1 is given by

$$PM_1 = PMF_1 + PMX_{11} \times X_1 + PMX_{12} \times X_2 \tag{3.9}$$

where PMF_1 is the fixed end moment at 1, PMX_{11} is the moment induced at 1 due to a unit value of the rotation at 1 and PMX_{12} is the moment induced at 1 due to a unit value of rotation at 2.

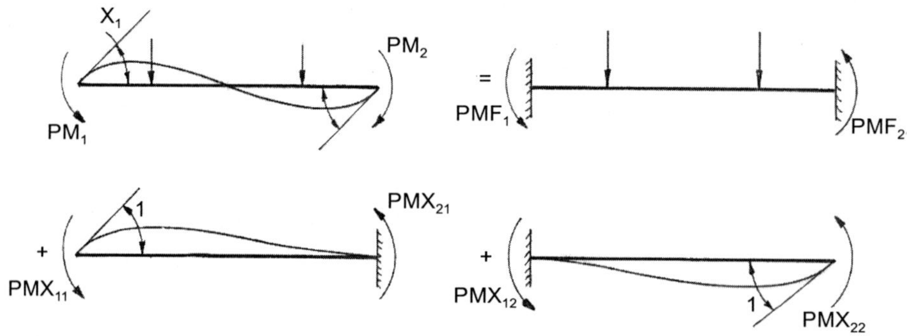

Fig. 3.14. Contribution to the end moment of a member of a continuous beam

Similarly,

$$PM_2 = PMF_2 + PMX_{21} \times X_1 + PMX_{22} \times X_2 \tag{3.10}$$

Equations (3.9) and (3.10) when written in matrix form become

$$\begin{Bmatrix} PM_1 \\ PM_2 \end{Bmatrix} = \begin{Bmatrix} PMF_1 \\ PMF_2 \end{Bmatrix} + \begin{bmatrix} PMX_{11} & PMX_{12} \\ PMX_{21} & PMX_{22} \end{bmatrix} \begin{Bmatrix} X_1 \\ X_2 \end{Bmatrix} \tag{3.11}$$

which in compact form becomes

$$\{PM\} = \{PMF\} + [PMX]\{X\} \tag{3.12}$$

Similarly, the end reactions can be shown to be equal to the superposition of reactions from the fixed end condition, to reactions developed due to nodal displacements. In matrix notations, this can be written as

$$\{PR\} = \{PRF\} + [PRX]\{X\} \tag{3.13}$$

In order to find out member forces and reactions for any structure, eqns. (3.12) and (3.13) are to be used. The construction of $\{PMF\}$, $\{PRF\}$, $[PMX]$ and $[PRX]$ are to be simultaneously developed with the formation of the stiffness matrix. However, final solution of eqn. (3.13) is awaited till the unknown displacements $\{X\}$ are evaluated.

Let us continue the problem of Fig. 3.1 and determine member end actions and support reactions. The member forces to be determined are the end moment and the shear force at each end of the member. The support reactions are four in number and they are consequently numbered. The end actions and reactions are shown with their positive sign convention in Fig 3.15. For fixed end condition, members *AB* and *BC* are considered separately and shown in Fig. 3.16.

$$\{PMF\} = \begin{bmatrix} \frac{PL}{4} \\ P \\ -\frac{PL}{4} \\ -P \\ \frac{PL}{8} \\ \frac{P}{2} \\ -\frac{PL}{8} \\ -\frac{P}{2} \end{bmatrix}$$

and $$\{PRF\} = \begin{Bmatrix} P \\ \frac{PL}{4} \\ \frac{3P}{2} \\ \frac{P}{2} \end{Bmatrix}$$

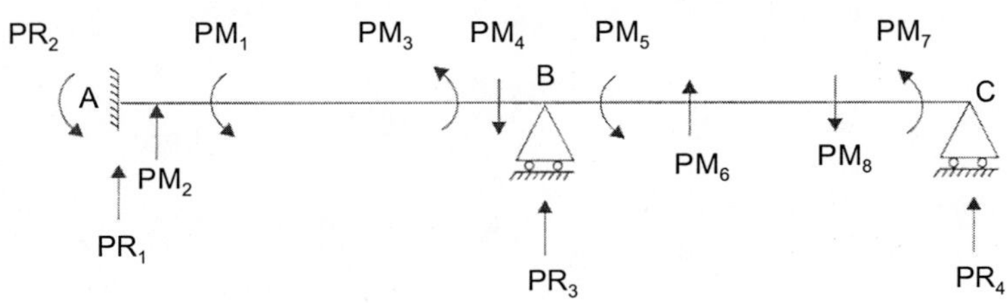

Fig. 3.15. The end moments and reactions of continuous beam of Fig. 3.1

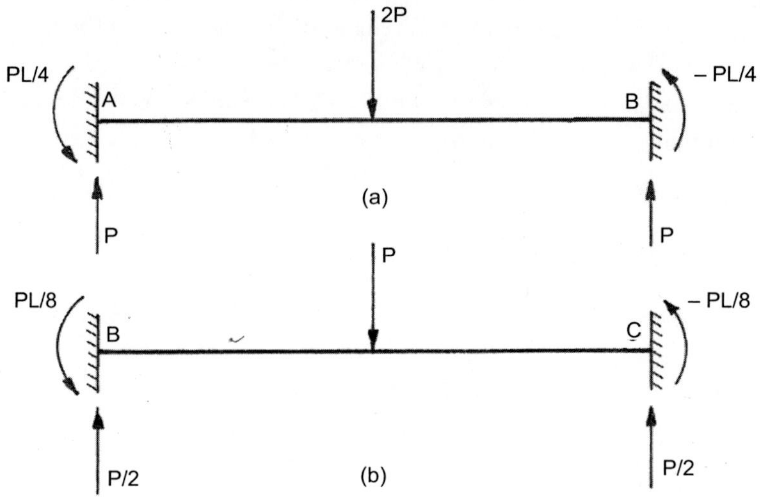

Fig. 3.16. Fixed end forces of members

Similarly, corresponding to a unit value of the displacement one at a time, the member end forces and reactions are shown in Fig. 3.17. Based on the values indicated in the Fig. 3.17, $\{PMX\}$ and $\{PRX\}$ can be formed. They are

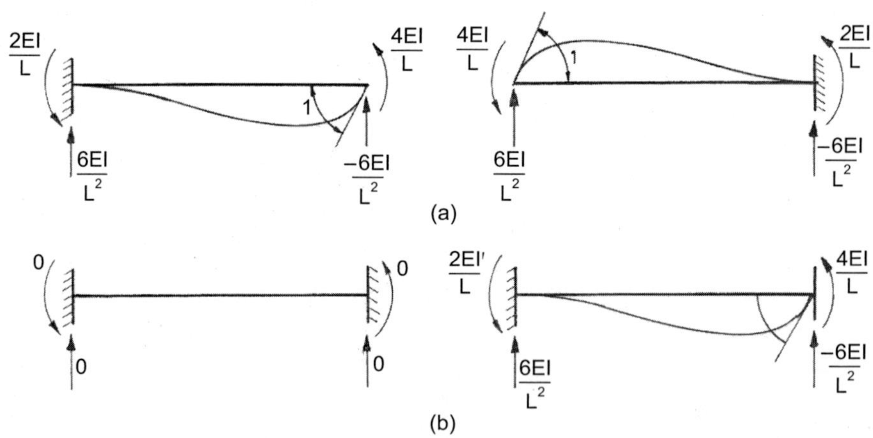

Fig. 3.17. Member end forces and reactions due to unit value of the displacement

$$[PMX] = \begin{bmatrix} \frac{2EI}{L} & 0 \\ \frac{6EI}{L^2} & 0 \\ \frac{4EI}{L} & 0 \\ \frac{6EI}{L^2} & 0 \\ \frac{4EI}{L} & \frac{2EI}{L} \\ \frac{6EI}{L^2} & \frac{6EI}{L^2} \\ \frac{2EI}{L} & \frac{4EI}{L} \\ \frac{6EI}{L^2} & \frac{6EI}{L^2} \end{bmatrix}$$

and $$[PRX] = \begin{bmatrix} \frac{6EI}{L^2} & 0 \\ \frac{2EI}{L} & 0 \\ 0 & \frac{6EI}{L^2} \\ -\frac{6EI}{L^2} & -\frac{6EI}{L^2} \end{bmatrix}$$

Applying eqn. (3.12) yields

$$\{PM\} = \frac{P}{8}\begin{Bmatrix} 2L \\ 8 \\ -2L \\ -8 \\ 1 \\ 4 \\ -L \\ -4 \end{Bmatrix} + \frac{EI}{L^2}\begin{bmatrix} 2L & 0 \\ 6 & 0 \\ 4L & 0 \\ 6 & 0 \\ 4L & 2L \\ 6 & 6 \\ 2L & 4L \\ 6 & 6 \end{bmatrix}\frac{PL^2}{112EI}\begin{Bmatrix} 1 \\ 3 \end{Bmatrix}$$

$$= \frac{P}{112}\begin{Bmatrix} 30L \\ 118 \\ -24L \\ -106 \\ 24L \\ 80 \\ 0 \\ -32 \end{Bmatrix}$$

This result agrees with those obtained from the flexibility method. Similarly, for determining the reactions, eqn. (3.13) yields

$$\{PR\} = \frac{P}{4}\begin{Bmatrix} 4 \\ L \\ 6 \\ 2 \end{Bmatrix} + \frac{EI}{L^2}\begin{bmatrix} 6 & 0 \\ 2L & 0 \\ 0 & 6 \\ -6 & -6 \end{bmatrix}\frac{PL^2}{112EI}\begin{Bmatrix} 1 \\ 3 \end{Bmatrix}$$

$$= \frac{P}{112}\begin{Bmatrix} 118 \\ 30L \\ 186 \\ 32 \end{Bmatrix}$$

It may be noted that whereas in the flexibility method, there are many possibilities of choosing the released structure, in stiffness method the analyst is left with no option, but to follow a rather set procedure. A few problems solved by the stiffness method are given in the following examples.

Example 3.4:

The plane frame of Fig. 3.18(*a*) is to be analysed by the displacement method. All members are uniform, and have flexural rigidity *EI*.

The axial deformation of the members will be neglected. The frame has three degrees of freedom-one horizontal displacement denoted by X_1 which indicates horizontal translation of the joint *B* or *C* which is same if the axial deformation is neglected and two rotations denoted by X_2 and X_3 at joints *B* and *C* respectively. The end moments of the members are also shown in Fig. 3.18(*b*). R_1, R_2 and R_3 are the reactions corresponding to the displacements X_1, X_2 and X_3 respectively.

Next, all joints are clamped and the fixed end moments at *B* and *C* of the members are determined. Referring to Fig. 3.18(*c*),

$$R_1 = -\frac{P}{2}$$

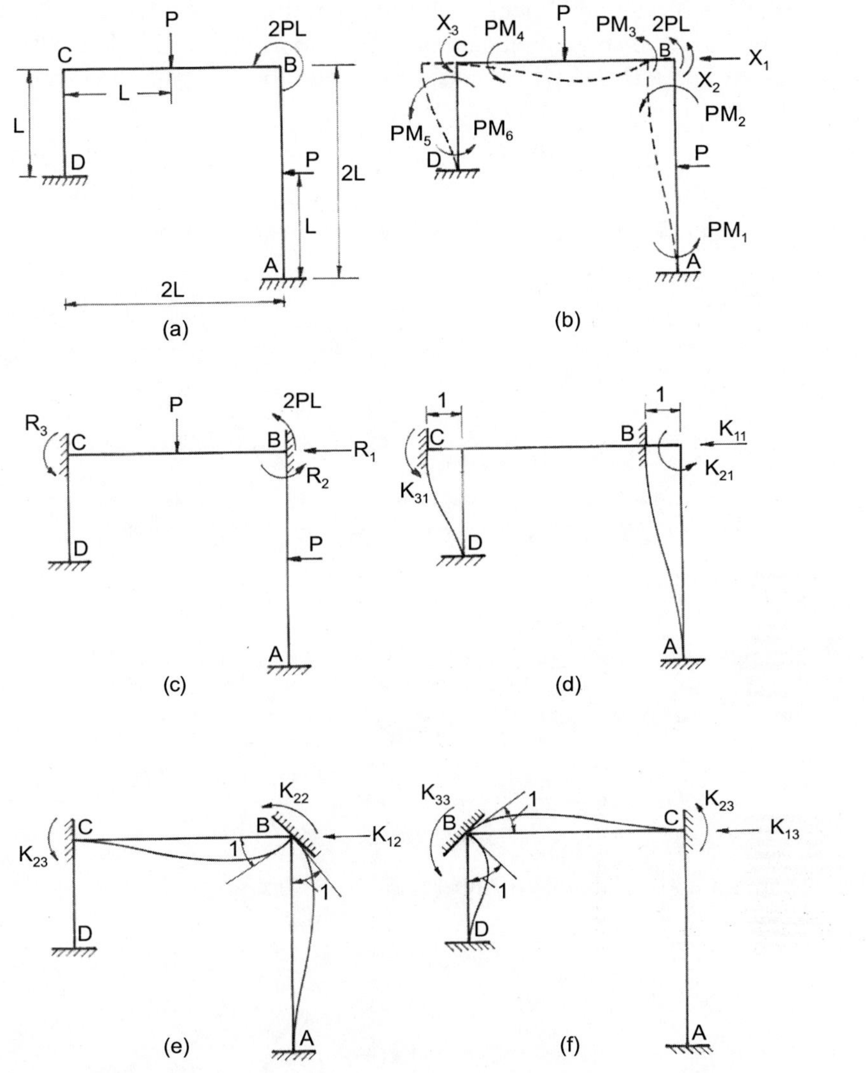

Fig. 3.18. Different steps to be followed in the rigid frame analysis by the stiffness method

$$R_2 = -\frac{PL}{4} + \frac{PL}{4} - 2PL = -2PL$$

$$R_3 = \frac{PL}{4}$$

Therefore,

$$\begin{Bmatrix} R_1 \\ R_2 \\ R_3 \end{Bmatrix} = \frac{P}{4}\begin{Bmatrix} -2 \\ -8L \\ L \end{Bmatrix}$$

Next, the stiffness coefficients are to be determined. The forces developed due to a unit horizontal displacement corresponding to X_1 is shown in Fig. 3.18(*d*) for members *AB* and *CD*. K_{11} will consist of the contribution from the members *AB* and *CD*. Thus

$$K_{11} = \frac{108}{8}\frac{EI}{L^3}$$

Referring to Fig. 3.18(*d*), K_{21} and K_{31} will assume the following values.

$$K_{21} = \frac{6EI}{4L^2}$$

$$K_{31} = \frac{6EI}{L^2}$$

In the same way, other stiffness coefficients may be determined. Referring to Fig. 3.18(*e*),

$$K_{12} = -\frac{6EI}{4L^2},\quad K_{22} = \frac{8EI}{2L},\quad K_{32} = \frac{2EI}{2L}$$

Similarly,

$$K_{13} = -\frac{6EI}{L^2},\quad K_{23} = \frac{2EI}{2L},\quad K_{33} = \frac{12EI}{2L}$$

Therefore, the stiffness matrix is

$$[K] = \frac{EI}{8L^3}\begin{bmatrix} 108 & -12L & -48L \\ -12L & 32L^2 & 8L^2 \\ -48L & 8L^2 & 48L^2 \end{bmatrix}$$

Finally applying the principle of superposition of forces, yields

$$\frac{EI}{8L^3}\begin{bmatrix} 108 & -12L & -48L \\ -12L & 32L^2 & 8L^2 \\ -48L & 8L^2 & 48L^2 \end{bmatrix}\begin{Bmatrix} X_1 \\ X_2 \\ X_3 \end{Bmatrix} + \frac{P}{4}\begin{Bmatrix} -2 \\ -8L \\ L \end{Bmatrix} = \begin{Bmatrix} 0 \\ 0 \\ 0 \end{Bmatrix}$$

$$\{X\} = [K]^{-1}\{R\} = \frac{L^3}{684EI}\begin{bmatrix} 92 & \frac{12}{L} & \frac{90}{L} \\ \frac{12}{L} & \frac{180}{L^2} & -\frac{18}{L^2} \\ \frac{90}{L} & -\frac{18}{L^2} & \frac{207}{L^2} \end{bmatrix}\frac{P}{4}\begin{Bmatrix} 2 \\ 8L \\ -L \end{Bmatrix}$$

Therefore,

$$\begin{Bmatrix} X_1 \\ X_2 \\ X_3 \end{Bmatrix} = \frac{PL^3}{EI}\begin{Bmatrix} 0.0694 \\ 0.5416/L \\ -0.0625/L \end{Bmatrix}$$

The end forces in the members numbered PM_1 to PM_6 are to be determined. Referring to Fig. 3.18(b) and Fig. 3.18(c), yields

$$[PMX] = \begin{bmatrix} -\dfrac{6EI}{4L^2} & \dfrac{2EI}{2L} & 0 \\ -\dfrac{6EI}{4L^2} & \dfrac{4EI}{2L} & 0 \\ 0 & \dfrac{4EI}{2L} & \dfrac{2EI}{2L} \\ 0 & \dfrac{2EI}{2L} & \dfrac{4EI}{2L} \\ -\dfrac{6EI}{L^2} & 0 & \dfrac{4EI}{L} \\ -\dfrac{6EI}{L^2} & 0 & \dfrac{2EI}{L} \end{bmatrix}$$

and

$$\{PMF\} = \begin{Bmatrix} -\dfrac{PL}{4} \\ \dfrac{PL}{4} \\ -\dfrac{PL}{4} \\ \dfrac{PL}{4} \\ 0 \\ 0 \end{Bmatrix}$$

Therefore, applying eqn. (3.12), yields

$$\{PM\} = \frac{PL}{4}\begin{Bmatrix} -1 \\ 1 \\ -1 \\ 1 \\ 0 \\ 0 \end{Bmatrix} + \frac{EI}{4L^2}\begin{bmatrix} -6 & 4L & 0 \\ -6 & 8L & 0 \\ 0 & 8L & 4L \\ 0 & 4L & 8L \\ -24 & 0 & 16L \\ -24 & 0 & 8L \end{bmatrix} \frac{PL^3}{EI}\begin{Bmatrix} 0.0694 \\ 0.5416/L \\ -0.0625/L \end{Bmatrix}$$

or,

$$\{PM\} = PL\begin{Bmatrix} 0.1875 \\ 1.2290 \\ 0.7707 \\ 0.6666 \\ -0.6664 \\ -0.5415 \end{Bmatrix}$$

Example 3.5:

Let us now see the application of the displacement method in the case of a truss of Fig. 3.19. The forces in each of the members are to be determined. *EA* is same for all the members.

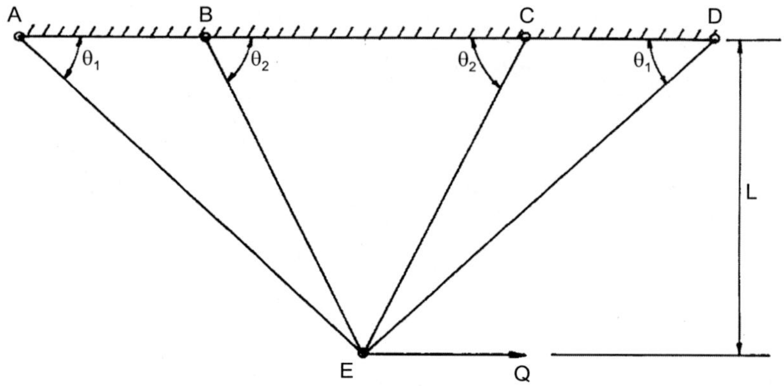

Fig. 3.19. Truss problem

The only joint that can be displaced is joint *E* and for this plane truss, joint *E* can have two translational displacements indicated by X_1 and X_2 in Fig. 3.20. R_1 and R_2 are the reactions acting at *E* corresponding to displacements X_1 and X_2.

Joint *E* is kept fixed and reactions at *E* due to the external load in this case will be the horizontal reaction acting to the left due to the force *Q* acting to the right. Therefore,

$$\begin{Bmatrix} R_1 \\ R_2 \end{Bmatrix} = \begin{Bmatrix} 0 \\ -Q \end{Bmatrix}$$

Next, the stiffness coefficients are to be determined. Let us first look into the member *AE*. The length of the member is given by

$$L_1 = \frac{L}{\sin\theta_1}$$

Due to unit movement in the vertical direction of joint *E*, member *AE* elongates [Fig. 3.20(*b*)] and the tensile force developed is $T_1 = \frac{EA}{L}\sin^2\theta_1$ and the vertical component of the tensile force is $\frac{EA}{L}\sin^3\theta_1$, and the horizontal component is $\frac{EA}{L}\sin^2\theta_1\cos\theta_1$ acting to the left for this member. Therefore, in order to determine K_{11}, the contribution of the vertical components of the forces from all members meeting at that joint is to be taken into account.

$$K_{11} = 2\,(\sin^3\theta_1 + \sin^3\theta_2)\frac{EA}{L}$$

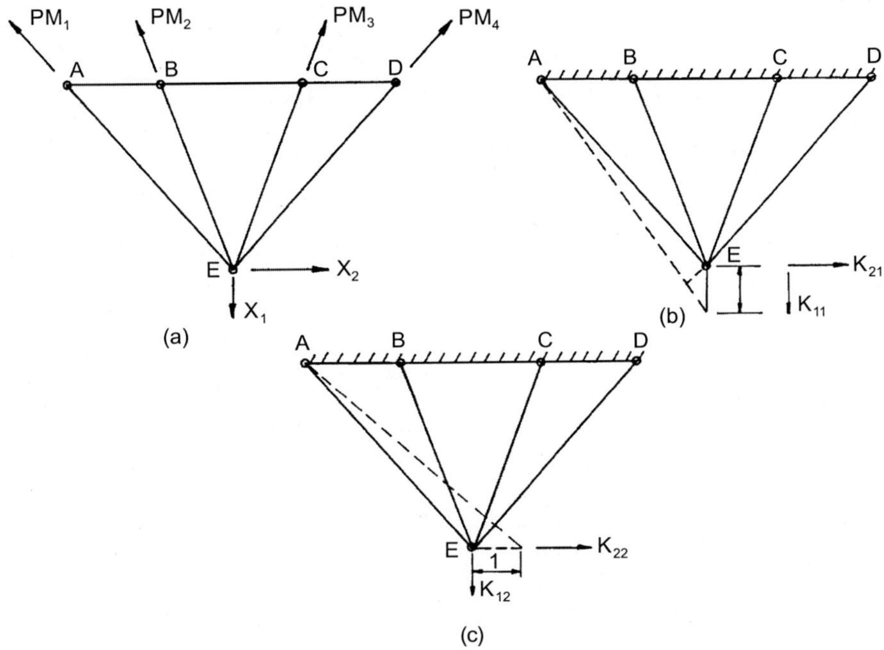

Fig. 3.20. Different steps in the solution of truss problem by the stiffness method

Similarly, sum of the horizontal, components of the forces in the members will give K_{21}. In this case it may be verified that this becomes zero. Thus, $K_{21} = 0$.

For stiffness coefficients corresponding, to unit horizontal displacement, let us again focus our attention to the member *AE*. Tensile force developed in *AE* is $\frac{EA}{L}\cos\theta_1\sin\theta_1$ and horizontal component of this force is $\frac{EA}{L}\cos^2\theta_1\sin\theta_1$. It may be noticed here that the sum of the vertical component of the forces in all members meeting at *E* due to a unit horizontal movement is equal to zero. Therefore,

$$K_{22} = 2\,(\cos^2\theta_1\sin\theta_1 + \cos^2\theta_2\sin\theta_2)\frac{EA}{L}$$

$$K_{21} = 0$$

Before proceeding with the final evaluation of the unknown displacement at joint *E*, let us have a look into the member forces. We know

$$\{PM\} = \{PMF\} + [PMX]\{X\}$$

In this case $\{PMF\}$ is a null matrix and referring to Fig. 3:20(b) and 3.20(c), it can be shown that

$$[PMX] = \frac{EA}{L}\begin{bmatrix} \sin^2\theta_1 & \cos\theta_1\sin\theta_1 \\ \sin^2\theta_2 & \cos\theta_2\sin\theta_2 \\ \sin^2\theta_2 & -\cos\theta_2\sin\theta_2 \\ \sin^2\theta_1 & -\cos\theta_1\sin\theta_1 \end{bmatrix}$$

Let us assume that $\theta_1 = 45°$ and $\theta_2 = 60°$. Substituting these values in stiffness coefficients, and applying the principle of superposition,.

$$\frac{EA}{L}\begin{bmatrix} 2.006 & 0 \\ 0 & 1.14 \end{bmatrix}\begin{Bmatrix} X_1 \\ X_2 \end{Bmatrix} + \begin{Bmatrix} 0 \\ -Q \end{Bmatrix} = \begin{Bmatrix} 0 \\ 0 \end{Bmatrix}$$

Solving the above equations,

$$X_1 = 0$$

$$X_2 = \frac{QL}{1.14AE} = 0.8771\frac{QL}{AE}$$

The member forces then are

$$\{PM\} = \frac{EA}{L}\begin{bmatrix} \frac{1}{2} & \frac{1}{2} \\ \frac{3}{4} & \sqrt{\frac{3}{4}} \\ \frac{3}{4} & -\sqrt{\frac{3}{4}} \\ \frac{1}{2} & -\frac{1}{2} \end{bmatrix}\begin{Bmatrix} 0 \\ \frac{1}{1.14} \end{Bmatrix}\frac{QL}{AE}$$

$$= Q\begin{Bmatrix} 0.438 \\ 0.380 \\ -0.380 \\ -0.438 \end{Bmatrix}$$

Example 3.6:

The grid of Fig. 3.21 will be solved. The grid is a planar structure and the difference between a grid and a rigid frame lies in the application of the load. The loading lies in the plane of the frame in the case of rigid frame, whereas the grid is subjected to loads perpendicular to its plane. Members are primarily subjected to flexural action in the case of a rigid frame, whereas in a grid, members are subjected to bending as well as twisting. Assume $GJ/EI = 0.8$ for both the members AB and BC. The joint B is assumed to be rigid. The degrees of freedom for the grid of Fig. 3.21 is 3, of which two are in rotation about two orthogonal axes in the horizontal plane and the third is the vertical displacement. The internal forces of the members of the grid are shown in Fig. 3.22(a).

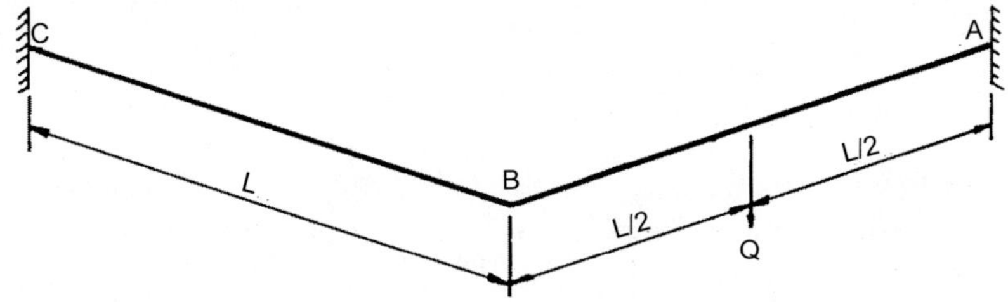

Fig; 3.21. A grid

The structure lies in the x-y plane and all applied forces act parallel to z-axis. Each member experiences torsion as well as shear and bending. The internal forces acting on a member in a plane grid are completely defined by two flexural end moments and the torsional moment in the member. PM_5 and PM_6 are the twisting moments of the members AB and BC respectively and their direction is positive when the arrows indicating the moments act outwards. $\{R\}$ matrix corresponding to the displacements in this case is

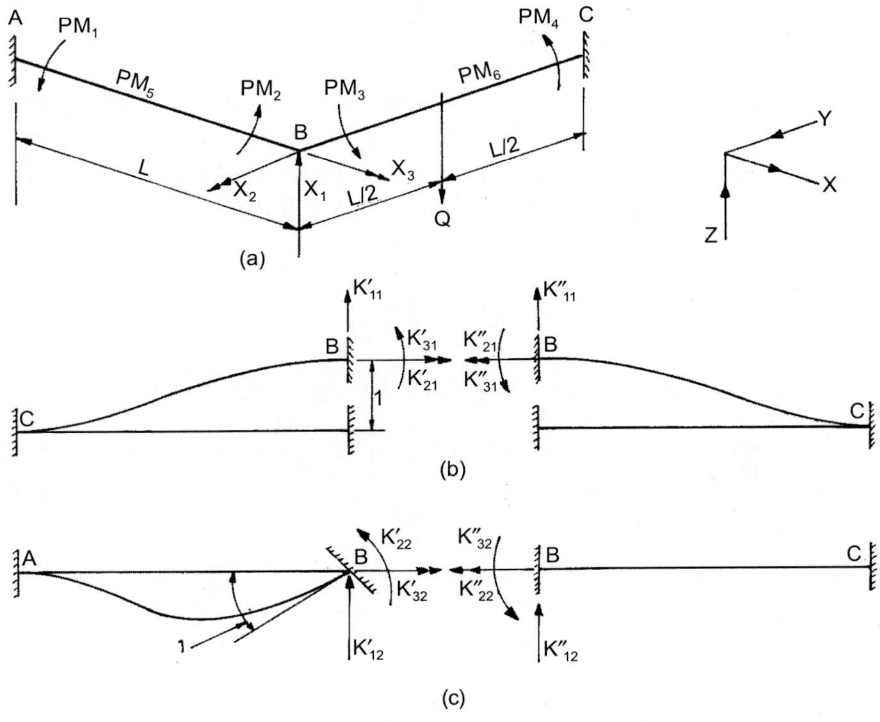

Fig. 3.22. Different steps in the analysis of the grid structure

$$\{R\} = \begin{Bmatrix} \dfrac{Q}{2} \\ 0 \\ \dfrac{QL}{8} \end{Bmatrix} = \frac{Q}{8}\begin{Bmatrix} 4 \\ 0 \\ L \end{Bmatrix}$$

The stiffness matrix for the grid is next to be formed. The stiffness coefficients corresponding to a unit value of X_1 are determined as follows. Referring to Fig. 3.22(*b*), two members *AB* and *BC* are shown separately for this unit displacement. The values of the different quantities indicated in the Fig. 3.22(*b*) are the, following:

$$K'_{11} = \frac{12EI}{L^3},\ K'_{21} = -\frac{6EI}{L^2},\ K'_{31} = 0$$

Similarly,

$$K''_{11} = \frac{12EI}{L^3},\ K''_{21} = 0,\ K''_{31} = \frac{6EI}{L^2}$$

The elements of the first column of the stiffness matrix will be obtained by taking into account the contribution of both the members. Therefore,

$$K_{11} = \frac{24EI}{L^3},\ K_{21} = -\frac{6EI}{L^2},\ K_{31} = \frac{6EI}{L^2}$$

Next, let us find out the stiffness coefficients due to a unit displacement corresponding to X_2. Again, two members *AB* and *BC* are shown separately for this unit displacement. The end *B* of the member *AB* is rotated flexurally. The end *B* of the member *BC* is twisted. The different quantities of the Fig. 3.22(*c*) are:

$$K'_{12} = -\frac{6EI}{L^2},\ K''_{22} = \frac{4EI}{L},\ K'_{32} = 0$$

$$K''_{12} = 0,\ K''_{22} = \frac{GJ}{L},\ K''_{32} = 0$$

Therefore, combining the contribution of both the members meeting at the joint, yields

$$K_{12} = -\frac{6EI}{L^2},\ K_{22} = \frac{4EI}{L} + \frac{GJ}{L},\ K_{32} = 0$$

In the same way, remaining stiffness coefficients may be determined., They are

$$K_{13} = \frac{6EI}{L^2},\ K_{23} = 0,\ K_{33} = \frac{GJ}{L} + \frac{4EI}{L}$$

Once the stiffness coefficients are evaluated, the stiffness matrix becomes

$$[K] = \frac{EI}{L^3}\begin{bmatrix} 24 & -6L & 6L \\ -6L & (4+\beta)L^2 & 0 \\ 6L & 0 & (4+\beta)L^2 \end{bmatrix}$$

Substituting $\beta = \dfrac{GJ}{EI} = 0.8$ in the above equation, yields,

$$[K] = \frac{EI}{L^3}\begin{bmatrix} 24 & -6L & 6L \\ -6L & 4.8L^2 & 0 \\ 6L & 0 & 4.8L^2 \end{bmatrix}$$

Applying the principle of superposition of forces yields

$$\frac{Q}{8}\begin{Bmatrix} -4 \\ 0 \\ -L \end{Bmatrix} = \frac{EI}{L^3}\begin{bmatrix} 24 & -6L & 6L \\ -6L & 4.8L^2 & 0 \\ 6L & 0 & 4.8L^2 \end{bmatrix}\begin{Bmatrix} X_1 \\ X_2 \\ X_3 \end{Bmatrix}$$

Solution of the above equation, gives

$$\begin{Bmatrix} X_1 \\ X_2 \\ X_3 \end{Bmatrix} = \frac{QL}{1658.88EI}\begin{Bmatrix} 6.36L \\ 79.20 \\ 36.00 \end{Bmatrix}$$

$$= \frac{QL}{829.44EI}\begin{Bmatrix} -31.68L \\ -39.60 \\ 18.00 \end{Bmatrix}$$

The member forces can now be determined

$$\begin{Bmatrix} PMF_1 \\ PMF_2 \\ PMF_3 \\ PMF_4 \\ PMF_5 \\ PMF_6 \end{Bmatrix} = \begin{Bmatrix} 0 \\ 0 \\ \dfrac{QL}{8} \\ -\dfrac{QL}{8} \\ 0 \\ 0 \end{Bmatrix}$$

Likewise, $[PMX]$ may be determined.

$$[PMX] = \begin{bmatrix} \frac{-6EI}{L^2} & \frac{2EI}{L} & 0 \\ \frac{-6EI}{L^2} & \frac{4EI}{L} & 0 \\ \frac{6EI}{L^2} & 0 & \frac{4EI}{L} \\ \frac{6EI}{L^2} & 0 & \frac{2EI}{L} \\ 0 & 0 & \frac{GJ}{L} \\ 0 & \frac{GJ}{L} & 0 \end{bmatrix}$$

Therefore,

$$\begin{Bmatrix} PM_1 \\ PM_2 \\ PM_3 \\ PM_4 \\ PM_5 \\ PM_6 \end{Bmatrix} = \begin{Bmatrix} 0 \\ 0 \\ \frac{QL}{8} \\ \frac{-QL}{8} \\ 0 \\ 0 \end{Bmatrix} + \begin{bmatrix} \frac{-6EI}{L^2} & \frac{2EI}{L} & 0 \\ \frac{-6EI}{L^2} & \frac{4EI}{L} & 0 \\ \frac{6EI}{L^2} & 0 & \frac{4EI}{L} \\ \frac{6EI}{L^2} & 0 & \frac{2EI}{L} \\ 0 & 0 & \frac{GJ}{L} \\ 0 & \frac{GJ}{L} & 0 \end{bmatrix} \frac{QL^2}{829.44EI} \begin{Bmatrix} -31.68L \\ -39.60 \\ 18.00 \end{Bmatrix}$$

$$= QL \begin{Bmatrix} 0.133 \\ 0.038 \\ -0.017 \\ -0.310 \\ 0.017 \\ -0.038 \end{Bmatrix}$$

The equilibrium of joint B can be checked.

3.4 INCORPORATION OF DIFFERENT LOADING CONDITIONS

It has been seen in the previous section that the flexibility matrix and the stiffness matrix of a structure are independent of the loading and are properties of the structure. A structural engineer has to deal with different types of loading in analysing the structure. If a structure is analysed by taking one loading condition at a time, then the same flexibility and stiffness matrices are to be inverted depending on the method chosen for the analysis, every time for a separate load. This is time consuming and in fact of is not necessary, as all the loading conditions can be incorporated in one loading matrix - different columns of the matrix representing different loading conditions. So eqns. (3.5) and (3.8) are modified as follows:

$$\underset{NR \times NLC}{[R]} = \underset{NR \times NR}{-[F]^{-1}} \; \underset{NR \times NLC}{[X]} \tag{3.14}$$

and

$$\underset{NP \times NLC}{[X]} = \underset{NP \times NP}{-[K]^{-1}} \; \underset{NP \times NLC}{[R]} \tag{3.15}$$

where

NR is the number of redundants of the structure.

NP is the number of degrees of freedom.

NLC is the number of loading conditions for which results are required..

Each column of [*P*] and [*X*] refers to the corresponding values of the forces and displacements of the structure for a particular condition.

Similarly, the eqns. (3.12) and (3.13) for determining reactions and member forces by the displacement method can be modified to incorporate different loading conditions. They are done as follows:

$$\underset{NF \times NLC}{[PM]} = \underset{NF \times NLC}{[PMF]} + \underset{NF \times NP}{[PMX]} \; \underset{NP \times NLC}{[X]} \tag{3.16}$$

$$\underset{NRE \times NLC}{[PR]} = \underset{NRE \times NLC}{[PRF]} + \underset{NRE \times NP}{[PRX]} \; \underset{NP \times NLC}{[X]} \tag{3.17}$$

where

NF is the number of unknown member forces.

and *NRE* is the number of unknown reactions.

3.5 OTHER TYPES OF LOADINGS

In addition to the conventional loading which act in the form of concentrated, uniform and varying loads that act on a structure, an engineer has to deal with other effects such as the variation in temperature, lack of fit, prestrain and settlement of supports *etc*.

For many reasons, supports of the structure yield. Though the value of the settlement is small, the geometry is appreciably affected in the case of statically indeterminate structure and considerable stresses and strains are introduced.

Certain structures may be subjected to severe variation in temperature. These temperature changes tend to alter the length or curvature of the member of the structure; If these members are not allowed to undergo free movements, then they will be strained.

Due to the error in fabrication, members sometimes become too long or too short to be properly fitted in its place. The members in such cases are forced to fit during erection. This results in members being stressed before any load is applied. A member may be prestrained as is done in prestressed concrete beams which results in internal forces in the members. .

Matrix methods presented so far can be readily extended to incorporate the above effects.

3.5.1 Treatment by the Flexibility Matrix Method

The effects of the variation in temperature, lack of fit, prestrain, settlement of support *etc.* can be catered in the flexibility matrix method by making eqn. (3.4) more general. It becomes

$$[F]\{R\} + \{X\} + \{X_S\} + \{XLT\} + \ldots = \{0\} \tag{3.18}$$

where $\{X_S\}$, $\{X_T\}\ldots$ represent the displacements at the releases of the primary structure due to prestrain, temperature changes *etc.* Eqn. (3.18) suggests that matrices $\{X_S\}$, $\{X_T\}$ *etc.* are to be added to the matrix $\{X\}$, displacements due to external load acting on the released structure. The procedure has been explained with the help of the following example.

Example 3.7:

Analyse the continuous beam of Fig 3.23 for (*a*) a unit vertical movement, (*b*) a unit rotation of support *B* in the clockwise direction. The flexural rigidity *EI* is constant throughout the beam. The beam has 2 degrees of statical indeterminacy. The internal support moments at *B* and *C* are considered as redundants. The continuous beam is thus made statically determinate by introducing hinges at supports *B* and *C*. The loading matrix generation will be dealt after the development of the flexibility matrix. The steps followed to determine flexibility coefficients will not be given here, as it has already been given in details. The flexibility matrix for this problem with the chosen redundants is given by

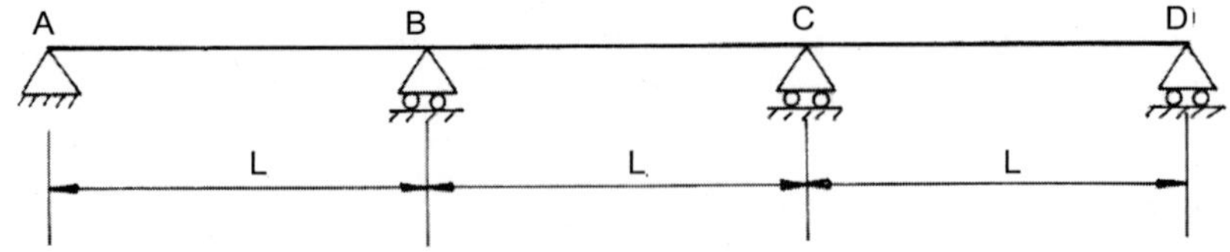

Fig. 3.23. Example 3.7

$$[F] = \frac{L}{6EI}\begin{bmatrix} 4 & 1 \\ 1 & 4 \end{bmatrix}$$

and $$[F]^{-1} = \frac{2EI}{5L}\begin{bmatrix} 4 & -1 \\ -1 & 4 \end{bmatrix}$$

For the first loading case, displacements corresponding to the redundants are denoted as X_{11} and X_{21}. Referring to Fig. 3.24(*b*), they assume the following values:

$$X_{11} = -\frac{2}{L} \text{ and } X_{21} = \frac{1}{L}$$

For this case, the solution is simple

$$[F] = [F]\{R\} = -\{X\}$$

$$\{R\} = -[F]^{-1}\{X\}$$

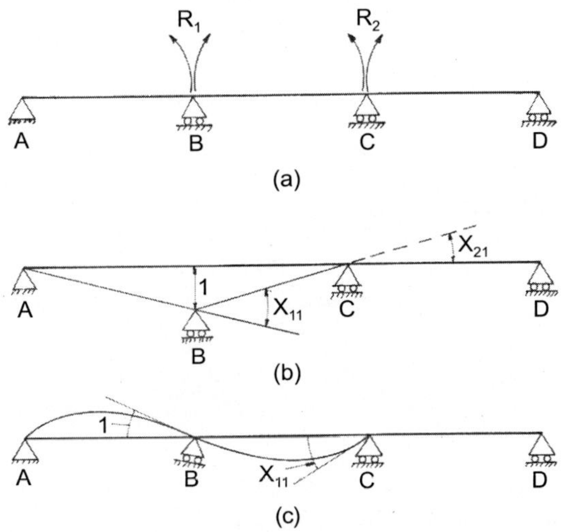

Fig. 3.24. Loading of Example 3.7

Therefore,

$$\begin{Bmatrix} R_1 \\ R_2 \end{Bmatrix} = -\frac{2EI}{5L}\begin{bmatrix} 4 & -1 \\ -4 & 4 \end{bmatrix}\frac{1}{L}\begin{Bmatrix} -2 \\ 1 \end{Bmatrix} = \frac{2EI}{5L^2}\begin{Bmatrix} 9 \\ -6 \end{Bmatrix}$$

In the second loading case, the joint *B* is rotated by unit angle. Therefore, relative rotation of support *B* is zero, whereas the relative rotation of support *C* be X_{22} [Fig. 3.24(*c*)].

Therefore,

$$X_{12} = 0 \text{ and } X_{22} = \theta \text{ (say)}$$

The final equation becomes

$$\begin{Bmatrix} R_1 \\ R_2 \end{Bmatrix} = -\frac{2EI}{5L}\begin{bmatrix} 4 & 1 \\ -1 & 4 \end{bmatrix}\begin{Bmatrix} 0 \\ \theta \end{Bmatrix}$$

The above equation gives

$$R_1 = \frac{2EI\theta}{5L} \text{ and } R_2 = -\frac{8EI\theta}{5L}$$

The values of R_1 and R_2 are given in terms of θ. So the, value of θ is to be determined. Considering the freebody diagram of the span *AB* (Fig. 3.25) of the continuous beam

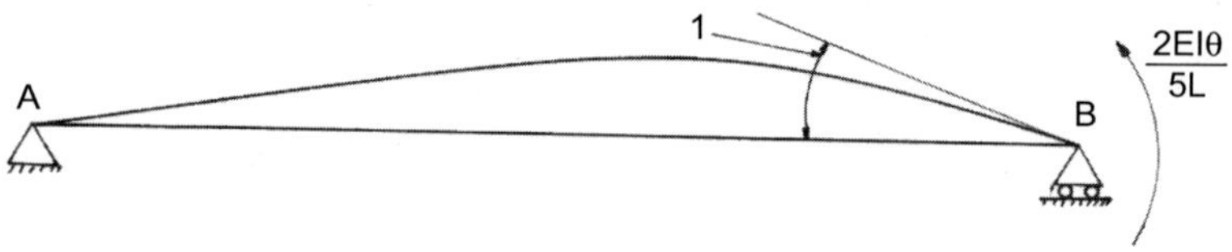

Fig. 3.25. Freebody of *AB*

where the joint *B* rotates by unity, the rotation due to couple at *B* should be equal to unity

$$\frac{2EI\theta}{5L}\cdot\frac{L}{3EI} = -1 \text{ or } \theta = -\frac{15}{2}$$

On back substitution of θ, the final values of the redundant forces are

$$R_1 = -\frac{3EI}{L} \text{ and } R_2 = \frac{12EI}{L}$$

3.5.2 Treatment by the Stiffness Method

Like the flexibility thethod, separate or combined effects of temperature variation, shrinkage, settlement of supports *etc.* can be readily incorporated into the analysis by the stiffness method. These effects are considered to act in the restrained structure. Resulting restraining forces are evaluated and are added to the [*R*] matrix produced by the loads. Thus, eqn. (3.8) becomes

$$[K][X] + [R] + [R_S] + [R_T] + \ldots = [0] \tag{3.19}$$

where $[R_S]$, $[R_T]$ *etc.* are the forces in the restrained structure due to settlement, temperature variation *etc.* The procedure is explained with the help of the following examples.

Example 3.8:

For the frame of Fig. 3.26, support *D* has vertical settlement of Δ simultaneously with horizontal movement to the right of 2Δ. *EI* is constant for all members.

The example is solved by the displacement method. The stiffness matrix and its inverse have been determined in Example 3.4.

$$[K]^{-1} = \frac{L^2}{684EI}\begin{bmatrix} 92 & \frac{12}{L} & -\frac{90}{L} \\ \frac{12}{L} & \frac{180}{L^2} & -\frac{18}{L^2} \\ -\frac{90}{L} & -\frac{18}{L^2} & \frac{20}{L^2} \end{bmatrix}$$

For given joint displacements, the forces in the restrained structure are to be evaluated. Referring to Fig. 3.27(*a*), the restraining forces corresponding to X_1, X_2 and X_3 for the vertical settlement Δ of support D, are

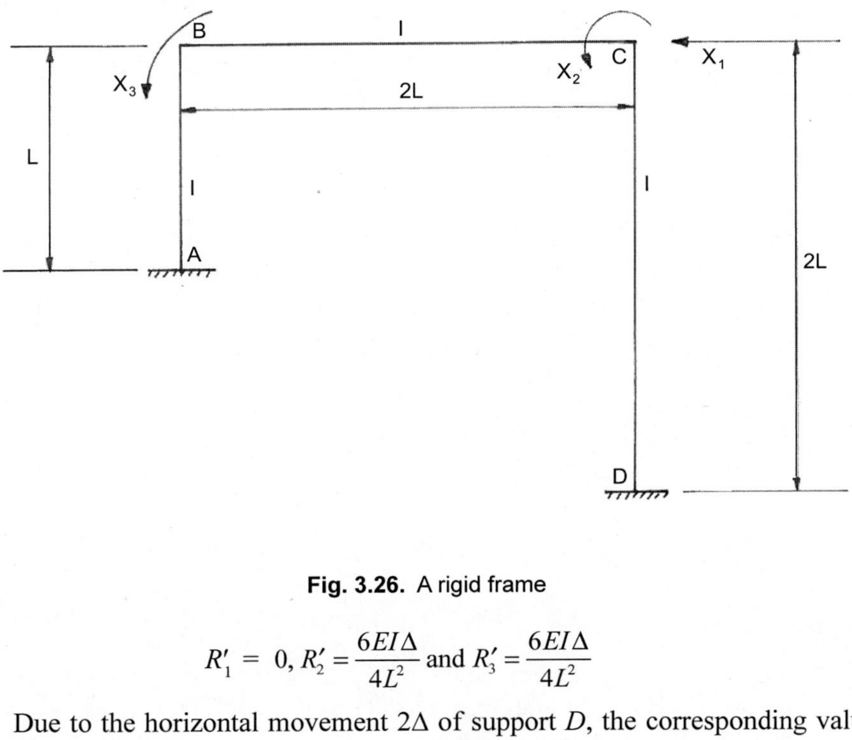

Fig. 3.26. A rigid frame

$$R_1' = 0,\ R_2' = \frac{6EI\Delta}{4L^2} \text{ and } R_3' = \frac{6EI\Delta}{4L^2}$$

Due to the horizontal movement 2Δ of support D, the corresponding values are [Fig. 3.27(*b*)].

$$R_1'' = \frac{24EI\Delta}{8L^3},\ R_2'' = -\frac{12EI\Delta}{4L^2},\ R_3'' = 0$$

Therefore, the final values of restoring forces are obtained by adding the two cases dealt separately as above.

$$R_1 = R_1' + R_1'' = \frac{24EI\Delta}{8L^3}$$

$$R_2 = R_2' + R_2'' = -\frac{6EI\Delta}{4L^2}$$

$$R_3 = R_3' + R_3'' = \frac{6EI\Delta}{4L^2}$$

Final equations are

$$\begin{Bmatrix} X_1 \\ X_2 \\ X_3 \end{Bmatrix} = \frac{L^3}{684EI}\begin{bmatrix} 92 & \frac{12}{L} & -\frac{90}{L} \\ \frac{12}{L} & \frac{180}{L^2} & -\frac{18}{L^2} \\ -\frac{90}{L} & -\frac{18}{L^2} & \frac{207}{L^2} \end{bmatrix} \frac{6EI\Delta}{4L^2}\begin{Bmatrix} \frac{2}{L} \\ -1 \\ 1 \end{Bmatrix} = -\frac{\Delta L}{456}\begin{Bmatrix} 82/L \\ -174/L^2 \\ 45/L^2 \end{Bmatrix}$$

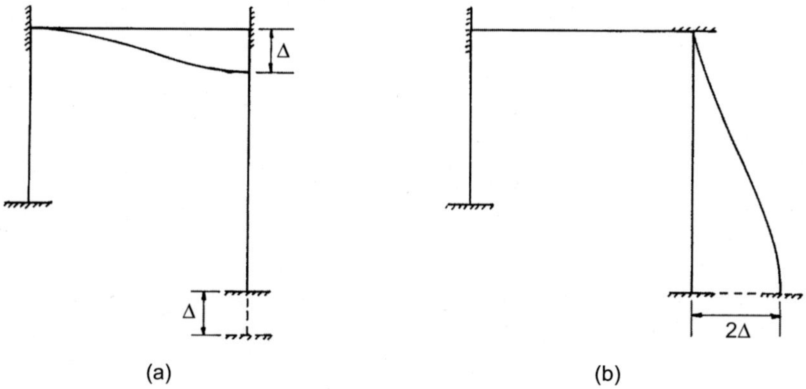

Fig. 3.27. Given point displacements of Example 3.8

After joint displacements are determined, the forces in the members can then be evaluated.

Example 3.9:

For the truss of Fig. 3.28, determine the forces in the members if the members *BE* is Δ too short. *EA* is same for all members. The inverse of the stiffness matrix for this truss worked out in Example 4.5 is

$$[K]^{-1} = \frac{L}{2.28EA}\begin{bmatrix} 1.14 & 0 \\ 0 & 2.004 \end{bmatrix}$$

In order to bring the member *BE* to joint *E*, a tensile force of $\frac{\Delta EA \sin 60^\circ}{L} = \frac{\sqrt{3}\,\Delta EA}{2L}$ is to be applied. This tensile force is applied at the joint *E* as shown in Fig. 3.29. The reactions have to be in opposite directions.

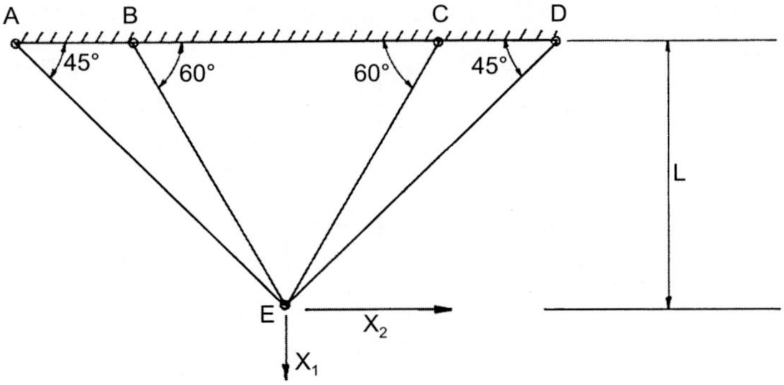

Fig. 3.28. Example 3.9

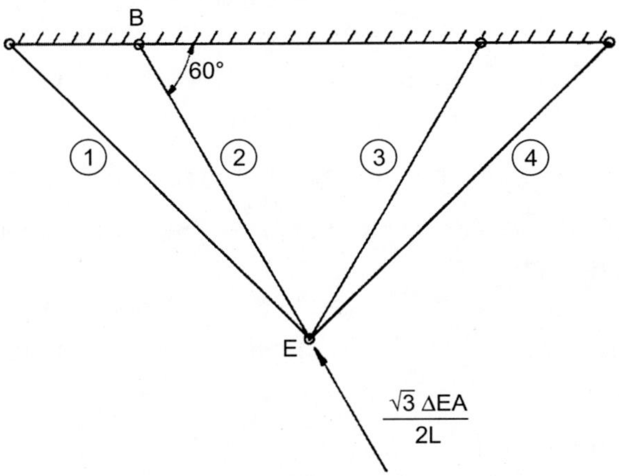

Fig. 3.29. External force concept in Example 3.9

Therefore,

$$\begin{Bmatrix} R_1 \\ R_2 \end{Bmatrix} = -\frac{\sqrt{3}\,\Delta\,EA}{2L}\begin{Bmatrix} \dfrac{\sqrt{3}}{2} \\ \dfrac{1}{2} \end{Bmatrix}$$

Final equations then become

$$\begin{Bmatrix} X_1 \\ X_2 \end{Bmatrix} = \frac{-L}{2.28\,EA}\begin{bmatrix} 1.14 & 0 \\ 0 & 2.004 \end{bmatrix}\frac{\sqrt{3}\,\Delta\,EA}{4L}\begin{Bmatrix} \sqrt{3} \\ \dfrac{1}{2} \end{Bmatrix}$$

$$= -\frac{\sqrt{3}\Delta}{4\times 2.28}\begin{bmatrix} 1.975 \\ 2.004 \end{bmatrix} = -\frac{\Delta}{2.28}\begin{bmatrix} 0.855 \\ 0.868 \end{bmatrix}$$

Once the displacements at the joints are determined, the forces in the members of the truss (based on Example 3.5) are

$$\{PM\} = \frac{EA}{L}\begin{bmatrix} \frac{1}{2} & \frac{1}{2} \\ \frac{3}{4} & \frac{\sqrt{3}}{4} \\ \frac{3}{4} & -\frac{\sqrt{3}}{4} \\ \frac{1}{2} & -\frac{1}{2} \end{bmatrix}\left(\frac{-\Delta}{2.28}\right)\begin{Bmatrix} 0.855 \\ 0.868 \end{Bmatrix} + \frac{EA}{2L}\begin{Bmatrix} 0 \\ \sqrt{3}\Delta \\ 0 \\ 0 \end{Bmatrix}$$

$$= \frac{EA\Delta}{L}\begin{Bmatrix} 0.8615/2.28 \\ 0.4203 \\ -0.265/2.28 \\ 5.5\times10^{-3} \end{Bmatrix} = \frac{EA\Delta}{L}\begin{Bmatrix} -0.3778 \\ 0.4203 \\ -0.1160 \\ -0.0002 \end{Bmatrix}$$

3.6 INCORPORATION OF SHEAR DEFORMATION

The shear deformation can be included in the stiffness matrix. This is done on the basis of flexibility coefficient.

Consider a cantilever beam *AB* of Fig. 3.30(*a*). Two coordinates 1 and 2 have been indicated. Applying eqn. (2.23), the flexibility coefficients after incorporating the shear deformation are

$$\begin{Bmatrix} f_{11} = \dfrac{L^3}{3EI} + \dfrac{L}{\mu AG}, \; f_{22} = \dfrac{L}{EI} \\ f_{12} = f_{21} = L^2/2EI \end{Bmatrix} \qquad (3.20)$$

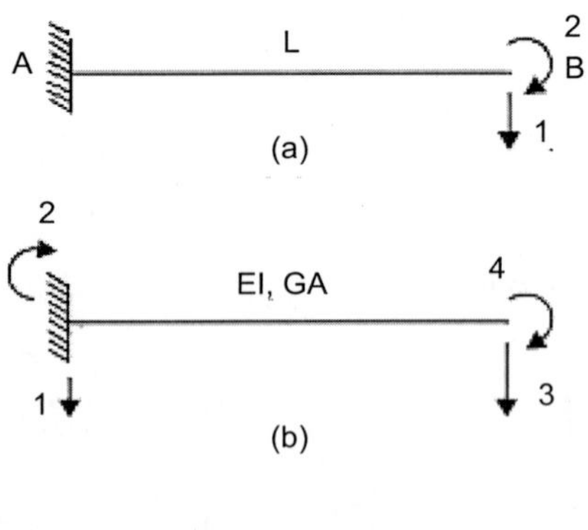

Fig. 3.30. Consideration for shear

where μ is a dimensionless coefficient and is dependent on the shape of the cross-section and is given by

$$\mu = I^2 / \left(A \int \frac{Q^2 dA}{b} \right) \tag{3.21}$$

where, A = cross-sectional area

b = width of the section

Q = first moment of area of the part of the cross-section above the section considered, about the neutral axis.

The flexibility matrix then, is

$$[F] = \begin{bmatrix} \dfrac{L^3}{3EI} + \dfrac{L}{\mu AG} & \dfrac{L^2}{2EI} \\ \dfrac{L^2}{2EI} & \dfrac{L}{EI} \end{bmatrix} \tag{3.22}$$

The stiffness matrix corresponding to coordinates of Fig. 3.30(a) is

$$[F]_e = [F]^{-1} = \frac{1}{1+\alpha} \begin{bmatrix} \dfrac{12EI}{L^3} & -\dfrac{6EI}{L^2} \\ -\dfrac{6EI}{L^2} & (4+\alpha)\dfrac{EI}{L} \end{bmatrix} \tag{3.23}$$

where $\alpha = \dfrac{12EI}{\mu AGL^2}$

In a similar way, the stiffness matrix corresponding to coordinates of Fig. 3.30(b) with the shear deformation considered can be determined. The final values for this case is

$$[K]_e = \frac{EI}{1+\alpha} \begin{bmatrix} \dfrac{12}{L^3} & & & symmetric \\ \dfrac{6}{L^2} & \dfrac{(4+\alpha)}{L} & & \\ -\dfrac{12}{L^3} & -\dfrac{6}{L^2} & \dfrac{12}{L^3} & \\ \dfrac{6}{L^2} & \dfrac{2-\alpha}{L} & -\dfrac{6}{L^2} & \dfrac{(4+\alpha)}{L} \end{bmatrix} \tag{3.24}$$

3.7 RELATION BETWEEN FLEXIBILITY AND STIFFNESS MATRICES

In Fig. 3.31 a beam with two different moments of inertia at two portions are shown. The unknown displacements at the intersection of the two parts denoted as X_1 and X_2 corresponding to the deflection and the rotation and the forces corresponding to X_1 and X_2 are R_1 and R_2. So, total displacement can be written as

$$\begin{Bmatrix} X_1 = f_{11}R_1 + f_{12}R_2 \\ X_2 = f_{21}R_1 + f_{22}R_2 \end{Bmatrix} \tag{3.25}$$

Here f_{11}, f_{12} *etc.* are flexibility coefficients. So eqn. (3.20) can be writtenn in compact form as

$$\{X\} = [F]\{R\}$$

or
$$\{R\} = [F]^{-1}\{X\} \tag{3.26}$$

$[F]$ is the flexibility matrix.

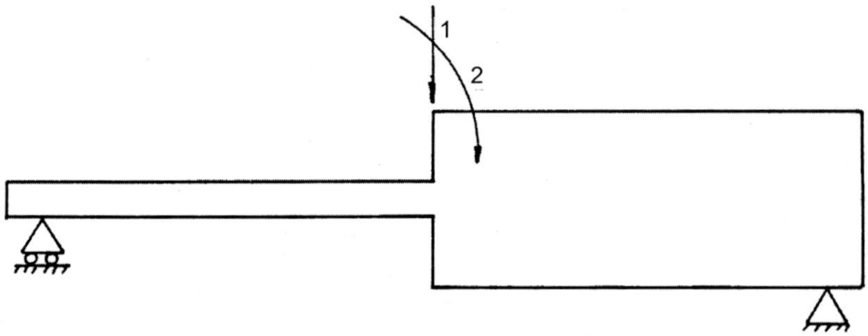

Fig. 3.31. A stepped beam

Based on the same coordinate system, let us form the stiffness matrix. If forces K_{11}, K_{21} are applied at 1 and 2 such that $X_1 = 1$ and $X_2 = 0$, then

$$\begin{Bmatrix} K_{11} \\ K_{21} \end{Bmatrix} = [F]^{-1}\begin{Bmatrix} 1 \\ 0 \end{Bmatrix} \tag{3.27}$$

Similarly, if forces K_{12}, K_{22} are applied at 1 and 2 such that $X_1 = 0$, but $X_2 = 1$, then

$$\begin{Bmatrix} K_{12} \\ K_{22} \end{Bmatrix} = [F]^{-1}\begin{Bmatrix} 0 \\ 1 \end{Bmatrix} \tag{3.28}$$

Combining eqns. (3.22) and (3.23), yields

$$\begin{bmatrix} K_{11} & K_{12} \\ K_{21} & K_{22} \end{bmatrix} = [F]^{-1}\begin{bmatrix} 1 & 0 \\ 0 & 1 \end{bmatrix}$$

or
$$[K] = [F]^{-1} \tag{3.29}$$

Hence, based on the same coordinate system, the stiffness matrix is the inverse of the flexibility matrix and *vice versa.*

However, it must be noted that for the solution of the same problem by the flexibility method and the stiffness method, different coordinate systems are usually chosen - as in the first case the unknowns are redundant forces whereas in the other case the unknowns are the degrees of freedom. So, for the same structure two coordinate systems will not be same. Therefore, inverse of a stiffness matrix for the solution of a structural problem is indeed a flexibility matrix, but it is not the same flexibility matrix as one would obtain for the solution of the same problem.

3.8 EQUIVALENT JOINT LOADS

Equation (3.8) for the displacement method is reproduced.

$$[K]\,[X] = -[R] \tag{3.30}$$

Reacting force matrix $[R]$ is replaced by $[P]$ such that

$$[P] = -[R] \tag{3.31}$$

Equation (3.30) is thus modified with the help of eqn. (3.31) as

$$[K]\,[X] = [P] \tag{3.32}$$

The procedure of calculating $[P]$ is same as that of $[R]$ with the sign changed. $[P]$ matrix is the equivalent joint load matrix. The construction of $[P]$ matrix is explained with help of the following example.

The frame of Fig. 3.32(*a*) is restrained at joints *B* and *C*. The reactions at the restrained joints are determined. Different fixed end forces are indicated in Fig. 3.32(*b*). The sum of the reacting forces at joints *B* and *C* is shown in Fig. 3.32(*c*). This is what is termed as equivalent joint loading. The part corresponding to the degrees of freedom will be associated with the $\{P\}$ matrix. If the frame has three degrees of freedom shown as X_1, X_2 and X_3 in Fig. 3.32(*c*), the corresponding $\{P\}$ matrix then will be

$$\begin{Bmatrix} P_1 \\ P_2 \\ P_3 \end{Bmatrix} = \begin{Bmatrix} 4 \\ -2 \\ -4 \end{Bmatrix}$$

3.9 CHOICE OF THE METHOD OF ANALYSIS

Matrix methods of structural analysis are primarily meant for structures of complex shapes which are large in size and for which hand computation is prohibitive. Matrix methods are computer-oriented methods and their relative merits and demerits should not be viewed with small sized problems. It has already been seen that there are two basic approaches - the flexibility method and the stiffness method. Looking into the fundamental philosophy of both these approaches, it is revealed that they are practically same and in both a set of simultaneous equations is generated. The choice of the method as such should, depend primarily on the ease with which the problem can be formulated. For a particular problem, one approach may appear easier than the other.

With the flexibility method, the analyst is free to make his own choice of the redundants and he has at his disposal a variety of choices. But in the stiffness method there is no option on the selection of the restrained structure and as such the analysis follows a rather set procedure. The analysis by the stiffness method is thus more mechanical than the flexibility method. The economy of analysis has, however, to be viewed on the basis of the time taken by the computer for the solution of equations. If the time taken by other operations are assumed to be same, the main time in the computer is consumed in the solution of the simultaneous equations. Generally, the elements of flexibility matrix method are sparsely populated, whereas the problem formulated by the stiffness method is banded in form. If advantage is taken of the branded matrix, then less storage is required in the computer, as well as, less solution time. For many problems, the stiffness method is thus better suited than the flexibility method. In certain problems like truss type structures such as transmission tower, lattice girders *etc*., where degrees of statical indeterminacy is much lesser than the degrees of kinematic indeterminacy, the flexibility method may appear more economical than the stiffness method in spite of taking advantage of its banded nature. The displacement method is ,however, easy to formulate. In this book more emphasis is laid on the development and application of the stiffness method for the solution of structural problems.

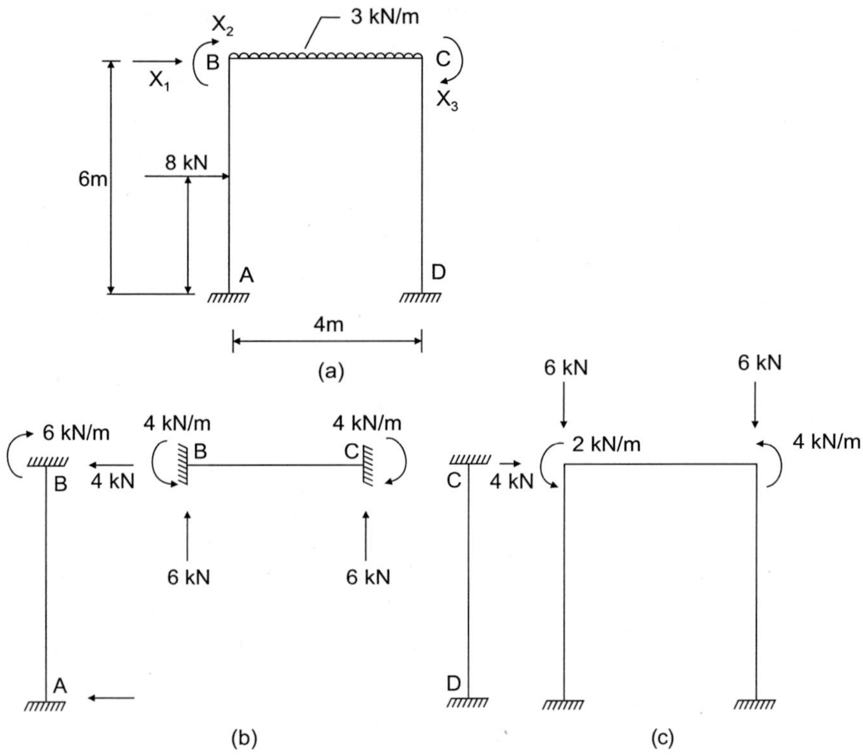

Fig. 3.32. Calculation of equivalent joint loads

REFERENCES AND SUGGESTED READINGS

3.1. J.M. Gere and W. Weaver Jr., Analysis of Framed Structures, 2nd Edition, D. Van Nostrand, USA, 1982.

3.2. A.S. Hall and R.W. Woodhead, Frame Analysis, John Wiley and Sons, USA, 1961.

3.3. K.H. Gerstle, Basic Structural Analysis, Prentice Hall, USA, 1974.

3.4. H. Kadestuncer, Elementary Matrix Analysis of Structures, McGraw-Hill Book Co., 1974.

3.5. W. MacGuire, R.H. Gallaghar & R. D. Zieman, Matrix Structural Analysis, John Wiley & Sons, USA, 2nd Edition, 2000.

3.6. J.S. Przmieniecki, Theory of Matrix Structural Analysis, McGraw-Hill, 1968.

3.7. M.F. Rubinstein, Matrix Computer Analysis, Prentice Hall, USA, 1969.

3.8. C.K. Wang, Introductory Structural Analysis with Matrix Methods, Prentice Hall, USA, 1973.

EXERCISE 3

Solve the problems 3.1 to 3.13 by the flexibility method. Select suitable redundants and determine their values.

3.1. Write the flexibility matrix corresponding to the coordinate indicated.

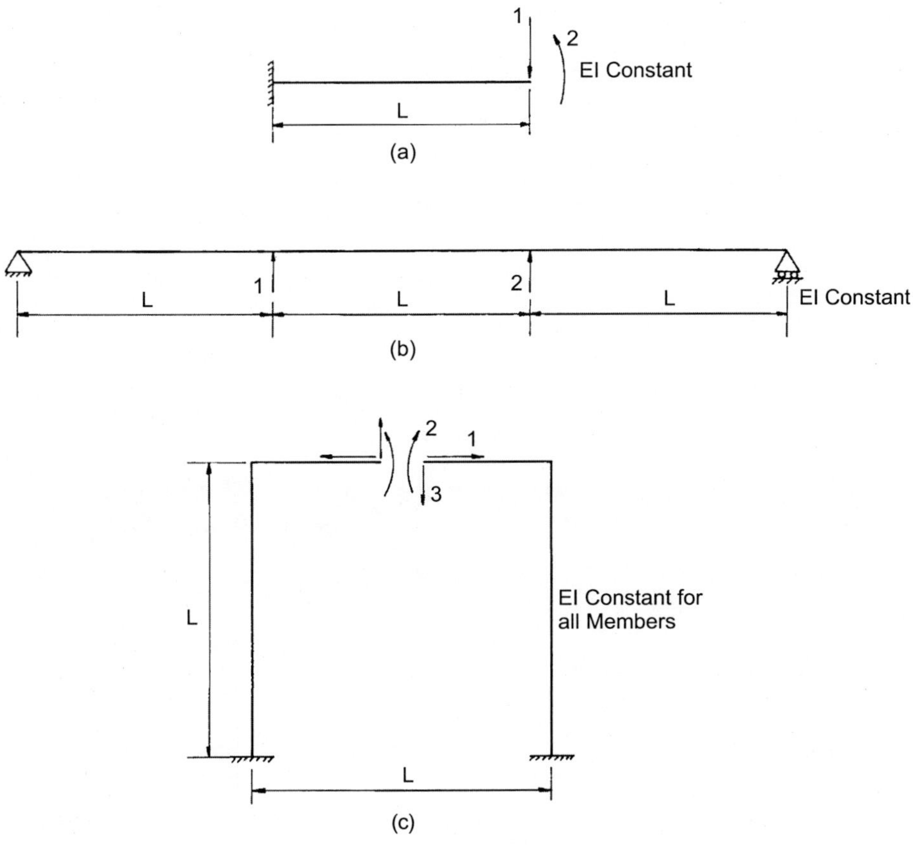

Prob. 3.1.

3.2. Determine the forces in the two springs at B and C. The beam AD has a constant flexural rigidity EI and both the springs have stiffness $100/L^3$.

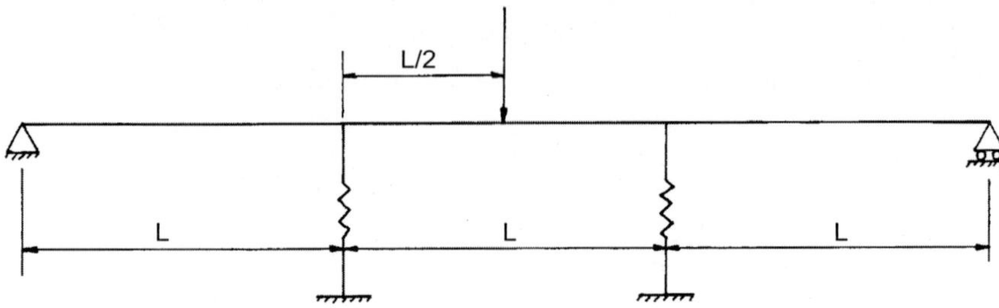

Prob. 3.2.

3.3. Determine the support reactions and draw the bending moment diagram for the portal frame subjected to a point load of 120 kN, acting at the mid-point of BC.

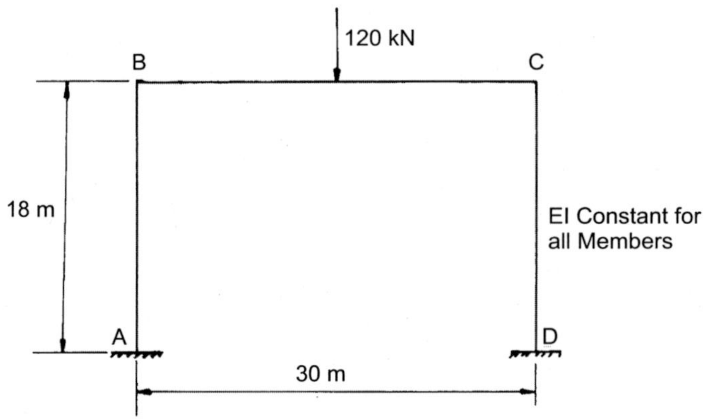

Prob. 3.3.

3.4. Determine the end moments of the members and draw the bending moment diagram for the gable frame. EI is same for all members.

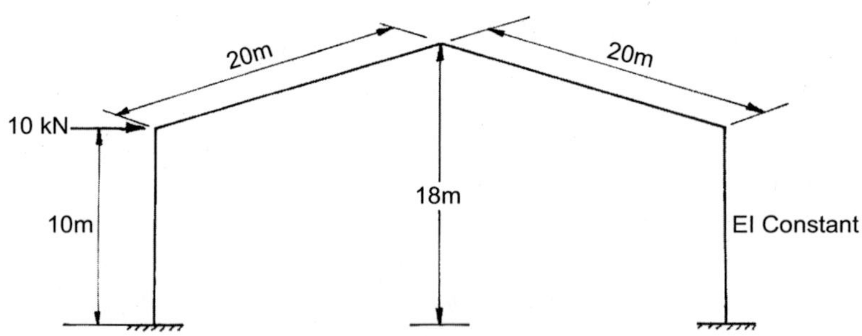

Prob. 3.4.

3.5. Determine the end-moments of the members and draw the bending moment diagram for the closed rectangular portal frame. *EI* is same for all members.

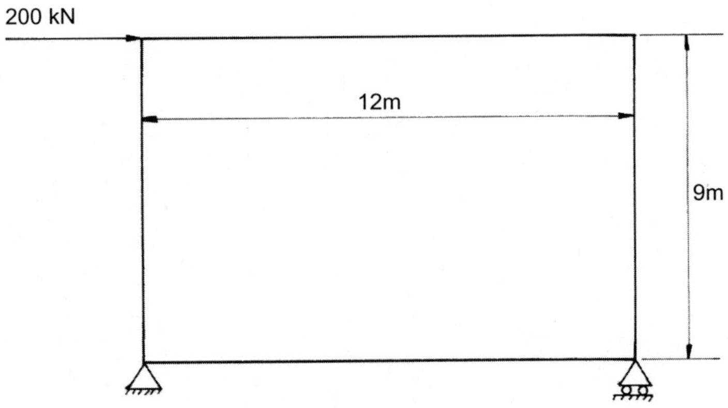

Prob. 3.5.

3.6. Determine the reactions at the supports. Also calculate the forces in the members and deflection at the load point. *EI* is same for all members.

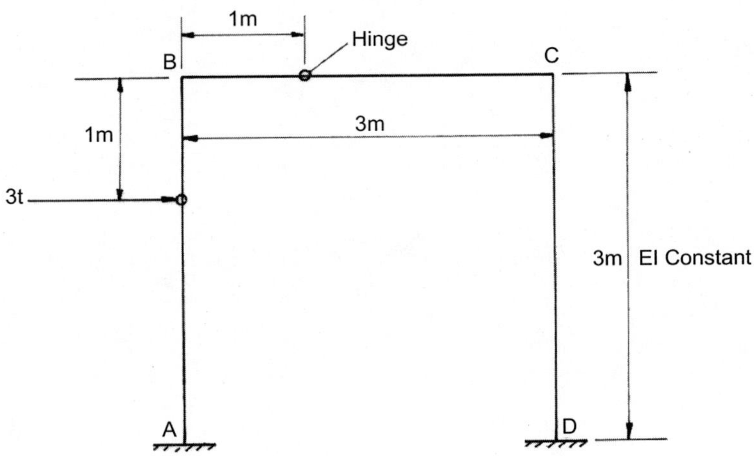

Prob. 3.6.

3.7. Determine the member forces of the trussed beam. Assume *I* as the moment of inertia of the members and *A* as the area for all members.

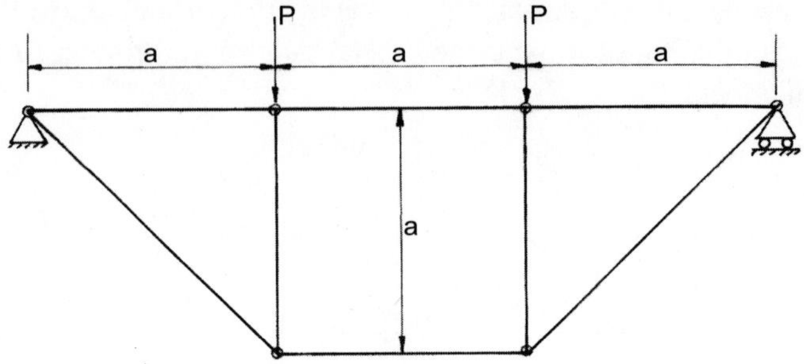

Prob. 3.7.

3.8. Determine the redundant forces.

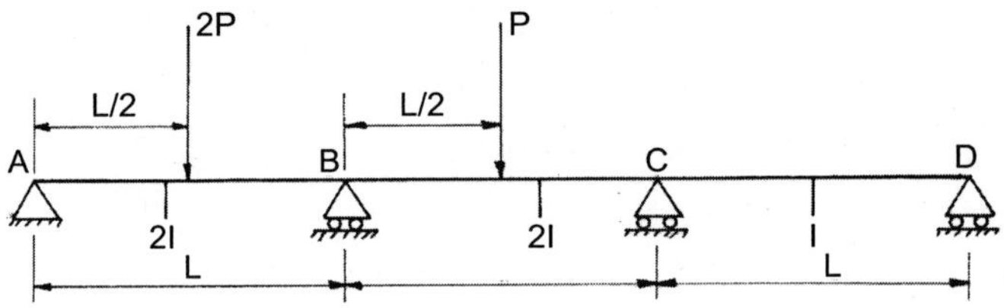

Prob. 3.8.

3.9. Find out the forces in all the members of the truss. The axial rigidity for the horizontal and vertical members is *2EA* and for the diagonal members *EA*.

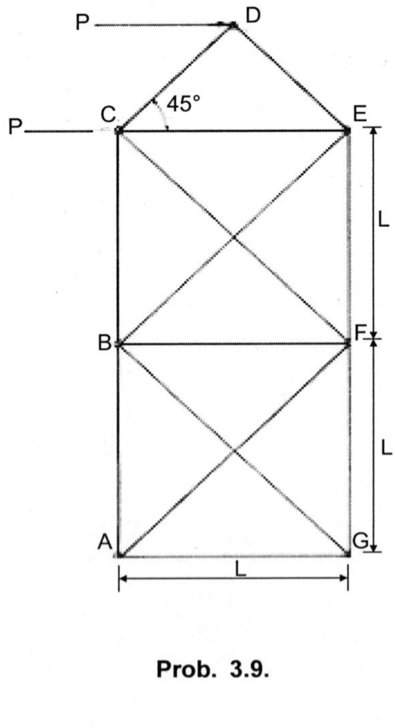

Prob. 3.9.

3.10. Evaluate the forces in all the members of the truss.

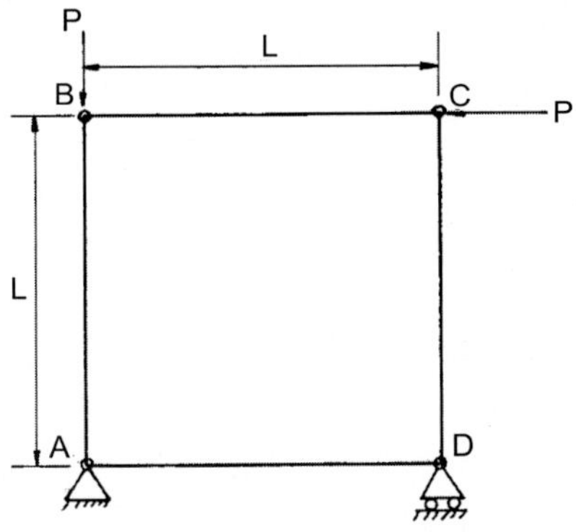

Prob. 3.10.

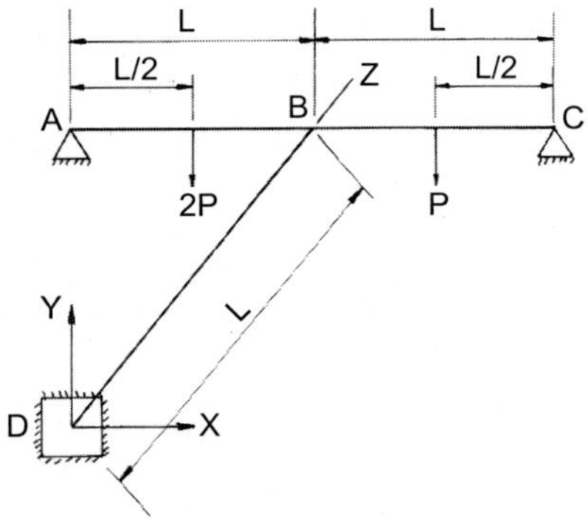

Prob. 3.11.

3.11. *A* and *C* are simple supports. Two moments about *x* and *z* axes may be considered as redundants. Joint *B* is rigid. $EI/GJ = 0.8$ for all the members. Determine the redundants.

3.12. For the problem 3.10, if the support settles by Δ, determine the member forces in the truss.

3.13. Find the reactions at the supports of the continuous beam if joint *C* is displaced by unity along with clockwise rotation of joint *A* of magnitude unity.

Solve the following problems by the stiffness method.

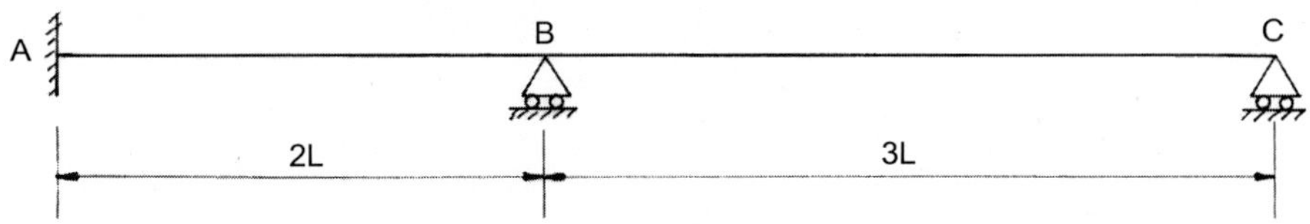

Prob. 3.13.

3.14. Form the stiffness matrix based on the coordinate system indicated.

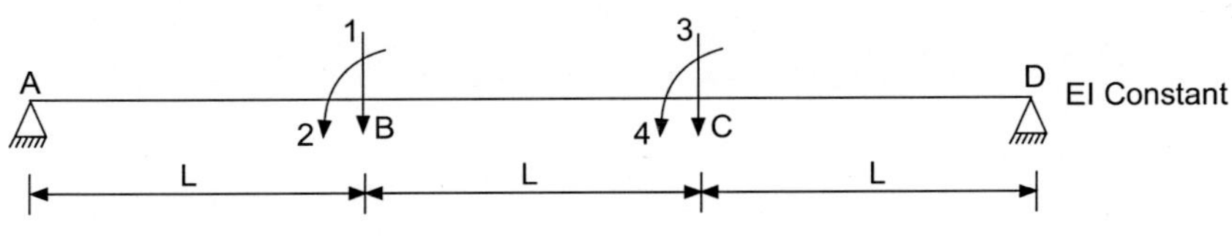

(a)

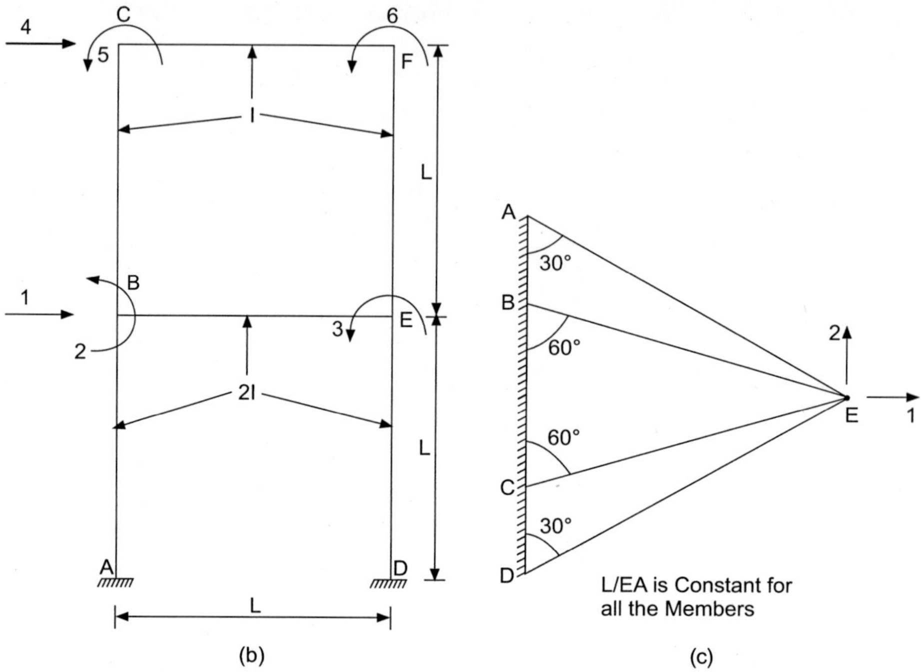

Prob. 3.14.

3.15. Find all support rotations.

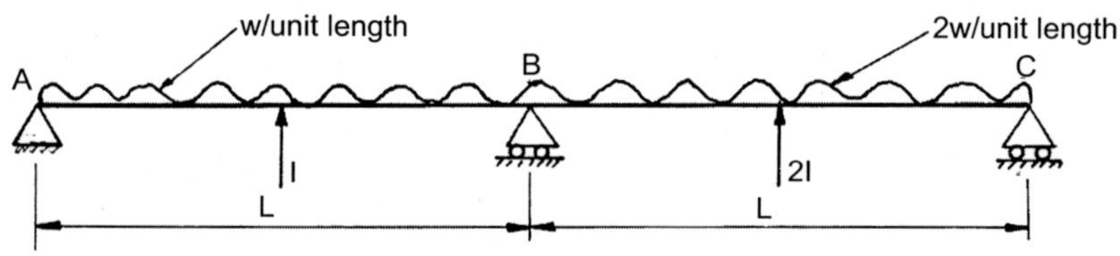

Prob. 3.15.

3.16. Find all the member forces.

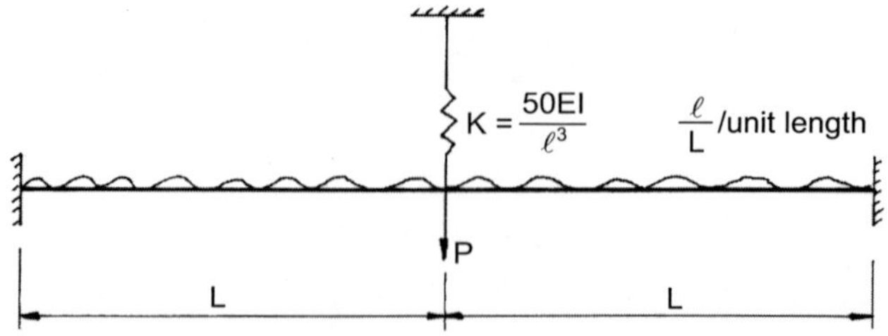

Prob. 3.16.

3.17. For the grid shown, find the unknown joint displacements and the member forces.

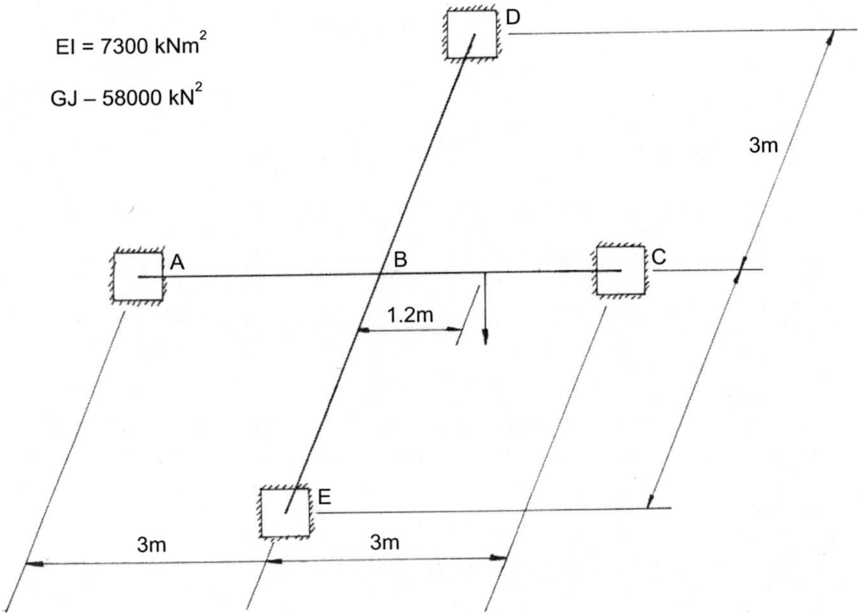

Prob. 3.17.

3.18. Determine the member forces for the truss.

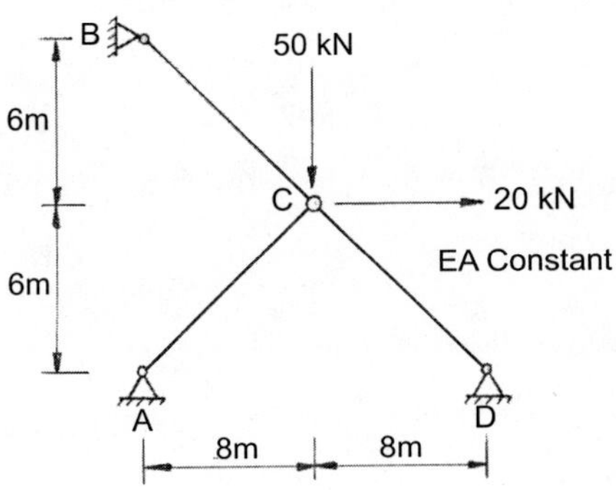

Prob. 3.18.

3.19. Determine the member forces for the truss.

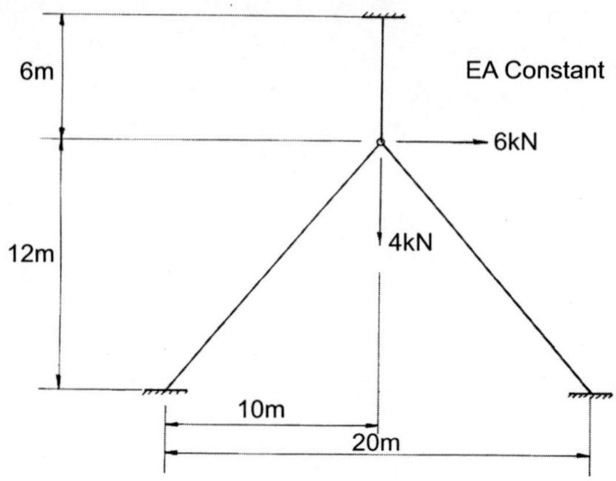

Prob. 3.19.

3.20. Find out the joint displacements and the member end forces for the rigid frame, *EI* constant for all members.

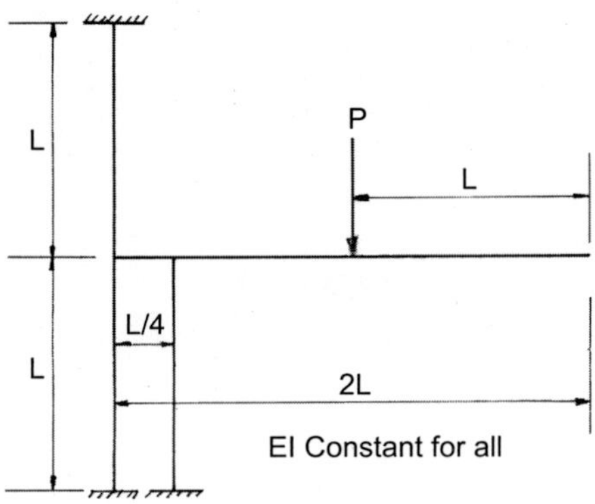

Prob. 3.20.

3.21. Determine the reactions and draw the B.M. diagram.

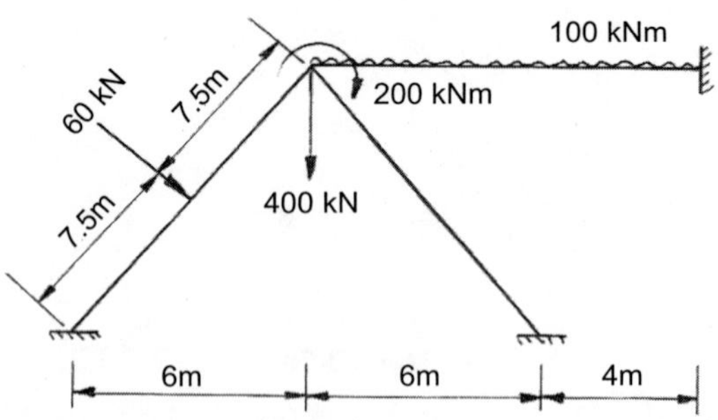

Prob. 3.21.

3.22. For the truss shown, determine the forces in the members, if AC is Δ too long and the support D settles by Δ. The axial rigidity is $2AE$ for horizontal and vertical members and EA for diagonal members.

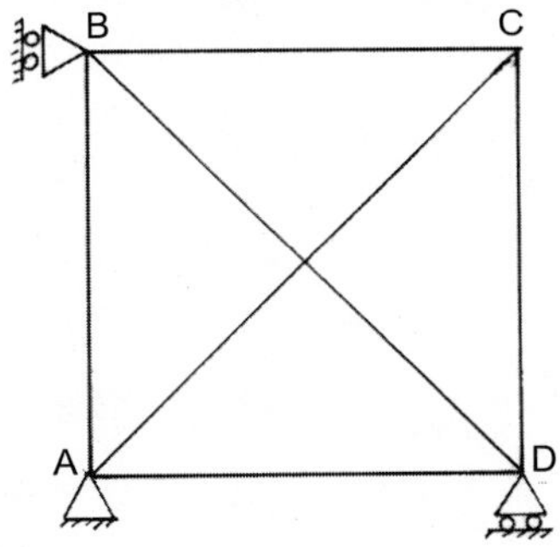

Prob. 3.22.

3.23. Determine the support reactions when the beam at B is rotated through an angle θ in the counterclockwise direction.

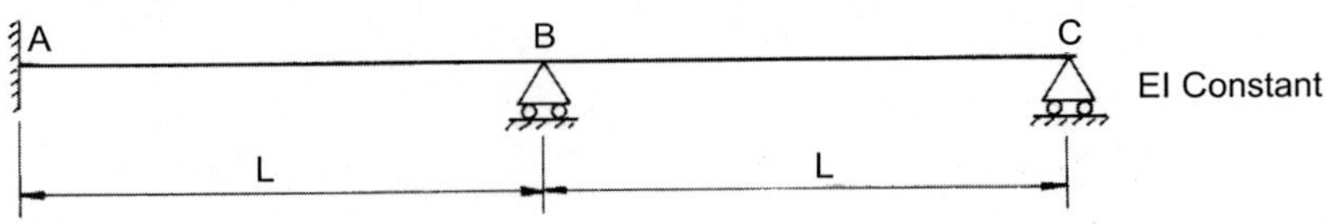

Prob. 3.23.

❑❑❑

4

DIRECT STIFFNESS METHOD

4.1 INTRODUCTION

A more formalised approach of the stiffness method has been presented in this chapter. It will be shown here that the overall stiffness matrix of the structure will be generated automatically by the computer, based on the geometric and elastic properties of the individual element. The unknown displacements at the joints are obtained from the overall solution of the problem. Once the displacements at the joints are known, the stress resultants for any member can be obtained. In this approach, the procedure becomes mechanical, the technique can be easily coded in the computer and it results in a saving of storage space.

4.2 LOCAL AND GLOBAL COORDINATE SYSTEM

In this procedure, the stiffness matrix of an individual element is to be evaluated. In order to do this, a clear concept of local and global axis system and their relations are necessary.

A local coordinate system is one that is defined for a particular element and not necessarily for the entire body or structure. This may be explained with reference to Fig. 4.1 where the bar of a truss has been indicated. This member develops axial force due to the external load and as such the local axis system is considered along the axis of the member. The stiffness matrix of this member in the local axis system is given by

$$[K]_e = \frac{EA}{L}\begin{bmatrix} 1 & -1 \\ -1 & 1 \end{bmatrix} \tag{4.1}$$

where EA is the axial rigidity of the uniform member and L, its length. In a truss, the members are oriented at different inclinations and as such, the stiffness matrix written as above for different members, will have to be based on different axes system. The difficulty which will result in such a case is that the member stiffness matrices at the joints cannot be added. This is possible only when it is referred to a common axes system. Global axis system in this common coordinate system is valid for the entire structure. The element stiffness matrix must be written in the global coordinate system before assembly. The displacements that will be obtained from the overall solution will be in global coordinate system. It is, however, desirable to determine the member forces in local coordinate system.

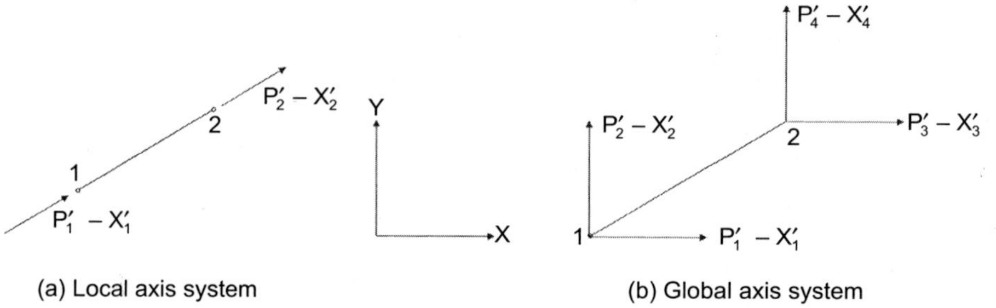

Fig. 4.1. Element of a truss

In order to understand how the conversion from local to global axis system is done in the case of stiffness matrix and similar conversion associated with the loading matrix and stress resultants, the knowledge of transformation of axis system is needed. In fact it is of fundamental importance in structural analysis to obtain transformations relating to the set of variables in one system to an equivalent set in the second system.

4.3 TRANSFORMATION OF VARIABLES

The discussion is initiated with the transformation of member coordinate axes.

4.3.1 Transformation of Member Coordinate Axes

Two sets of cartesian coordinate system are shown in Fig. 4.2. *O*, *X*′, *Y*′ and *Z*′ be the local axis system and *O*, *X*, *Y* and *Z* be the global axis system. Both sets are right handed. The direction cosines of *X*′ with respect to *X*, *Y* and *Z* are l_1, m_1 and n_1. Similarly, direction cosines of *Y*′ and *Z*′ with respect to *X*, *Y* and *Z* are l_2, m_2, n_2 and l_3, m_3, n_3 respectively.

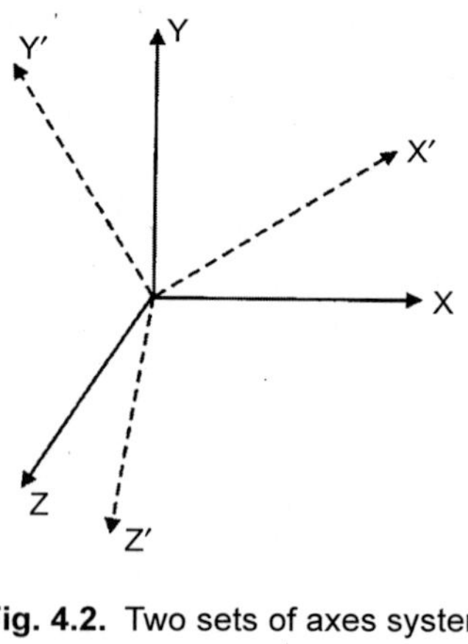

Fig. 4.2. Two sets of axes system

If a displacement vector $\{w'\}$ has components w'_x, w'_y and w'_z along the local axes system, the same vector denoted as $\{w\}$ in the global axes system has components w_x, w_y and w_z respectively along that axes system, then it can be shown that.

$$\begin{Bmatrix} w'_x = l_1 w_x + m_1 w_y + n_1 w_z \\ w'_y = l_2 w_x + m_2 w_y + n_2 w_z \\ w'_z = l_3 w_x + m_3 w_y + n_3 w_z \end{Bmatrix} \qquad (4.2)$$

Equation (4.2) written in matrix form becomes

$$\begin{Bmatrix} w'_x \\ w'_y \\ w'_z \end{Bmatrix} = \begin{bmatrix} l_1 & m_1 & n_1 \\ l_2 & m_2 & n_2 \\ l_3 & m_3 & n_3 \end{bmatrix} \begin{Bmatrix} w_x \\ w_y \\ w_z \end{Bmatrix}$$

or,

$$\{w'\} = [\lambda]\ \{w\} \tag{4.3}$$

where $[\lambda]$ is called rotation matrix. If both sets of reference axes are orthogonal, then

$$[\lambda]^{-1} = [\lambda]^T \tag{4.4}$$

which saves the time for computation, because the transpose of a matrix is easier to obtain than its inverse.

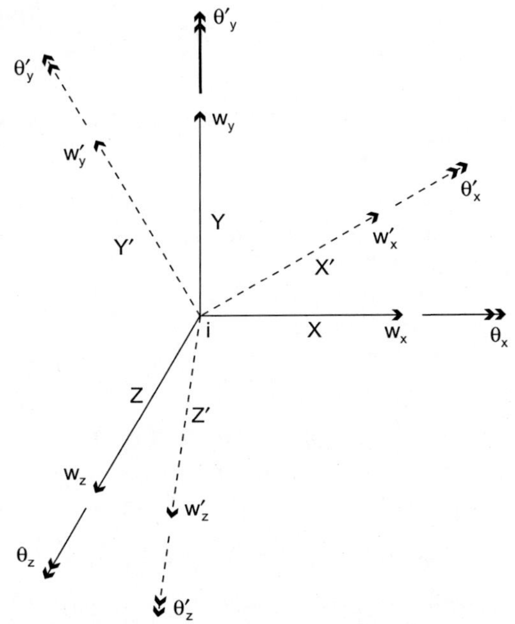

Fig. 4.3. Generalised displacements along local and global axes system

Usually, there are six displacement components at a point of an elastic structure-three translations and three rotations. Just as the translational components discussed above, the rotational components of the two axes system can be related.

4.3.2 Transformation of Member Displacement Matrix

A generalised displacement consisting of three components of translation and three components of rotation, is given by

$$\{X'\}^T = \{w'_x \quad w'_y \quad w'_z \quad \theta'_x \quad \theta'_y \quad \theta'_z\}$$

or,

$$\{X'\}^T = \{\{\Delta'\}_l \ \vdots \ \{\theta'\}_i\} \tag{4.5}$$

The generalised displacement matrix is partitioned into translational and rotational components and the subscript 'i' refers to the displacement parameters at point 'i' (generally they are, at the nodes *i.e.* joints). The generalised displacement component in a local axis system can be related to the global axis system at a node i by the following relation.

$$\begin{Bmatrix} \{\Delta'\}_i \\ \hline \{\theta'\}_i \end{Bmatrix} = \left[\begin{array}{c|c} [\lambda] & [0] \\ \hline [0] & [\lambda] \end{array}\right] \begin{Bmatrix} \{\Delta\}_i \\ \hline \{\theta\}_i \end{Bmatrix} \tag{4.6}$$

A straight member has two nodes at its two ends. If the axis system is considered parallel at two nodes, then we have

$$\begin{Bmatrix} \{\Delta'\}_1 \\ \{\theta'\}_1 \\ \{\Delta'\}_2 \\ \{\theta'\}_2 \end{Bmatrix} = \begin{bmatrix} [\lambda] & & \text{submatrices not} & \\ & [\lambda] & \text{shown are zeros} & \\ & & [\lambda] & \\ & & & [\lambda] \end{bmatrix} \begin{Bmatrix} \{\Delta\}_1 \\ \{\theta\}_1 \\ \{\Delta\}_2 \\ \{\theta\}_2 \end{Bmatrix} \tag{4.7}$$

Subscript 1 refers to the displacement at node 1 and subscript 2 to that at node 2. In compact form, eqn. (4.7) becomes

$$\{X'\} = [T]\{X\} \tag{4.8}$$

where $[T]$ is referred to as member transformation matrix. $[T]$ is a square matrix of order 12×12. It can be shown that

$$[T][T]^{-1} = [T][T]^T = [I]$$

i.e.

$$[T]^{-1} = [T]^T \tag{4.9}$$

It has already been seen that for the solution of a particular structure, the degrees of freedom need not be six at each node and will be reduced depending on the problem at hand. In that case the size of $[T]$ is reduced.

4.3.3 Transformation of the Member Force Matrix

Let a member be in equilibrium under the action of forces $\{P'\}$ which results in displacement $\{X'\}$ in local coordinate system. Let $\{P\}$ be the corresponding forces in the global coordinate system.

For a virtual displacement, eqn. (4.8) becomes

$$\delta\{X'\} = [T]\,\delta\,\{X\} \tag{4.10}$$

Taking transpose of eqn. (4.10),

$$\delta\{X'\}^T = \delta\,\{X\}^T\,[T]^T \tag{4.11}$$

Due to this virtual displacement, the virtual work done by the forces in local coordinate system is

$$\delta\{X'\}^T\,\{P'\}$$

and the corresponding virtual work in the global coordinate system is

$$\delta\{X\}^T\,\{P\}$$

Equating the virtual work in two coordinate systems,

$$\delta\{X'\}^T\{P'\} = \delta\{X\}^T\{P\} \tag{4.12}$$

Combining eqns. (4.11) and (4.12), yields

$$\delta\{X\}^T[T]^T\{P'\} = \delta\{X\}^T\{P\}$$

or,

$$\{P\} = [T]^T\{P'\} \tag{4.13}$$

Also,

$$\{P'\} = [T]\{P\} \tag{4.14}$$

As

$$[T][T]^T = [I]$$

4.3.4 Transformation of the Member Stiffness Matrix

It is known that the element stiffness matrix in local coordinates is connected by the following force displacement relationship

$$[P'] = [k]_e\{X\}_e \tag{4.15}$$

$[X']_e$ represents the displacements associated with that member. Premultiplying both sides of eqn. (4.15) by $[T]^T$ and substituting $[X]_e$ from eqn. (4.10) into eqn. (4.15), results

$$[T]^T[P']_e = [T]^T[k]_e[T][X]_e \tag{4.16}$$

From eqns. (4.14) and (4.16), yields,

$$[P]_e = [T]^T[k]_e[T][X]_e \tag{4.17}$$

If the element stiffness matrix in global coordinates is defined as

$$[K]_e = [T]^T[k]_e[T] \tag{4.18}$$

then, the force-displacement relation in the global coordinate system can be written as

$$[P]_e = [K]_e[X]_e \tag{4.19}$$

Equations (4.13) and (4.18) give the relation of transforming the loading and the stiffness matrix from, the local to the global axis system.

The derivation done so far will be illustrated with the elements of different structures.

4.4 TRANSFORMATION OF THE STIFFNESS MATRIX OF THE MEMBER OF A TRUSS

A typical member of a truss is shown in Fig. 4.4. As the member carries only the axial force, the local axis system has X'-axis only. The local and global numbering of the displacements has also been indicated in Fig. 4.4. The direction cosines of X'-axis with X and Y-axes are $\cos\theta$ and $\sin\theta$ respectively. The member is assumed to have uniform axial rigidity AE and length L.

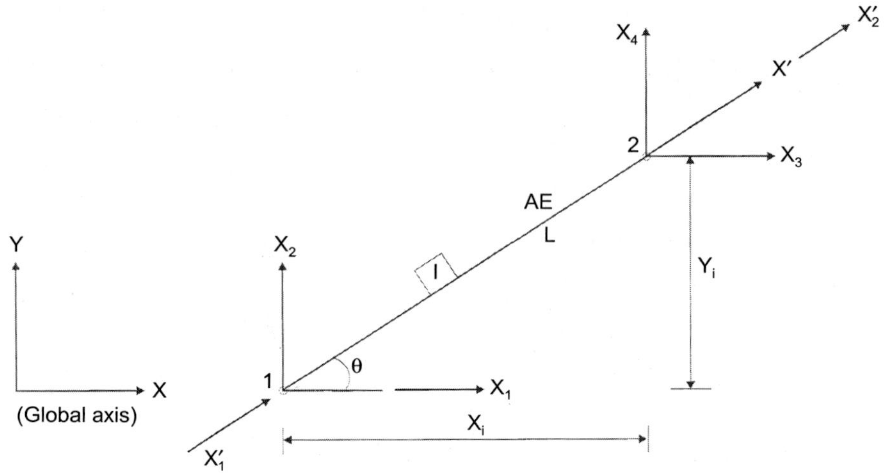

Fig. 4.4. A typical member of a truss with local and global axis system

By elementary considerations, the stiffness matrix of the member in local coordinates is

$$[k]_e = \frac{EA}{L}\begin{bmatrix} 1 & -1 \\ -1 & 1 \end{bmatrix} \tag{4.20}$$

In order to obtain the transformation matrix, X'_1 is to be related to X_1, and X_2 and X'_2 are to be related to X_3 and X_4. X_1 and X_2 are the displacements at node 1 parallel to the global-axis system. X_3 and X_4 are the corresponding displacements along the global axis at node 2. Thus

$$\begin{Bmatrix} X'_1 \\ \hline X'_2 \end{Bmatrix} = \begin{bmatrix} l_1 & m_1 & 0 & 0 \\ \hline 0 & 0 & l_1 & m_1 \end{bmatrix} \begin{Bmatrix} X_1 \\ X_2 \\ X_3 \\ X_4 \end{Bmatrix} \tag{4.21}$$

where $l_1 = \cos\theta$ and $m_1 = \sin\theta$

Therefore,

$$[T] = \begin{bmatrix} l_1 & m_1 & 0 & 0 \\ 0 & 0 & l_1 & m_1 \end{bmatrix}$$

The element stiffness matrix in global coordinates as given in eqn. (4.18) can now be calculated. For the truss it is done as shown below.

$$[K]_e = \begin{bmatrix} l_1 & 0 \\ m_1 & 0 \\ 0 & l_1 \\ 0 & m_1 \end{bmatrix} \frac{EA}{L}\begin{bmatrix} 1 & -1 \\ -1 & 1 \end{bmatrix}\begin{bmatrix} l_1 & m_1 & 0 & 0 \\ 0 & 0 & l_1 & m_1 \end{bmatrix}$$

or,

$$[K]_e = \frac{EA}{L}\begin{bmatrix} l_1^2 & & Symmetrical & \\ l_1 m_1 & m_1^2 & & \\ -l_1^2 & -l_1 m_1 & l_1^2 & \\ -l_1 m_1 & -m_1^2 & l_1 m_1 & m_1^2 \end{bmatrix} \tag{4.22}$$

4.5 TRANSFORMATION OF THE STIFFNESS MATRIX OF THE MEMBER OF A RIGID FRAME

A typical member of a frame is, shown in Fig. 4.5. The local axis system is shown as X', Y' and Z' and global axis system as X, Y and Z. The member is assumed to have uniform flexural rigidity EI and uniform axial rigidity EA.

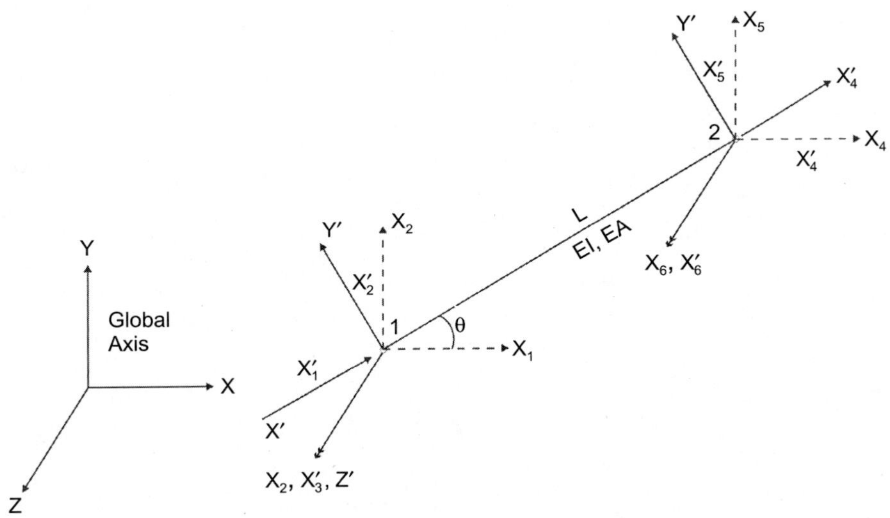

Fig. 4.5. A typical member of a frame with local and global coordinate system

The axial deformation of the member will be considered. The other deformation in the local axis system will be the displacement in the Y'-direction and rotation about Z'-axis. The deformations in the global coordinates are the translations along X and Y-axes and rotation about Z'-axis. The direction cosines of X' with respect to X and Y-axes are cos θ and sin θ respectively. Six deformations-three at each node, are numbered as shown in Fig 4.5 for both the local and the global axis systems.

The stiffness matrix of the member of the frame in the local axis system is given by:

$$[k]_e = \begin{bmatrix} \frac{EA}{L} & 0 & 0 & -\frac{EA}{L} & 0 & 0 \\ & \frac{12EI}{L^3} & \frac{6EI}{L^2} & 0 & -\frac{12EI}{L^3} & \frac{6EI}{L^2} \\ & & \frac{4EI}{L} & 0 & -\frac{6EI}{L^2} & \frac{2EI}{L} \\ & & & \frac{EA}{L} & 0 & 0 \\ & \textit{Symmetrical} & & & \frac{12EI}{L^3} & -\frac{6EI}{L^2} \\ & & & & & \frac{4EI}{L} \end{bmatrix} \tag{4.23}$$

The relationship between local and global axis system is given by:

$$\begin{Bmatrix} X_1' \\ X_2' \\ X_3' \\ \hline X_4' \\ X_5' \\ X_6' \end{Bmatrix} = \left[\begin{array}{ccc|ccc} l_1 & m_1 & 0 & 0 & 0 & 0 \\ -m_1 & l_1 & 0 & 0 & 0 & 0 \\ 0 & 0 & 1 & 0 & 0 & 0 \\ \hline 0 & 0 & 0 & l_1 & m_1 & 0 \\ 0 & 0 & 0 & -m_1 & l_1 & 0 \\ 0 & 0 & 0 & 0 & 0 & 1 \end{array}\right] \begin{Bmatrix} X_1 \\ X_2 \\ X_3 \\ \hline X_4 \\ X_5 \\ X_6 \end{Bmatrix} \tag{4.24}$$

where $l_1 = \cos\theta$ and $m_1 = \sin\theta$

Therefore, for the member of the frame shown in Fig. 4.5, the transformation matrix is given by:

$$[T] = \begin{bmatrix} l_1 & m_1 & 0 & 0 & 0 & 0 \\ -m_1 & l_1 & 0 & 0 & 0 & 0 \\ 0 & 0 & 1 & 0 & 0 & 0 \\ 0 & 0 & 0 & l_1 & m_1 & 0 \\ 0 & 0 & 0 & -m_1 & l_1 & 0 \\ 0 & 0 & 0 & 0 & 0 & 1 \end{bmatrix} \tag{4.25}$$

The element stiffness match in the global axis system can now be calculated according to eqn. (4.18). Instead of explicitly determining it, the necessary result may be obtained by the matrix multiplication done by the computer.

If the axial force in the members of the rigid frame is neglected, there will be two deformations in the local axis system at each node, *i.e.* X_1' and X_2' in node 1 and X_3' and X_4' in node 2 (Fig. 4.6). However, global axis displacement labels remain the same as the case with axial load.

$$[k]_e = \begin{bmatrix} \frac{12EI}{L^3} & & \textit{Symmetrical} & \\ \frac{6EI}{L^2} & \frac{4EI}{L} & & \\ -\frac{12EI}{L^3} & -\frac{6EI}{L^2} & \frac{12EI}{L^3} & \\ \frac{6EI}{L^2} & \frac{2EI}{L} & -\frac{6EI}{L^2} & \frac{4EI}{L} \end{bmatrix} \tag{4.26}$$

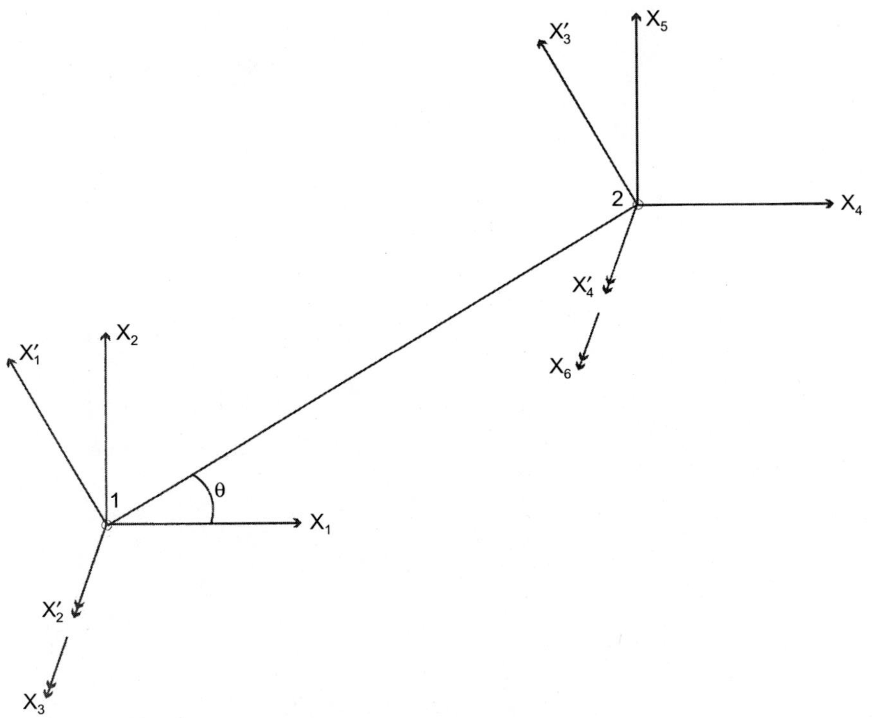

Fig. 4.6. A frame member neglecting axial deformation

and the transformation matrix for this case becomes

$$[T] = \begin{bmatrix} -m_1 & l_1 & 0 & 0 & 0 & 0 \\ 0 & 0 & 1 & 0 & 0 & 0 \\ 0 & 0 & 0 & -m_1 & l_1 & 0 \\ 0 & 0 & 0 & 0 & 0 & 1 \end{bmatrix} \tag{4.27}$$

4.6 TRANSFORMATION OF THE STIFFNESS MATRIX OF THE MEMBER OF A GRILLAGE

The plan of a typical member of a grid or grillage is shown in Fig. 4.7. The flexural and torsional rigidities of the uniform members are *EI* and *GJ* repectively. In local axis system, the deformations at each node are a vertical displacement and two rotational displacements which are numbered in Fig. 4.7(a) The deformation labels in global coordinates are shown in Fig. 47(b).

The stiffness matrix of the member of the grid in local axis system is given by:

$$[k]_e = \begin{bmatrix} \frac{12EI}{L^3} & & & & & \\ 0 & \frac{GJ}{L} & & \textit{Symmetrical} & & \\ \frac{6EI}{L^2} & 0 & \frac{4EI}{L} & & & \\ -\frac{12EI}{L^3} & 0 & -\frac{6EI}{L^2} & \frac{12EI}{L^3} & & \\ 0 & -\frac{GJ}{L} & 0 & 0 & \frac{GJ}{L} & \\ \frac{6EI}{L^2} & 0 & \frac{2EI}{L} & -\frac{6EI}{L^2} & 0 & \frac{4EI}{L} \end{bmatrix} \tag{4.28}$$

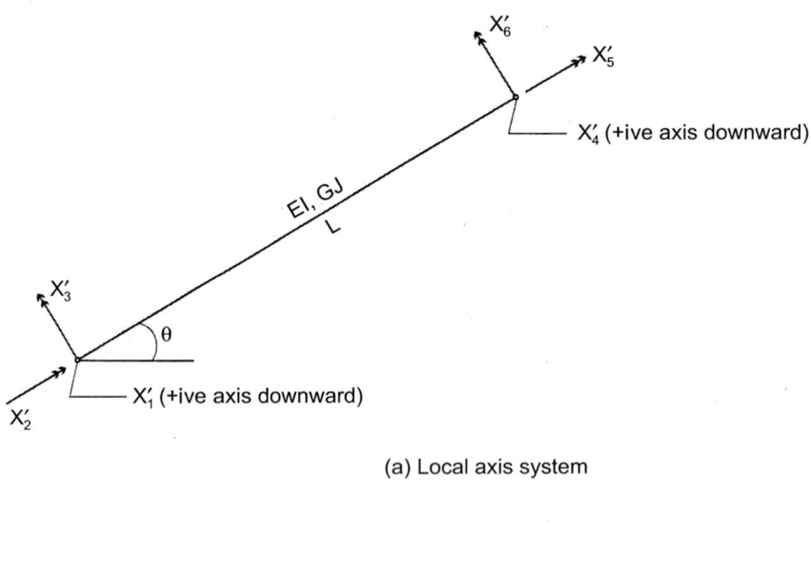

(a) Local axis system

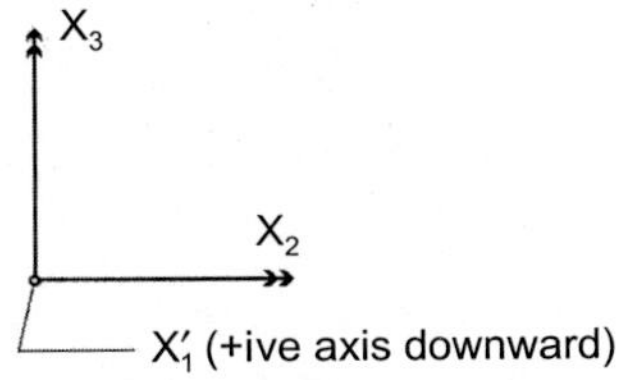

(b) Global axis

Fig. 4.7. A typical member of a grid

The relationship of the deformation in local and global axis system is given by

$$\begin{Bmatrix} X_1' \\ X_2' \\ X_3' \\ X_4' \\ X_5' \\ X_6' \end{Bmatrix} = \begin{bmatrix} 1 & 0 & 0 & 0 & 0 & 0 \\ 0 & \cos\theta & \sin\theta & 0 & 0 & 0 \\ 0 & -\sin\theta & \cos\theta & 0 & 0 & 0 \\ 0 & 0 & 0 & 1 & 0 & 0 \\ 0 & 0 & 0 & 0 & \cos\theta & \sin\theta \\ 0 & 0 & 0 & 0 & -\sin\theta & \cos\theta \end{bmatrix} \begin{Bmatrix} X_1 \\ X_2 \\ X_3 \\ X_4 \\ X_5 \\ X_6 \end{Bmatrix} \qquad (4.29)$$

In compact form, eqn. (4.29) is

$$\{X\} = [T]\,\{X\} \qquad (4.30)$$

Once the transformation matrix is obtained, the stiffness matrix of the member of the global axis system can be obtained in the same way stated earlier.

4.7 TRANSFORMATION OF THE STIFFNESS MATRIX OF THE MEMBER OF A SPACE FRAME

The possible displacements at the ends of a typical space frame member *i* are indicated in Fig. 4.8(a) for local axis system and Fig. 4.8(b) for global axis system.

The member stiffness matrix of the space frame member in local axis system is given by eqn. (4.31). In this equation A_x is the cross-sectional area of the member, L is the length of the member, I_x and I_z are the principal moments of inertia of the cross-section of the member with respect to y_m and z_m axes respectively.

If *i* and *j*, the two ends of the member has coordinates x_i, y_i, z_i and x_j, y_j, z_j, then direction cosines of the member is given by

$$C_x = \frac{x_j - x_i}{L},\ C_y = \frac{y_j - y_i}{L} \text{ and } C_z = \frac{z_j - z_i}{L} \qquad (4.32)$$

The length L of the member is given by

$$L = \sqrt{(x_j - x_i)^2 + (y_j - y_i)^2 + (z_j - z_i)^2} \qquad (4.33)$$

$$
[k]_e = \left[\begin{array}{cccccc|cccccc}
\frac{EA_x}{L} & & & & & & & & & & & \\
0 & \frac{12EI_z}{L^3} & & & & & & & & & & \\
0 & 0 & \frac{12EI_y}{L^3} & & & \textit{Symmetrical} & & & & & & \\
0 & 0 & 0 & \frac{GI_x}{L} & & & & & & & & \\
0 & 0 & -\frac{6EI_y}{L^2} & 0 & \frac{4EI_y}{L} & & & & & & & \\
0 & \frac{6EI_z}{L^2} & 0 & 0 & 0 & \frac{4EI_z}{L^2} & & & & & & \\
\hline
-\frac{EA_x}{L} & 0 & 0 & 0 & 0 & 0 & \frac{EA_x}{L} & & & & & \\
0 & -\frac{12EI_y}{L^3} & 0 & 0 & 0 & -\frac{6EI_z}{L^2} & 0 & \frac{12EI_z}{L^3} & & & & \\
0 & 0 & -\frac{12EI_y}{L^3} & 0 & \frac{6EI_y}{L^2} & 0 & 0 & 0 & \frac{12EI_y}{L^3} & & & \\
0 & 0 & 0 & -\frac{GI_x}{L} & 0 & 0 & 0 & 0 & 0 & \frac{GI_x}{L} & & \\
0 & 0 & -\frac{6EI_y}{L^2} & 0 & \frac{2EI_y}{L} & 0 & 0 & 0 & \frac{6EI_y}{L^2} & 0 & \frac{4EI_y}{L} & \\
0 & \frac{6EI_z}{L^2} & 0 & 0 & 0 & \frac{2EI_z}{L} & 0 & -\frac{6EI_z}{L^2} & 0 & 0 & 0 & \frac{4EI_z}{L}
\end{array}\right] \qquad (4.31)
$$

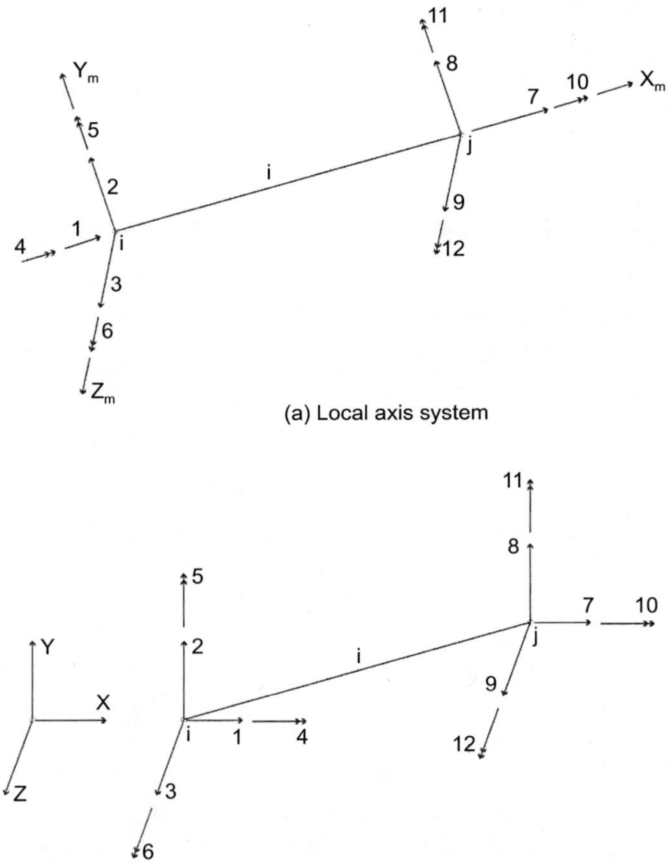

(a) Local axis system

(b) Global axis system

Fig. 4.8. A typical member of a space frame

The rotation matrix $[\lambda]$ (eqn. (4.3) is given by

$$[\lambda] = \begin{bmatrix} C_x & C_y & C_z \\ -\dfrac{C_x C_y}{\sqrt{C_x^2 + C_y^2}} & \sqrt{C_x^2 + C_y^2} & -\dfrac{C_y C_z}{\sqrt{C_y^2 + C_z^2}} \\ -\dfrac{C_z}{\sqrt{C_x^2 + C_z^2}} & 0 & \dfrac{C_x}{\sqrt{C_x^2 + C_z^2}} \end{bmatrix} \qquad (4.34)$$

The transformation matrix of the space frame member can be formed accordingly.

The rotational matrix $[\lambda]$ is valid for all positions of the member except when it is vertical. $[\lambda]$ matrix for this case is

$$[\lambda] = \begin{bmatrix} 0 & C_y & 0 \\ -C_y & 0 & 0 \\ 0 & 0 & 1 \end{bmatrix} \tag{4.35}$$

4.8 HORIZONTALLY CIRCULAR CURVED BEAM ELEMENT

Consider a structural member circularly curved in plan and is uniform. The member will develop two moments P_1 and P_2 and a force P_3 (the member is part of a grid structure) as shown in Fig. 4.9. The end *i* is fixed, but the end *j* is free. Let us take an infinitesimal arc length *ds* at an angle θ from *ij* subtending $d\theta$ at the centre.

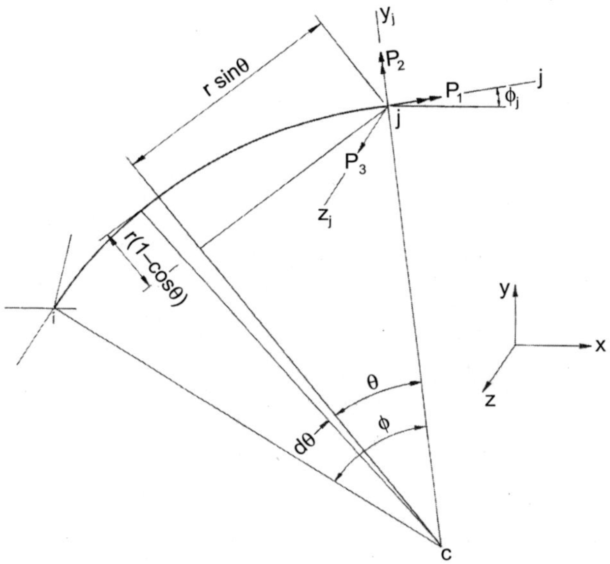

Fig. 4.9. Curved beam element

Due to unit values of the forces at *j* the bending moments and torques developed at *ds* is

$$\begin{aligned} M_1 &= \sin\theta, \quad M_2 = -\cos\theta, \quad M_3 = r\sin\theta \\ T_1 &= \cos\theta, \quad T_2 = -\sin\theta, \quad T_3 = r(1-\cos^2\theta) \end{aligned} \tag{4.36}$$

The flexibility coefficient $(F_{jj})_{11}$ is obtained as [4.1].

$$(F_{jj})_{11} = \int_s \frac{M_1^2}{EI_y} ds + \int_s \frac{T_1^2}{GI_x} ds \tag{4.37}$$

or,

$$(F_{jj})_{11} = \frac{r}{2EI_y}(\phi - \sin\phi\cos\phi) + \frac{r}{2GI_x}(\phi + \sin\phi\cos\phi) \tag{4.38}$$

Other flexibility coefficients can be similarly obtained. Each element of the flexbility matrix consists of two parts – flexure and torsion, and each element is sum of the two.

$$[F_{jj}] = [F_{jj}]_{flex} + [F_{jj}]_{tor} \tag{4.39}$$

where $$[F_{jj}]_{flex} = \frac{r}{2EI_y}\begin{bmatrix} \phi - \sin\phi\cos\phi & & symmetrical \\ \sin^2\phi & \phi + \sin\phi\cos\phi & \\ r(\phi - \sin\phi\cos\phi) & -r\sin^2\phi & r^2(\phi - \sin\phi\cos\phi) \end{bmatrix} \tag{4.40a}$$

$$[F_{jj}]_{tor} = \frac{r}{2GI_x}\begin{bmatrix} \phi + \sin\phi\cos\phi & & symmetrical \\ \sin^2\phi & \phi + \sin\phi\cos\phi & \\ r(\phi - 2\sin\phi + \sin\phi\cos\phi) & -r(2 - 2\cos\phi - \sin^2\phi) & r^2(\phi - \sin\phi\cos\phi) \end{bmatrix} \tag{4.40b}$$

Equations (4.40a) and (4.40b) are matrices which indicate displacements in the direction x_j, y_j and z_j due to unit values corresponding to P_1, P_2 and P_3 respectively.

The stiffness matrix at the jth end can be obtained by inversion. Thus,

$$[k]_{jj} = [F]_{jj}^{-1} \tag{4.41}$$

The stiffness matrix is to be expressed in structural directions by

$$[K_{jj}] = [R]^T [k]_{jj} [R] \tag{4.42}$$

where $$[R]_j = \begin{bmatrix} \cos\phi_j & \sin\phi_j & 0 \\ -\sin\phi_j & \cos\phi_j & 0 \\ 0 & 0 & 1 \end{bmatrix} \tag{4.43}$$

Other parts of the member stiffness matrix is given by

$$[K_{ij}] = -[T_{ij}][K_{jj}] \tag{4.44}$$

where $$[T_{ij}] = \begin{bmatrix} 1 & 0 & y_{ij} \\ 0 & 1 & -x_{ij} \\ 0 & 0 & 1 \end{bmatrix} \tag{4.45}$$

Similarly,

$$[K_{ji}] = -[K]_{jj} [T_{ij}]^T \tag{4.46}$$

and

$$[K_{ij}] = [T_{ij}]^T [K]_{jj} [T_{ij}] \tag{4.47}$$

4.9 OVERALL STIFFNESS MATRIX

The stiffness matrix of an individual member of the structure forms the basic component. Element stiffness matrices of all members are to be assembled to relate the form and the displacement of the complete structure. The subsequent discussion is undertaken to show as to how the contribution of the element stiffness matrix is taken care of in the formation of the overall stiffness matrix, also known as the structure stiffness matrix.

It may be noticed by looking into problems of Chapter 3 that the stiffness at a joint is obtained by adding the stiffnesses of all the members meeting at the joint.

In the first step, the degrees of freedom of the structure are to be numbered-starting with 1 and ending with *NP* where *NP* is the total degrees of freedom. The restraints are then numbered beyond *NP*. These numberings are referred to as degrees of freedom corresponding to global numbering or simply the global degrees of freedom. If the unknown displacements and the internal forces in the members are only to be determined, then all restraints can be given the number (*NP*+1). This procedure will save some storage space. However, in addition to the above two, if the reactions at the supports are to be evaluated, then the restraints (*i.e.* the known displacements) are to be numbered sequentially. The reaction aspect is put up till the next chapter. In the mean time the part of the stiffness matrix associated with the degrees of freedom is discussed.

The force-displacement relationship for the complete structure is given by

$$[K][X] = [P] \tag{4.48}$$

Both $[K]$ and $[P]$ consist of contributions from individual element. Assembling the contribution of element stiffness matrix of individual member to form overall stiffness matrix is now discussed. This is best demonstrated with an example.

Before further discussion is initiated, a revision of some of the terms is felt necessary. The chapter has been started by introducing the concept of local axis system and global axis system. We shall henceforth deal with two different numbering in global axis system for the displacements as the member ends. The first set of numbering is referred to as local numbering which will remain identical for every member. This numbering refers to the typical displacement labels of the member. The members meet at a joint in the structure. The same member ends will have different displacement labels when the total structure is concerned.

The continuous beam of Fig. 4.10 has four degrees of freedom-rotation at four supports. The horizontal movement of the supports is ignored. In Fig. 4.10(a), the numbering from left-hand side corresponding to global coordinates has been indicated. In Fig. 4.10(b), the local numbering of a typical member and in Fig. 4.10(c), the positive direction of the displacements have been shown.

Calculation of the element stiffness matrix is next to be done. This is done for a typical element on the basis of local axis system. For the problem at hand, the size of the element stiffness matrix is 4 × 4. For the first element, they are, say,

Global	5	1	6	2
Local	1	2	3	4

$$[K]_1 = \begin{matrix} 5 & 1 \\ 1 & 2 \\ 6 & 3 \\ 2 & 4 \end{matrix} \begin{bmatrix} a_{11} & a_{12} & a_{13} & a_{14} \\ a_{21} & a_{22} & a_{23} & a_{24} \\ a_{31} & a_{32} & a_{33} & a_{34} \\ a_{41} & a_{42} & a_{43} & a_{44} \end{bmatrix} \tag{4.49}$$

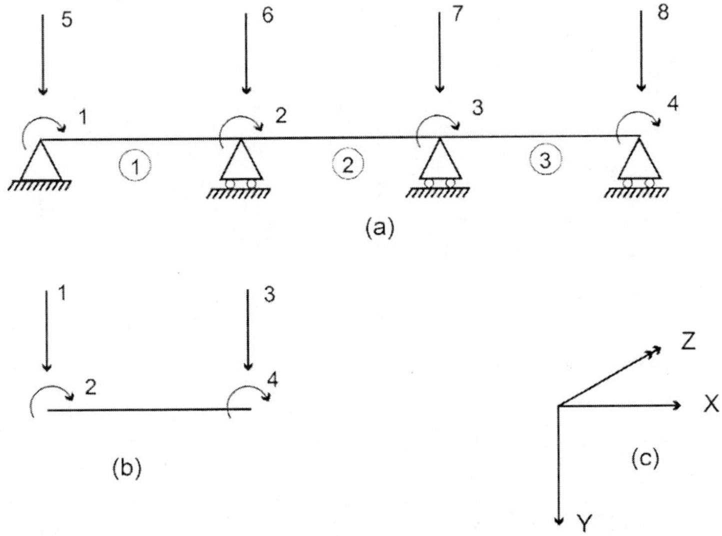

Fig. 4.10. Local and global numbering of a continuous beam

For members which are inclined and where local and global coordinates are not coincident, the element stiffness matrix is to be evaluated on the basis of transformation as per eqn. (4.18). There is one to one correspondence between the joints of the elements and that of the assemblage. The elements of the element stiffness matrix of eqn. (4.49) should now be put in their proper place in the overall stiffness matrix. Equation (4.32) indicates that local numbering 1 corresponds to global numbering 5 which means that a_{11} should occupy 5th row 5th column (*i.e.* K_{55}) of the overall stiffness matrix. Local member 2 corresponds to global number 1. Therefore, a_{12}, a_{21}, a_{22} will occupy the positions of K_{51}, K_{15} and K_{11} of the overall stiffness matrix. The proper location of all elements of eqn. (4.49) in the overall stiffness matrix has been shown in Fig. 4.11.

The process is to be repeated for all the elements. It is to be borne in mind that individual stiffness at a particular location in the overall stiffness matrix is to be added. If the element stiffness matrices for the element members 2 and 3 be then, the overall stiffness matrix has been shown in Fig. 4.12. The steps to be followed in the computer program have been shown in the flow chart of Fig. 4.13.

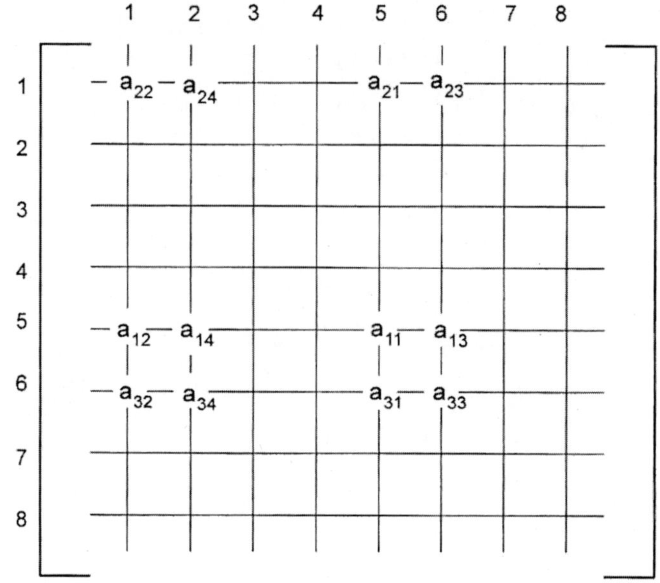

Fig. 4.11. Assemblage into global system of element of Fig. 4.10

	1	2	3	4	5	6	7	8
1	a_{22}	a_{24}	0	0	a_{21}	a_{23}	0	0
2	a_{42}	$a_{44}+b_{22}$	b_{24}	0	a_{41}	$a_{43}+b_{21}$	b_{23}	0
3	0	b_{42}	$b_{44}+c_{22}$	c_{24}	0	b_{41}	$b_{43}+c_{21}$	c_{23}
4	0	0	c_{42}	c_{44}	0	0	c_{41}	c_{43}
5	a_{12}	a_{14}	0	0	a_{11}	a_{13}	0	0
6	a_{32}	$a_{34}+b_{12}$	b_{14}	0	a_{31}	$a_{33}+b_{11}$	b_{13}	0
7	0	b_{32}	$b_{34}+c_{12}$	c_{14}	0	b_{31}	$b_{33}+c_{11}$	c_{13}
8	0	0	c_{32}	c_{34}	0	0	c_{31}	c_{33}

Fig. 4.12. Overall stiffness matrix

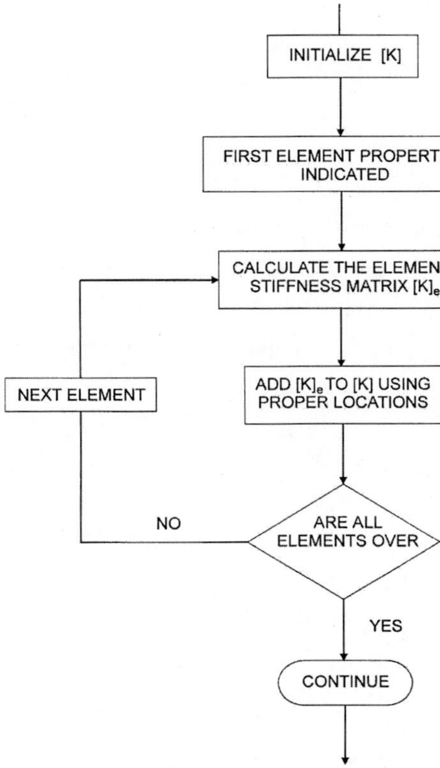

Fig. 4.13. Flowchart of assembly from the element stiffness matrix to the overall stiffness matrix

$$\begin{array}{llcccc} & \textit{Global} & 6 & 2 & 7 & 3 \\ & \textit{Local} & 1 & 2 & 3 & 4 \end{array}$$

$$[K]_2 = \begin{array}{cc} 6 & 1 \\ 2 & 2 \\ 7 & 3 \\ 3 & 4 \end{array} \begin{bmatrix} b_{11} & b_{12} & b_{13} & b_{14} \\ b_{21} & b_{22} & b_{23} & b_{24} \\ b_{31} & b_{32} & b_{33} & b_{34} \\ b_{41} & b_{42} & b_{43} & b_{44} \end{bmatrix}$$

$$\begin{array}{llcccc} & \textit{Global} & 7 & 3 & 8 & 4 \\ & \textit{Local} & 1 & 2 & 3 & 4 \end{array}$$

$$[K]_3 = \begin{array}{cc} 7 & 1 \\ 3 & 2 \\ 8 & 3 \\ 4 & 4 \end{array} \begin{bmatrix} c_{11} & c_{12} & c_{13} & c_{14} \\ c_{21} & c_{22} & c_{23} & c_{24} \\ c_{31} & c_{32} & c_{33} & c_{34} \\ c_{41} & c_{42} & c_{43} & c_{44} \end{bmatrix} \tag{4.50}$$

The overall loading matrix can be similarly formulated as an assemblage of the contribution from the element loading matrix. For the formation of the overall loading matrix, however, the load applied to the joint directly is to be taken care of.

4.10 BOUNDARY CONDITIONS

One practice of labelling is to put the number corresponding to the unknown displacements first and then, the restrained displacements. This facilitates taking into account the boundary conditions in an effective manner.

There are a number of ways of catering the boundary conditions. A few of the approaches have been discussed here.

If the overall stiffness matrix is to be formed in half-band form, then the numbering of displacements should be such that the band width is minimum. For this case displacement labels are put in a systematic manner irrespective of whether the joint displacements are unknowns or restraints. The restrained displacements are taken care of by assigning a high value (say 10^{25}) in the diagonal element corresponding to that displacement.

If the unknown displacements are labelled first, then the matrix operations can be restricted up to unknown displacement labels and beyond that, the overall stiffness matrix may be ignored.

Another approach of catering the boundary conditions will be to make the diagonal element corresponding to that displacement label equal to unity while all other elements in the corresponding rows and columns as also the loading vector in that row, are put equal to zero.

Another approach is not to assemble in the overall stiffness matrix, the contributions from the nodes which have restrained displacements *i.e.* row and column corresponding to fixed displacement is eliminated.

4.10.1 Boundary Conditions Corresponding to Skewed Supports

If the constraints applied at the nodes are in the direction of global axis system, they can be taken care of as stated above. The same concept can be extended to the problems where the boundary condition corresponding to a skew support is to be incorporated.

A beam element is of Fig. 4.14 has a skewed roller support. Therefore, in the overall formulation, the vertical displacement corresponding to the roller support is zero. At *i*th node, where the roller support is placed, displacement from the global axis system is transformed to skew coordinates.

$$[X_i]_G = [\lambda]\,[X]_{skew} \tag{4.51}$$

$[\lambda]_i$ is a simple point transformation involving the direction cosines which relate the global and skewed axis system. The transformation matrix for that element then becomes

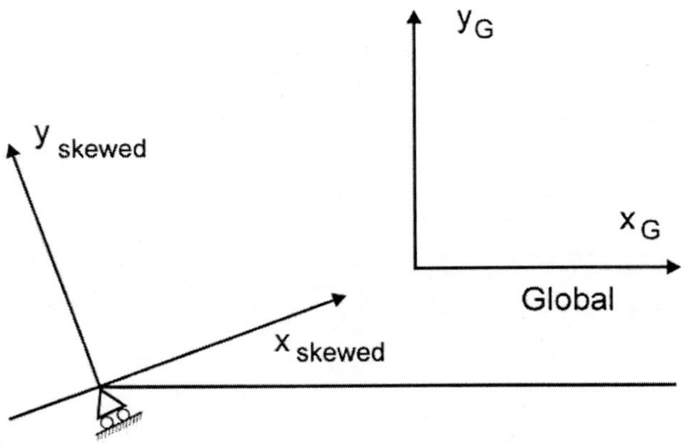

Fig. 4.14. Skewed support

$$[T] = \begin{bmatrix} [I] & & & & & & \\ & [I] & & & & & \\ & & [\lambda]_i & & & & \\ & & & \cdot & & & \\ & & & & \cdot & & \\ & & & & & \cdot & \\ & & & & & & [I] \end{bmatrix} \tag{4.52}$$

where $[I]$ is an identity matrix. There is one submatrix $[\lambda]_i$ in $[T]$ for each node with skewed constraints.

The resulting stiffness matrix then becomes,

$$[K] = [T]^T [k]_e [T] \tag{4.53}$$

This transformation is done at the individual stiffness matrix generation level *i.e.* before they are assembled to form the overall stiffness matrix.

After the unknown displacements are known for the overall solution, the skewed displacements are to be converted back by global coordinates by using eqn. (4.51), for obtaining the necessary stresses in the elements.

The incorporation of the boundary conditions has been discussed in details in the computer programs and the examples presented later.

4.11 COMPUTATION OF INTERNAL FORCES

The solution of eqn. (4.48) will yield the displacements at all joints. The end actions of the members are next to be determined. This is possible on the basis of eqn. (3.12).

$$[PM]_e = [PMF]_e + [PMF]_e\,[X]_e \tag{4.54}$$

or,

$$[PM]_e = [PMF]_e + [k]_e\,[X]_e \tag{4.55}$$

Only change to be noted is that $[PMX]$ is replaced by $[k]_e$. $[PMF]_e$ matrix for each element is given as an input to the computer.

4.12 COMPUTER PROGRAM FOR THE TRUSS ANALYSIS BY THE DIRECT STIFFNESS METHOD

Table 4.1 gives the listing of the computer program in 'C' developed for the analysis of a plane truss by the direct stiffness method. It is also given in the CD (file name: dir_tr.c) attached with this book. Different notations used and different steps followed have been indicated with comment statement in the computer program. The computer program for the same in FORTRAN (file name: DIR_TR.F) is also given in the CD.

Input data required for the computer program consists of the material and geometric properties of the members and their global numbering. The x and y coordinates of one end of the member with respect to the other end are specified as input. The data of the loading matrix is also fed in the computer.

Example 4.1:

The truss of Fig. 4.15 is to be analysed with the help of the computer program of Table 4.1. The horizontal and vertical members have area 1 and the inclined members 1.414. The loading has been shown in Fig. 4.15.

Global numbering has been indicated in Fig. 4.15(a). For this problem $NT = 9$, $NP = 8$, $NF = 10$ and $NLC = 1$. It is assumed that $E = 1$. It may be noted that the total degrees of freedom for this problem is 8. All the restraints have been assigned the number 9. The preparation of the input data corresponding to one typical member is explained here.

The inclined member 4 is considered. A is assumed as the initial point and B as the terminal point. The data for this member then is

1 2 7 8 1.0 –1.0 1.414

The local numbering for this element has been shown as dotted line in the figure. B also could have been assumed as the initial point and A as the terminal point, Then the data will be

7 8 1 2 –1.0 1.0 1.414

The input data for the problem of Fig. 4.13 is given in Table 4.2. It may be noted that two loading conditions have been considered.

The computer output for the problem is given in Table 4.3. The forces in the members can be checked with those obtained by Wang's displacement method* for the same problem [4.2, 4.3]. They have been found to be of the same order.

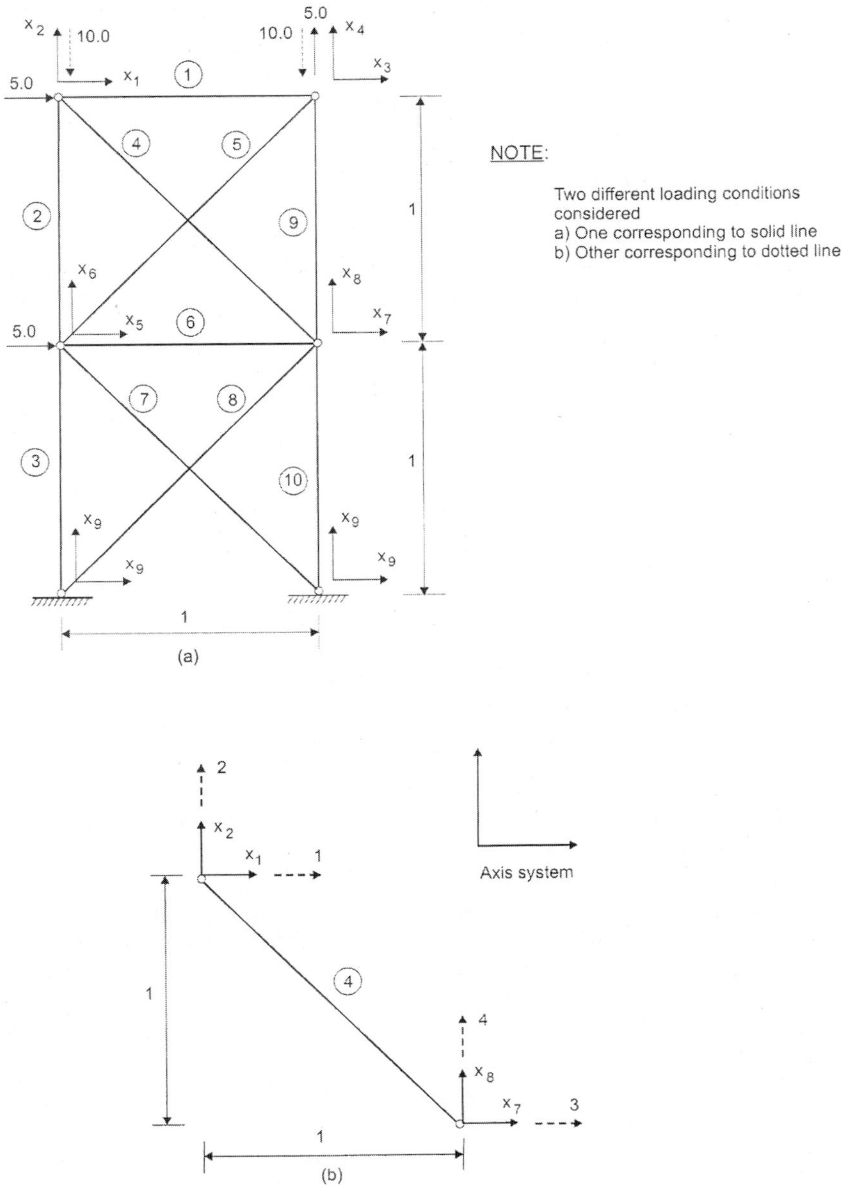

Fig. 4.15. Truss for solution by direct stiffness method

* It is a method where the stiffness matrix is formed in a different manner and this has not been included in this book.

```
/*
      Table 4.1. Computer Program for the Analysis of Plane Truss by Direct
                              Stiffness Method
*/
#include <stdio.h>
#include <math.h>
#include <dos.h>
#include <conio.h>
// FUNCTION INVER FOR MATRIX INVERSION
void inver(float a[][30],int n){
// VARIABLE DECLARATION
        float max,temp,pivot;
        int col,i,j,flag=1,index[20];
        for(i=0;i<n;i++) index[i] = 0;
        while(flag != 0){
                max=-1.0;
                for(i=0;i<n;i++){
                        if(index[i]==0){
                                        temp = fabs(a[i][i]);
                                        if(temp > max){
                                                col=i;
                                                max=temp;
                                        }
                                }
                        }
                        if(max > 0){
                                index[col] = 1;
                                pivot = a[col][col];
                                a[col][col] = 1.0;
                                pivot = 1/pivot;
                                for(j=0;j<n;j++) a[col][j] *= pivot;
                                for(i=0;i<n;i++){
                                        if((i-col) != 0){
                                                temp = a[i][col];
                                                a[i][col] = 0.0;
                                                for(j=0;j<n;j++) a[i][j]
-= a[col][j] * temp;
                                        }
                                }
                        }
                        else flag=0;
                }
        }
void planetruss()
{
/* NP TOTAL DEGREES OF FREEDOM
   NT TOTAL DEGREES OF FREEDOM INCLUDING RESTRAINTS
   NF NUMBER OF MEMBERS
   NLC NUMBER OF LOADING CONDITIONS
   E  YOUNG'S MODULUS
   STIFO THE OVERALL STIFFNESS MATRIX
   */
float stifo[30][30],x[30][4],fi[50][4],load[30][4];
float v[50],h[50],aa[50],xcos,xsin,xl,e,dirc,dirs,dircs;
int np1,np2,np3,np4,i,j,np,nt,nf,nlc,k,npg[40][4];
FILE *fpt1,*fpt2;
```

```
fpt1=fopen("DIR_TR_IN","r");
fpt2=fopen("DIR_TR_OUT","w");
fprintf(fpt2,"\tTruss Analysis\n");
fscanf(fpt1,"%d%d%d%d%f",&nt,&np,&nf,&nlc,&e);
fprintf(fpt2,"\tNT= %d\tNP = %d\tNF = %d\tNLC = %d\tE =
%.3f\n",nt,np,nf,nlc,e);
// INITIALIZE STIFO
for(i=0;i<nt;i++)
{
for(j=0;j<nt;j++) stifo[i][j] = 0.0;
}
/* CALCULATE THE ELEMENT STIFFNESS MATRIX AND PUT THEM IN
   THEIR PROPER LOCATION IN STIFO
   NPG[I][1],NPG[I][2],NPG[I][3],NPG[I][4] ARE THE GLOBAL
   NUMBERINGS CORRESPONDING TO THE LOCAL NUMBERING 1,2,3,4
   H[I],V[I] ARE THE PROJECTIONS ALONG X AND Y AXIS OF THE
   MEMBER
   AA[I] IS THE CROSS-SECTIONAL AREA OF THE MEMBER */
fprintf(fpt2,
" MEMBER  NP1 NP2 NP3 NP4   H    V    A    L    COS    SIN\n");
for(i=0;i<nf;i++)
{
for(j=0;j<4;j++) fscanf(fpt1,"%d",&npg[i][j]);
fscanf(fpt1,"%f%f%f",&h[i],&v[i],&aa[i]);
xl = sqrt((h[i]*h[i])+(v[i]*v[i]));
xcos = h[i] / xl;
xsin = v[i] / xl;
np1 = npg[i][0]-1;
np2 = npg[i][1]-1;
np3 = npg[i][2]-1;
np4 = npg[i][3]-1;
fprintf(fpt2,"%3d%3d%3d%3d%3d",i+1,np1,np2,np3,np4);
fprintf(fpt2,"%7.3f%7.3f%7.3f%7.3f%7.3f%7.3f\n",h[i],
v[i],aa[i],xl,xcos,xsin);
dirc = e * aa[i] * xcos * xcos / xl;
dirs = e * aa[i] * xsin * xsin / xl;
dircs = e * aa[i] * xcos * xsin / xl;
stifo[np1][np1] = stifo[np1][np1] + dirc;
stifo[np1][np2] = stifo[np1][np2] + dircs;
stifo[np1][np3] = stifo[np1][np3] - dirc;
stifo[np1][np4] = stifo[np1][np4] - dircs;
stifo[np2][np2] = stifo[np2][np2] + dirs;
stifo[np2][np3] = stifo[np2][np3] - dircs;
stifo[np2][np4] = stifo[np2][np4] - dirs;
stifo[np3][np3] = stifo[np3][np3] + dirc;
stifo[np3][np4] = stifo[np3][np4] + dircs;
stifo[np4][np4] = stifo[np4][np4] + dirs;
stifo[np2][np1] = stifo[np1][np2];
stifo[np3][np1] = stifo[np1][np3];
stifo[np4][np1] = stifo[np1][np4];
stifo[np3][np2] = stifo[np2][np3];
stifo[np4][np2] = stifo[np2][np4];
stifo[np4][np3] = stifo[np3][np4];
}
for(j=0;j<nlc;j++)
{
for(i=0;i<np;i++) fscanf(fpt1,"%f",&load[i][j]);
}
```

```
// OUTPUT THE STIFFNESS MATRIX
fprintf(fpt2,"\tThe Stiffness Matrix\n");
for(i=0;i<np;i++)
{
        for(j=0;j<np;j++) {
                fprintf(fpt2,"%14.5e",stifo[i][j]);
if((j+1)%5==0)
                        fprintf(fpt2,"\n");
        }
fprintf(fpt2,"\n");
}
// INVERT MATRIX STIFO
inver(stifo,np);
// OUTPUT THE LOAD MATRIX
fprintf(fpt2,"\tThe Load Matrix\n");
for(j=0;j<nlc;j++)
{
        for(i=0;i<np;i++) {
                fprintf(fpt2,"%14.5e",load[i][j]);
if((i+1)%5==0)
                        fprintf(fpt2,"\n");
        }
fprintf(fpt2,"\n");
}
// CALCULATE THE JOINT DISPLACEMNTS
for(i=0;i<np;i++)
{
for(j=0;j<nlc;j++)
{
x[i][j] = 0.0;
for(k=0;k<np;k++) x[i][j] += stifo[i][k] * load[k][j];
}
}
// OUTPUT MATRIX X
fprintf(fpt2,"\tThe X Matrix\n");
for(j=0;j<nlc;j++)
{
        for(i=0;i<np;i++){
                fprintf(fpt2,"%14.5e",x[i][j]);
                if((i+1)%5==0)
                        fprintf(fpt2,"\n");
        }
fprintf(fpt2,"\n");
}
// RESTRAINT DISPLACEMENTS ARE MADE ZERO IN NEXT 4 STATEMENTS
for(i=np;i<nt;i++)
{
for(j=0;j<nlc;j++) x[i][j] = 0.0;
}
// COMPUTE INTERNAL FORCES IN THE MEMBERS OF THE TRUSS
for(i=0;i<nf;i++)
{
xl = sqrt(h[i] * h[i] + v[i] * v[i]);
xcos = h[i] / xl;
xsin = v[i] / xl;
np1 = npg[i][0]-1;
np2 = npg[i][1]-1;
np3 = npg[i][2]-1;
```

```
np4 = npg[i][3]-1;
for(j=0;j<nlc;j++)
fi[i][j]=e*aa[i]/xl*(xcos*(x[np3][j]-x[np1][j])+xsin*
(x[np4][j]-x[np2][j]));

}
// OUTPUT MATRIX FI
fprintf(fpt2,"\tThe FI Matrix\n");
for(j=0;j<nlc;j++)
{
        for(i=0;i<nf;i++) {
                fprintf(fpt2,"%14.5e",fi[i][j]);
                if((i+1)%5==0)
                        fprintf(fpt2,"\n");
        }
fprintf(fpt2,"\n");
}

fclose(fpt1);
fclose(fpt2);


}

void main()
{planetruss();
}
```

Table 4.2. Input Data of Truss

```
9 8 10 2 1.0
1 2 3 4 1.0 0.0 1.0
1 2 5 6 0.0 -1.0 1.0
5 6 9 9 0.0 -1.0 1.0
1 2 7 8 1.0 -1.0 1.414
5 6 3 4 1.0 1.0 1.414
5 6 7 8 1.0 0.0 1.0
5 6 9 9 1.0 -1.0 1.414
9 9 7 8 1.0 1.0 1.414
3 4 7 8 0.0 -1.0 1.0
7 8 9 9 0.0 -1.0 1.0
5.0 0.0 0.0 5.0 5.0 0.0 0.0 0.0
0.0 -10.0 0.0 -10.0 0.0 0.0 0.0 0.0
```

Table 4.3 Output of the Truss by Direct Stiffness Method

```
        Truss Analysis
        NT= 9   NP = 8  NF = 10 NLC = 2 E = 1.000
MEMBER   NP1 NP2 NP3 NP4   H       V       A       L       COS     SIN
  1   0   1   2   3   1.000   0.000   1.000   1.000   1.000   0.000
  2   0   1   4   5   0.000  -1.000   1.000   1.000   0.000  -1.000
  3   4   5   8   8   0.000  -1.000   1.000   1.000   0.000  -1.000
  4   0   1   6   7   1.000  -1.000   1.414   1.414   0.707  -0.707
  5   4   5   2   3   1.000   1.000   1.414   1.414   0.707   0.707
  6   4   5   6   7   1.000   0.000   1.000   1.000   1.000   0.000
  7   4   5   8   8   1.000  -1.000   1.414   1.414   0.707  -0.707
  8   8   8   6   7   1.000   1.000   1.414   1.414   0.707   0.707
  9   2   3   6   7   0.000  -1.000   1.000   1.000   0.000  -1.000
 10   6   7   8   8   0.000  -1.000   1.000   1.000   0.000  -1.000
```

```
        The Stiffness Matrix
  1.49992e+000 -4.99925e-001 -1.00000e+000  0.00000e+000  0.00000e+000
  0.00000e+000 -4.99925e-001  4.99925e-001
 -4.99925e-001  1.49992e+000  0.00000e+000  0.00000e+000  0.00000e+000
 -1.00000e+000  4.99925e-001 -4.99925e-001
 -1.00000e+000  0.00000e+000  1.49992e+000  4.99925e-001 -4.99925e-001
 -4.99925e-001  0.00000e+000  0.00000e+000
  0.00000e+000  0.00000e+000  4.99925e-001  1.49992e+000 -4.99925e-001
 -4.99925e-001  0.00000e+000 -1.00000e+000
  0.00000e+000  0.00000e+000 -4.99925e-001 -4.99925e-001  1.99985e+000
  0.00000e+000 -1.00000e+000  0.00000e+000
  0.00000e+000 -1.00000e+000 -4.99925e-001 -4.99925e-001  0.00000e+000
  2.99985e+000  0.00000e+000  0.00000e+000
 -4.99925e-001  4.99925e-001  0.00000e+000  0.00000e+000 -1.00000e+000
  0.00000e+000  1.99985e+000  0.00000e+000
  4.99925e-001 -4.99925e-001  0.00000e+000 -1.00000e+000  0.00000e+000
  0.00000e+000  0.00000e+000  2.99985e+000
        The Load Matrix
  5.00000e+000  0.00000e+000  0.00000e+000  5.00000e+000  5.00000e+000
  0.00000e+000  0.00000e+000  0.00000e+000
  0.00000e+000 -1.00000e+001  0.00000e+000 -1.00000e+001  0.00000e+000
  0.00000e+000  0.00000e+000  0.00000e+000
        The X Matrix
  3.87295e+001  1.21818e+001  3.62750e+001 -2.81816e+000  1.89106e+001
  9.63640e+000  1.60924e+001 -5.36360e+000
 -1.09085e+000 -1.52731e+001  1.09083e+000 -1.52731e+001 -2.36347e+000
 -7.45475e+000  2.36347e+000 -7.45474e+000
        The FI Matrix
 -2.45456e+000  2.54545e+000  9.63640e+000 -3.59981e+000  3.47126e+000
 -2.81816e+000 -6.55685e+000  7.58528e+000  2.54545e+000 -5.36360e+000

  2.18168e+000 -7.81832e+000 -7.45475e+000 -3.08536e+000 -3.08536e+000
  4.72693e+000 -3.59953e+000 -3.59953e+000 -7.81832e+000 -7.45474e+000
```

4.13 COMPUTER PROGRAM FOR THE FRAME ANALYSIS BY DIRECT STIFFNESS METHOD

The computer program for the plane frame analysis by the direct stiffness method in both the languages - C (file name: dir_fr.c) and FORTRAN (DIR_FR.F), are given in the attached CD. Notations have been explained in the listing of the computer program for both cases. Different steps followed in the computer code have been explained with 'comment' statements.

In comparison to the computer program of the truss presented in the earlier section, the present one has several new features. For the first time, we are encountering a computer program where the overall stiffness matrix is stored in halfband form. The displacement matrix has been obtained from a solution of simultaneous equations. This is done by subroutine GAUSS.

The main program contains a number of subroutines - a subroutine for calculating the element stiffness matrix in local coordinate system, a subroutine for calculating the transformation matrix and another subroutine for computing the member forces. The input data (Table 4.4) is simple to prepare. It consists of pertinent local and global information. In addition to the material and geometric properties of the members, the information regarding total degrees of freedom, total number of nodes, number of restraints, number of loading conditions *etc.* must be specified. This computer program does not generate the loading matrix which however, is fed directly as the input data. The computer output has been indicated in Table 4.5.

Example 4.2:

The frame shown in Fig. 4.16 is to be analysed. For this problem $NT = 6$, $NLC = 1$, $NNODE = 4$, $NMEM = 3$, $NN = 2$, $NBH = 7$. It may be noted that all the restraints have been given a displacement label 7. For a small problem of the type shown in Fig. 4.16, it is not possible to show the advantages of storing the overall stiffness-matrix in half band form. For bigger size problems storing the overall stiffness matrix in halfband form will be obvious.

The input data for the problem is given in Table 4.4. The computer output has been given in Table 4.5.

Table 4.4. Input Data of the Plane Frame Problem

```
4 3 3 2 6 6 1 1
0.0 0.0 5.0 8.0 13.0 8.0 18.0 0.0
3 1 1 2 2 4
7 7 7 1 2 3 1 2 3 4 5 6 4 5 6 7 7 7
7 200000000.
0.00036 0.00036 0.00036
0.6 0.6 0.6
0.0 -80.0 106.7 0.0 -80.0 -106.7
0 0 0 0 0 0
0.0 80.0 -106.7 0.0 80.0 106.7
0 0 0 0 0 0
```

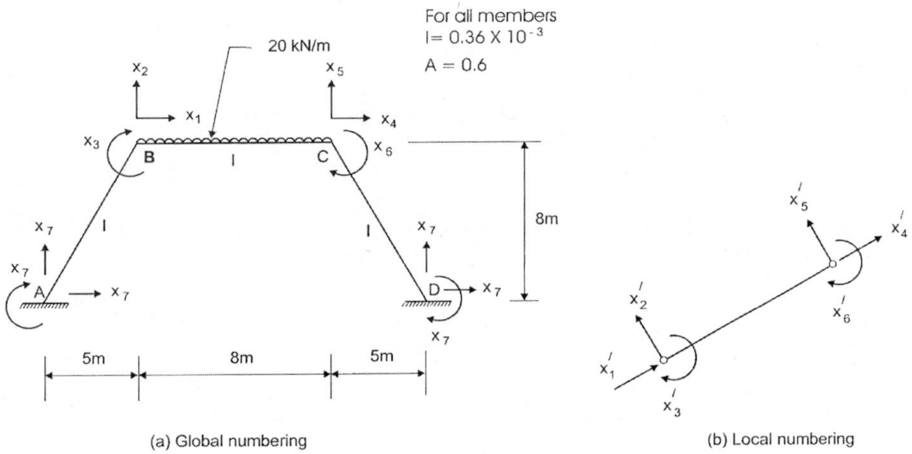

Fig. 4.16. Frame for solution of plane frame by direct stiffness method

Table 4.5. Computer Output of the Plane Frame Problem

```
    NNODE = 4   NMEM = 3   NFREE = 3   NN = 2   NPM = 6   NT = 6   NLC = 1
 The Coordinates of the Nodes
      0.0000        0.0000        5.0000        8.0000
     13.0000        8.0000       18.0000        0.0000

 Node Numbering
   3   1   1   2   2   4
 The memberwise NPG Values
          Member No. 1
   7   7   7   1   2   3
          Member No. 2
   1   2   3   4   5   6
          Member No. 3
   4   5   6   7   7   7
         NBH = 7 E =          2e+008
         Moment of Inertia of Members
  3.60000e-004  3.60000e-004  3.60000e-004
         Cross-Sectional Areas
  6.00000e-001  6.00000e-001  6.00000e-001
         The OSTIF Matrix
  9.276e+006  9.225e+006 -5.088e+003 -3.574e+006 -5.716e+006
 -4.116e+003  0.000e+000  1.131e+007  4.152e+003 -5.716e+006
 -9.147e+006  2.573e+003 -5.667e+006  0.000e+000  4.940e+004
  4.116e+003 -2.573e+003  1.526e+004  8.826e+003  0.000e+000
  0.000e+000  9.276e+006  2.208e+006  5.088e+003 -2.192e+006
  0.000e+000  0.000e+000  0.000e+000  1.131e+007 -9.935e+002
  1.351e+006  0.000e+000  0.000e+000  0.000e+000  0.000e+000
  4.940e+004  6.883e+003  0.000e+000  0.000e+000  0.000e+000
  0.000e+000  0.000e+000  1.575e+007  0.000e+000  0.000e+000
  0.000e+000  0.000e+000  0.000e+000  0.000e+000
         The LOAD Matrix
   0.00000 -80.00000 106.70000    0.00000 -80.00000-106.70000

         The DISPLACEMENT Matrix
   0.53122   -0.86328    0.10734   -0.23611   -0.38370    0.07050

          Load condition: 1
          Member end forces of the member: 1
-152.044 -47.223-866.724 152.044  47.223 145.897
          Member end forces of the member: 2
  89.066  12.082-145.898 -89.066 147.918-494.837
          Member end forces of the member: 3
 110.253  21.265 494.837-110.253 -21.265-170.237
```

4.14 COMPUTER PROGRAM FOR THE GRILLAGE ANALYSIS BY THE DIRECT STIFFNESS METHOD

The computer program for the grillage analysis by the direct stiffness method in both the languages – C and FORTRAN are given in the attached CD. Notations have been explained in the listing of the computer program for both cases. Different steps followed in the computer code have been explained with 'comment' statements.

In comparison to the computer program of the truss presented in the earlier section, the present one has several new features. Much like the case of the frame analysis, the overall stiffness matrix is stored in halfband form. The displacement matrix has been obtained from a solution of simultaneous equations.

The main program contains a number of subroutines - a subroutine for calculating the element stiffness matrix in local coordinate system, a subroutine for calculating the transformation matrix and another subroutine for computing the member forces. The input data (Table 4.6) is simple to prepare. It consists of pertinent local and global information. In addition to the material and geometric properties of the members, the information regarding, total degrees of freedom, total number of nodes, number of restraints, number of loading conditions *etc.* must be specified. This computer program does not generate the loading matrix which however, is fed directly as the input data. The computer output has been indicated in Table 4.7.

The input data consists of among others, member geometries, their global numbering and the geometric properties. A code LDIS is used as an input. This takes care of the uniform load present in a member. For that member, the computer program calculates the element loading matrix and puts it in the proper location of overall loading matrix. The computer program can cater shear deformation of the member based on LGG value. If LGG = 0, the shear deformation is not included.

Example 4. 3:

The grid of Fig. 4.17 is to be analysed. It has been assumed that $E = 1$, $I = 1$, $L = 1$, $P = 1$, $G = 0.8$, $J = 1.0$ and $A = 1$. For this problem, $NT = 4$, $NP = 3$, $NLC = 1$, $NELM = 2$, $NPM = 6$, $NBH = 4$. All the restraints have been given a displacement label 4. Like the frame problem of the previous section, this problem is also of small size. As such the advantages of storing the overall stiffness matrix in halfband form cannot be felt. Shear deformation has not been considered in this problem.

The displacement labels for a typical member corresponding to local coordinates are shown in Fig. 4.17(c). The data corresponding to item No. 3 of Table 4.6 for member no. 1 is

1.0 0. 0. 1. 1. 4. 4. 4. 1 2 3 0 0

The input data for the problem is given in Table 4.6. The computer output for the problem is given in Table 4.7. The values are nearly identical to those obtained from Wang's displacement method.

Table 4.6. Input Data for the Grillage Problem

```
4 3 1 2 6 4 0 0
1. .8 0.
1. 0. 0. 1. 1. 0 4 4 4 1 2 3 0. 0.
0. 1. 0. 1. 1. 0 1 2 3 4 4 4 0. 0.
0.5 -0.125 0.
0. 0. 0. 0. 0. 0.
-0.5 0. -0.125 -0.5 0.0 0.125
```

Table 4.7. Computer Output of the Grillage Problem

```
   Direct Stiffness Method of Analysis of Grillage
   NT             NP       NLC      NELM     NPM      NBH          LGG     LDIS
4      3    1     2    6        4        0      0
   E = 1   G = 0.8        FF = 0
   XX = 1.000   XY = 0.000        AA = 0.000       XI = 1.000       XJ = 1.000
   SPDR = 0.000 Q = 0.000
   NPG Values
   4    4        4        1        2        3        LUND = 0
   XX = 0.000   XY = 1.000        AA = 0.000       XI = 1.000       XJ = 1.000
   SPDR = 0.000 Q = 0.000
   NPG Values
   1    2        3        4        4        4        LUND = 0
   The Overall Stiffness Matrix
  2.40000e+001 -6.00000e+000 -6.00000e+000 -3.60000e+001
  4.80000e+000  0.00000e+000  7.20000e+000  0.00000e+000
  4.80000e+000  7.20000e+000  0.00000e+000  0.00000e+000
  1.00000e+025  0.00000e+000  0.00000e+000  0.00000e+000
       The Overall Loading Matrix
       Loading case:     1
  5.00000e-001 -1.25000e-001  0.00000e+000
       The Displacement Matrix
             0.0381944             0.0217014             0.0477431
       Stress in Member Number = 1
 -1.7188e-001 -1.7361e-002 -1.3368e-001  1.7188e-001  1.7361e-002
-3.8194e-002
       Stress in Member Number = 2
 -1.7188e-001  3.8194e-002  1.7361e-002 -8.2813e-001 -3.8194e-002
3.1076e-001
```

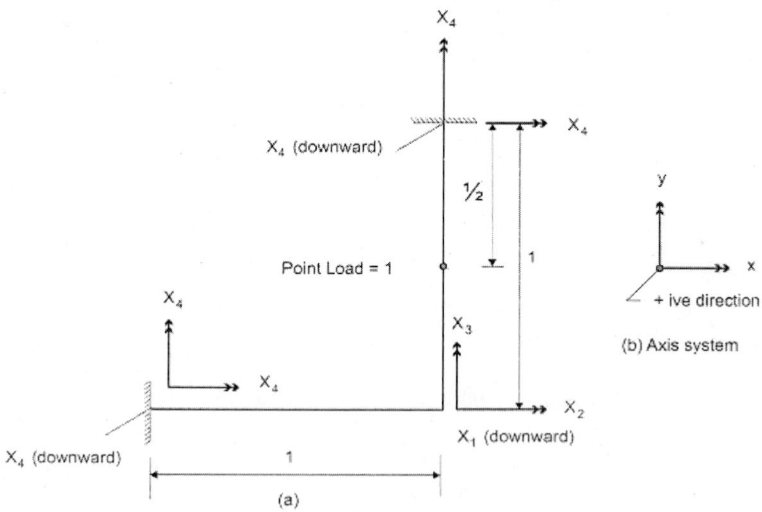

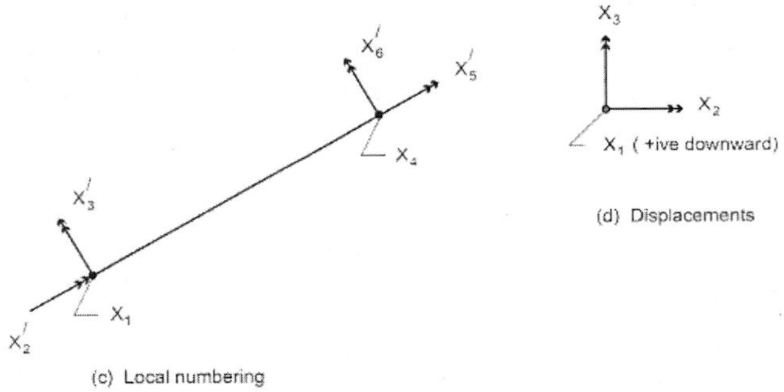

Fig. 4.17. Grillage for solution by the direct stiffness method

REFERENCES AND SUGGESTED READINGS

4.1. J.M. Gere and W. Weaver Jr., Analysis of Framed Structures, 2nd Edition, D. Van Nostrand, 1982.

4.2. C.K. Wang, Matrix Methods of Structural Analysis, International Textbook, Pasadena, USA, 1970.

4.3. C.K. Wang, Introductory Structural Analysis with Matrix Method, Prentice Hall, 1973.

4.4. W. McGuire, R.H. Gallagher and R.D. Ziemian, Matrix Structural Analysis, 2nd Edition, John Wiley & Sons Inc., USA, 2000.

EXERCISE 4

4.1. Write a computer program for the solution of a plane frame by the direct stiffness method. The computer program is to have the flexibility of considering both with or without axial force.

4.2. Prepare the input data for the plane truss of Fig. below according to the computer program of Table 4.1. Assume $E = 1.0$, the areas of the members of top chord and bottom chord are 2 units and that of all internal members 1.

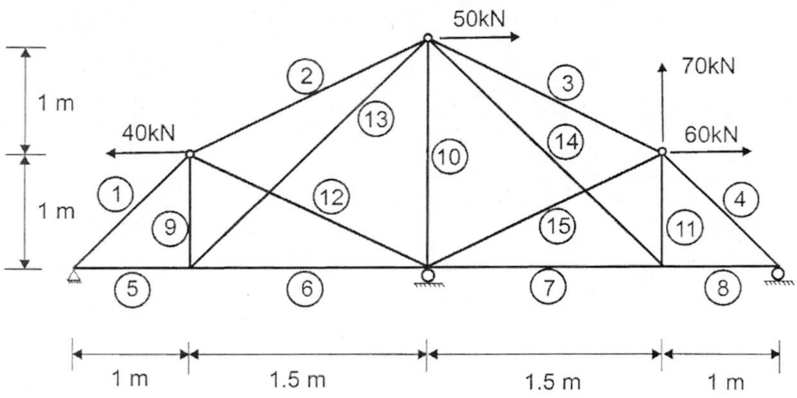

Prob. 4.2.

4.3. Prepare the input data for the grillage according to the program of Table 4.4. Assume $EI = 1.0$ and $GJ = 0.8$ for all members. The shear deformation is not to be included. The loads acting have been indicated in Prob. 4.3.

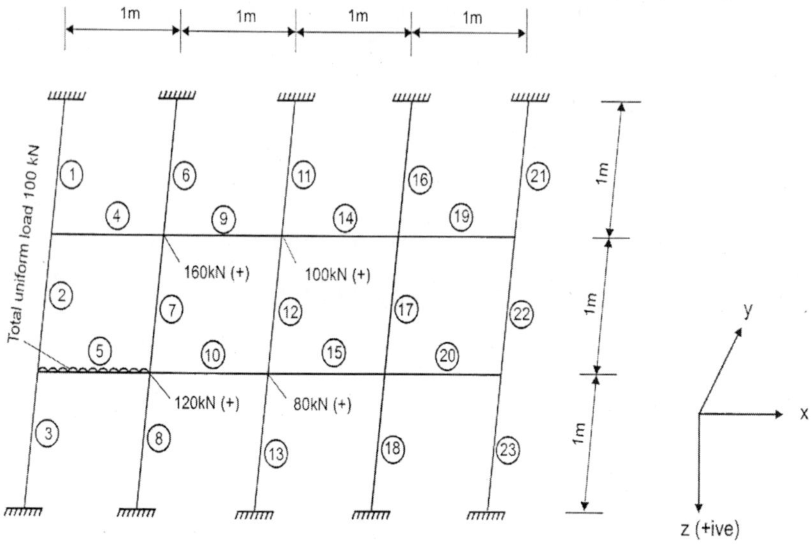

Prob. 4.3.

4.4. With the help of computer programs find out the forces in the members of the trusses shown. Cross-sectional areas of the outer members = 2 units and those of internal members = 1.

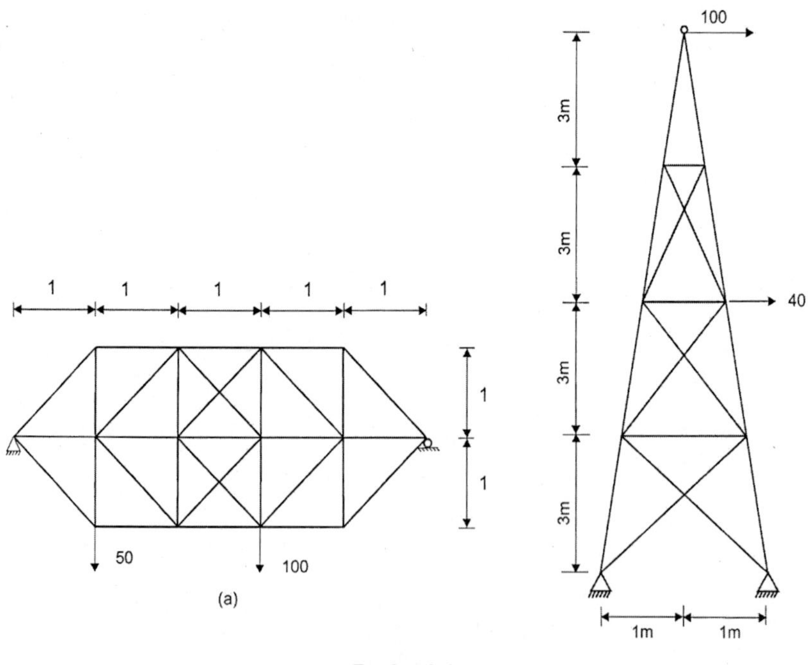

Prob. 4.4.

4.5. Solve the plane frames by the direct stiffness method; neglect axial force for (a) and (b), but incorporate axial force for (c). Assume $EA = 0.1$ *El* for (c).

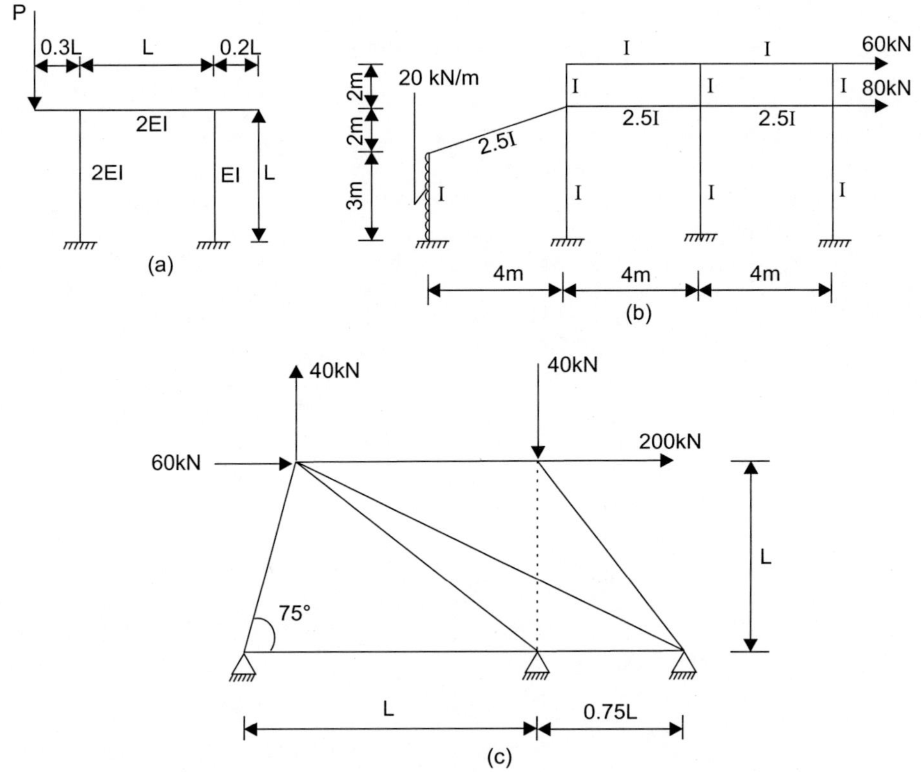

Prob. 4.5.

4.6. Determine the member force for the grillages shown. Assume $EI = 1.25 \times 10^9$ kNm², GJ=1 × 10^9 kNmm².

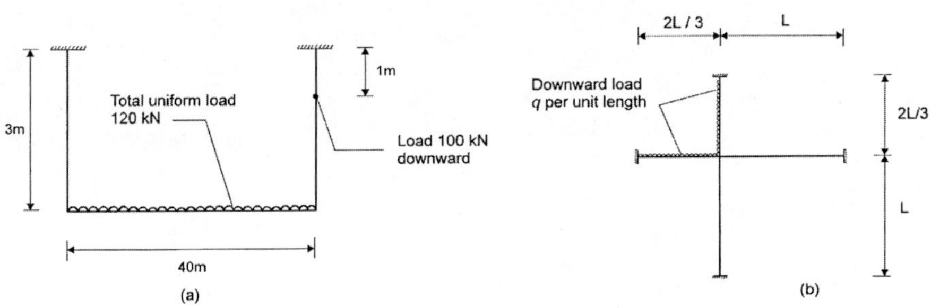

Prob. 4.6.

5

SUBSTRUCTURE TECHNIQUE FOR THE ANALYSIS OF STRUCTURAL SYSTEMS

5.1 INTRODUCTION

The analysis of structural problems by the stiffness matrix method results in the following system of simultaneous equations:

$$[K][X] = [P] \tag{5.1}$$

A major part of the computer time (CPU time) is involved in the solution of eqn. (5.1). The direct method of generating the structure stiffness matrix as a solid matrix is suited only to problems of small size. For big size problems, advantage is taken of the fact that the structure stiffness matrix $[K]$ is symmetric and banded. $[K]$ is stored in halfband and the solution of simultaneous equations is obtained from Gauss's method for forward elimination and then backward substitution. This half-band storage of $[K]$ is a preferred approach for problems where all storage remains within the core memory. The storage can be made more economical by storing the half-band of $[K]$ in skyline form.

For large sized problems where half-band solution no longer becomes economic, more sophisticated and specialized techniques are resorted to. Some of them are the frontal solution technique (5.1, 5.2), sparse matrix techniques (5.3) and substructure technique (5.4, 5.5, 5.6).

There are different approaches to the analysis of structural systems using substructure technique. In all of them, the total structure is divided into a number of substructures and the stiffness matrix of the substructure is expressed in terms of its boundary displacements. The overall stiffness matrix of the structure is the assembly of these reduced stiffness matrices. By dividing the total structure into substructures, the number of equations that may be solved at any one time for the analysis of the total system is considerably reduced.

5.2 BASIC CONCEPTS

The basic concept is explained with the plane truss of Fig. 5.1. To start with, the structure is to be conveniently partitioned into smaller structural systems. Each portion of these partitioned structural systems is known as a substructure. Each substructure consists of a group of elements.

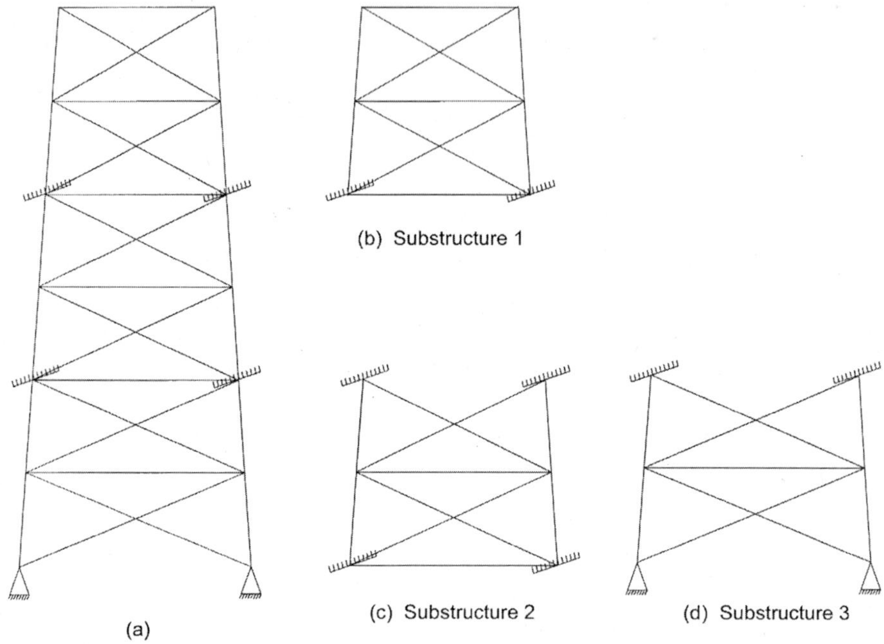

(a) (b) Substructure 1 (c) Substructure 2 (d) Substructure 3

Fig. 5.1. A plane truss with substructures

Each substructure is analysed first by assuming the boundaries between the substructures as restrained. This can be carried out by the direct stiffness method discussed in the previous chapter considering each substructure as if it were a complete system in itself. As the displacements at the substructure boundaries are artificially constrained, joint displacements from this analysis will not represent the true values of the displacements at the joints. Therefore, these displacements need to be corrected. This correction is obtained from the analysis of the total structure which is based on the displacements of the artificially restrained boundaries of the substructure. Once the actual displacements of all the joints of the complete structure have been evaluated, the member forces can then be determined.

The main advantage of analysing the structure using substructure technique can be seen with respect to the plane truss of Fig. 5.1. By the direct stiffness method, the number of unknown displacements for the truss is 24, as there are 12 free joints. When analysed by the substructure technique according to the substructures of Fig. 5.1, maximum number of unknown displacements to be catered at any one time is 8. For substructures 1, 2 and 3, the unrestrained components of joint displacements are 8, 4 and 4 respectively. For the analysis of whole structure, 8 displacements are there corresponding to the artificial constraints. Thus, in comparison to the direct stiffness method, the number of unknowns to be dealt with at any one time is considerably lower.

While dividing a structure into a number of substructures, care should be taken to partition the structure. Number of unknowns for the individual substructure and those of the total structure considering the unknowns at the substructure joints should preferably be of the same order for economical solution. Therefore, the partitioning of the total structure into substructure must be judiciously done.

5.3 DIRECT STIFFNESS METHOD RESTATED

Equation (4.48) is rewritten with slight modification

$$[K]_c\,[X]_c = [P]_c + [R]_c \tag{5.2}$$

where $[K]_c$ is the complete structure stiffness matrix (including all restraints).

$[X]_c$ is the complete joint displacement matrix.

$[P]_c$ is the complete joint load matrix.

$[R]_c$ is the complete support reaction matrix.

While writing the elements of eqn. (5.2), it may be noted that the unknown displacements are labelled first, then the boundary displacements $[K]_c$ are partitioned into groups associated with unrestrained and restrained joint displacements, as follows:

$$\begin{bmatrix} [K]_{uu} & [K]_{ur} \\ [K]_{ru} & [K]_{rr} \end{bmatrix} \begin{bmatrix} [X]_u \\ [X]_r \end{bmatrix} = \begin{bmatrix} [P]_u \\ [P]_r \end{bmatrix} + \begin{bmatrix} [O]_u \\ [R]_r \end{bmatrix} \tag{5.3}$$

where $[K]_{uu}$ is the stiffness matrix associated with unknown joint displacements.

$[K]_{ru}$ is the stiffness matrix associated with the restrained displacements of the support developed by the unit values of unknown joint displacements.

$[K]_{ur}$ is the stiffness matrix associated with the unknown joint displacements developed by unit values of the displacements corresponding to restrained support displacements.

$[K]_{rr}$ is the stiffness matrix associated with restrained support displacements developed by unit values of restrained support displacements.

Equation (5.3) can be resolved into two independent equations

$$[K]_{uu}\,[X]_u = [P]_u \tag{5.4}$$

$$[K]_{ru}\,[X]_u = [P]_r + [R]_r \tag{5.5}$$

It may be noted that $[X]_r = 0$

Unknown joint displacements can be evaluated from eqn. (5.4)

$$[X]_u = [K]_{uu}^{-1}\,[P]_u \tag{5.6}$$

Once $[X]_u$ has been obtained, support reactions can be determined by substituting $[X]_u$ into eqn. (5.5). Thus

$$[R]_r = [K]_{ru}\,[K]_{uu}^{-1}\,[P]_u - [P]_r \tag{5.7}$$

5.4 THE SUBSTRUCTURE TECHNIQUE

The complete structure is partitioned into a number of substructures. Each substructure need to be analysed independently for:

(*a*) the load applied within each substructure

(*b*) possible substructure boundary displacements

For the analysis of *n* th substructure of a given structural system, the equations describing the static equilibrium of the joints of the substructure can be expressed in the following form when its boundaries are fixed

$$\left[\begin{array}{c|c} [K]^n_{uu} & [K]^n_{ur} \\ \hline [K]^n_{ru} & [K]^n_{rr} \end{array}\right]\left[\begin{array}{c} [X]^n_{uf} \\ \hline [O] \end{array}\right] = \left[\begin{array}{c} [P]^n_{uf} \\ \hline [P]^n_{rf} \end{array}\right] + \left[\begin{array}{c} [O] \\ \hline [R]^n_{rf} \end{array}\right] \tag{5.8}$$

Subscript f refers to the case where substructure boundaries are fixed. Equation (5.8) results in two equations

$$[K]^n_{uu}\,[X]^n_{uf} = [P]^n_{uf} \tag{5.9}$$

$$[K]^n_{ru}\,[X]^n_{uf} = [P]^n_{rf} + [R]^n_{rf} \tag{5.10}$$

From eqn. (5.9), one gets

$$[K]^n_{ru}\,[X]^n_{uf} = [P]^n_{rf} + [R]^n_{rf} \tag{5.11}$$

Substituting the value of $[X]^n_{uf}$ from eqn. (5.11) into eqn. (5.10) yields,

$$[R]^n_{rf} = [K]^n_{ru}\,([K]^n_{uu})^{-1}\,[P]^n_{uf} - [P]^n_{rf} \tag{5.12}$$

$[R]^n_{rf}$, the reactions developed at the fixed substructure boundaries, are to be determined from eqn. (5.12). Therefore, in the first stage of analysis, the substructure boundaries are kept fixed and $[X]^n_{uf}$ and $[R]^n_{rf}$ for the *n* th substructure are determined from eqn. (5.11) and (5.12) respectively.

In the next stage of analysis, substructure boundaries are subjected to displacements. These substructure boundary displacements are to be obtained from the analysis of the total structure, and this is discussed a little later. In the meantime, it is assumed that substructure joints are given boundary displacements. The subscript *d* refers to this case. Therefore, the response of substructure *n* to the application of true boundary displacements with all others loadings removed are written. In this case

$$[P]^n_{ud} = [P]^n_{rd} = [0] \tag{5.13}$$

The equilibrium equations for this case are

$$\left[\begin{array}{c|c} [K]^n_{uu} & [K]^n_{ur} \\ \hline [K]^n_{ru} & [K]^n_{rr} \end{array}\right]\left[\begin{array}{c} [X]^n_{ud} \\ \hline [X]^n_{rd} \end{array}\right]\left[\begin{array}{c} [0] \\ \hline [R]^n_{rd} \end{array}\right] \tag{5.14}$$

where $[X]^n_{rd}$ represents the actual displacements of the substructure boundaries.

Equation (5.14) can be expressed into two equations

$$[K]^n_{uu}\,[X]^n_{ud} + [K]^n_{ur}\,[X]^n_{rd} = [0] \tag{5.15}$$

$$[K]^n_{ru}\,[X]^n_{ud} + [K]^n_{rr}\,[X]^n_{rd} = [R]^n_{rd} \tag{5.16}$$

Equation (5.15) yields

$$[X]^n_{ud} = -\,([K]^n_{ud})^{-1} + [K]^n_{ur}\,[X]^n_{rd} \tag{5.17}$$

Substituting $[X]^n_{ud}$ from eqn. (5.17) into eqn. (5.16) yields

$$-[K]^n_{ru}\,([K]^n_{uu})^{-1}\,[K]^n_{ur}\,[X]^n_{rd} + [K]^n_{rr}\,[X]^n_{rd} = [R]^n_{rd}$$

or,

$$([K]^n_{rr} - [K]^n_{ru}\,([K]^n_{uu})^{-1}\,[K]^n_{ur}\,[X]^n_{rd} = [R]^n_{rd} \tag{5.18}$$

Introducing,

$$[K]^n_{s} = [K]^n_{rr} - [K]^n_{ru}\,([K]^n_{uu})^{-1}\,[K]^n_{ur} \tag{5.19}$$

Equation (5.18) is written as

$$[K]^n_{s}\,[X]^n_{rd} = [R]^n_{rd} \tag{5.20}$$

It may be recalled that $[R]^n_{rd}$ refers to the reaction at the substructure boundaries to maintain equilibrium, when it is subjected to boundary displacements $[X]^n_{rd}$. Therefore, $[K]^n_{s}$ can be appropriately described as the substructure stiffness matrix.

Thus, for each substructure, the substructure stiffness matrix as given by eqn. (5.19) is to be obtained. These stiffness matrices are assembled to get the complete structure stiffness matrix $[K]_c$. The displacements associated with the complete structure stiffness matrix are those on the substructure boundaries along with those at the real boundaries. Therefore,

$$[K]_c\,[X]_c = [P]_c + [R]_c \tag{5.21}$$

The evaluation $[P]_c$ in eqn. (5.21) needs some discussion. $[R]^n_{rf}$, the reactions at the fixed boundaries of each substructure obtained from eqn. (5.12) with their sign changed constitute one part of $[P]_c$. To this is added the loads applied at the joints linking the substructure. It is to be borne in mind that loads applied at the joints of the substructure boundary will not be considered in the analysis of the substructure, but will be taken into account when analysing the complete structure.

In $[X]_c$ of eqn. (5.21) has $[X]^n_{rd}$ of each substructure. Once $[X]_c$ is evaluated on the basis of eqn. (5.21), then selecting appropriate values of $[X]^n_{rd}$ for each substructure, and using eqn. (5.17), $[X]^n_{ud}$ can be obtained. True displacements of the unrestrained joints of the substructure, then is

$$[X]^n_{u} = [X]^n_{uf} + [X]^n_{ud} \tag{5.22}$$

Once the true displacements of the joints are known, internal forces of the members can then be computed.

Therefore, true displacements at the joint of the structure are determined in four stages.

(*a*) After keeping the substructure boundaries fixed, the displacements at the unrestrained joints of the substructure are determined. All loads acting in this substructure are considered except those acting at the substructure boundaries. The displacements in this case are obtained by eqn. (5.11) and the reactions at the substructure boundaries by eqn. (5.12).

(*b*) Next, from the analysis of total structure based on the substructure stiffness matrix defined by eqn. (5.19), the unknown displacements at the substructure boundaries are obtained.

This is done by applying eqn. (5.21). Therefore, $[X]^n_{rd}$ values of each substructure are obtained in this step.

(*c*) Based on these $[X]^n_{rd}$ values, the displacements at all unrestrained joints of substructure can then be obtained by eqn. (5.17)

(*d*) Finally, the displacements obtained from steps (*a*) and (*c*) are added to get the true displacements.

In calculation at every step, it must be borne in mind that the unrestrained joints are labelled first, then the restraints.

5.5 AN ILLUSTRATIVE EXAMPLE

The truss of Fig. 5.2 is to be analysed by the substructure technique. Cross-sectional areas of the vertical and horizontal bars = 200 mm^2 and those of diagonal bars = 141.14 mm^2, $E = 5.0 \times 10^9$ N/mm^2. All loads acting at different locations have magnitude unity. The truss is splitted up into two substructures and they are shown in Fig. 5.3 and are denoted as substructures 1 and 2 respectively.

Analysis of substructure 1

The displacements of the substructure is labelled. The unrestrained displacements are denoted first by 1, 2, 3 and 4 and then the artificially restrained displacements are labelled.

The structure stiffness matrix of this substructure is then generated. This is done in the usual way by adding the member stiffness matrices with like labels. It assumes

$$[K]^1_c = \left[\begin{array}{cccc|cccc} 5 & 1 & 0 & 0 & -4 & 0 & -1 & -1 \\ 1 & 5 & 0 & -4 & 0 & 0 & -1 & -1 \\ 0 & 0 & 5 & -1 & -1 & 1 & -4 & 0 \\ 0 & -4 & -1 & 5 & 1 & -1 & 0 & 0 \\ \hline -4 & 0 & -1 & -1 & 5 & -1 & 0 & 0 \\ 0 & 0 & 1 & -1 & -1 & 1 & 0 & 0 \\ -1 & -1 & -4 & 0 & 0 & 0 & 5 & 1 \\ -1 & -1 & 0 & 0 & 0 & 0 & 1 & 1 \end{array}\right] \times 0.125 \times 10^9$$

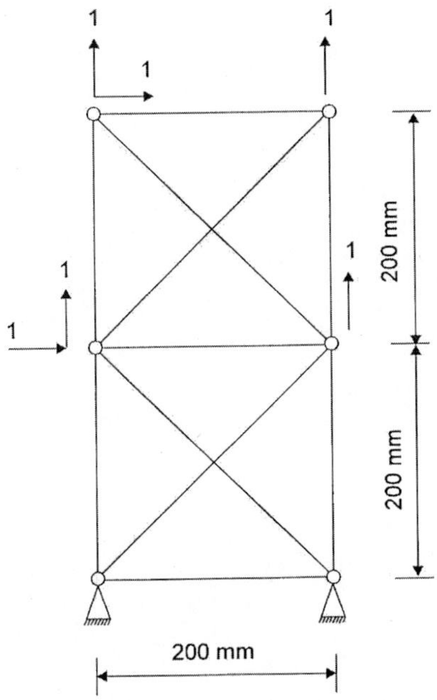

Fig. 5.2. An illustrative example

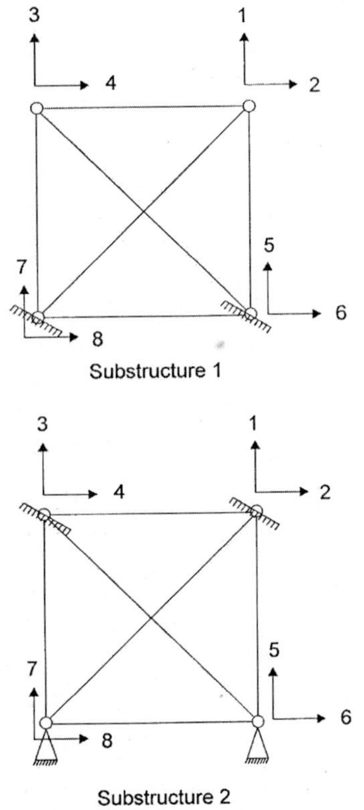

Fig. 5.3. Substructures of the example

This matrix is partitioned as given in eqn. (5.8) as

$$[K]_c^1 = \left[\begin{array}{c|c} [K]_{uu}^1 & [K]_{ur}^1 \\ \hline [K]_{ru}^1 & [K]_{rr}^1 \end{array}\right]$$

The joint load matrix for this substructure is given by

$$[P]_c^1 = \left[\begin{array}{c} 1 \\ 0 \\ 1 \\ 1 \\ \hline 0 \\ 0 \\ 0 \\ 0 \end{array}\right]$$

It may be noted that loads at the substructures boundaries have not been considered in the above $[P]_c^1$ matrix.

The internal joint displacements are given by

$$[X]^1_{uf} = ([K]^1_{uu})^{-1}\,[P]^1_{uf}$$

$$[X]^1_{uf} = \begin{bmatrix} 10 & -6 & -1 & -5 \\ -6 & 30 & 5 & 25 \\ -1 & 5 & 10 & 6 \\ -5 & 25 & 6 & 30 \end{bmatrix} \frac{2\times10^{-9}}{11} \begin{bmatrix} 1 \\ 0 \\ 1 \\ 1 \end{bmatrix}$$

$$= \frac{2\times10^{-9}}{11} \begin{bmatrix} 4 \\ 24 \\ 15 \\ 31 \end{bmatrix}$$

The fixed boundary actions for substructure 1 are equal to

$$[R]^1_{rf} = [K]^1_{ru}\,[X]^1_{uf} - [P]^1_{rf}$$

$$= \frac{10^9}{8} \begin{bmatrix} -4 & 0 & -1 & 1 \\ 0 & 0 & 1 & -1 \\ -1 & -1 & -4 & 0 \\ -1 & -1 & 0 & 0 \end{bmatrix} \frac{2\times10^{-9}}{11} \begin{bmatrix} 4 \\ 24 \\ 15 \\ 31 \end{bmatrix} - \begin{bmatrix} 0 \\ 0 \\ 0 \\ 0 \end{bmatrix}$$

$$[R]^1_{rf} = \frac{2}{88} \begin{bmatrix} 0 \\ 16 \\ 88 \\ 28 \end{bmatrix}$$

The substructure stiffness matrix for substructure 1 can now be evaluated

$$[K]^1_s = [K]^1_{rr} - [K]^1_{ru}\,([K]^1_{uu})^{-1}\,[K]^1_{ur}$$

$$= \begin{bmatrix} 5 & -1 & 0 & 0 \\ -1 & 1 & 0 & 0 \\ 0 & 0 & 5 & 1 \\ 0 & 0 & 1 & 1 \end{bmatrix} \frac{10^9}{8} - \begin{bmatrix} -4 & 0 & -1 & 1 \\ 0 & 0 & 1 & -1 \\ -1 & -1 & -4 & 0 \\ -1 & -1 & 0 & 0 \end{bmatrix} \frac{10^9}{8}$$

$$\begin{bmatrix} 10 & -6 & -1 & -5 \\ -6 & 30 & 5 & 25 \\ -1 & 5 & 10 & 6 \\ -5 & 25 & 6 & 30 \end{bmatrix} \frac{2\times10^{-9}}{11} \times \begin{bmatrix} -4 & 0 & -1 & -1 \\ 0 & 0 & -1 & -1 \\ -1 & 1 & -4 & 0 \\ 1 & -1 & 0 & 0 \end{bmatrix} \frac{10^9}{8}$$

(5 6 7 8 – boundary displacement labels)

$$= \begin{bmatrix} 0 & 0 & 0 & 0 \\ 0 & 1 & 0 & -1 \\ 0 & 0 & 0 & 0 \\ 0 & -1 & 0 & 1 \end{bmatrix} \frac{10^9}{22}$$

Analysis of substructure 2

There are no interior displacements for substructure 2. The structure stiffness matrix is same as that of $[K]_{uu}$ part of substructure 1. Further actual boundary displacements will have no effect on the substructure stiffness matrix. Therefore, the substructure stiffness matrix for substructure 2 is

$$[K]_s^2 = \begin{bmatrix} 5 & 1 & 0 & 0 \\ 1 & 5 & 0 & -4 \\ 0 & 0 & 5 & -1 \\ 0 & -4 & -1 & 5 \end{bmatrix} \frac{10^9}{8}$$

This type of substructure is a special case where these are no unrestrained displacements.

Analysis of the complete structure

The displacements at the substructure boundary *i.e.*, corresponding to the restrained displacements in the restrained structure are labelled and shown in Fig. 5.4

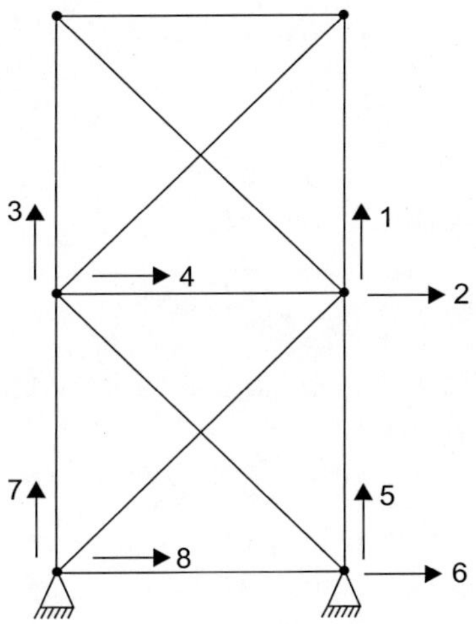

Fig. 5.4. Labeling of the boundary displacements of the complete structure

The substructure stiffness matrix for the two substructures has already been obtained. The labels of these substructure stiffness matrices are considered in this case.

However, in the general case, the labels are to be changed with respect to the total structure. The complete structure stiffness matrix is obtained by adding the elements of substructure stiffness matrix with the labels.

In this case

$$[K]_c = \begin{bmatrix} 55 & 11 & 0 & 0 \\ 11 & 59 & 0 & -48 \\ 0 & 0 & 55 & -11 \\ 0 & -48 & -11 & 59 \end{bmatrix} \frac{10^9}{88}$$

The joint load matrix for the total structure is obtained by adding the elements with like labels of the reactions at the substructure boundaries with sign changed when the substructure boundaries have been considered fixed with the loads at these boundaries. Therefore,

$$[P]_c = \frac{1}{44}\begin{bmatrix} 0 \\ 16 \\ 88 \\ 78 \end{bmatrix} + \begin{bmatrix} 1 \\ 0 \\ 1 \\ 1 \end{bmatrix} = \frac{1}{44}\begin{bmatrix} 44 \\ 16 \\ 132 \\ 122 \end{bmatrix}$$

$[X]_c$ is to be obtained from the solution of the complete structure.

$$[X]_c = [K]_c^{-1}[P]_c$$

$$= 10^{-9}\begin{bmatrix} 1.43 & -0.852 & -0.144 & -0.72 \\ -0.852 & 4.26 & 0.72 & 3.6 \\ -0.144 & 0.72 & 1.43 & 0.852 \\ -0.72 & 3.6 & 0.852 & 4.42 \end{bmatrix} \frac{1}{44}\begin{bmatrix} 44 \\ 16 \\ 132 \\ 122 \end{bmatrix}$$

$$= \frac{10^{-9}}{44}\begin{bmatrix} -57.56 \\ 564.91 \\ 297.88 \\ 677.62 \end{bmatrix}$$

The real displacements at the unrestrained joints of the substructure 1 are next to be determined. For substructure 1

$$[X]_{rd}^1 = \frac{10^{-9}}{44}\begin{bmatrix} -57.56 \\ 564.91 \\ 297.88 \\ 677.62 \end{bmatrix}$$

It may be noted that in this case $[X]_{rd}^1$ is same as $[X]_c^1$ and they are displacements corresponding to labels 5, 6, 7 and 8 of substructure 1 which are same as labels, 1, 2, 3 and 4 of the complete structure.

$$[X]_u^1 = [X]_{uf}^1 - ([K]_{uu}^1)^{-1}\,[K]_{ur}^1\,[X]_{rd}^1$$

$$= \frac{2\times10^{-9}}{11}\begin{bmatrix}4\\24\\15\\31\end{bmatrix} - \begin{bmatrix}10 & -6 & -1 & -5\\-6 & 30 & 5 & 25\\-1 & 5 & 10 & 6\\-5 & 25 & 6 & 30\end{bmatrix}\frac{2\times10^{-9}}{11}$$

$$= \begin{bmatrix}-4 & 0 & -1 & -1\\0 & 0 & -1 & -1\\-1 & 1 & -4 & 0\\1 & -1 & 0 & 0\end{bmatrix}\frac{10^9}{8}\times\frac{10^{-9}}{44}\begin{bmatrix}-57.56\\264.91\\297.88\\677.62\end{bmatrix}$$

$$= \frac{2\times10^{-9}}{11}\begin{bmatrix}2.29\\-16.48\\-23.52\\-13.45\end{bmatrix}$$

$$[X]^1_{rd} = \frac{2\times10^{-9}}{11}\begin{bmatrix}2.29\\-16.48\\-23.52\\-13.45\end{bmatrix}$$

For substructure 2, there are no unrestrained joints as the substructure boundaries are fixed.

All the information is now available for the determination of internal forces in the members. The procedure for determining internal forces is same as that treated in earlier chapters.

5.6 COMPUTER PROGRAM FOR THE TRUSS ANALYSIS BY THE SUBSTRUCTURE TECHNIQUE

The computer program for the plane truss analysis by the substructure technique in both the languages – C and FORTRAN, are given in the attached CD. Notations have been explained in the listing of the computer program in both cases. Different steps followed in the computer code have been explained with "comment" statements.

The computer program presented here is more advanced than earlier chapters, and as such it is to be carefully followed. There are many matrix operations to be carried out. This is best illustrated with an example.

Example 5.1:

The truss of Fig. 5.5 is to be analysed by the substructure technique. In Fig. 5.5(b), the truss is divided into two substructures and in Fig. 5.5(c), into 3. The preparation of the input data corresponding to Fig. 5.5(c) will be discussed.

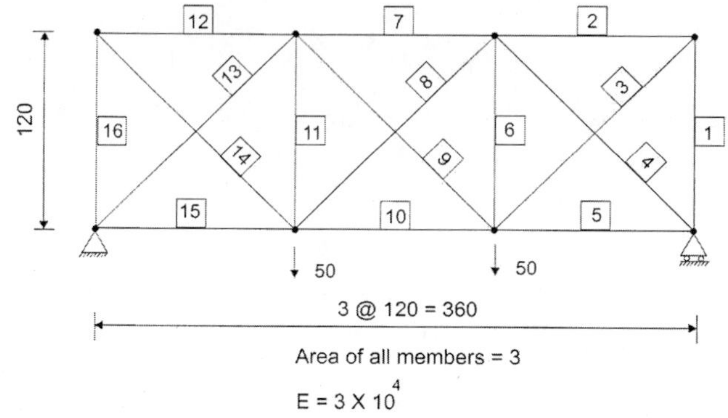

(a) Complete truss

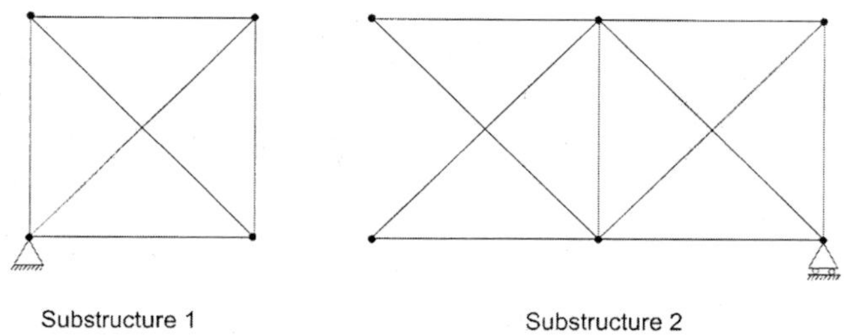

(b) Truss divided into two substructures

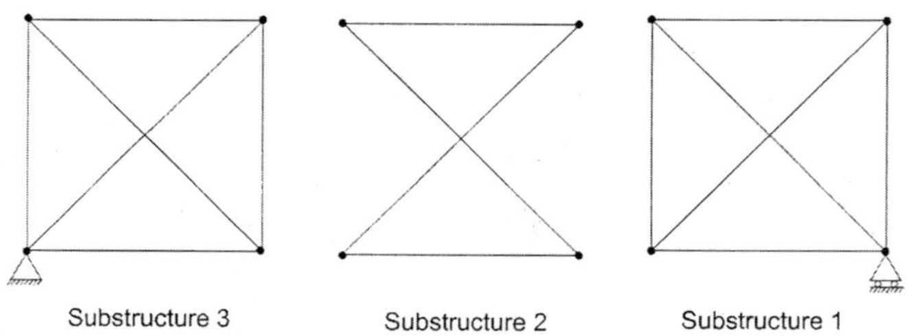

(c) Truss divided into three substructures

Fig. 5.5. Details of a truss with substructures

Out of the three substructures, there are two substructure boundaries where there are 4 unrestrained joints in all. Each joint has 2 degrees of freedom. Therefore, $NS = 8$ and $NPART = 3$. Only one loading condition is considered, *i.e.* $NLC = 1$.

For substructure 1, the displacement lables are put as shown in Fig. 5.6(a). The unrestrained joints are numbered first, then the displacements at the substructure boundary and lastly, at the real boundary. For this substructure, then $NI\ (IJ) = 7$, $NP\ (IJ) = 3$, $NR\ (H) = 4$ and $NF\ (IJ) = 6$. The member data corresponding to item no. 4 to be prepared is exactly same as those required by the direct stiffness method for the truss.

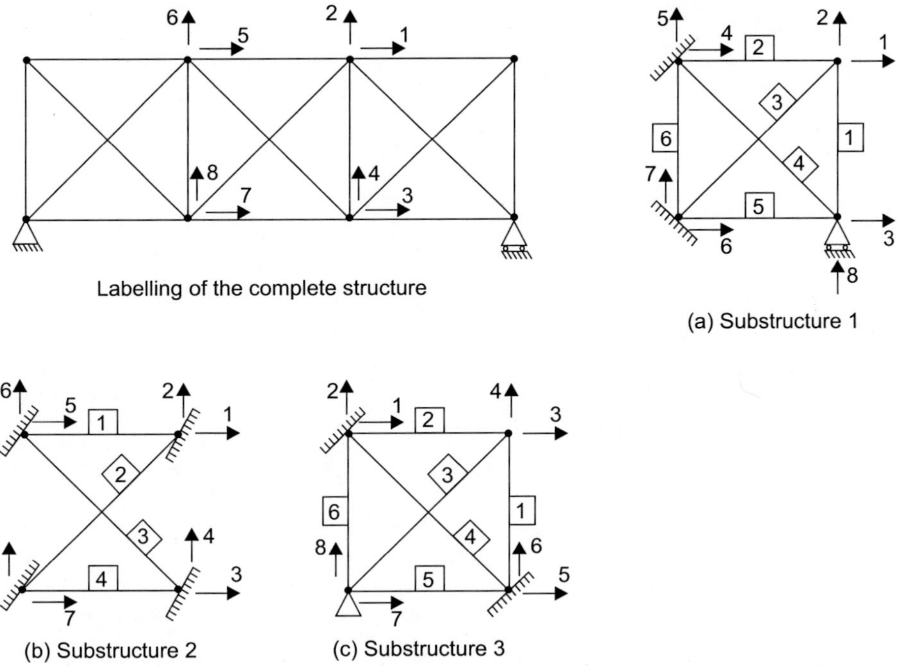

Fig. 5.6 Displacement labels of the individual substructure and the complete structure

Note that *NSUB(IJ)* for substructures 1 and 2 is zero while that for substructures 3 is 4 which is the last global number of degrees of freedom of substructure 2 based on the complete structure (*see* Fig. 5.6 (d)). Also to be noted is the fact that though there may not be any load acting at the joint, still *ND* should not be made zero, but be given a minimum value 1. The input data for the problem at hand is given in Table 5.1. The output of the truss problem is presented in Table 5.2.

The values of the forces in the members of the same truss obtained by the direct stiffness method as also by the substructure technique by dividing into two substructures (Fig. 5.5(b)] have been given in Table 5.3.

Table 5.1. Input Data for the Truss Analysis by Substructure Technique

```
8 1 3
30000.0
7 3 4 6
3 8 1 2 0.0 120.0 3.0
4 5 1 2 120.0 0.0 3.0
6 7 1 2 120.0 120.0 3.0
4 5 3 8 120.0 -120.0 3.0
6 7 3 8  120.0 0.0 3.0
6 7 4 5  0.0 120.0 3.0
1
1 1 0.0
0
8 0 8 4
5 6 1 2  120.0  0.0  3.0
```

```
7 8 1 2  120.0  120.0  3.0
5 6 3 4  120.0 -120.0 3.0
7  8  3  4  120.0 0.0  3.0
0
6  2  4  6
5  6  3  4  0.0  120.0  3.0
1  2  3  4  120.0  0.0  3.0
7  8  3  4  120.0  120.0  3.0
1  2  5  6  120.0  -120.0  3.0
7  8  5  6  120.0  0.0  3.0
7  8  1  2  0.0  120.0  3.0
1
1  1  0.0
4
2
4  1  -50.0
8  1  -50.0
```

Table 5.2. Computer Output for the Truss Analysis by Substructure Technique

```
Analysis of Truss by Substructure Technique
        NS = 8  NLC = 1 NPART = 3       E =    3.00000e+004
Substructure No. = 1
  MEMBER  NP1 NP2 NP3 NP4     H              V              A
   1    3    8    1    2  0.00000e+000  1.20000e+002  3.00000e+000
   2    4    5    1    2  1.20000e+002  0.00000e+000  3.00000e+000
   3    6    7    1    2  1.20000e+002  1.20000e+002  3.00000e+000
   4    4    5    3    8  1.20000e+002 -1.20000e+002  3.00000e+000
   5    6    7    3    8  1.20000e+002  0.00000e+000  3.00000e+000
   6    6    7    4    5  0.00000e+000  1.20000e+002  3.00000e+000
Substructure No. = 2
  MEMBER  NP1 NP2 NP3 NP4     H              V              A
   1    5    6    1    2  1.20000e+002  0.00000e+000  3.00000e+000
   2    7    8    1    2  1.20000e+002  1.20000e+002  3.00000e+000
   3    5    6    3    4  1.20000e+002 -1.20000e+002  3.00000e+000
   4    7    8    3    4  1.20000e+002  0.00000e+000  3.00000e+000
Substructure No. = 3
  MEMBER  NP1 NP2 NP3 NP4     H              V              A
   1    5    6    3    4  0.00000e+000  1.20000e+002  3.00000e+000
   2    1    2    3    4  1.20000e+002  0.00000e+000  3.00000e+000
   3    7    8    3    4  1.20000e+002  1.20000e+002  3.00000e+000
   4    1    2    5    6  1.20000e+002 -1.20000e+002  3.00000e+000
   5    7    8    5    6  1.20000e+002  0.00000e+000  3.00000e+000
   6    7    8    1    2  0.00000e+000  1.20000e+002  3.00000e+000
ND = 2
The Pload Matrix
  0.00000e+000  0.00000e+000  0.00000e+000 -5.00000e+001  0.00000e+000
  0.00000e+000  0.00000e+000 -5.00000e+001
The XX Matrix
  2.70091e-002 -1.97078e-001  9.36758e-002 -2.24087e-001  1.00731e-001
 -1.97078e-001  3.40639e-002 -2.24087e-001
Member Force in Substructure No. = 1
        NFSUB = 0
        NPJ = 3 NRJ = 4 NFF = 6 NTJ = 7
         Member Force in substructure No. 1
         Member No. 1 -2.44521e+001
         Member No. 2 -2.44521e+001
         Member No. 3  3.45805e+001
```

```
        Member No. 4 -3.61302e+001
        Member No. 5  2.55479e+001
        Member No. 6  2.02568e+001
Member Force in Substructure No. = 2
        Member No. 1 -5.52912e+001
        Member No. 2  7.48281e+000
        Member No. 3  7.48282e+000
        Member No. 4  4.47089e+001
Member Force in Substructure No. = 3
       NFSUB = 4
       NPJ = 2 NRJ = 4 NFF = 6 NTJ = 6
        Member Force in substructure No. 3
        Member No. 1  2.02568e+001
        Member No. 2 -2.44521e+001
        Member No. 3 -3.61302e+001
        Member No. 4  3.45805e+001
        Member No. 5  2.55479e+001
        Member No. 6 -2.44521e+001
```

Table 5.3. Comparison of the Results of the Truss of Fig. 5.5 Obtained by Various Methods of Analysis

Member No.	Direct Stiffness	Two Substructures (Fig. 5.5(b))	Three Substructures (Fig. 5.5(c))
1	–24.45208*	–24.45208	–24.45208
2	–24.45208	–24.45208	–24.45208
3	34.58047	34.58043	34.58047
4	–36.13021	–36.13019	–36.13021
5	25.54792	25.54790	25.54792
6	20.25677	20.25678	20.25677
7	–55.29115	–55.29113	–55.29115
8	7.48281	7.48281	7.48281
9	7.48281	7.48281	7.48281
10	44.70885	44.70882	44.70885
11	20.25677	20.25677	20.25677
12	–24.45208	–24.45206	–24.45208
13	–36.13021	–36.13021	–36.13021
14	34.58047	34.58046	34.58049
15	25.54792	25.54788	25.54792
16	–24.45208	–24.45207	–24.45208

*- *indicates compression*

REFERENCES AND SUGGESTED READINGS

5.1. E. Hinton and D.R.J. Owen, Finite Element Programming, Academic Press, 1977.

5.2. B.M. Irons, A Frontal Solution Program for the Finite Element Analysis, Int. J of Num. Meth. in Engg., v. 2, 1970, pp. 5–32.

5.3. W.R. Spillers, Techniques for the Analysis of Large Structures, J. of Strl. Divn., Proc. ASCE, v. 94, 1968, pp. 2521–2534.

5.4. J.S. Przmieniecki, Theory of Matrix Structures, McGraw-Hill, 1968.

5.5. M.F. Rubinstein, Matrix Computer Analysis of Structures, Prentice Hall, 1966.

5.6. M.F. Rubinstein, Combined Analysis of Structure and Recursion, J. Strl. Divn, Proc. ASCE, v. 94, 1967, pp. 231–235.

EXERCISE 5

Assume $E = 22.5 \times 10^6$ kN/mm^2

5.1. Write a computer program in FORTRAN for the solution of a plane frame by the substructure technique.

5.2. Extend the computer program given in Table 5.1 for the analysis of a space truss.

5.3. Neglecting axial deformation and considering the moment of inertia of all members to be same, determine the member forces by partitioning the structure into 3 substructures.

5.4. Treating each element as a substructure, determine the complete structure stiffness matrix and the member forces.

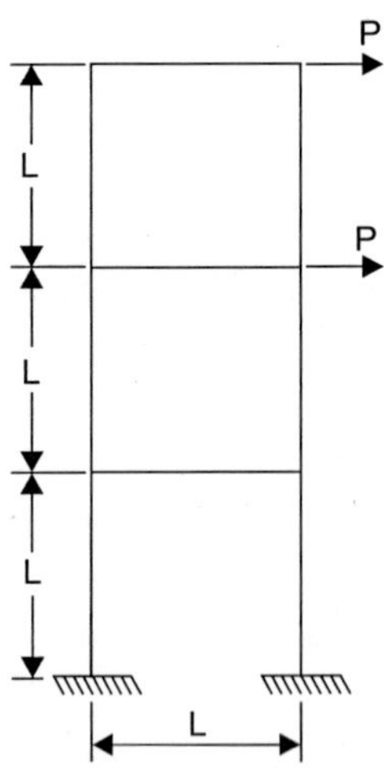

Prob. 5.3.

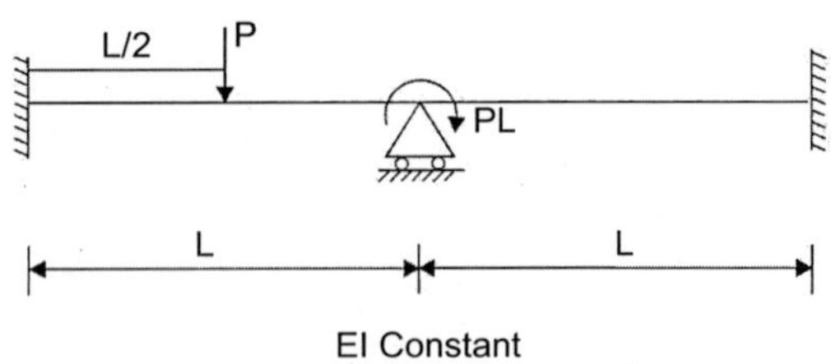

Prob. 5.4

5.5. Prepare the input data for the truss to be solved by the computer program of Table 5.1. Areas of all vertical and horizontal members = 5000 mm^2 and those of inclined members = 4000 mm^2.

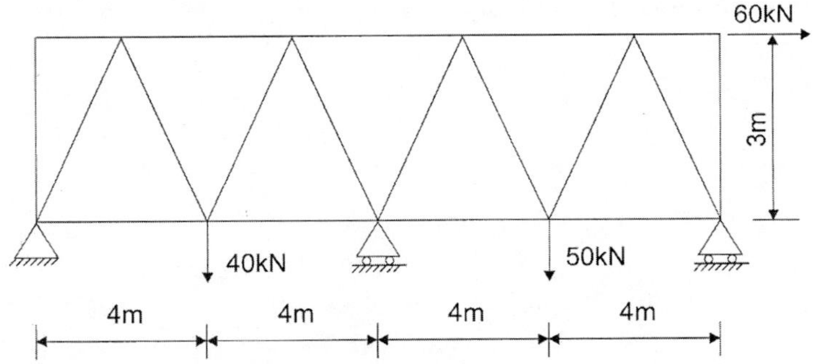

Prob. 5.5.

5.6. For the frame, determine the member forces after dividing the frame into two substructures *EI* is same for all.

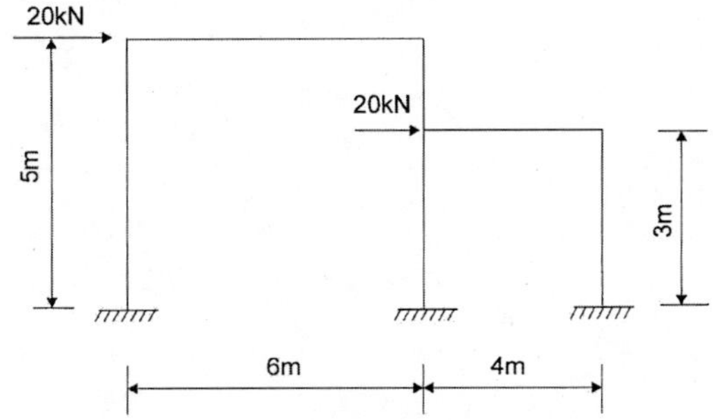

Prob. 5.6.

5.7. Analyse the grid by dividing into 2 substructures. The moment of inertia and polar moment of inertia of each member are 1000 mm^4 and 7000 mm^4 respectively. Assume $G = E/3$.

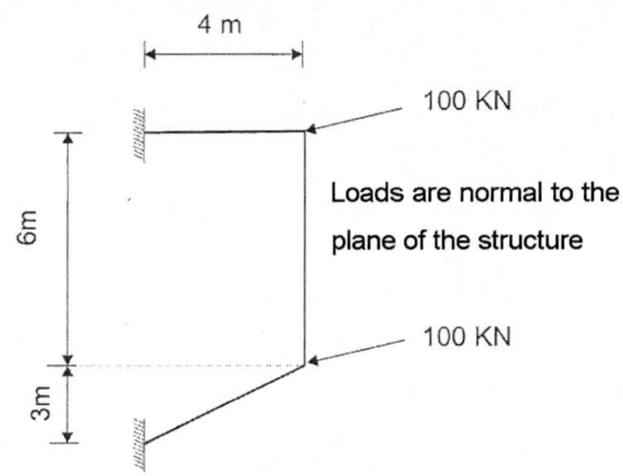

Prob. 5.7.

5.8. The space truss is to be divided into two substructures. For the modified computer program according to problem 5.2, prepare the input data for the truss. The cross-sectional area of each member of the space truss in mm^2 is indicated adjacent to the member.

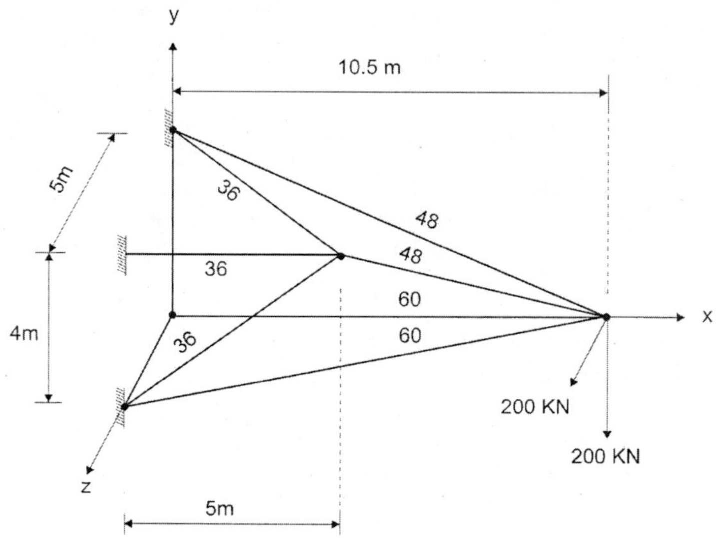

Prob. 5.8.

❑❑❑

6

THE FLEXIBILITY MATRIX METHOD

6.1 INTRODUCTION

The basic approach to the flexibility matrix method has been introduced in Chapter 3 where no attention has been paid to orient it towards the needs of the digital computer. The purpose of this chapter is to introduce a more formalised approach for the analysis performed by the flexibility matrix method. Unlike the stiffness method, it does not follow a set procedure which is primarily due to the existence of various choices of the primary structure. However, the complete formulation of the flexibility method can be done in terms of matrices and matrix operations.

6.2 ELEMENT FLEXIBILITY MATRIX

Due to the external loading, the element of the structure deforms in its own "standard" way, *e.g.*, the members of a truss can only have axial deformation, the deformation of the member of frames or continuous beams where axial rigidity is not considered, is generally denoted by the slopes at the member ends and so on. Defining the deformation in this fashion results in the definition of flexibility matrices which become typical of that structure and the approach becomes more mechanical as will be shortly seen.

The flexibility coefficient of the truss member of Fig. 6.1 (a) (which is say element i) is

$$[FM]_i = [f_{11}]_i = \frac{L}{AE} \tag{6.1}$$

The flexibility matrix of the beam element i of Fig. 6.1 (b) corresponding to the labels shown are

$$[FM]_i = \begin{bmatrix} f_{11} & f_{12} \\ f_{21} & f_{22} \end{bmatrix}_i = \frac{L}{6EI}\begin{bmatrix} 2 & 1 \\ 1 & 2 \end{bmatrix} \tag{6.2}$$

The element of Fig. 6.1(c) is subjected to axial force, bending moment and twisting moment. The flexibility matrix of the member corresponding to the labels indicated is

$$[FM]_i = \begin{bmatrix} f_{11} & f_{12} & f_{13} & f_{14} \\ f_{21} & f_{22} & f_{23} & f_{24} \\ f_{31} & f_{32} & f_{33} & f_{34} \\ f_{41} & f_{42} & f_{43} & f_{44} \end{bmatrix} = \begin{bmatrix} \frac{L}{3EI} & \frac{L}{6EI} & 0 & 0 \\ \frac{L}{6EI} & \frac{L}{3EI} & 0 & 0 \\ 0 & 0 & \frac{L}{EA} & 0 \\ 0 & 0 & 0 & \frac{L}{GJ} \end{bmatrix} \tag{6.3}$$

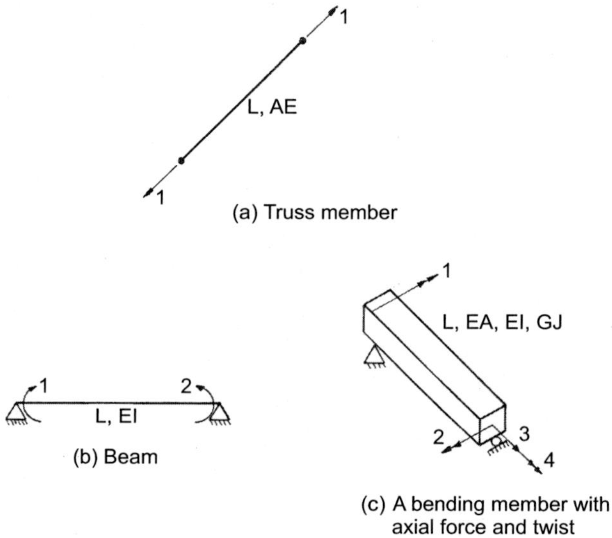

Fig. 6.1. FM Matrix

6.3 PRINCIPLE OF CONTRAGREDIENCE

This principle is utilised in the construction of the flexibility matrix of the structure in an automatic way.

A set of forces PM_1, PM_2 and PM_3 and the corresponding displacements XM_1, XM_2 and XM_3 exist at point O in a structure (Fig. 6.2). The effect of this force system on another point O' of the structure, the coordinates of O is (x, y) with O' as origin, is to be investigated.

Applying equations of statics.

$$\left.\begin{aligned} PM_1 &= P_1 \\ PM_2 &= P_2 \\ PM_3 &= P_3 + P_2 x - P_1 y \end{aligned}\right\}$$

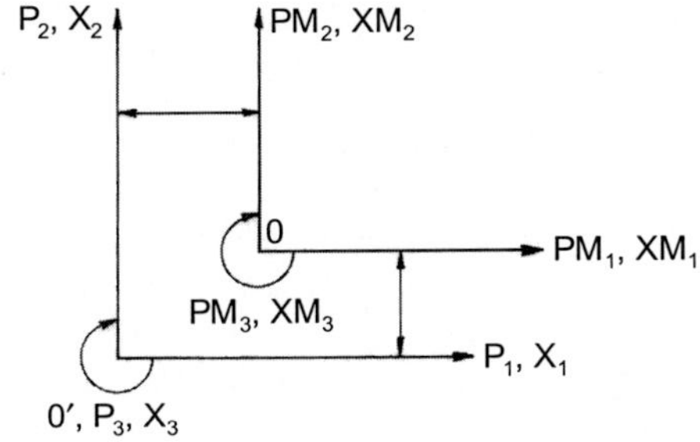

Fig. 6.2. Forces and displacements in the parallel axis system

or, $$\begin{Bmatrix} PM_1 \\ PM_2 \\ PM_3 \end{Bmatrix} = \begin{bmatrix} 1 & 0 & 0 \\ 0 & 1 & 0 \\ -y & x & 1 \end{bmatrix} \begin{Bmatrix} P_1 \\ P_2 \\ P_3 \end{Bmatrix} \tag{6.4}$$

or, $$\{PM\} = [T_1]\,\{P\} \tag{6.5}$$

Two systems of displacements are related thus

$$\begin{Bmatrix} XM_1 \\ XM_2 \\ XM_3 \end{Bmatrix} = \begin{bmatrix} 1 & 0 & y \\ 0 & 1 & -x \\ 0 & 0 & 1 \end{bmatrix} \begin{Bmatrix} X_1 \\ X_2 \\ X_3 \end{Bmatrix} \tag{6.6}$$

$$\{XM\} = [T_2]\,\{X\} \tag{6.7}$$

It can be easily shown that

$$[T_1]\,[T_2]^{\mathrm{T}} = [I]$$

also, $$[T_2]\,[T_1]^{\mathrm{T}} = [I] \tag{6.8}$$

Substituting $[T_2]$ from the second relation of eqn. (6.8) into eqn. (6.7), yields

$$\{XM\} = ([T_1]^T)^{-1}\,\{X\}$$

or, $$\{X\} = [T_1]^T\,\{XM\} \tag{6.9}$$

Thus, the principle of contragredience is illustrated by the following equations:

$$\{PM\} = [T_1]\,\{P\}$$

and $$\{X\} = [T_1]\,\{XM\}$$

The forces and displacements are said to transform contragrediently.

6.4 THE EQUILIBRIUM MATRIX

Consider the frame of Fig. 6.3. Suppose this frame needs to be analysed by the flexibility matrix method. P_1 and P_2 are chosen as the redundants and PM_1, PM_2, PM_3 and PM_4 are the end moments of the member. The relationship between P_1 and P_2 and PM_1, PM_2, PM_3 and PM_4 is to be determined. In order to do that, unit values corresponding to P_1 and P_2 are applied separately and the necessary equilibrium diagrams are drawn as shown in Fig. 6.3 (d) and (e). Corresponding to Fig. 6.3(d), the forces at member references are

$$PM_1 = 1, PM_2 = 0, PM_3 = 0 \text{ and } PM_4 = 0$$

Similarly, corresponding to Fig. 6.3(e), the forces at member references are

$$PM_1 = 0, PM_2 = -V, PM_3 = -V \text{ and } PM_4 = 0$$

Therefore, the following relationship can be written

$$\begin{Bmatrix} PM_1 \\ PM_2 \\ PM_3 \\ PM_4 \end{Bmatrix} = \begin{bmatrix} 1 & 0 \\ 0 & -V \\ 0 & -V \\ 0 & 0 \end{bmatrix} \begin{Bmatrix} P_1 \\ P_2 \end{Bmatrix} \tag{6.10}$$

$$\{PM\} = [E]\,\{P\} \tag{6.11}$$

$[E]$ matrix is known as the equilibrium matrix. Each column of the equilibrium matrix is generated by applying unit value of the redundant and finding out the forces at member references. Comparing eqns. (6.11) and (6.5) reveals that $[E] = [T_l]$. Therefore, $[E]$ is nothing but force transformation matrix.

6.5 CONSTRUCTION OF THE FLEXIBILITY MATRIX OF THE STRUCTURE

Suppose for the frame of Fig. 6.3, $H = V = L$. The flexibility matrix of the system consisting of the flexibility matrix of the member in their unconnected state is written as

$$[FM] = \left[\begin{array}{c|c} [FM]_1 & [0] \\ \hline [0] & [FM]_2 \end{array}\right]$$

or,

$$[FM] = \frac{L}{6EI}\begin{bmatrix} \begin{bmatrix} 2 & 1 \\ 1 & 2 \end{bmatrix} & \begin{bmatrix} 0 & 0 \\ 0 & 0 \end{bmatrix} \\ \begin{bmatrix} 0 & 0 \\ 0 & 0 \end{bmatrix} & \begin{bmatrix} 2 & 1 \\ 1 & 2 \end{bmatrix} \end{bmatrix} \tag{6.12}$$

The displacements at the member ends are given by

$$\{XM\} = [FM]\,\{PM\} \tag{6.13}$$

Similarly, the displacements corresponding to the redundants of the total structure are

$$\{X\} = [F]\,\{P\} \tag{6.14}$$

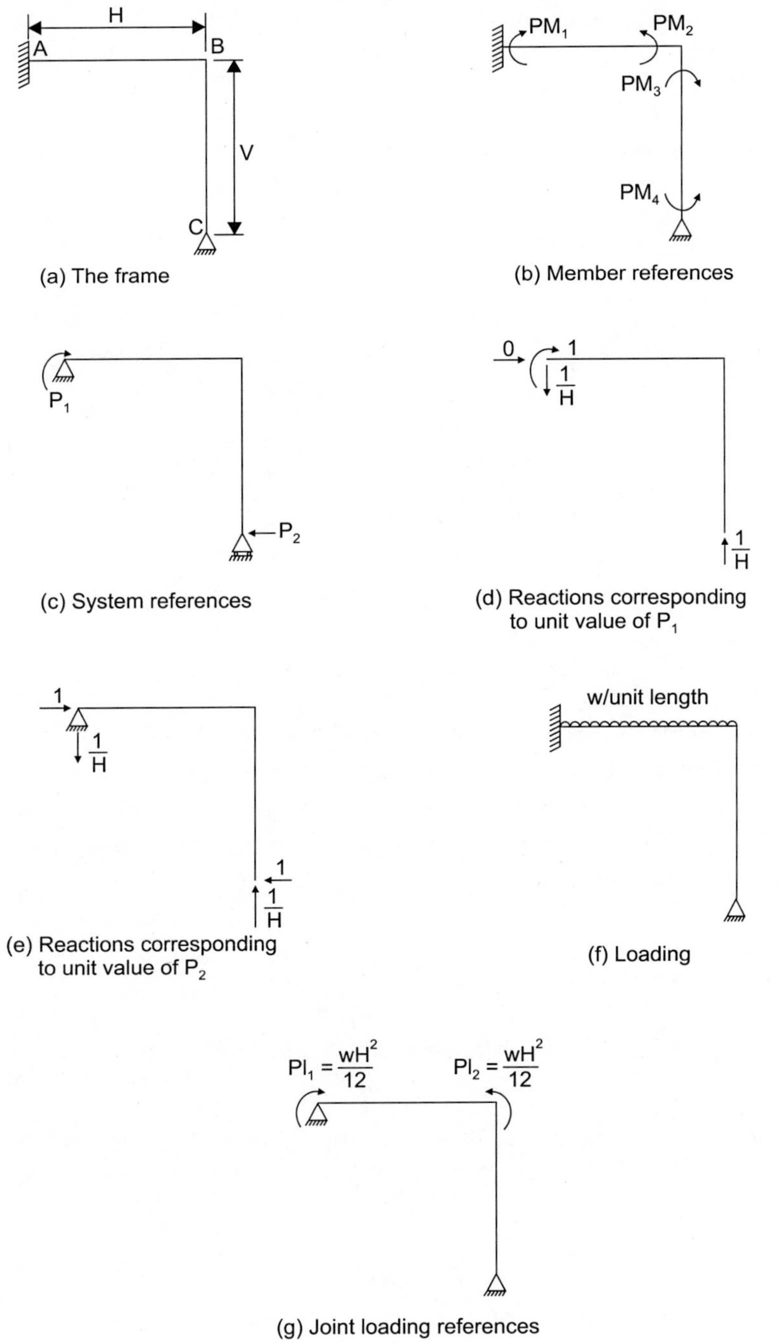

Fig. 6.3. A frame with member and system references

Substituting $\{PM\}$ from eqn. (6.5) and changing $[E]$ for $[T_l]$, and replacing the value of $\{XM\}$ from eqn. (6.9), eqn. (6.13) becomes

$$([E]^T)^{-1}\{X\} = [FM][E]\{P\} \tag{6.15}$$

Premultiplying both sides of eqn. (6.15) by $[E]^T$, results

$$\{X\} = [E]^T[FM][E]\{P\} \tag{6.16}$$

On comparing eqn. (6.16) with eqn. (6.14) immediately reveals that the structure flexibility matrix is given by

$$[F] = [E]^T[FM][E] \tag{6.17}$$

Therefore, for the frame of Fig. 6.3, the structure flexibility matrix is

$$[F] = \begin{bmatrix} 1 & 0 & 0 & 0 \\ 0 & -L & L & 0 \end{bmatrix} \frac{L}{6EI} \begin{bmatrix} 2 & 1 & 0 & 0 \\ 1 & 2 & 0 & 0 \\ 0 & 0 & 2 & 1 \\ 0 & 0 & 1 & 2 \end{bmatrix} \begin{bmatrix} 1 & 0 \\ 0 & -L \\ 0 & -L \\ 0 & 0 \end{bmatrix} = \frac{L}{6EI} \begin{bmatrix} 2 & -L \\ -L & 4L^2 \end{bmatrix}$$

6.6 MATRIX DETERMINATION OF THE DISPLACEMENT VECTOR

The flexibility coefficients indicate displacements. Therefore, following similar procedure as above, the displacement vector $\{X\}$ of eqn. (6.1) can be obtained.

The loading of the frame is indicated in Fig. 6.3 (f). X_1, and X_2 are the displacements corresponding to redundants (*see* Fig. 6.3 (c)) of the primary frame due to external loading. Let $\{PL\}$ represent a vector of external loading usually representing the equivalent joint loading at salient points for the frame of Fig. 6.3. The $\{PL\}$ matrix is

$$\begin{Bmatrix} PL_1 \\ PL_2 \end{Bmatrix} = \frac{wL^2}{12} \begin{Bmatrix} 1 \\ 1 \end{Bmatrix}$$

Let $[E_0]$ be the matrix which relates $\{PM\}$ to $\{PL\}$ such that

$$\{PM\} = [E_0]\{PL\} \tag{6.18}$$

It has already been shown that (*i.e.*, eqns. (6.13) and (6.9) are rewritten)

$$\{XM\} = [FM]\{PM\} \tag{6.19}$$

Also

$$\{X\} = [E]^T\{XM\} \tag{6.20}$$

Combining the three eqns. (6.18), (6.19) and (6.20), yields

$$([E]^T)^{-1}\{X\} = [FM][E_0]\{PL\} \tag{6.21}$$

Premultiplying both sides of eqn. (6.21) by $[E]^T$, yields

$$\{X\} = [E]^T[FM][E_0]\{PL\} \tag{6.22}$$

$\{X\}$ represents the displacements caused by the redundants. For the frame of Fig. 6.3, corresponding to PL_1 and PL_2 indicated, eqn. (6.18) becomes

$$\begin{Bmatrix} PM_1 \\ PM_2 \\ PM_3 \\ PL_4 \end{Bmatrix} = \begin{bmatrix} 1 & 0 \\ 0 & 1 \\ 0 & 1 \\ 0 & 0 \end{bmatrix} \begin{Bmatrix} PL_1 \\ PL_2 \end{Bmatrix}$$

i.e.,

$$[E_0] = \begin{bmatrix} 1 & 0 \\ 0 & 1 \\ 0 & 1 \\ 0 & 0 \end{bmatrix}$$

Therefore,

$$\{X\} = \begin{bmatrix} 1 & 0 & 0 & 0 \\ 0 & -L & -L & 0 \end{bmatrix} \frac{L}{6EI} \begin{bmatrix} 2 & 1 & 0 & 0 \\ 1 & 2 & 0 & 0 \\ 0 & 0 & 2 & 1 \\ 0 & 0 & 1 & 2 \end{bmatrix} \begin{bmatrix} 1 & 0 \\ 0 & 1 \\ 0 & 1 \\ 0 & 0 \end{bmatrix} \frac{wL^2}{12} \begin{Bmatrix} 1 \\ 1 \end{Bmatrix}$$

$$= \frac{wL^3}{72EI} \begin{bmatrix} 2 & 1 \\ -L & -2L \end{bmatrix} \begin{Bmatrix} 1 \\ 1 \end{Bmatrix} = \frac{wL^3}{72EI} \begin{Bmatrix} 3 \\ -3L \end{Bmatrix} = \frac{wL^3}{24EI} \begin{Bmatrix} 1 \\ -1 \end{Bmatrix}$$

It may be noted that only the non-zero componentss of equivalent joint loading and not all possible components be considered in the construction of $\{PL\}$ matrix. Further, the sign of $\{PL\}$ matrix need not coincide with those components of member reference system.

6.7 DETERMINATION OF MEMBER FORCES

Once the redundants $\{P\}$ are obtained from the solution of simultaneous equations [eqn. (6.14)], the forces in the members are to be obtained. The member forces are determined by the following equation:

$$\{PM\} = [E_0]\{PL\} + [E]\{P\} + \{PMF\} \tag{6.23}$$

$\{PMF\}$ constitutes the constraining forces corresponding to the member references. For a bending member, the first part of eqn. (6.23) represents the forces in the members due to the loading in the primary structure and the second part, that due to redundants acting on the primary structure.

6.8 PROCEDURE OF THE ANALYSIS OF STATICALLY INDETERMINATE STRUCTURES

The procedure for analysis of statically indeterminate structure is as follows:

1. The structure is converted into a statically determinate structure and the redundants are accordingly identified.
2. Select element references for which member flexibility matrix exists.

3. Select system references.
4. Select references corresponding to equivalent joint loading.
5. Generate $[E]$, $[FM]$, $[E_0]$, $[PMF]$ and $[PL]$ matrices.
6. Form $[F] = [E]^T [FM] [E]$
7. Form $[X] = [E]^T [FM] [E_0] [PL]$
8. Solve for $[P]$ from eqn. (6.14)
9. Determine $[PM]$ from eqn. (6.23)

6.9 ILLUSTRATED EXAMPLES

Example 6.1:

The continuous beam of Fig. 6.4 is to be analysed. The particulars of the beam have been indicated in the figure.

The beam is made statically determinate by removing the reactions of the roller supports. All the necessary references have been shown in Fig. 6.4 (b), (c) and (d). All the necessary matrices generated are given below:

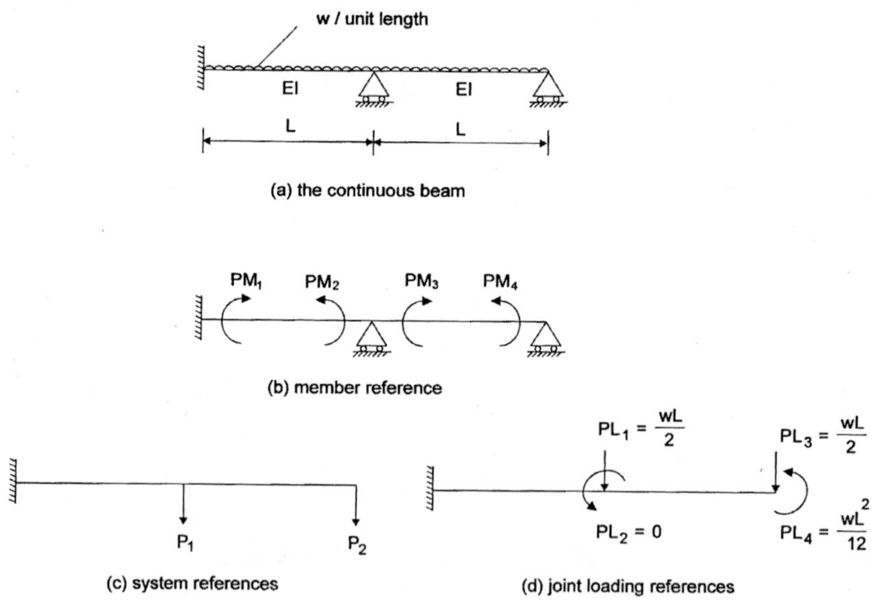

Fig. 6.4. The beam of Example 6.1

$$[E] = \begin{bmatrix} -L & -2L \\ 0 & -L \\ 0 & -L \\ 0 & 0 \end{bmatrix}, \qquad [FM] = \frac{1}{6EI}\begin{bmatrix} 2 & 1 & 0 & 0 \\ 1 & 2 & 0 & 0 \\ 0 & 0 & 2 & 1 \\ 0 & 0 & 1 & 2 \end{bmatrix}, \qquad \{PL\} = \begin{Bmatrix} wL \\ 0 \\ \dfrac{wL}{2} \\ \dfrac{wL^2}{12} \end{Bmatrix},$$

$$\{PMF\} = \frac{wL^2}{12}\begin{Bmatrix} -1 \\ -1 \\ -1 \\ -1 \end{Bmatrix}, \qquad [E_0] = \begin{bmatrix} -L & 1 & -2L & 1 \\ 0 & 1 & -L & 1 \\ 0 & 0 & -L & 1 \\ 0 & 0 & 0 & 1 \end{bmatrix}$$

Therefore,

$$[F] = \begin{bmatrix} -L & 0 & 0 & 0 \\ -2L & -L & -L & 0 \end{bmatrix}\frac{L}{6EI}\begin{bmatrix} 2 & 1 & 0 & 0 \\ 1 & 2 & 0 & 0 \\ 0 & 0 & 2 & 1 \\ 0 & 0 & 1 & 2 \end{bmatrix}\begin{bmatrix} -L & -2L \\ 0 & -L \\ 0 & -L \\ 0 & 0 \end{bmatrix}$$

$$= \frac{L}{6EI}\begin{bmatrix} 2 & 5 \\ 5 & 16 \end{bmatrix}$$

$$\{X\} = \begin{bmatrix} -L & 0 & 0 & 0 \\ -2L & -L & -L & 0 \end{bmatrix}\frac{L}{6EI}\begin{bmatrix} 2 & 1 & 0 & 0 \\ 1 & 2 & 0 & 0 \\ 0 & 0 & 2 & 1 \\ 0 & 0 & 1 & 2 \end{bmatrix}\begin{bmatrix} -L & 1 & -2L & 1 \\ 0 & 1 & -L & 1 \\ 0 & 0 & -L & 1 \\ 0 & 0 & 0 & 1 \end{bmatrix}\begin{Bmatrix} wL \\ 0 \\ \dfrac{wL}{2} \\ \dfrac{wL^2}{12} \end{Bmatrix}$$

$$= \frac{wL^4}{72EI}\begin{Bmatrix} 51 \\ 144 \end{Bmatrix}$$

Applying eqn. (6.14),

$$\frac{L^3}{6EI}\begin{bmatrix} 2 & 5 \\ 5 & 16 \end{bmatrix}\begin{Bmatrix} P_1 \\ P_2 \end{Bmatrix} = \frac{wL^4}{72EI}\begin{Bmatrix} 51 \\ 144 \end{Bmatrix}$$

or,

$$\begin{Bmatrix} P_1 \\ P_2 \end{Bmatrix} = -\frac{wL}{28}\begin{Bmatrix} 32 \\ 11 \end{Bmatrix}$$

The member forces are given by$_2$

$$\begin{Bmatrix} PM_1 \\ PM_2 \\ PM_3 \\ PM_4 \end{Bmatrix} = \begin{bmatrix} -L & 1 & -2L & 1 \\ 0 & 1 & -L & 1 \\ 0 & 0 & -L & 1 \\ 0 & 0 & 0 & 1 \end{bmatrix} \frac{wL}{12} \begin{Bmatrix} 12 \\ 0 \\ 6 \\ L \end{Bmatrix} + \begin{bmatrix} -L & -2L \\ 0 & -L \\ 0 & -L \\ 0 & 0 \end{bmatrix} \frac{-wL}{28} \begin{Bmatrix} 32 \\ 11 \end{Bmatrix} + \frac{wL^2}{12} \begin{Bmatrix} -1 \\ -1 \\ -1 \\ -1 \end{Bmatrix}$$

$$= \frac{wL^2}{28} \begin{Bmatrix} 2L \\ 3L \\ 3L \\ 0 \end{Bmatrix}$$

Example 6. 2:

Analyse the truss of Fig. 6.5. All the particulars are given in the diagram. Relative axial rigidities are given in brackets.

The member and system references have been indicated in Fig. 6.5(b) and (c). Therefore, based on these references, the matrices which are input to the computer are as follows (tensile force is considered as positive).

$$[E] = \begin{bmatrix} 0 & -3/4 \\ 0 & 0 \\ 0 & 0 \\ 0 & 5/4 \\ -0.8 & 0 \\ -0.6 & 0 \\ 1.0 & 0 \\ 1.0 & 0 \\ -0.8 & 0 \\ -0.6 & 0 \end{bmatrix}$$

Diagonals of $[FM] = [3 \quad 8 \quad 4 \quad 4 \quad 2 \quad 3 \quad 3 \quad 5 \quad 3 \quad 4]$

Off-diagonals of $[FM]$ are zero.

$$\{PL\} = \{10\}$$

$$[E_0]^T = [3/4 \quad 0 \quad 1 \quad -5/4 \quad 0 \quad 1 \quad 0 \quad 0 \quad 0 \quad 0]$$

The flexibility matrix of the truss

$$[F] = [E]^T [FM] [E] = \begin{bmatrix} 13.72 & 0 \\ 0 & 7.9375 \end{bmatrix}$$

$$\{X\} = [E]^T [FM] [E_0] \{PL\} = \begin{Bmatrix} 18.000 \\ 79.375 \end{Bmatrix}$$

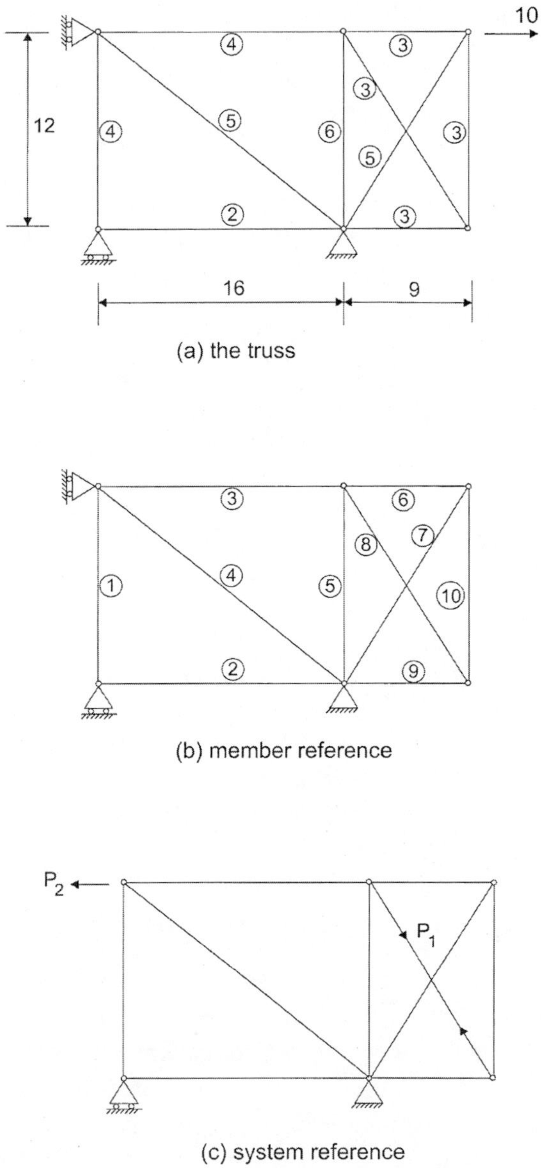

Fig. 6.5. The truss of Example 6.2

On solution of eqn. (6.14)

$$\{P\} = \begin{Bmatrix} 1.312 \\ 10.000 \end{Bmatrix}$$

The axial forces in the member of the truss, then are

$$\{PM\} = \begin{bmatrix} 0 & -3/4 \\ 0 & 0 \\ 0 & 0 \\ 0 & 5/4 \\ -0.8 & 0 \\ -0.6 & 0 \\ 1.0 & 0 \\ 1.0 & 0 \\ -0.8 & 0 \\ -0.6 & 0 \end{bmatrix} \begin{Bmatrix} 1.312 \\ 10.000 \end{Bmatrix} + \begin{bmatrix} 3/4 \\ 0 \\ 1 \\ -5/4 \\ 0 \\ 1 \\ 0 \\ 0 \\ 0 \\ 0 \end{bmatrix} \{10\}$$

$$= \begin{Bmatrix} 0 \\ 0 \\ 10 \\ 0 \\ -1.050 \\ 4.000 \\ 1.312 \\ 1.312 \\ -1.050 \\ -0.787 \end{Bmatrix}$$

6.10 CHOICE OF THE RELEASED STRUCTURE

It is now realised that there is no unique choice regarding the primary structure. The main disadvantage of the flexibility method is that the construction of the transformation matrices does not follow a set pattern. Both $[E]$ and $[E_0]$ matrices are input to the computer and they are to be generated by hand calculation. For big sized problems, this generation becomes a tedious affair and takes away much of the charm of this computer-oriented method. For this reason, the stiffness method is more popularly applied for the solution of large sized problems. However, in certain problems where degree of statical indeterminacy is much lesser than kinematical indeterminacy, the flexibility method may be preferred. In the flexibility matrix method, the choice of the primary structure is of paramount importance. While making the choice, the following points must be borne in mind:

(*a*) The complete primary structure should be statically determinate.

(*b*) The choice should be such that the generation of the necessary transformation matrices $[E]$ and $[E_0]$ is not tedious.

(*c*) It may happen that depending on the choice of the released structure, the flexibility matrix may become ill-conditioned. The primary structures which do not result in well-conditioned flexibility matrix of the structure should be avoided.

REFERENCES AND SUGGESTED READINGS

6.1. A. S. Hall and R. W. Woodhead, Frame Analysis, John Wiley & Sons, 1961.

6.2. H. C. Martin, Introduction to Matrix Methods of Structural Analysis, McGraw-Hill, 1966.

6.3. J. H. Argyris and S. Kesley, Energy Theorems and Structural Analysis, Butterworths, London, 1960.

6.4. C. K. Wang, Introductory Structural Analysis with Matrix Methods, Prentice Hall, 1973.

6.5. W. McGuire, R. H. Gallagher and R. D. Ziemian, Matrix Structural Analysis, Second Edition, John Wiley & Sons, 2000.

6.6. J. M. Gere and W. Weaver Jr, Analysis of Framed Structures, Second Edition, D. van Nostrand, 1982.

EXERCISE 6

The problems given are to be solved by the method presented in this chapter.

6.1. Analyse the continuous beam

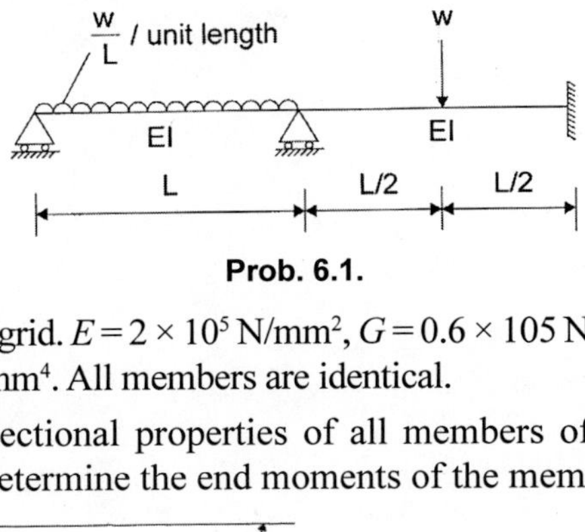

Prob. 6.1.

6.2. Analyse the grid. $E = 2 \times 10^5$ N/mm², $G = 0.6 \times 105$ N/mm², $I = 4 \times 10^8$ mm⁴, $J = 4 \times 10^7$ mm⁴. All members are identical.

6.3. The cross-sectional properties of all members of the space frame are identical. Determine the end moments of the members.

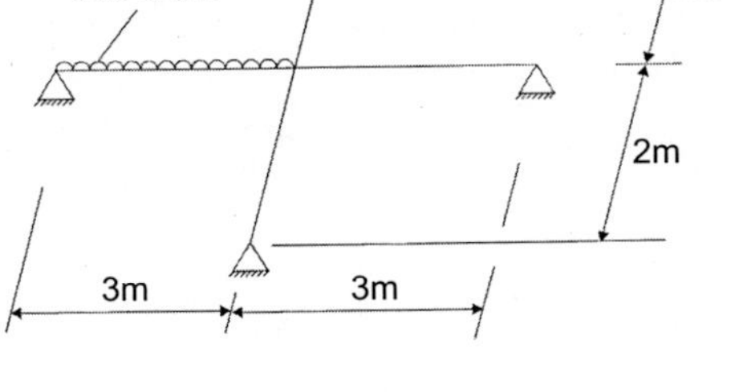

Prob. 6.2.

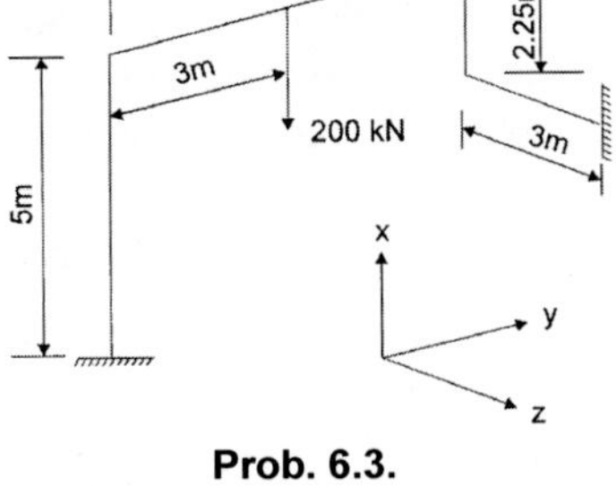

Prob. 6.3.

6.4. Develop a computer program for the analysis of any framed structure by the method presented in this chapter.

6.5. Analyse the frame.

6.6. Analyse the plane truss. The axial rigidity of the members of the truss is identical.

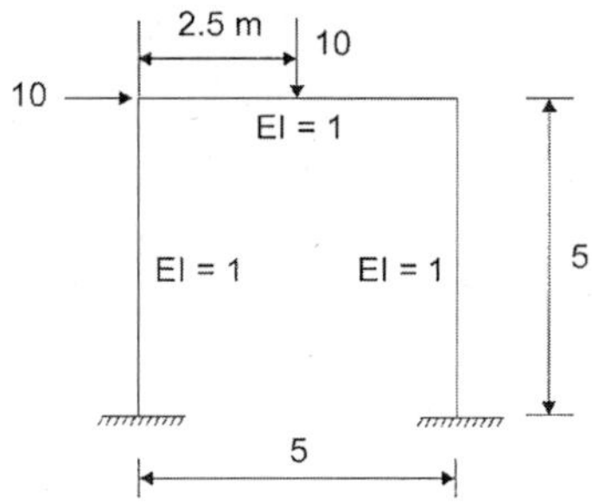

Prob. 6.5.

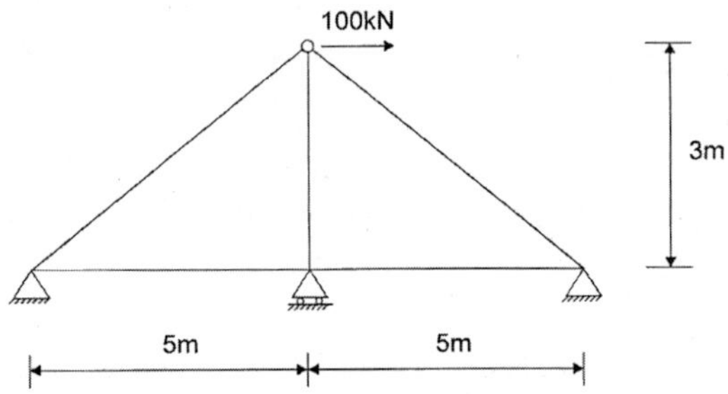

Prob. 6.6.

6.7. Analyse the beam. Draw the bending moment and shear force diagrams.

6.8. For the grillage as shown, $E = 2.5\ G$. For the loading indicated, determine the variation of bending and twisting moment along the length of the member.

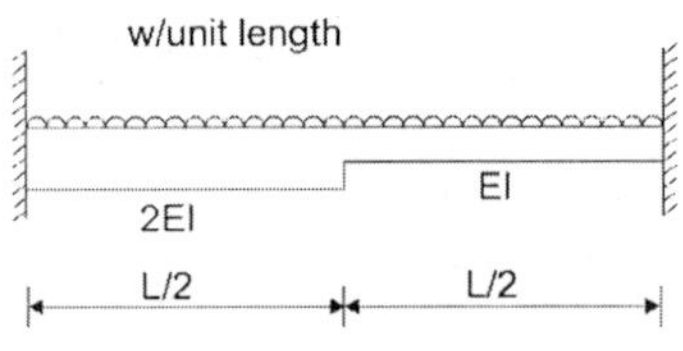

Prob. 6.7.

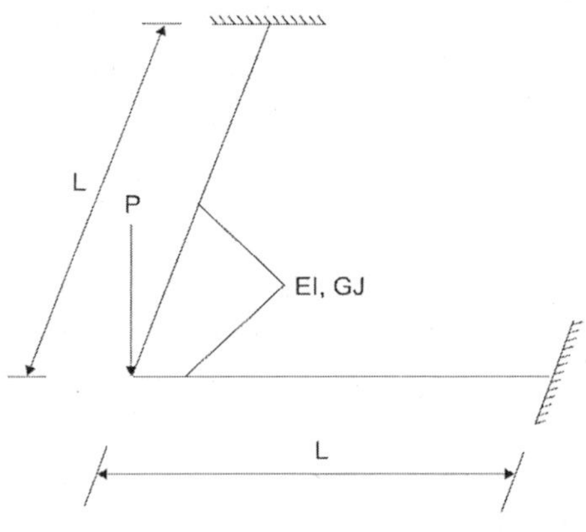

Prob. 6.8.

❑❑❑

7

ELEMENTS OF ELASTICITY

7.1 INTRODUCTION

Theory of elasticity provides a more sophisticated tool of stress analysis than the strength of materials approach. It tackles the problems in a more rigorous way. Though the analysis caters more detailed aspects of the problem, but the solution of many of the problems is anything but simple. With the availability of high-speed computers, it becomes somewhat a luxury to solve most of the structural problems by the theory of elasticity approach.

For the basic formulation of plate and shell structures, the knowledge of the theory of elasticity, however, is needed. Even any numerical analysis of plates and shells requires those basic equations. Many books exist where the theory of elasticity, plates and shells have been treated in details to which an interested reader can refer [1–5}. This chapter, however, is limited only to the introduction of the basic relationships which will be needed for the finite element analysis.

7.2 SOME NOTATIONS AND RELATIONS IN THE THEORY OF ELASTICITY

7.2.1 Surface and Body Forces

The force distributed over the surface of the body is called surface force, *e.g.*, hydrostatic pressure, frictional force *etc.*

The force which is distributed over the volume of the body is called the body force, *e.g.*, self-weight of a structure, magnetic force, inertia force, *etc.*

7.2.2 Components of Stresses

The stress acting on an area can be splitted into a normal component and a tangential component. If at point O of an element of sides dx, dy and dz which are parallel to reference x, y and z axis system is taken out of a stressed body, one normal stress component and two shear stress components act all parallel to the reference axis system. The components of the stresses along with the axis system are shown in Fig. 7.1.

The sign convention adopted for the normal stress is that it is positive when it produces tension; therefore, if it produces compression it is negative. Shear stresses shown in Fig. 7.1 is accompanied with two subscripts — the first indicating the plane given by its normal on which it acts and the second gives the direction of the component of the stress. To explain it further, if we consider the two shearing stress components acting on the plane which is perpendicular to x-axis, then the component which is in the direction of y-axis will be τ_{xy}, and that in the z-direction will be τ_{xz}. The positive directions of the components of shearing stress on any side of the element are taken as the positive directions of the coordinate axes if a tensile stress on the same plane would have the positive direction of the corresponding axis. Thus, if on any side, the tensile stress is opposite to the positive direction of the reference axis, the positive directions of the shearing stress components will then be reversed. All the stress components shown in Fig. 7.1 are positive.

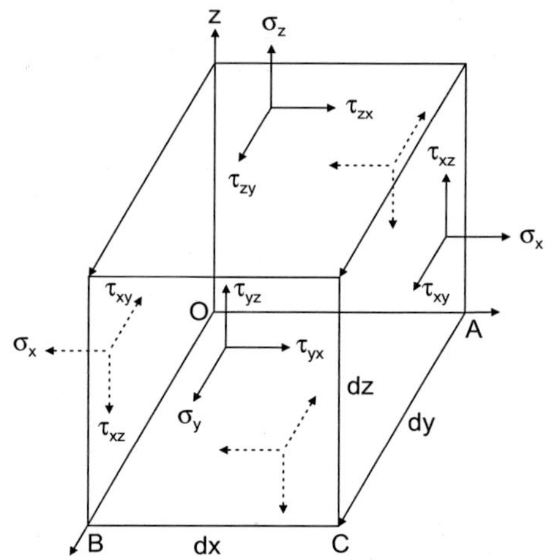

Fig. 7.1. Stressed element

It is evident from Fig. 7.1 that there are three components of normal stress σ_x, σ_y and σ_z and six components of shear stress τ_{xy}, τ_{yz}, τ_{zx}, τ_{yx}, τ_{zy} and τ_{xz}. Further it is known that $\tau_{xy} = \tau_{yx}$, $\tau_{xz} = \tau_{zx}$ and $\tau_{yz} = \tau_{zy}$. Therefore, to describe the stress at a point, the information of six quantities σ_x, σ_y, σ_z, τ_{xy}, τ_{yz}, τ_{zx} are all that are needed.

7.2.3 Components of Strain

The deformation of the face *OACB* is shown in Fig. 7.2. The displacements of point *O* of the deformed body is resolved into components u, v and w parallel to the reference axes x, y and z respectively, for the infinitesimal element of Fig. 7.1. The side will have two types of deformation. Point *A* will have displacement in the x-direction which is $u + \dfrac{\partial u}{\partial x} dx$

Therefore, the increase in length of the element whose length is dx, is $\dfrac{\partial u}{\partial x} dx$. Hence, the strain in x-direction is $\dfrac{\partial u}{\partial x}$. Similarly, it can be shown that the strains in y and z-directions are $\dfrac{\partial v}{\partial y}$ and $\dfrac{\partial w}{\partial z}$ respectively.

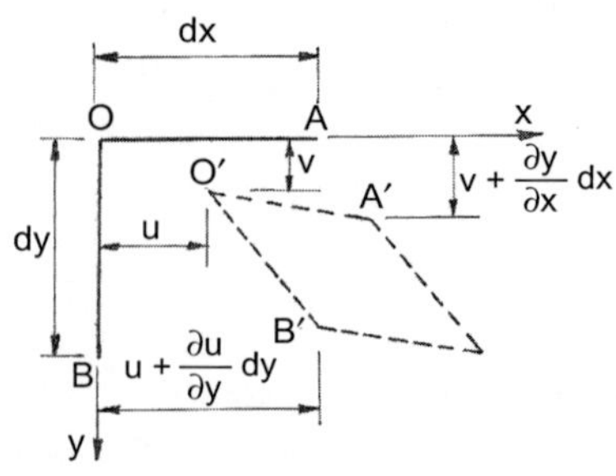

Fig. 7.2. Deformation of the element

Point A also have a component of the displacement in the y-direction which is $v + \frac{\partial v}{\partial x} dx$. Similarly, the displacement of point B in x-direction is $u + \frac{\partial u}{\partial y} dy$.

Therefore, inclination of $O'A'$ from OA is $\frac{\partial v}{\partial x}$ and that of $O'B'$ from OB is $\frac{\partial u}{\partial y}$.

Therefore, change in the $\angle AOB$ after deformation is $\left(\frac{\partial v}{\partial x} + \frac{\partial u}{\partial y}\right)$. By definition, this is the shearing strain between the plane xz and yz. Similarly, other shearing strain components can be determined.

Introducing the letter ε indicate the normal strain and γ for the shearing strain, the following then are the strain-displacement relationship.

$$\left.\begin{aligned} \varepsilon_x &= \frac{\partial u}{\partial x},\ \varepsilon_y = \frac{\partial v}{\partial y},\ \varepsilon_z = \frac{\partial w}{\partial z} \\ \gamma_{xy} &= \frac{\partial u}{\partial y} + \frac{\partial v}{\partial x},\ \gamma_{yz} = \frac{\partial v}{\partial z} + \frac{\partial w}{\partial y},\ \gamma_{zx} = \frac{\partial w}{\partial x} + \frac{\partial u}{\partial z} \end{aligned}\right\} \tag{7.1}$$

7.2.4 Stress-Strain Relationship

For a linearly elastic material, the following are the stress-strain relationships.

$$\left.\begin{aligned} \varepsilon_x &= \frac{1}{E}[\sigma_x - \nu(\sigma_y + \sigma_z)] \\ \varepsilon_y &= \frac{1}{E}[\sigma_y - \nu(\sigma_z + \sigma_x)] \\ \varepsilon_z &= \frac{1}{E}[\sigma_z - \nu(\sigma_x + \sigma_y)] \\ \gamma_{xy} &= \frac{\tau_{xy}}{G},\ \gamma_{yz} = \frac{\tau_{yz}}{G},\ \gamma_{zx} = \frac{\tau_{zx}}{G} \end{aligned}\right\}$$

where ν is the Poisson's ratio of the material, E is the Young's modulus of elasticity and G is the shear modulus of elasticity. Again E and G are related as follows:

$$G = \frac{E}{2(1+\nu)}$$

7.3 TWO-DIMENSIONAL PROBLEMS

7.3.1 Plane Stress

If a thin plate is loaded in its own plane in such a way that the forces are distributed uniformly over the thickness, then the plate is said to be in a state of plane stress. For the plate of Fig. 7.3, the stress components σ_x, σ_y, and τ_{xy} represent the state of stress and stress components $\sigma_z = \tau_{yz} = \tau_{zx} = 0$.

Beams, buttress of dams, web frame of tankers, superstructure deck interaction problems of ships are some of the examples of plane stress problems. The stress-strain relations for the plane stress problem then become

$$\begin{Bmatrix} \varepsilon_x \\ \varepsilon_y \\ \gamma_{xy} \end{Bmatrix} = \frac{1}{E}\begin{bmatrix} 1 & -\nu & 0 \\ -\nu & 1 & 0 \\ 0 & 0 & 2(1+\nu) \end{bmatrix}\begin{Bmatrix} \sigma_x \\ \sigma_y \\ \tau_{xy} \end{Bmatrix} \tag{7.4}$$

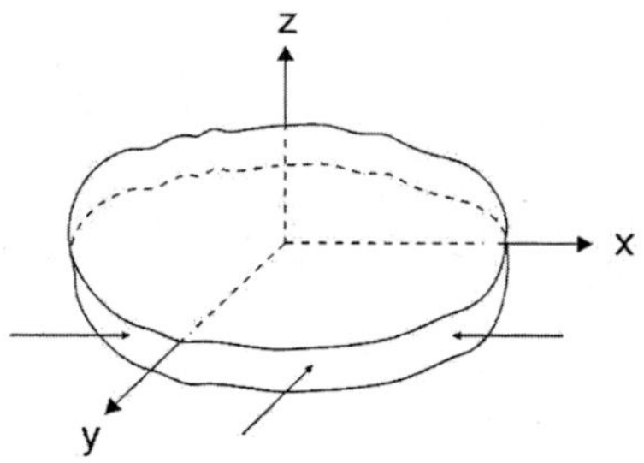

Fig. 7.3. Thin plate under plane stress

Equation (7.4) has resulted after making necessary substitution of plane stress conditions into eqn. (7.2) and replacing G in terms of E from eqn. (7.3). In compact form, eqn. (7.4) can be written as

$$\{\varepsilon\} = [S]\,\{\sigma\} \tag{7.5}$$

Therefore,

$$\{\sigma\} = [D]\,\{\varepsilon\} \tag{7.6}$$

where $$[D] = [S]^{-1} = \frac{E}{1-\nu^2}\begin{bmatrix} 1 & \nu & 0 \\ \nu & 1 & 0 \\ 0 & 0 & \dfrac{1-\nu}{2} \end{bmatrix} \tag{7.7}$$

7.3.2 Plane Strain

If the dimension of the body in the z-direction is very much larger in comparison to other two dimensions, the section at any point of the body perpendicular to z-axis is constant and the necessary forces applied to the body do not vary along the length, it is assumed that the deformation of the plane which is perpendicular to the z-axis becomes independent of its distance from z-axis, *i.e.*, all cross-sections are assumed to be in the same condition. For this condition then, longitudinal displacement w is zero. Further, $\gamma_{yz} = \gamma_{zx} = \varepsilon_z = 0$. For the analysis of this body, two sections at unit distance apart are then considered.

Culvert or tunnel, dam, retaining wall, cylindrical tubes subjected to internal pressure arc some of the examples of plane strain problems.

Since $\gamma_{yz} = \gamma_{zx} = 0$, then from eqn. (7.2), $\tau_{yz} = \tau_{zx} = 0$. σ_z however is not zero. Since $\varepsilon_z = 0$ then from eqn. (7.2)

$$\sigma_z \times \nu(\sigma_x + \sigma_y) = 0$$

or,

$$\sigma_z = \nu(\sigma_x + \sigma_y) \tag{7.8}$$

Thus, the plane strain problem like the plane stress problem, reduces to the determinaton of σ_x, σ_y and τ_{yx} as a function of x and y only. Substituting the values of σ_z from eqn. (7.8) into first two equations of eqn. (7.2) results in

$$\left.\begin{aligned} \varepsilon_x &= \frac{1-\nu^2}{E}\sigma_x - \frac{\nu(1+\nu)}{E}\sigma_y \\ \varepsilon_y &= \frac{1-\nu^2}{E}\sigma_y - \nu\frac{(1+\nu)}{E}\sigma_x \end{aligned}\right\} \tag{7.9}$$

Therefore, the stress-strain relation of a plane strain problem is

$$\begin{Bmatrix} \varepsilon_x \\ \varepsilon_y \\ \tau_{xy} \end{Bmatrix} = \frac{1-\nu^2}{E}\begin{bmatrix} 1 & -\dfrac{\nu}{1-\nu} & 0 \\ -\dfrac{\nu}{1-\nu} & 1 & 0 \\ 0 & 0 & \dfrac{2}{1-\nu} \end{bmatrix}\begin{Bmatrix} \sigma_x \\ \sigma_y \\ \tau_{xy} \end{Bmatrix} \tag{7.10}$$

or,

$$\{\varepsilon\} = [S]\{\sigma\} \tag{7.11}$$

Equation (7.11) can be rewritten as

$$\{\sigma\} = [D]\{\varepsilon\} \tag{7.12}$$

where

$$[D] = \frac{E(1-\nu)}{(1-2\nu)(1+\nu)}\begin{bmatrix} 1 & \dfrac{\nu}{1-\nu} & 0 \\ \dfrac{\nu}{1-\nu} & 1 & 0 \\ 0 & 0 & \dfrac{1-2\nu}{2(1-\nu)} \end{bmatrix} \tag{7.13}$$

Therefore, it is seen that there exists great similarity between plane stress and plane strain problems. It can further be verified that $[D]$ matrix of plane stress can be reduced to $[D]$ matrix of plane strain by replacing E and ν in the former expression by

$$E' = \frac{E}{1-\nu^2} \text{ and } \upsilon' = \frac{\nu}{1-\nu}$$

7.3.3 Differential Equations of Equilibrium

The rectangular element of Fig. 7.4 is in equilibrium in the stressed condition.

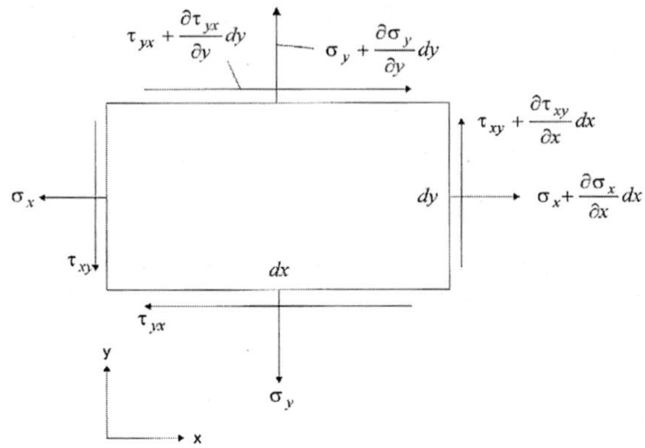

Fig. 7.4. Element in equilibrium

Stresses acting at the different faces of the element are shown in Fig. 7.4. Body forces per unit volume acting in X and Y directions are F_x and F_y respectively. The equation of equilibrium of the forces in the x-direction is

$$\left(\sigma_x + \frac{\partial \sigma_x}{\partial x} dx\right) dy - \sigma_x\, dx + \left(\tau_{xy} + \frac{\partial \tau_{xy}}{\partial y} dy\right) dx - \tau_{xy} dx + F_x\, dx\, dy = 0$$

or,
$$\frac{\partial \sigma_x}{\partial x} + \frac{\partial \tau_{xy}}{dy} + F_x = 0 \tag{7.14}$$

Similarly, the equilibrium of the forces in the y-direction will result in the other equation

$$\frac{\partial \tau_{xy}}{dx} + \frac{\partial \sigma_y}{dy} + F_y = 0 \tag{7.15}$$

It may be noted that the third equilibrium equation, *i.e.*, the moment of the forces about any point in space is zero, results in $\tau_{xy} = \tau_{yx}$. This has already been considered and accordingly, no distinction between τ_{xy} and τ_{yx} has been made as is evident in Fig. 7.4.

It may, further, be seen that a two-dimensional theory of elasticity problem involves three unknowns - σ_x, σ_y and τ_{xy}. Only two equilibrium equations are available, therefore, the problem is statically indeterminate. Further discussions regarding the solution of plane problems will not be made, as they are not needed as far as the finite element analysis is concerned. However, before the topic is closed, the equilibrium equations of three-dimensional problems are given which are nothing, but the extension of eqns. (7.14) and (7.15).

$$\left.\begin{aligned}\frac{\partial \sigma_x}{\partial x}+\frac{\partial \tau_{xy}}{\partial y}+\frac{\partial \tau_{xz}}{\partial z}+F_x=0\\ \frac{\partial \tau_{yx}}{\partial x}+\frac{\partial \sigma_y}{\partial y}+\frac{\partial \tau_{yz}}{\partial z}+F_y=0\\ \frac{\partial \tau_{zx}}{\partial x}+\frac{\partial \tau_{zy}}{\partial y}+\frac{\partial \sigma_z}{\partial z}+F_z=0\end{aligned}\right\} \tag{7.16}$$

7.4 BENDING OF THIN PLATES

The theory of bending of plates which assumes the material of the plate to be linearly elastic and the deformation of the plate to be small is known as the small deflection theory of plates or simply as classical theory of thin plates.

The plane parallel to the two faces of the plate and bisecting the thickness of the plate in the undeformed state is called the middle plane or middle surface of the plate. A body having the middle surface in the form of a plane and whose thickness is sufficiently small compared to its other two dimensions is called a thin plate.

7.4.1 Basic Assumptions

The classical theory of thin plates is based on the following assumptions:

1. A normal to the middle plane of the plate before bending transforms into a normal to the middle plane after bending.
2. The normal stress component perpendicular to the plane of the plate is small compared with other stress components and is neglected in the stress-strain relationship.
3. The middle plane of the plate remains unstrained after bending.

7.4.2 Deformation of the Plate

Coordinate system shown for the plate is given as x and y-axes in the middle plane of the plate and z-axis perpendicular to the middle plane.

Figure 7.5 indicates the section of a plate parallel to xz plane. After bending, a point P on the middle plane is deflected to P' with a deflection w. Another point Q, at a distance z from the undeformed middle plane is displaced to Q' which is on the normal to the middle plane after bending. The displacement of the point Q in x-direction is given by

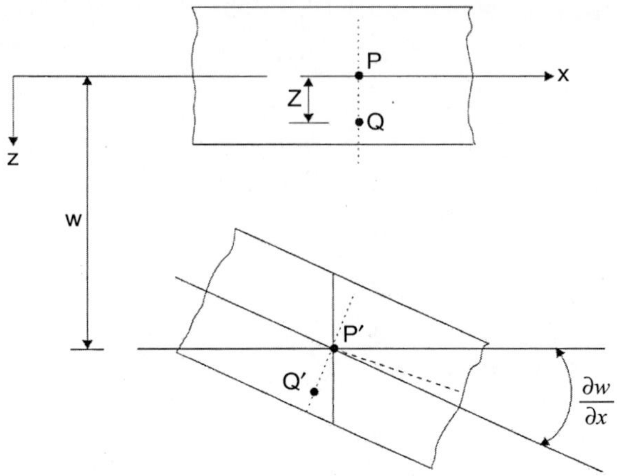

Fig. 7.5. Deformation of the plate in xz plane

$$u = -z\frac{\partial w}{\partial x} \tag{7.17}$$

Similarly,
$$v = -z\frac{\partial w}{\partial y} \tag{7.18}$$

7.4.3 Strain-Displacement Relationship

Referring to eqn. (7.1) and combining it with eqns. (7.17) and (7.18), yields

$$\left.\begin{aligned} \varepsilon_x &= \frac{\partial u}{\partial x} = -z\frac{\partial^2 w}{\partial x^2} \\ \varepsilon_y &= \frac{\partial v}{\partial y} = -z\frac{\partial^2 w}{\partial y^2} \\ \gamma_{xy} &= \frac{\partial u}{\partial y} + \frac{\partial v}{\partial x} = -2z\frac{\partial^2 w}{\partial x\,\partial y} \end{aligned}\right\} \tag{7.19}$$

7.4.4 Stress-Strain Relationship

Assumption (2) states that $\sigma_z = 0$. Further, $\tau_{xz} = \tau_{yz} = 0$. Therefore, eqn. (7.2) becomes

$$\left.\begin{aligned} \varepsilon_x &= \frac{1}{E}(\sigma_x - \nu\sigma_y) \\ \varepsilon_y &= \frac{1}{E}(\sigma_y - \nu\sigma_x) \\ \gamma_{xy} &= \frac{\tau_{xy}}{G} \end{aligned}\right\} \tag{7.20}$$

Solution of eqn. (7.20), yields

$$\left.\begin{aligned} \sigma_x &= \frac{E}{1-\nu^2}(\varepsilon_x + \nu\varepsilon_y) \\ \sigma_y &= \frac{E}{1-\nu^2}(\varepsilon_y + \nu\varepsilon_x) \\ \tau_{xy} &= \frac{E}{1-\nu^2}\left(\frac{1-\nu}{2}\gamma_{xy}\right) \end{aligned}\right\} \tag{7.21}$$

Substituting ε_x, ε_y and γ_{xy} from eqn. (7.19) into eqn. (7.21), yields

$$\left.\begin{aligned} \sigma_x &= -\frac{Ez}{1-\nu^2}\left(\frac{\partial^2 w}{\partial x^2} + \nu\frac{\partial^2 w}{\partial y^2}\right) \\ \sigma_y &= -\frac{Ez}{1-\nu^2}\left(\frac{\partial^2 w}{\partial y^2} + \nu\frac{\partial^2 w}{\partial x^2}\right) \\ \tau_{xy} &= \frac{-Ez}{1-\nu^2}(1-\nu)\frac{\partial^2 w}{\partial x\,\partial y} \end{aligned}\right\} \tag{7.22}$$

Once the stress components have been expressed in terms of the curvatures, the bending moment and twisting moment per unit length of the plate can be obtained. The bending mements have two components parallel to *xz*-plane and *yz*-plane respectively. Thus,

$$M_x = \int_{-h/2}^{h/2} \sigma_x \, z dz \tag{7.23}$$

Substituting σ_x from eqn. (7.22) into eqn. (7.23) and noting that the deflection w is not a function of z, eqn., (7.23) becomes

$$M_x = -\frac{E}{1-\nu^2}\left(\frac{\partial^2 w}{\partial x^2} + \nu\frac{\partial^2 w}{\partial y^2}\right)\int_{-h/2}^{h/2} z^2\, dz = -D\left(\frac{\partial^2 w}{\partial x^2} + \nu\frac{\partial^2 w}{\partial y^2}\right) \tag{7.24}$$

where D denotes $[Eh^3/12\,(1-\nu^2)]$ and is called the flexural rigidity of the plate. Similarly, it can be shown that

$$M_y = -D\left(\frac{\partial^2 w}{\partial y^2} + \nu\frac{\partial^2 w}{\partial x^2}\right) \tag{7.25}$$

The twisting moment is given by

$$M_{xy} = -M_{yx} = -\int_{-h/2}^{h/2} \tau_{xy} \cdot z dz$$

or,

$$M_{xy} = D(1-\nu)\frac{\partial^2 w}{dx\, dy} \tag{7.26}$$

The negative sign of the integral has been put because of the fact that for positive τ_{xy} and positive z, dM_{xy} is negative.

The moment curvature relationship in matrix notations then is

$$\begin{Bmatrix} M_x \\ M_y \\ M_{xy} \end{Bmatrix} = D\begin{bmatrix} 1 & \nu & 0 \\ \nu & 1 & 0 \\ 0 & 0 & \dfrac{1-\nu}{2} \end{bmatrix}\begin{Bmatrix} -\dfrac{\partial^2 w}{\partial x^2} \\ -\dfrac{\partial^2 w}{\partial y^2} \\ 2\dfrac{\partial^2 w}{\partial x\,\partial y} \end{Bmatrix} \tag{7.27}$$

In compact form, it is written as $\{M\} = [D]\,\{k\}$. (7.28)

It may be mentioned here that eqn. (7.27) is valid for isotropic plate only. For orthotropic plates, where the plate has two different properties in two orthogonal directions, the moment curvature relationship becomes

$$\begin{Bmatrix} M_x \\ M_y \\ M_{xy} \end{Bmatrix} = \begin{bmatrix} D_x & D_1 & 0 \\ D_1 & D_y & 0 \\ 0 & 0 & D_{xy} \end{bmatrix}\begin{Bmatrix} -\dfrac{\partial^2 w}{\partial x^2} \\ -\dfrac{\partial^2 w}{\partial y^2} \\ 2\dfrac{\partial^2 w}{\partial x\,\partial y} \end{Bmatrix} \tag{7.29}$$

7.4.5 Equilibrium Equations

Though it has been assumed that plane sections normal to the middle plane before bending remain plane after bending which signifies that τ_{xz} and τ_{yz} are negligible, yet the shear forces acting at the different edges of the element will be considered. All the necessary forces in the element of a plate are shown in Fig. 7.6. A lateral load p per unit area acts on the plate.

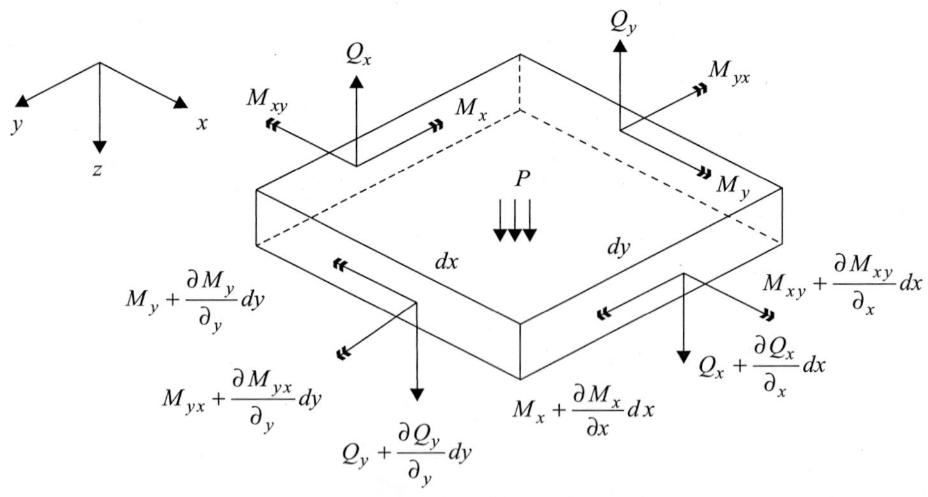

Fig. 7.6 Forces acting on the plate

Considering the equilibrium of all the forces in the z-direction, yields.

$$\left(Q_x + \frac{\partial Q_x}{\partial x}dx\right)dy - Q_x\,dy + \left(Q_y + \frac{\partial Q_y}{\partial y}dy\right)dx - Q_y\,dx + p\,dx\,dy = 0$$

or,

$$\frac{\partial Q_x}{\partial x} + \frac{\partial Q_y}{dy} + p = 0 \tag{7.30}$$

Taking moments of all the forces acting on the element with respect to x-axis and neglecting higher order terms

$$\left(M_{xy} + \frac{\partial M_{xy}}{\partial x}dx\right)dy - M_{xy}\,dy + M_y\,dx - \left(M_y + \frac{\partial M_y}{\partial y}dy\right)dx + Q_y\,dx\,dy = 0$$

or,

$$\frac{\partial M_{xy}}{\partial x} - \frac{\partial M_y}{\partial y} + Q_y = 0 \tag{7.31}$$

Similarly, taking moment of all the forces about y-axis will yield

$$\frac{\partial M_x}{\partial x} + \frac{\partial M_{yx}}{\partial y} - Q_x = 0 \tag{7.32}$$

Equations (7.30), (7.31) and (7.32) can be combined into a single equation given by

$$\frac{\partial^2 M_x}{\partial x^2} - 2\frac{\partial^2 M_{xy}}{\partial x\,\partial y} + \frac{\partial^2 M_y}{\partial y^2} + p = 0 \tag{7.33}$$

7.4.6 Differential Equation for Deflection

Substituting M_x, M_y, and M_{xy} in terms of curvatures as given by eqn. (7.27) into equation (7.33) results

$$\frac{\partial^2}{\partial x^2}\left[-D\left(\frac{\partial^2 w}{\partial x^2} + \nu\frac{\partial^2 w}{\partial y^2}\right)\right] - \frac{2\partial^2}{\partial x\,\partial y}\left[D(1-\nu)\frac{\partial^2 w}{\partial x\,\partial y}\right] + \frac{\partial^2}{\partial y^2}\left[-D\left(\frac{\partial^2 w}{\partial y^2} + \nu\frac{\partial^2 w}{\partial x^2}\right)\right] + p = 0$$

or,

$$\frac{\partial^4 w}{\partial x^2} + 2\frac{\partial^4 w}{\partial x^2\,\partial y^2} + \frac{\partial^4 w}{\partial y^4} = \frac{p}{D} \tag{7.34}$$

Equation (7.34) is known as the plate equation.

7.4.7 Shearing Forces

Equation (7.32) is rewritten as

$$Q_x = \frac{\partial M_x}{\partial x} - \frac{\partial M_{xy}}{\partial y} \tag{7.35}$$

Substituting M_x and M_y from eqn. (7.27) into eqn. (7.35) yields

$$Q_x = -D\left(\frac{\partial^3 w}{\partial x^3} + \frac{\partial^3 w}{\partial x\,\partial y^3}\right) \tag{7.36}$$

Similarly, it can be shown that

$$Q_y = -D\left(\frac{\partial^3 w}{\partial x^2\,\partial y} + \frac{\partial^3 w}{\partial y^3}\right) \tag{7.37}$$

7.5 BOUNDARY CONDITIONS

The boundary conditions discussed below are with respect to the rectangular plate.

7.5.1 Simply Supported Edge

If the edge $x = a$ is simply supported, the deflection and bending moment along the edge must be zero. That is

$$(w)_{x=a} = 0 \text{ and } (M_x)_{x=a} = -D\left(\frac{\partial^2 w}{\partial x^2} + \nu\frac{\partial^2 w}{\partial y^2}\right) = 0$$

But $w = 0$ along the edge $x = a$ implies that

$$\frac{\partial w}{\partial y} = \frac{\partial^2 w}{\partial y^2} = 0 \text{ along this edge}$$

The boundary conditions for a simply supported edge are therefore equivalent to

$$(w)_{x=a} = 0 \text{ and } \left(\frac{\partial^2 w}{\partial x^2}\right)_{x=a} = 0 \tag{7.38}$$

7.5.2 Clamped Edge

If an edge at $x = a$ is clamped, the deflection as well as the slope of the middle plane is zero, that is

$$(w)_{x=a} = 0 \text{ and } \left(\frac{\partial w}{\partial x}\right)_{x=a} = 0 \tag{7.39}$$

7.5.3 Free Edge

If the edge $x = a$ is free, then the bending moment, the twisting moment and the vertical shearing force at that edge is zero, *i.e.*,

$$(M_x)_{x=a} = 0,\ (M_{xy})_{x=a} = 0 \text{ and } (Q_x)_{x=a} = 0 \tag{7.40}$$

It was shown by Kirchoff that three boundary conditions are too many and that two conditions are sufficient for complete determination of deflection. Kirchoff replaced two conditions prescribing M_{xy} and Q_x by a single equation. This is explained below.

On the edge $x = a$, for an element of length dy, the twisting moment is $M_{xy}\,dy$, which may be replaced by two vertical forces of magnitude M_{xy}, at a distance dy apart. For the immediate next element of length dy, the twisting moment assumes a magnitude $\left[M_{xy} + \frac{\partial M_{xy}}{\partial y}dy\right]dy$, which may be replaced by vertical forces of magnitude $M_{xy} + \frac{\partial M_{xy}}{\partial y}dy$ as shown in Fig. 7.7 by dotted lines. At point *H*, there acts, therefore, a net vertical force of magnitude $\frac{\partial M_{xy}}{\partial y}$. For positive M_{xy}, this statically equivalent vertical force is acting upward which is negative according to the sign convention. Distribution of M_{xy} is, therefore, equivalent to a distribution of shearing forces of intensity.

$$Q'_x = -\left(\frac{\partial M_{xy}}{\partial y}\right)_{x=a} \tag{7.41}$$

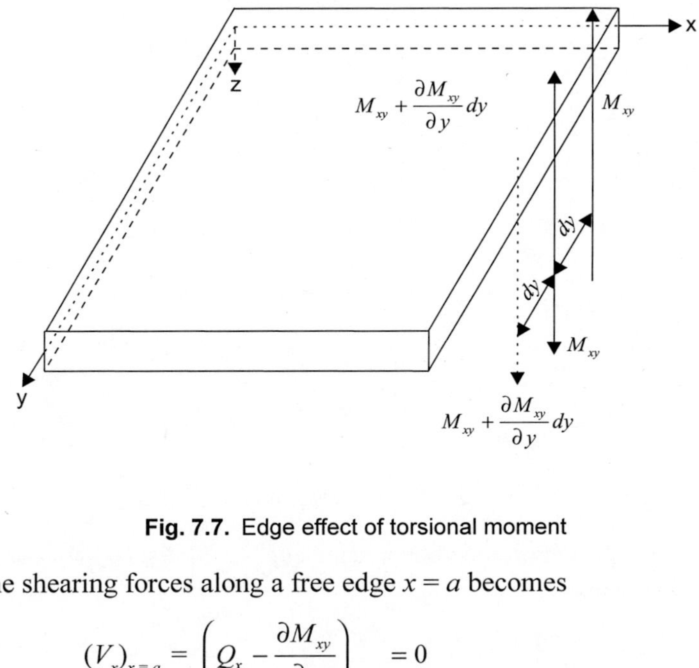

Fig. 7.7. Edge effect of torsional moment

Hence, the shearing forces along a free edge $x = a$ becomes

$$(V_x)_{x=a} = \left(Q_x - \frac{\partial M_{xy}}{\partial y}\right)_{x=a} = 0 \tag{7.42}$$

Replacing the values of Q_x and M_{xy} from eqn. (7.36) and eqn. (7.27) respectively in eqn. (7.42), yields

$$\left[\frac{\partial^3 w}{\partial x^3} + (2-\nu)\frac{\partial^3 w}{\partial x\,\partial y^2}\right]_{x=a} = 0 \tag{7.43}$$

The other boundary condition to be satisfied at a free edge is

$$(M_x)_{x=a} = 0$$

$$\left[\frac{\partial^2 w}{\partial x^2} + \nu\frac{\partial^2 w}{\partial y^2}\right]_{x=a} = 0 \tag{7.44}$$

Equations (7.43) and (7.44) may be taken as boundary conditions at a free edge.

7.5.4 Elastically Supported Edge

If an edge $x = a$ is elastically supported and the translational stiffness or spring constant is K_s, shown in Fig. 7.8(a), the boundary conditions for this edge will be

$$(w)_{x=a} = \pm\frac{D}{K_s}\left[\frac{\partial^3 w}{\partial x^3} + (2-\nu)\frac{\partial^3 w}{\partial x\,\partial y^2}\right]_{x=a} \tag{7.45}$$

and
$$(M_x)_{x=a} = \left[\frac{\partial^2 w}{\partial x^2} + \nu\frac{\partial^2 w}{\partial y^2}\right]_{x=a} = 0 \tag{7.46}$$

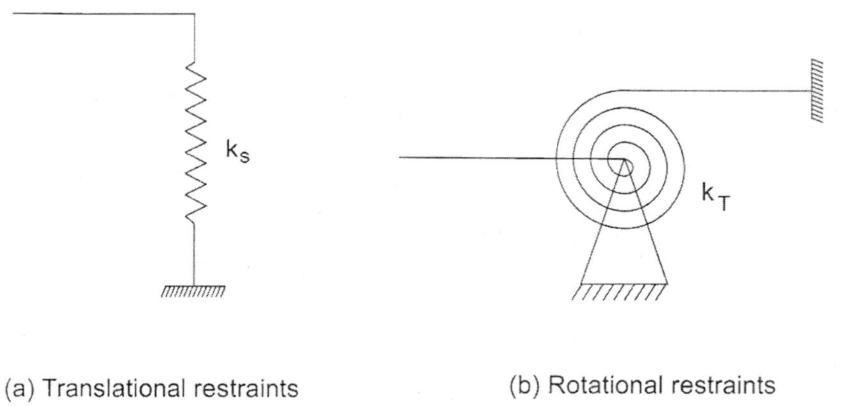

Fig. 7.8. Elastically translational and rotational supports

If this elastic support is provided by a beam supporting the edge of the plate, then eqn. (7.43) becomes

$$\left[EI_b\frac{\partial^4 w}{\partial y^4}\right]_{x=a} = \mp D\left[\frac{\partial^3 w}{\partial x^3} + (2-\nu)\frac{\partial^3 w}{\partial x\,\partial y^2}\right]_{x=a} \tag{7.47}$$

where I_b represents the moment of inertia of the beam.

7.5.5 Edge Having Elastic Rotational Restraint

If an edge has an elastic rotational restraint k_r, then the boundary conditions will be

$$(w)_{x=a} = 0$$

and
$$\left(\frac{\partial w}{\partial x}\right)_{x=a} = \mp\frac{D}{k_r}\left[\frac{\partial^2 w}{\partial x^2} + \nu\frac{\partial^2 w}{\partial x^2}\right]_{x=a} \tag{7.48}$$

7.6 CONCLUDING REMARKS

No attempt will be made to solve the plate differential equation (7.34). The closed bound solutions for various types of boundary conditions for some plates of certain shapes under typical loading conditions have been obtained and results have been presented in tabulated or graphical form [7.4,7.6].

Here only the derivation of basic relationship of the plane stress and plate bending problems of the plate have been carried out which will be required in the subsequent chapters when the finite element analysis is presented. For the same reason, no exercise is given in this chapter.

REFERENCES AND SUGGESTED READINGS

7.1. A. Green and W. Zerna, Theoretical Elasticity, Second Edition, Oxford University Press, 1968.

7.2. Fifth Conference on Electronic Computation, Proc. ASCE, J. Strl. Divn., v. 97, Jan., 1971.

7.3. I. S. Sokolnikoff, Mathematical Theory of Elasticity, McGraw-Hill, 1966.

7.4. R. Szilard, Theory and Analysis of Plates: Classical and Numerical Methods, Prentice Hall, 1974.

7.5. S. Timoshenko and J. N. Goodier, Theory of Elasticity, Second Edition, McGraw Hill, 1951.

7.6. S. Timoshenko and S. Woinowsky-Krieger, Theory of Plates and Shells, Second Edition, McGraw-Hill, 1959.

❑❑❑

8

INTRODUCTION TO THE FINITE ELEMENT METHOD

8.1 INTRODUCTION

In all the previous chapters, we have seen how the matrix methods are applied to one-dimensional or skeletal structures. For the analysis, we consider one member of the structure at a time. The choice of the element is obvious — the member is the discrete element considered. However, when we focus our attention to a continuum, such as two-dimensional structures like plates and shells, we are immediately faced with the difficulty regarding the choice of discretization, as no obvious choice exists. In such cases, like conventional one-dimensional structures which are considered as an assemblage of structural elements interconnected at their intersections, the continuum will be assumed to consist of imaginary divisions. Thus, the real continuum is divided into a finite number of discrete elements as shown in Fig. 8.1.

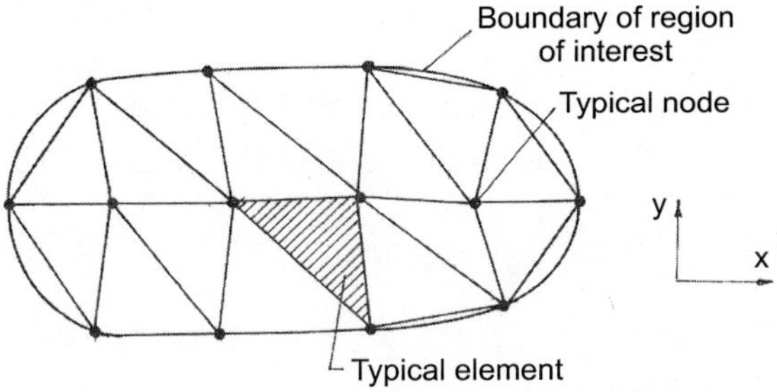

Fig. 8.1. A continuum divided into finite elements

8.2 THE FINITE ELEMENT METHOD

In a continuum that is divided into a mesh, two adjacent elements placed side by side will have a common edge (Fig.8.1). Its true representation will result in a more complicated analysis. Therefore, in order to make the analysis simpler, it is assumed that the elements are connected at the nodal points and it is only there that the continuity requirement is to be satisfied. Once the discretization is made, the analysis follows a rather set procedure. The analysis presented here regarding the finite element method is throughout based on the displacement approach. The stiffness matrix of the individual elements needs to be formulated. The forces actually distributed in the real structure are transformed to act at nodal points. In the finite element analysis, the continuum is divided into a 'finite' number of elements, having 'finite' dimensions and reducing the continuum having 'infinite' degrees of freedom to 'finite' degrees of freedom. The term 'finite element' was coined by Clough (8.1).

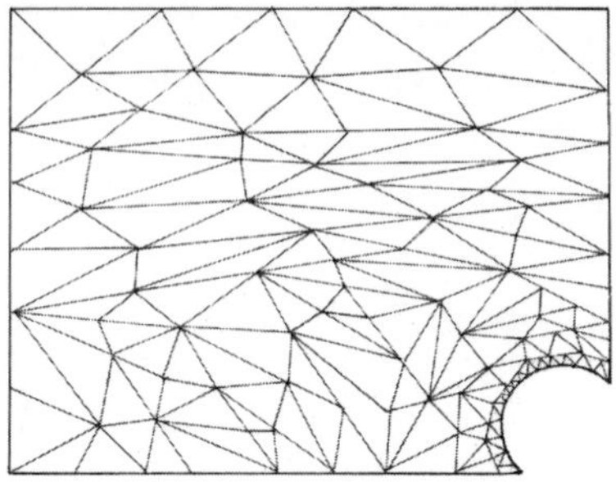

Fig. 8.2. Mesh division of a plate with hole

It has been found that the accuracy of the solution in general increases with the increase in the number of elements considered. However, involving more number of elements will take more computer time for obtaining the solution. A question that arises in the mind of the analyst is that what should be the optimum number of divisions that would give a satisfactory solution. The answer given by the experts to this query is that (*a*) the required subdivision of the continuum is to be based on past experience, (*b*) if an analysis is attempted for the first time, then convergence is to be tested for varied mesh gradings. For regions where higher concentration of stresses is expected, such as around the openings, a finer mesh division is to be made (Fig. 8.2). For last few years, adaptive finite element procedures have been developed to generate a mesh which would indicate accurate value.

In the finite element method, the problem is formulated in two stages:

(*a*) the element formulation

(*b*) the system formulation

The first stage involves the derivation of the element stiffness matrix that yields a relationship between nodal forces and nodal displacements. The next stage is the formulation of the stiffness and loads of the entire structure. The resulting solution of it yields the deformation and the internal forces at the points desired.

8.3 BRIEF HISTORY OF THE DEVELOPMENT OF THE FINITE ELEMENT METHOD

Although the name "finite element" is of recent origin, the concept has been used for centuries. The basic philosophy is to replace the actual problem by a simpler model which will closely approximate the solution of the problem at hand.

One of the earlier problems dealt with this concept was by ancient mathematicians, who found the circumference of the circle by approximating it as a polygon. Each side is a finite element according to present-day notations. Depending on whether the approximating polygon is inscribing or circumscribing the circle, a lower or upper bound solution to the true circumference can be obtained. Further, as the number of sides of the polygon are increased, the approximate value of the circumference will converge towards the true value. The finite element concept led to an accurate evaluation of the unknown quantities in those early days as can be found in ancient records.

The development of the finite element method has been carried out in two stages. In the initial stages, the theory was formulated on the basis of applying elementary theory of structural analysis with intuitive reasoning. Argyris and Kesley (8.2) have pioneered the concept of applying energy principles to the formulation of structural analysis problems. The starting point of the finite element method is a paper by Turner, Clough, Martin and Topp (8.3). It presented the application of simple finite elements (pin-joined bar and triangular plate with inplane load) for the analysis of aircraft structures.

Second stage of the development started with a more rigorous study of the convergence aspect of the finite element method. One such study by Melosh (8.4) led to the formulation of the finite element method based on the principle of minimum potential energy. Soon after that, de Veubeke (8.5) introduced stress or equilibrium elements based on the principle of minimum complementary energy. More or less at the same time, the concept of hybrid element was established by using the dual principles of minimum potential energy and minimum complementary energy was put forward by Pian (8.6).

The concept that the formulation of the finite element method can be set up by the variational principles, led to a revolution in its application. The structural analysis problems could now be placed on a more general footing. Load acting on the structure, be it a body force or a surface force can now be elegantly expressed in terms of generalized nodal forces. Further, the conversion of the physical nature of the problem into a mathematical model freed the method from its only field of application, the structural mechanics and paved the way for its application to other fields. In fact, any physical problem for which a variational formulation is possible, can now be solved by the finite element analysis.

According to the type of models of the element which exist, they are of three types-displacement model based on the principle of minimum potential energy, the stress or equilibrium model based on the principle of minimum complementary energy, and the hybrid model which takes into account the dual nature of the variational principles.

During the last two decades, the finite element method has made rapid strides. The theories are now placed on a firm mathematical footing. The generality and versatility of the method are now thoroughly exposed. A number of international conferences have been held in different parts of the world (8.7-8.21, 8.40-8.44). As far as the structural analysis is concerned, from the initial attention focused on the elastic analysis of plane stress and plate bending problems, the method has been successfully extended to the cases of analysis of three-dimensional structures (8.5, 8.45,8.46), curved structures (8.22, 8.47-8.49), stability and vibration problems (8.23-8.28, 8.50-8.53), non-linear analysis taking into account the large deflection and material non-linearity either individually or combined (8.29-8.31, 8.54-8.56).

8.4 BASIC STEPS IN THE FINITE ELEMENT METHOD FOR THE SOLUTION OF STATIC PROBLEMS

The following are the steps adopted for analyzing a structural engineering static problem by the finite element method (8.32-8.33).

Step 1. Discretisation of the domain: The continuum is divided into a number of finite elements by imaginary lines or surfaces. The interconnected elements may have different sizes and shapes. The success of this idealization lies in how closely this discretised continuum represents the actual continuum. The choice of the simple element or higher order element, straight or curved, its shape, refinement is to be decided before the mathematical formulation starts.

Step 2. Identification of variables: It is assumed that the elements are connected at their intersecting points referred to as nodal points. At each node, unknown displacements are to be prescribed. They are dependent on the problem at hand. For example, in a plane stress problem the minimum unknowns are two linear translations at each node. In a simpler model of the plate bending problem, the unknowns at a node are the deflection and two rotations. In ordinary cases, to every unknown displacement at a node there corresponds a nodal force which can be physically visualized. But it has already been discussed that the finite element method transcends the physical based reasoning. The problem may be identified in such a way that in addition to the displacement which occurs at the nodes depending on the physical nature of the problem, certain other quantities such as 'strain', 'curvature' *etc.* may need be specified as nodal unknowns for the element, which however, may not have a corresponding physical quantity in the generalized forces. The value of these quantities can, however, be obtained from variational principles.

Step 3. Choice of approximating functions: Once the variables and local coordinate system have been chosen, the next step is the choice of the displacement function. In fact, it is the displacement function that is the starting point of the mathematical analysis. This function represents the variation of the displacement within the element. The function can be approximated in a number of ways. The displacement function may be approximated in the form of a linear function or higher-order functions. A convenient way to express it is by using polynomial expressions. A typical question is posed in the finite element analysis, that is, the solution may converge to the exact values either by increasing the degrees of the polynomial or by decreasing the element size. This aspect will be discussed later.

The shape of the element or the geometry may also be approximated. The coordinates of corner nodes define the element shape accurately if the element is actually made of straight lines or planes. This simplification of geometry may sometimes lead to wrong results, *e.g.*, flat elements in a shell. The analyst will also have to decide on the weightage to be given to the geometry and displacements for a particular problem.

Step 4. Formation of the element stiffness matrix: After the continuum is discretised with desired element shapes, the element stiffness matrix is formulated. This can be done in a number of ways. Basically it is a minimization procedure whatever may be the approach adopted. For certain elements, the form involves a great deal of sophistication. With the exception of a few simple elements, the element stiffness matrix for majority of elements is not available in explicit form. As such, they require numerical integration for their evaluation.

The geometry of the element is defined with reference to the global frame. In many problems such as those of rectangular plates, the global and local axis system are coincident and for them no further calculation is needed at the element level beyond computation of element stiffness matrix in local coordinates. Coordinate transformation must be done for all elements where it is needed.

Step 5. Formation of the overall stiffness matrix: After the element stiffness matrices in global coordinates are formed, they are assembled to form of the overall stiffness matrix. The assembly is done through the nodes which are common to adjacent elements. At the nodes, the continuity of the displacement function and possibly their derivatives are established. The overall stiffness matrix is symmetric and banded.

Step 6. Incorporation of boundary conditions: The boundary restraint conditions are to be imposed in the stiffness matrix. There are various techniques available to satisfy the boundary conditions. In some of these approaches, the size of the stiffness matrix may be reduced or condensed in its final form. To ease the computer programming aspect and to elegantly incorporate the boundary conditions, the size of the overall stiffness matrix is kept the same.

Step 7. Formation of the element loading matrix: The loading forms an essential parameter in any structural engineering problem. The loading inside an element is transferred at the nodal points and consistent element loading matrix is formed. Sometimes, based on the typicality of the problem, the loading matrix may be simplified.

Step 8. Formation of the overall loading matrix: Like the overall stiffness matrix, the element loading matrices are assembled to form the overall loading matrix. This matrix has one column per loading case and it is either a column vector or a rectangular matrix depending on the number of loading conditions.

Step 9. Solution of simultaneous equations: All the equations required for the solution of the problem are now developed. In the displacement method, the unknowns are the nodal displacements. Rarely is the stiffness matrix stored as a solid matrix. Advantages are taken of the symmetric nature of the problem and banded properties. The Gauss elimination and Cholsky's factorization are the most commonly used procedures for the solution of simultaneous equations. These methods are well suited to a small or moderate number of equations. For large sized problems, frontal technique is one of the methods of obtaining solutions. For systems of large order, Gauss-Seidel or Jacobi iterations are more suited.

Step 10. Calculation of stresses or stress-resultants: In the previous step, nodal displacements are calculated and these values are utilized for the calculation of stresses or stress-resultants. This may be done for all elements of the continuum or it may be limited only to some predetermined elements.

Results may be obtained by graphical means. It may be desirable to plot the contour of the deformed shape of the continuum. The contour of the principal stresses may be one of the sought after items for certain category of problems.

8.5 ADVANTAGES AND DISADVANTAGES OF THE FINITE ELEMENT METHOD

Physical problems result in differential equations. To obtain exact solution of them in closed-bound form poses formidable challenge. Numerical methods are employed to obtain approximate solution. There are many approximate methods such as Rayleigh-Ritz method, Galerkin method, the least square method and the finite difference method are often used to obtain necessary solution. The finite element method stands superior to all of them. The main advantage of the finite element method is that in this method, approximations are confined to relatively small subdomains whereas in other methods the admissible function satisfies the boundary condition of the entire domain which becomes extremely difficult when the domain has irregular shape. In the finite element method, the admissible functions are valid over the simple domain and have nothing to do with the boundary, however simple or complex it may be.

Thus, the main advantage of the finite element method can be put in one sentence – the physical problems which were so far intractable and complex for any closed-bound solution can now be analysed by the finite element method. The advantages in relation to the complexity of the problem are stated below:

1. The method can be efficiently applied to cater irregular geometry. It can take care of any type of boundary.
2. Material anisotropy and nonhomogeneity can be catered without much difficulty.
3. Any type of loading can be handled.

Based on the above discussion one may form the idea that the finite element method is the most efficient for the analysis of any type of structural engineering or physical problem. There are many types of problems where some other method of analysis which may prove to be more efficient than the finite element method. One of the major disadvantages of the method is the time involved in the solution of the problem. In many problems related to vibration and stability, the time consumed in carrying out the finite element analysis may turn out to be very high.

It may therefore be a luxury to undertake vibration and stability analysis of simpler structures where the application of even simpler computer methods such as the finite strip (8.34) or other semi-analytical methods (8.35, 8.36, 8.37) will lead to more economical solution. But those methods will work within their limitations and will not be as versatile as the finite element method.

In many cases one is so much carried away by the versatility of the finite element method that one loses sight of its limitations. It must be remembered that in whatever sophisticated manner the problem might have been formulated and solved, it has been done so within the frame of its assumptions. Proper engineering judgment is to be exercised in interpreting the results.

In the finite element method, if the size of the problem is relatively large, it may lead to round-off errors. A computer works with a limited number of digits. Solving the problem on the basis of restricted numbers of digits may not yield the desired degree of accuracy or it may give erroneous results in certain cases. The magnitude of these round-off errors varies with the problem, the problem description and computer configuration. For many problems, however, the increase in number of digits for the purpose of calculation improves the accuracy. An error analysis can be made to this effect (8.38, 8.39).

When a continuum is discretized, an infinite degrees of freedom system is converted into a model having finite number of degrees of freedom. In a continuum function which are continuous are now replaced by ones which are piece-wise continuous within individual elements. Thus, the actual continuum is represented by a set of approximations.

Injudicious choice of the shape of the element will lead to a considerable error in the solution. Triangular elements in the shape of equilateral or rectangular elements in the shape of a square will always perform better than those having unequal length of the sides. For very long shapes the attainment of convergence will be extremely slow.

Simplification of the actual boundary is another source of error. The domain may be reduced to the shape of a polygon. If the mesh is refined, then the error involved in the discretized boundary may be reduced.

REFERENCES AND SUGGESTED READINGS

8.1. R. W. Clough, The Finite Element Method in Plane Stress Analysis, Proc. Second Conf. Electronic Computation, ASCE, Pittsburg, 1960.

8.2. J. H. Argyris and S. Kesley, Energy Theorems and Structural Analysis, Butterworths, London, 1960.

8.3. M. J. Turner, R. W. Clough, H. C. Martin and L. J. Topp, Stiffness and deflection analysis of complex structures, J. Aero. Sc., v. 23, Sept., 1956, pp. 805–824.

8.4. R. J. Melosh, Basis of Derivation of Matrices for Direct Stiffness Method, J. AIAA, v. 1, no. 7, 1963.

8.5. B. F. de Veubeke, Upper and Lower Bounds in Matrix Structural Analysis, in "Matrix Methods of Structural Analysis", AGARD, v. 72, 1964, pp. 165–201.

8.6. Y. K. Cheung, The Finite Strip Method in Structural Analysis, Pergamon Press, 1976.

8.7. Conference on "The Use of Digital Computers in Structural Engineering", The University of New Castle, England, July, 1966.

8.8. S. J. Fenves, N. Perrone, A. R. Robinson and W. C. Schnobrich (Editors), Numerical and Computer Methods in Structural Mechanics, Academic Press, 1972.

8.9. Fifth Conference on Electronic Computation, J. Strl. Divn., Proc. ASCE, v. 97, Jan., 1971.

8.10. Finite Element Review, Advanced Engineering Corporation, Linkuping, Sweden, 1979.

8.11. R. H. Gallaghar, Y. Yamada and J. T. Oden (Editors), Recent Advances in Matrix Methods of Structural Analysis and Design, Tokyo, Aug. 25–30, 1979, University of Alabama Press, 1980.

8.12. R. Glowinski, E. Y. Rodin and O. C. Zienkiewicz, Energy Methods in Finite Element Analysis, John Wiley and Sons, 1979.

8.13. ISD/ISSC Symposium on Finite Element Techniques, Stuttgart, West Germany, June, 1969.

8.14. ISSC Symposium on "High Speed Computing in Elasticity", Liege, Belgium, 1970.

8.15. J. C. McCutcheon, M. S. Mirza and A. A. Mufti (Editors), Proceedings on Specialized Conference on the Finite Element Method in Civil Engineering, (held at McGill University, Canada 1972), Engineering Institute of Canada, Ottawa, London, 1972.

8.16. J. T. Oden, O. C. Zienkiewicz, R. H. Gallaghar and C. Taylor (Editors), Finite Elements in Fluids, v. I and II, John Wiley and Sons, 1979.

8.17. V. Palmano and A. Kabila (Editors), International Conference on Finite Element Methods, University of New South Wales, Australia, 1975.

8.18. Proceedings of the First Conference on Matrix Methods of Structural Mechanics, Wright-Patterson Air Force Base, Dayton, Ohio, AFFDL-TR-66-80, October 26–28, 1965.

8.19. Proceedings of the Second International Symposium in Computing Methods in Applied Science and Engineering, Iria, Versailles, France, 1975.

8.20. Symposium on "Applied Finite Element in Civil Engineering", Vanderbilt University, Tennessee, USA, Nov. 1969.

8.21. O. C. Zienkiewicz, R. Lewis and E. Stagg (Editors), Numerical Methods in Offshore Engineering, John Wiley & Sons, 1977.

8.22. G. A. Mohr, Numerically Integrated Triangular Element for Doubly Curved Thin Shells, Computers and Structures, v. 11, June 1980, pp. 565–573.

8.23. K. J. Bathe, Finite Element Procedures, Prentice-Hall, 1996.

8.24. A. Chajes, Principles of Structural Stability Theory, Prentice-Hall, 1974.

8.25. A. N. Chen, A Finite Element Study and Bifurcation and Limit Point Buckling of Elasto-plastic Arches, Computers and Structures, v. 60, no. 2, 1996, pp. 189–196.

8.26. R. H. Gallaghar, Finite Element Analysis Fundamentals, Prentice-Hall, 1975.

8.27. J. Padavan, Non-linear Vibrations of General Structures, J. of Sound and Vibration, v. 72, Oct. 1980, pp. 427–442.

8.28. G. B. Warburton, The Dynamical Behaviour of Structures, Pergamon Press, 1976.

8.29. R. Szilard, Theory and Analysis of Plates, Prentice-Hall, 1974.

8.30. M. M. Shokrieh and L. B. Lessard, Effects of Material Non-linearity on the Three-dimensional Stress State of Pin-loaded Composite Laminate, J. Comp. Materials, v. 30, no. 7, 1996, pp. 839–861.

8.31. D. Venugopal Rao, A. H. Sheikh and M. Mukhopadhyay, A Finite Element Large Deflection Analysis of Stiffened Plates, Computers and Structures, v. 47, no.6, 1993, pp. 987–993.

8.32. M. Mukhopadhyay, Free Vibrations of Rectangular Plates Having Different Degrees of Elastic Restraint of Rotation, J. Sound and Vibration, v. 67, no. 4, 1979, pp. 459–468.

8.33. M. Mukhopadhyay, Vibrations, Dynamics and Structural Systems, Oxford and IBH, 2nd Edition, 2000.

8.34. D. D. Milasinovic, The Finite Strip in Computational Mechanics, Faculties of Civil Engineering, Subotica, Belgrade, 2003.

8.35. M. Mukhopadhyay, Response of Rectangular Plates Having Different Degrees in Elastic Restraint in Harmonic Excitation, Ocean Engineering, Pergamon Press, v. 8, no. 5, 1981, pp. 497–505.

8.36. M. Mukhopadhyay, Buckling Analysis of Rectangular Ship Plating, International Shipbuilding Progress, v. 26, no. 297, 1979, pp. 89–97.

8.37. M. Mukhopadhyay, A Semi-analytic Solution for the Free Vibration of Annular Sector Plates, Journal of Sound and Vibration, v. 63, no. 1, 1979, pp.87–95.

8.38. R. J. Melosh, Structural Engineering Analysis by Finite Elements, Prentice-Hall, 1990.

8.39. R. J. Melosh, Estimating Manipulation Errors for Finite Element Analysis, Finite Elements in Analysis and Design, v. 3, no. 3, Oct. 1987.

8.40. 11th Annual Workshop on "the Finite Element Method in Biomedical Engineering, Biomechanics and Related Fields", University of Ulm, Germany, 11–14 July, 2004.

8.41. Finite Element Circus, Fall 2002, Penn State University, Oct. 25–26, 2002.

8.42. International Conference on Finite Element Methods: Superconvergence, Post-processing and a Post-priori Estimates, University of Jyvaskyla, Finland, July 1–6, 1996.

8.43. International Conference on the Finite Element Modelling, The Luxembourgh “Henry Tudor” Research Laboratory, Luxembourgh, 13–14 November, 2003.

8.44. International Conference on “ p and hp Finite Element Methods: Mathematics and Engineering Practice”, Washington University, May 31 – June 2, 2000.

8.45. G. Dhondt, The Finite Element Method for Three-dimensional Thermo-Mechanical Applications, Wiley Europe, 2004.

8.46. Y. C. Yip and R. C. Averill, A Three-dimensional Laminated Plate Finite Element with High Order Zig-zag Sublimate, Journal of Computational Engineering and Science, v. 2,no. 1, 2001, pp. 137–180.

8.47. N. Rattanawangcharoen, H. Bai and A. H. Shah, A 3D Cylindrical Finite Element Model for Thick Curved Beam Stress Element, IJNME, v. 59, 2004, pp. 511–531.

8.48. T. Y. Yang, Finite Element Structural Analysis, Prentice-Hall, Englewoods, Cliffs, 1986.

8.49. Jing Li, A Geometrically Exact Curved Beam Theory and its Finite Element Formulation/implementation, Master of Science Thesis in Aerospace Engineering, Virginia Polytechnique Institute, USA, 2000.

8.50. M. Petyt, Introduction to Finite Element Vibration Analysis, Cambridge University Press, Cambridge, U.K., 2nd edition, 1998.

8.51. M. Mukhopadhyay, Vibrations, Dynamics and Structural Systems, Oxford & IBH, 2nd edition, 2000.

8.52. S. Y. Wang, S. T. Quek and N. K. Ang, Dynamic Stability Analysis of Finite Element Modelling of Piezo-electric Composite, International Journal of Solids and Structures, v. 41, no. 3, 1987, pp. 745–764.

8.53. K. Chandrasekhara and K. Bhatia, Active Buckling Control of Smart Composite Plates – A Finite Element Analysis, Smart Materials and Structures, v. 2, March 1993, pp. 31–39.

8.54. M. A. Crisfield, Non-linear Finite Element Analysis of Solids and Structures, v. I and II, John Wiley and Sons, 1991 and 1997.

8.55. P. K. Kythe and D. Wee, An Introduction to Linear and Non-linear Finite Element Analysis, Birkhauser, Boston, 2002.

8.56. J. T. Oden, Finite Elements of Non-linear Continua, McGraw-Hill, 1972.

❑❑❑

9

FINITE ELEMENT ANALYSIS OF PLANE ELASTICITY PROBLEMS

9.1 INTRODUCTION

The present chapter deals with the element representation of plane stress problems. The derivation of element stiffness matrices is demonstrated for two types of elements – three-noded triangular element and four-noded rectangular element. Some of the higher order elements are then discussed. Isoparametric element will, however, be presented in the next chapter. A computer program for the solution of plane stress problems using triangular element has been presented. An example worked out with the help of the computer program has also been given.

The membrane part of the stress condition of some structures such as stiffened plate, folded plate, shells *etc.* can be represented by the plane stress elements. The difference between plane stress and plane strain problems has been discussed in Chapter 7. The finite element analysis of these two problems is identical excepting the [D] matrix, which relates the stress and the corresponding strain.

9.2 THREE-NODED TRIANGULAR ELEMENT

The simplest element available in the analysis of plane elasticity problems is the 3-noded triangular element. This element is most versatile in that it can fit into any irregular boundary – the curved boundaries are to be approximated by a series of straight lines. The element is also referred to as constant strain triangle, because it will be seen in the derivation of element stiffness matrix that the strain remains constant at any point within the element.

A triangular element of arbitrary shape with nodes at the corners denoted as 1, 2 and 3 is shown in Fig 9.1. The co-ordinates of 1,2 and 3 are (x_1, y_1), (x_2, y_2) and (x_3, y_3). The thickness of the element is t. The displacements parallel to the axis systems at points 1, 2 and 3 are (u_1, v_1), (u_2, v_2) and (u_3, v_3). Forces corresponding to the displacements are (Px_1, Py_1), (Px_2, Py_2) and (Px_3, Py_3). Therefore, the complete displacements and force vectors for these elements are

$$\{X\}_e^T = \{u_1\ v_1\ u_2\ v_2\ u_3\ v_3\} \tag{9.1}$$

$$\{P\}_e^T = \{Px_1\ Py_1\ Px_2\ Py_2\ Px_3\ Py_3\} \tag{9.2}$$

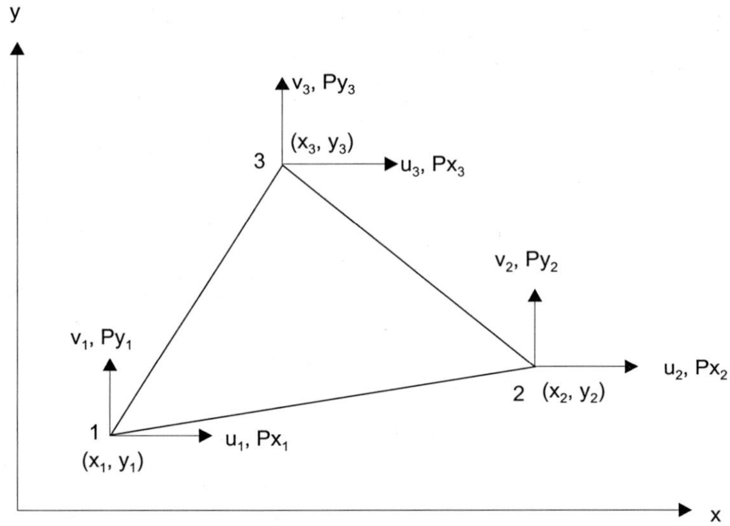

Fig. 9.1. Three-noded triangular plane stress element

9.2.1 Displacement Function

Each node is free to move in x and y directions so that there are two degrees of freedom per node. Therefore, for 3 nodes there are in all six degrees of freedom. The displacement functions for u and v are represented by polynomials. Total number of constant terms to be associated with the polynomial will be six, because the number of constant terms equals the total degrees of freedom for the element. Therefore, for this case the displacement function is represented by two linear polynomials

$$\begin{Bmatrix} u = \alpha_1 + \alpha_2 x + \alpha_3 y \\ v = \alpha_4 + \alpha_5 x + \alpha_6 y \end{Bmatrix} \tag{9.3}$$

or

$$\begin{Bmatrix} u \\ v \end{Bmatrix} = \begin{bmatrix} 1 & x & y & 0 & 0 & 0 \\ 0 & 0 & 0 & 1 & x & y \end{bmatrix} \begin{Bmatrix} \alpha_1 \\ \alpha_2 \\ \alpha_3 \\ \alpha_4 \\ \alpha_5 \\ \alpha_6 \end{Bmatrix} \tag{9.4}$$

or

$$\{f\} = [C]\,\{\alpha\} \tag{9.5}$$

9.2.2 Displacement Function Expressed in Terms of Nodal Displacements

Using eqn. (9.4) nodal displacements can be expressed in terms of respective nodal coordinates. Thus,

$$\begin{Bmatrix} u_1 \\ v_1 \\ u_2 \\ v_2 \\ u_3 \\ v_3 \end{Bmatrix} = \begin{bmatrix} 1 & x_1 & y_1 & 0 & 0 & 0 \\ 0 & 0 & 0 & 1 & x_1 & y_1 \\ 1 & x_2 & y_2 & 0 & 0 & 0 \\ 0 & 0 & 0 & 1 & x_2 & y_2 \\ 1 & x_3 & y_3 & 0 & 0 & 0 \\ 0 & 0 & 0 & 1 & x_3 & y_3 \end{bmatrix} \begin{Bmatrix} \alpha_1 \\ \alpha_2 \\ \alpha_3 \\ \alpha_4 \\ \alpha_5 \\ \alpha_6 \end{Bmatrix} \tag{9.6}$$

or,

$$\{X\}_e = [A]\{\alpha\} \tag{9.7}$$

The constant terms associated with the polynomial can now be determined from eqn. (9.7). Thus

$$\{\alpha\} = [A]^{-1}\{X\}_e \tag{9.8}$$

Substituting the value of $\{\alpha\}$ from eqn. (9.8) into eqn. (9.5), yields

$$\{f\} = [C][A]^{-1}\{X\}_e \tag{9.9}$$

or

$$\{f\} = [(I)N_1\ (I)N_2\ (I)N_3]\ \{X\}_e \tag{9.10}$$

where $[I]$ is a 2×2 identity matrix

$$N_1 = \frac{a_1 + b_1 x + c_1 y}{2\Delta},$$

$$N_2 = \frac{a_2 + b_2 x + c_2 y}{2\Delta}$$

and

$$N_3 = \frac{a_3 + b_3 x + c_3 y}{2\Delta}$$

where

$$a_1 = x_2 y_3 - y_2 x_3, \quad b_1 = y_2 - y_3, \quad \text{and} \quad c_1 = x_3 - x_2$$

$$a_2 = x_3 y_1 - x_1 y_3, \quad b_2 = y_3 - y_1, \quad \text{and} \quad c_2 = x_1 - x_3$$

$$a_3 = x_1 y_2 - x_2 y_1, \quad b_3 = y_1 - y_2, \quad \text{and} \quad c_3 = x_2 - x_1$$

$$2\Delta = \det \begin{bmatrix} 1 & x_1 & y_1 \\ 1 & x_2 & y_2 \\ 1 & x_3 & y_3 \end{bmatrix}$$

$= 2 \times$ area of the triangular elements 1 2 3

It may be mentioned that the absolute value of the area of the triangle is to be considered. Equation (9.9) can be written as

$$\{f\} = [N]\{X\}_e \tag{9.11}$$

It will be shown in later chapter that $[N]$ denotes the shape function matrix and $[N] = [C][A]^{-1}$. In the present case, the size being small, $[A]$ has been inverted algebraically. For elements where $[A]$ is larger, the inversion has to be done by the computer.

Whether from eqn. (9.4) or eqn. (9.11), it is seen that displacements along the edge vary linearly and its magnitude is dependent on the coordinates at the two nodes connecting that edge. Therefore, the compatibility of displacements of adjacent triangular elements with a common boundary is ensured.

9.2.3 Strain–Nodal Parameter Relationship

Strain components for a plane stress problem are given by

$$[\varepsilon] = \begin{Bmatrix} \varepsilon_x \\ \varepsilon_y \\ \gamma_{xy} \end{Bmatrix} = \begin{Bmatrix} \dfrac{\partial u}{\partial x} \\ \dfrac{\partial v}{\partial y} \\ \dfrac{\partial u}{\partial y} + \dfrac{\partial v}{\partial x} \end{Bmatrix} \tag{9.12}$$

Equation (9.12) can be written as

$$\{\varepsilon\} = \begin{bmatrix} \dfrac{\partial}{\partial x} & 0 \\ 0 & \dfrac{\partial}{\partial y} \\ \dfrac{\partial}{\partial y} & \dfrac{\partial}{\partial x} \end{bmatrix} \begin{Bmatrix} u \\ v \end{Bmatrix} \tag{9.13}$$

Combining eqns. (9.11) and (9.13) yields

$$\{\varepsilon\} = \begin{bmatrix} \dfrac{\partial N_1}{\partial x} & 0 & \dfrac{\partial N_2}{\partial x} & 0 & \dfrac{\partial N_3}{\partial x} & 0 \\ 0 & \dfrac{\partial N_1}{\partial y} & 0 & \dfrac{\partial N_2}{\partial y} & 0 & \dfrac{\partial N_3}{\partial y} \\ \dfrac{\partial N_1}{\partial y} & \dfrac{\partial N_1}{\partial x} & \dfrac{\partial N_2}{\partial y} & \dfrac{\partial N_2}{\partial x} & \dfrac{\partial N_3}{\partial y} & \dfrac{\partial N_3}{\partial x} \end{bmatrix} \{X\}_e \tag{9.14}$$

Performing all the necessary differentiation, eqn. (9.14) reduces to

$$\{\varepsilon\} = \frac{1}{2\Delta} \begin{bmatrix} b_1 & 0 & b_2 & 0 & b_3 & 0 \\ 0 & c_1 & 0 & c_2 & 0 & c_3 \\ c_1 & b_1 & c_2 & b_2 & c_3 & b_3 \end{bmatrix} \tag{9.15}$$

In compact form, eqn. (9.15) becomes

$$\{\varepsilon\} = [B]\,\{X\}_e \tag{9.16}$$

A careful look into $[B]$ matrix will immediately reveal that all the elements of $[B]$ are constants, which indicates that the strain is constant at any point within the element. Strains are first derivatives of displacements. So it is obvious that when displacement functions are linear polynomial, the strain will be constant.

9.2.4 Stress-Strain Relationship

It has been seen in eqn. (9.12) that the stress-strain relationship for the plane stress problem is given by

$$\{\sigma\} = [D]\{\varepsilon\} \tag{9.17}$$

Replacing $\{\varepsilon\}$ from eqn. (9.16) into eqn. (9.17) yields

$$\{\sigma\} = [D][B]\{X\}_e \tag{9.18}$$

9.2.5 Derivation of the Element Stiffness Matrix

The element stiffness matrix is derived here by applying Castigliano's theorem. Total strain energy of the plate element is given by

$$U = \frac{1}{2}\int_v \{\sigma_x \varepsilon_x \quad \sigma_y \varepsilon_y \quad \tau_{xy} \gamma_{xy}\} dv \tag{9.19}$$

or,

$$U = \frac{1}{2}\int_v \{\varepsilon_x\ \varepsilon_y\ \gamma_{xy}\} \begin{Bmatrix} \sigma_x \\ \sigma_y \\ \tau_{xy} \end{Bmatrix} dv$$

or,

$$U = \frac{1}{2}\int_v \{\varepsilon\}^T \{\sigma\} dv \tag{9.20}$$

Substituting $\{\varepsilon\}$ and $\{\sigma\}$ from eqns. (9.16) and (9.18) into eqn. (9.20) yields

$$U = \frac{1}{2}\int_v ([B]\{X\}_e)^T [D][B]\{X\}_e\, dv$$

or,

$$U = \frac{1}{2}\int_v \{X\}_e^T [B]^T [D][B]\{X\}_e\, dv \tag{9.21}$$

The integrand is constant. As such eqn. (9.21) becomes

$$U = \frac{1}{2}\{X\}_e^T [B]^T [D][B]\{X\}_e\, \Delta.t \tag{9.22}$$

where t is the thickness of the plate.

Now applying Castigliano's theorem (eqn. 9.16),

$$\frac{\partial U}{\partial \{X\}_e} = [B]^T [D][B]\, \Delta \cdot t \cdot \{X\}_e$$

or,

$$\{P\}_e = [K]_e \{X\}_e \tag{9.23}$$

where

$$[K]_e = [B]^T [D][B]\, \Delta \cdot t \tag{9.24}$$

$[K]_e$ has been explicitly evaluated and the final values for a plate element are given as eqn. (9.25).

$$[k]_e = \frac{t}{4\Delta}\begin{bmatrix} D_1b_1^2 + D_{12}c_1^2 & & & & & \\ D_1D_2b_1c_1 + D_{12}b_1c_1 & D_1c_1^2 + D_{12}b_1^2 & & & & \\ D_1b_2b_1 + D_{12}c_1c_2 & D_1D_2b_2c_1 + D_{12}b_1c_2 & D_1b_2^2 + D_{12}c_2^2 & & & \\ D_1D_2b_1c_2 + D_{12}b_2c_1 & D_1c_1c_2 + D_{12}b_1b_2 & D_1D_2b_2c_2 + D_{12}b_2c_2 & D_1c_2^2 + D_{12}b_2^2 & & \\ D_1b_1b_3 + D_{12}c_1c_3 & D_1D_2b_3c_1 + D_{12}b_1c_3 & D_1b_2b_3 + D_{12}c_2c_3 & D_1D_2b_3c_2 + D_{12}b_2c_3 & D_1b_3^2 + D_{12}c_3^2 & \\ D_1D_2b_1c_3 + D_{12}b_3c_1 & D_1c_1c_3 + D_{12}b_1b_3 & D_1D_2b_2c_3 + D_{12}b_3c_2 & D_1c_2c_3 + D_{12}b_2b_3 & D_1D_2b_3c_3 + D_{12}b_3c_3 & D_1c_3^2 + D_{12}b_3^2 \end{bmatrix} \qquad (9.25)$$

For plane stress

$$D_1 = E/(1 - \nu^2), \qquad D_2 = \nu, \qquad D_{12} = D_1(1 - D_2)/2$$

For plane strain

$$D_1 = \frac{E(1-\nu)}{(1+\nu)(1-2\nu)}, \qquad D_2 = \frac{\nu}{1-\nu}, \qquad D_{12} = \frac{D_1(1-D_2)}{2}$$

9.2.6 Determination of Element Stresses

After all the unknown displacements are obtained, the stresses in the elements are to be obtained as the final step. Equation (9.18) is rewritten as

$$\{\sigma\} = [D][B]\{X\}_e$$

For this element, eqn. (9.18) becomes

$$\{\sigma\} = \frac{1}{2\Delta}\begin{bmatrix} D_1b_1 & D_1D_2c_1 & D_1b_2 & D_1D_2c_2 & D_1b_3 & D_1D_2c_3 \\ D_1D_2b_1 & D_1c_1 & D_1D_3b_2 & D_1c_2 & D_1D_2b_3 & D_1c_3 \\ D_{12}c_1 & D_{12}b_1 & D_{12}c_2 & D_{12}b_2 & D_{12}c_3 & D_{12}b_3 \end{bmatrix}\{X\}_e \qquad (9.26)$$

The loading on the plate or more precisely the evaluation of the loads acting at the nodes can be done for the plane stress problem by simply lumping as shown in the example presented later in the chapter. The general treatment of the load acting on the element will be done in the next chapter. The remaining steps required for the complete solution of the problem are done by the computer. But those steps are more or less similar to the approach presented by the direct stiffness method as will be evident in the computer program presented later.

9.3 CRITERIA FOR THE CHOICE OF THE DISPLACEMENT FUNCTION

As already mentioned, the choice of the displacement function is the starting point of the mathematical analysis, and its selection has a lot of bearing on the accuracy of the results. On the basis of this assumed displacement function, the problem reduces from infinite degrees of freedom to some finite number. If the choice is improper then the result will not converge to the exact solution irrespective of the fineness of the subdivision. In order that simulated structure deforms as closely as possible to the actual structure, the displacement function chosen should satisfy the following three minimum requirements.

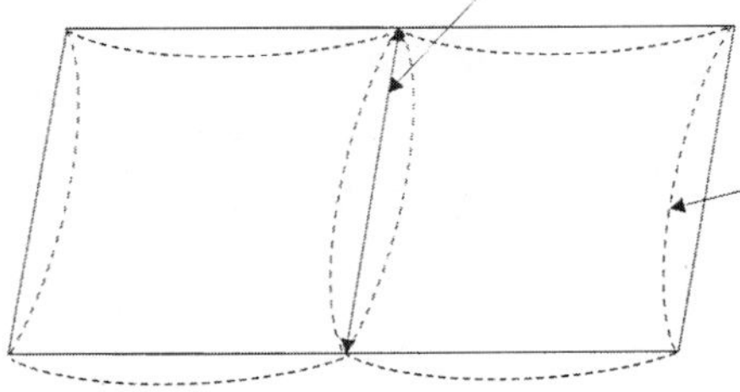

Fig. 9.2. Discontinuity at the element boundary

1. The chosen displacement function should ensure continuity within the element and also the compatibility between the adjacent elements.

   The polynomial expressions representing the displacement function obviously ensure the continuity of displacements within the element. There should not be any kinking, gaps or overlaps at the edge common to adjacent elements (Fig. 9.2). This condition is referred to as 'inter-elemental continuity' condition. This condition is satisfied only when displacements along the edge are dependent only on the nodal displacements on that edge.
2. The chosen displacement function should include the rigid body modes of the displacement of the element. This suggests that rigid body displacements at the nodes must not induce any straining of the element.

   In order to avoid complicated expressions, the terms pertaining to the rigid body modes have not been sometimes considered in the displacement function. Such formulation in curvilinear coordinates revealed that due to the violation of this condition, the convergence of the result has only slowed down, but the results have indeed converged to the correct value (9.1).
3. The chosen displacement function must represent the constant strain state of the element.

   This requirement suggests that as the subdivisions of a body are increased so that elements get smaller so as to reach infinitesimal size, the strains in each element assume constant value. More serious consequences result from the violation of this condition – the result converges to some incorrect value, which may be significantly different from the correct one.

9.4 POLYNOMIAL DISPLACEMENT FUNCTIONS

One of the prime considerations to be given for the choice of the displacement function is that it should not be biased about local coordinate system. This condition is referred to as geometric invariance, geometric isotropy or spatial isotropy. This suggests that for two-dimensional problems, the x and y terms should be balanced in the expression or there should not be more weightage of either x or y terms. The geometric isotropy condition can be met by choosing terms from Pascal's triangle given as Fig. 9.3.

1	(constant – 1 term)
$x \quad y$	(constant – 2 term)
$x^2 \quad xy \quad y^2$	(constant – 3 term)
$x^3 \quad x^2y \quad xy^2 \quad y^3$	(constant – 4 term)
$x^4 \quad x^3y \quad x^2y^2 \quad xy^3 \quad y^4$	(constant – 5 term)
$x^5 \quad x^4y \quad x^3y^2 \quad x^2y^3 \quad xy^4 \quad y^5$	(constant – 1 term)

Fig. 9.3. Pascal triangle

This immediately indicates the number of terms present in a complete polynomial of a particular order.

The terms of the displacement function should always be chosen from lower order polynomials and gradually moved towards next higher order and so on. These lower order terms help in satisfying the convergence criteria.

9.5 VERIFICATION OF THE CONVERGENCE CRITERIA OF THE DISPLACEMENT FUNCTION OF 3-NODED TRIANGULAR ELEMENT

The displacement function of the 3-noded triangular element given by eqn. (9.3) is reproduced below.

$$\begin{Bmatrix} u = \alpha_1 + \alpha_2 x + \alpha_3 y \\ v = \alpha_4 + \alpha_5 x + \alpha_6 y \end{Bmatrix} \tag{9.3}$$

Let the three conditions required to satisfy the convergence criteria be studied with respect to u and v polynomial expression of eqn. (9.3).

Condition 1 : Both u and v being polynomial series satisfy the continuity requirement within the element. It has been shown in Art. 9.2 that the inter-elemental continuity along any edge also exists.

Condition 2 : The rigid body u and v displacements terms are α_1 and α_4. The rigid body rotation terms are α_3 and α_5. All these constant terms are not zero. Therefore, the displacement function contains terms depicting the rigid body modes.

Condition 3 : For the plane stress problem, three strain terms are ε_x, ε_y and γ_{xy}. The constant strain terms in the displacement function are α_2, α_3, α_5 and α_6, which are non-zero. So, condition 3 is also satisfied.

9.6 NUMBER OF TERMS IN A POLYNOMIAL

It has been mentioned that the number of constants to be associated with a polynomial equals the total degrees of freedom of the element. Let the background of it be seen with respect to the constant strain triangular element.

The direct strain in x-direction is constant and is given by

$$\varepsilon_x = \frac{\partial u}{\partial x} = \alpha_2 \tag{9.27}$$

Integrating eqn. (9.27) with respect to x yields

$$u = \alpha_2 x + f(y) \tag{9.28}$$

Similarly, $$v = \alpha_6 y + g(x) \tag{9.29}$$

Again $$\gamma_{xy} = \frac{\partial u}{\partial y} + \frac{\partial v}{\partial x} = c \tag{9.30}$$

Combining eqns. (9.28), (9.29) and (9.30), yields

$$\frac{dg(x)}{dx} + \frac{df(y)}{dy} = c \tag{9.31}$$

Sum of the derivations of two independent functions in x and y can be equal to a constant if both are separately equal to constants.

Let $$\frac{dg(x)}{dx} = \alpha_5 \tag{9.32}$$

Integrating eqn. (9.32) with respect to x, yields

$$g(x) = \alpha_5 x + \alpha_4 \tag{9.33}$$

Similarly, $$\frac{df(y)}{dy} = \alpha_3 \tag{9.34}$$

Integrating eqn. (9.34) with respect to y yields,

$$f(y) = \alpha_3 y + \alpha_1 \tag{9.35}$$

Substituting the values of $g(x)$ and $f(y)$ from eqns (9.33) and (9.35) into eqns. (9.28) and (9.29), result in

$$\left.\begin{aligned} u &= \alpha_1 + \alpha_2 x + \alpha_3 y \\ v &= \alpha_4 + \alpha_5 x + \alpha_6 y \end{aligned}\right\} \tag{9.36}$$

9.7 FOUR-NODED RECTANGULAR ELEMENT

For structures having regular shaped boundaries, rectangular elements can be effectively employed for the analysis. The plate is divided into a suitable number of rectangular elements. It may be associated with other elements depending on the nature of the structure. A skew plate of Fig. 9.4 is divided into triangular and rectangular elements. In the following the derivation of element stiffness matrix of the rectangular element will be presented. The basic steps will be same as that adopted for constant strain triangular element.

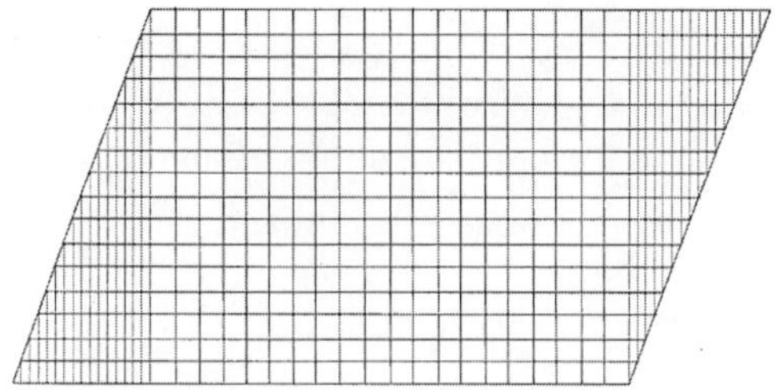

Fig. 9.4. Mesh division of a skew plate

A typical rectangular element having sides *a* and *b* is shown in Fig. 9.5. The thickness of the element is *t*. The nodes are numbered such that the coordinates of 1, 2, 3 and 4 are (0,0), (0, *b*), (*a*, 0) respectively. The displacements at nodes 1, 2 *etc*. parallel to the axis system are (u_1, v_1), (u_2, v_2) *etc*. and the corresponding forces are (Px_1, Py_1), (Px_2, Py_2) *etc*. The displacement vector and the force vector are then given by

$$\left.\begin{aligned} \{X\}_e^T &= \{u_1 \quad v_1 \quad u_2 \quad v_2 \quad u_3 \quad v_3 \quad u_4 \quad v_4\} \\ \text{and} \qquad & \\ \{P\}_e^T &= \{Px_1 \quad Py_1 \quad Px_2 \quad Py_2 \quad Px_3 \quad Py_3 \quad Px_4 \quad Py_4\} \end{aligned}\right\} \tag{9.37}$$

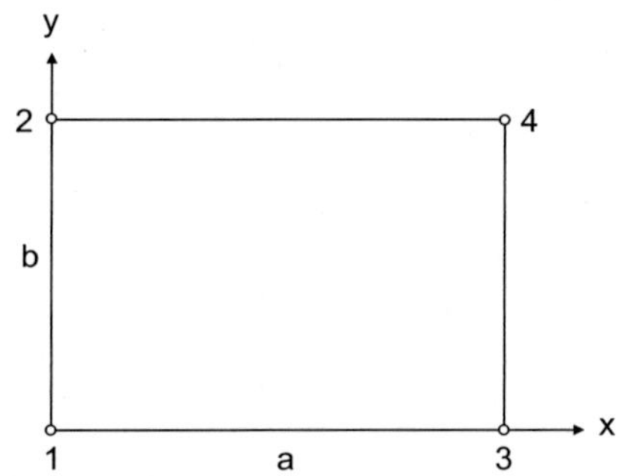

Fig. 9.5. Rectangular element (with corner nodes)

9.7.1 Displacement Function

Like triangular plane stress element, the state of displacement at any point (*x*, *y*) can be represented by two components *u* and *v*. To be consistent with 8 nodal displacements at four corners of the rectangle, the displacement equation to be chosen should contain 8 arbitrary constants. The functions expressed as a polynomial then are

$$\left.\begin{aligned} u &= \alpha_1 + \alpha_2 x + \alpha_3 y + \alpha_4 xy \\ v &= \alpha_5 + \alpha_6 x + \alpha_7 y + \alpha_8 xy \end{aligned}\right\} \tag{9.38}$$

$$\{f\} = \begin{Bmatrix} u \\ v \end{Bmatrix} = \begin{bmatrix} 1 & x & y & xy & 0 & 0 & 0 & 0 \\ 0 & 0 & 0 & 0 & 1 & x & y & xy \end{bmatrix} \begin{Bmatrix} \alpha_1 \\ \alpha_2 \\ \alpha_3 \\ \alpha_4 \\ \alpha_5 \\ \alpha_6 \\ \alpha_7 \\ \alpha_8 \end{Bmatrix} \tag{9.39}$$

or
$$\{f\} = [C]\,\{a\} \tag{9.40}$$

It may be verified that the displacement function satisfies all the requirements for convergence criteria. The functions being linear satisfy inter-elemental continuity.

9.7.2 Displacement Function in Terms of Nodal Displacements

Nodal displacements can be expressed in terms of its nodal coordinates with the help of eqn. (9.39)

$$\begin{Bmatrix} u_1 \\ v_1 \\ u_2 \\ v_2 \\ u_3 \\ v_3 \\ u_4 \\ v_4 \end{Bmatrix} = \begin{bmatrix} 1 & 0 & 0 & 0 & 0 & 0 & 0 & 0 \\ 0 & 0 & 0 & 0 & 1 & 0 & 0 & 0 \\ 1 & 0 & b & 0 & 0 & 0 & 0 & 0 \\ 0 & 0 & 0 & 0 & 1 & 0 & b & 0 \\ 1 & a & 0 & 0 & 0 & 0 & 0 & 0 \\ 0 & 0 & 0 & 0 & 1 & a & 0 & 0 \\ 1 & a & b & ab & 0 & 0 & 0 & 0 \\ 0 & 0 & 0 & 0 & 1 & a & b & ab \end{bmatrix} \begin{Bmatrix} \alpha_1 \\ \alpha_2 \\ \alpha_3 \\ \alpha_4 \\ \alpha_5 \\ \alpha_6 \\ \alpha_7 \\ \alpha_8 \end{Bmatrix} \tag{9.41}$$

or
$$\{X\}_e = [A]\,\{\alpha\} \tag{9.42}$$

From eqn. (9.42), one can write

$$\{\alpha\} = [A]^{-1}\,\{X\}_e \tag{9.43}$$

The algebraic equations given by eqn. (9.41) are very simple and therefore $[A^{-1}]$ can be easily obtained. It is given below

$$[A]^{-1} = \begin{bmatrix} 1 & 0 & 0 & 0 & 0 & 0 & 0 & 0 \\ -1/a & 0 & 0 & 0 & 1/a & 0 & 0 & 0 \\ -1/b & 0 & 1/b & 0 & 0 & 0 & 0 & 0 \\ 1/ab & 0 & -1/ab & 0 & -1/ab & 0 & 1/ab & 0 \\ 0 & 1 & 0 & 0 & 0 & 0 & 0 & 0 \\ 0 & -1/a & 0 & 0 & 0 & 1/a & 0 & 0 \\ 0 & -1/b & 0 & 1/b & 0 & 0 & 0 & 0 \\ 0 & -1/ab & 0 & -1/ab & 0 & -1/ab & 0 & 1/ab \end{bmatrix} \tag{9.44}$$

Replacing (α) from eqn. (9.43) into eqn. (9.40) yields

$$\{f\} = [C][A]^{-1}\{X\}_e \tag{9.45}$$

or

$$\{f\} = [N]\{X\}_e \tag{9.46}$$

9.7.3 Strain-Nodal Displacement Relationship

Strain components for a plane stress problem are given by

$$\{\varepsilon\} = \begin{Bmatrix} \varepsilon_x \\ \varepsilon_y \\ \varepsilon_{xy} \end{Bmatrix} = \begin{bmatrix} \frac{\partial}{\partial x} & 0 \\ 0 & \frac{\partial}{\partial y} \\ \frac{\partial}{\partial y} & \frac{\partial}{\partial x} \end{bmatrix} \begin{Bmatrix} u \\ v \end{Bmatrix} \tag{9.47}$$

Performing the necessary differentiation on eqn. (9.38), eqn. (9.47) can be written as

$$\begin{Bmatrix} \varepsilon_x \\ \varepsilon_y \\ \gamma_{xy} \end{Bmatrix} = \begin{bmatrix} 0 & 1 & 0 & y & 0 & 0 & 0 & 0 \\ 0 & 0 & 0 & 0 & 0 & 0 & 1 & x \\ 0 & 0 & 1 & x & 0 & 1 & 0 & y \end{bmatrix} \{\alpha\} \tag{9.48}$$

or

$$\{\varepsilon\} = [G]\{\alpha\} \tag{9.49}$$

Combining eqn. (9.48) within eqn. (9.43), yields

$$\{\varepsilon\} = [G][A]^{-1}\{X\}_e$$

or

$$\{\varepsilon\} = [B]\{X\}_e \tag{9.50}$$

The strain variation within the elements is linear as given by eqn. (9.49).

9.7.4 Stress-Strain Relationship

Stresses and strains are related as follows

$$\{\sigma\} = [D]\{\varepsilon\} \tag{9.51}$$

Replacing {ε} from eqn. (9.50) into eqn. (9.51), yields

$$\{\sigma\} = [D][B]\{X\}_e \tag{9.52}$$

9.7.5 Derivation of the Element Stiffness Matrix

It has already been seen that the stiffness matrix of the element is given by

$$[k]_e = \int_v [B]^T[D][B]\, dv \tag{9.53}$$

So for an element of constant thickness, eqn. (9.53) becomes

$$[k]_e = \int_A [B]^T[D][B]\, dx\, dy \tag{9.54}$$

In order to evaluate the integral, $[B]^T$ $[D]$ $[B]$ are first to be calculated and then the elements of the resulting matrix are to be evaluated. $[K]_e$ has been explicity evaluated and is given by eqn. (9.55) for an orthtropic plate element.

9.7.6 Evaluation of Element Stresses

After the unknown displacements are evaluated, the stresses in the element can be determined by eqn. (9.56)

9.8 A NOTE ON THE RECTANGULAR ELEMENT

Strains varying linearly within the element suggest that stresses also vary linearly within the element. It may be remembered that stresses are constant within a triangular element. If the same structure is to be solved by using first triangular elements and then by rectangular elements, it has been found that in order to attain a particular level of accuracy, the plate is to be divided into much lesser number of meshes in the case of rectangular elements.

9.9 A NOTE ON ELEMENT STRESSES

It has been shown that element stresses do not satisfy the equilibrium conditions for the individual element. This suggests that the applied loads do not balance the internal stresses at a nodal point. As a consequence, it has been found that the stresses at a common node obtained from different elements do not attain a unique value.

What should then be the correct value of the stress at the node? In one approach, the stresses are calculated at all the nodes of the element and the average value is assigned at the centroid of the element. In some another approach, the stresses at a particular node are taken as the average values obtained from all the adjacent elements common to that node. For a more detailed discussion about element stresses the reader may look into the chapter on Adaptive Finite Elements analysis.

$$[K]_e = \frac{t}{12}\begin{bmatrix}
\begin{array}{l}4D_x p^{-1}\\+4D_{xy}p\end{array} & & & & & & & p=\dfrac{a}{b} \\
\begin{array}{l}3D_1\\+3D_{xy}\end{array} & \begin{array}{l}4D_y p\\+4D_{xy}p^{-1}\end{array} & & & & & & \\
\begin{array}{l}2D_x p^{-1}\\-4D_{xy}p\end{array} & \begin{array}{l}-3D_1\\+3D_{xy}\end{array} & \begin{array}{l}4D_x p^{-1}\\+4D_{xy}p\end{array} & & & & & \\
\begin{array}{l}-3D_1\\+3D_{xy}\end{array} & \begin{array}{l}-4D_y p\\+2D_{xy}p^{-1}\end{array} & \begin{array}{l}-3D_1\\-3D_{xy}\end{array} & \begin{array}{l}4D_y p\\+4D_{xy}p^{-1}\end{array} & & & & \\
\begin{array}{l}-4D_x p^{-1}\\+2D_{xy}p\end{array} & \begin{array}{l}3D_1\\+3D_{xy}\end{array} & \begin{array}{l}-2D_x p^{-1}\\-2D_{xy}p\end{array} & \begin{array}{l}3D_1\\+3D_{xy}\end{array} & \begin{array}{l}4D_1 p^{-1}\\4D_{xy}p\end{array} & & & \\
\begin{array}{l}3D_1\\-3D_{xy}\end{array} & \begin{array}{l}2D_y p^{-1}\\4D_{xy}p^{-1}\end{array} & \begin{array}{l}3D_1\\+3D_{xy}\end{array} & \begin{array}{l}-2D_y p\\-2D_{xy}p^{-1}\end{array} & \begin{array}{l}-3D_1\\-3D_{xy}\end{array} & \begin{array}{l}4D_x p\\4D_{xy}p^{-1}\end{array} & & \\
\begin{array}{l}-2D_x p^{-1}\\-2D_{xy}p\end{array} & \begin{array}{l}-3D_1\\+3D_{xy}\end{array} & \begin{array}{l}-4D_x p^{-1}\\+2D_{xy}p\end{array} & \begin{array}{l}-3D_1\\-3D_{xy}\end{array} & \begin{array}{l}2D_x p^{-1}\\-4D_{xy}p\end{array} & \begin{array}{l}3D_1\\+3D_{xy}\end{array} & \begin{array}{l}4D_x p^{-1}\\+4D_{xy}p\end{array} & \\
\begin{array}{l}-3D_1\\-3D_{xy}\end{array} & \begin{array}{l}-2D_y p\\-2D_{xy}p^{-1}\end{array} & \begin{array}{l}-3D_1\\+3D_{xy}\end{array} & \begin{array}{l}2D_y p\\-4D_{xy}p^{-1}\end{array} & \begin{array}{l}3D_1\\-3D_{xy}\end{array} & \begin{array}{l}-4D_y p\\+2D_{xy}p^{-1}\end{array} & \begin{array}{l}3D_1\\+3D_{xy}\end{array} & \begin{array}{l}-4D_y p\\+4D_{xy}p^{-1}\end{array}
\end{bmatrix} \tag{9.55}$$

$$\{\sigma\}_e =$$

$$\frac{1}{ab}\begin{bmatrix} -D_1(b-y) & -D_1D_2(a-x) & -D_1y & D_1D_2(a-x) & D_1(b-y) & -D_1D_2x & D_1y & D_1D_2x \\ -D_1D_2(b-y) & -D_1(a-x) & -D_1D_2y & D_1(a-x) & D_1D_2(b-y) & -D_1x & D_1D_2y & D_1x \\ -D_\eta(a-x) & -D_\eta(b-y) & +D_\eta(a-x) & -D_\eta y & -D_\eta x & D_\eta(b-y) & D_\eta x & D_\eta y \end{bmatrix}\{X\}_e \quad (9.56)$$

9.10 COMPUTER PROGRAM FOR THE PLANE STRESS ANALYSIS USING THREE–NODED TRIANGULAR ELEMENT

The computer program in both C-language and FORTRAN for the solution of plane stress problem using constant strain triangular elements has been given in the attached CD. The notations have been explained at the beginning of the computer programs and different steps followed there have been explained.

To use this computer program, every element data is to be prepared, *i.e.*, the coordinates of each node of the element and global numbering of the nodes are input to the computer program.

The computer program for CST elements consists of a main program and four subroutines. The entire input data is to be prepared on the basis of the main program. It will then call the subroutines to perform the necessary functions. In the computer program, the element stiffness matrix is generated in subroutine TELMS. The assembly of element stiffness matrices to form overall stiffness matrix is obtained in subroutine ASSEM. The overall stiffness matrix is stored in half band form. The simultaneous equations are solved by using subroutine GAUSS. The element stresses are generated in subroutine TSTRM.

The preparation of the data is explained with the help of an example problem shown in Fig. 9.6. Due to symmetry, only the bottom half of the plate is analysed. The boundary conditions indicated in the computer program are corresponding to this problem and they have not been presented in a general manner.

The element numbers have been indicated within small circles shown in Fig. 9.6(b). In all there are 9 nodes and each node has two degrees of freedom so that NT = 18. The half bandwidth NBH = 10, NLC = 1. The preparation of data for a typical element say 5 is explained here. The node numbers corresponding to element 5 are 2, 3 and 5. Any one node can be taken as the origin. Suppose node 2 is taken as the origin. The positive direction of the axis system has been shown in Fig. 9.6(b) The data for element coordinates and node numbers corresponding to the READ statement of the computer program will be

0. 0.5 0. 0. 0. 0.25 1.0

2 3 5

The input data corresponding to '*A*' and '*B*' are irrevalent and so are '*NX*' and '*NY*' for a more general problem. The input data of the problem is given in Table 9.1 and the computer output has been presented in Table 9.2.

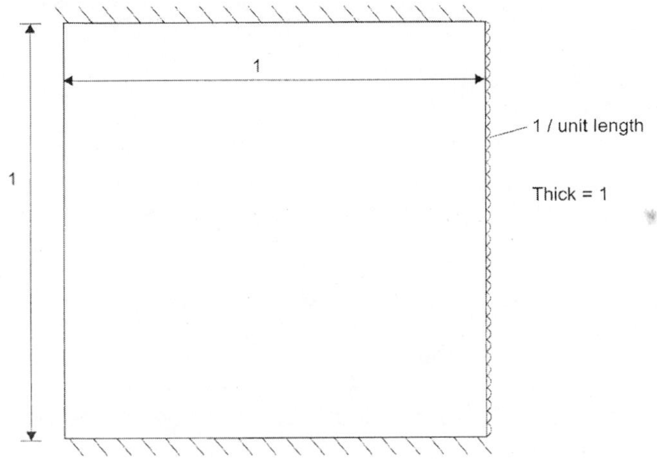

(a) The given plate

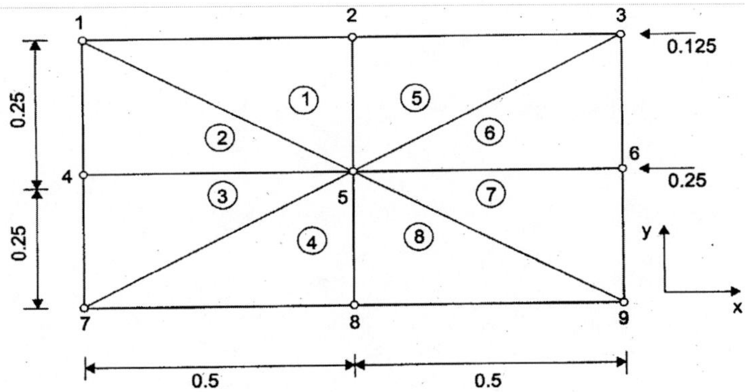

(b) The half plate

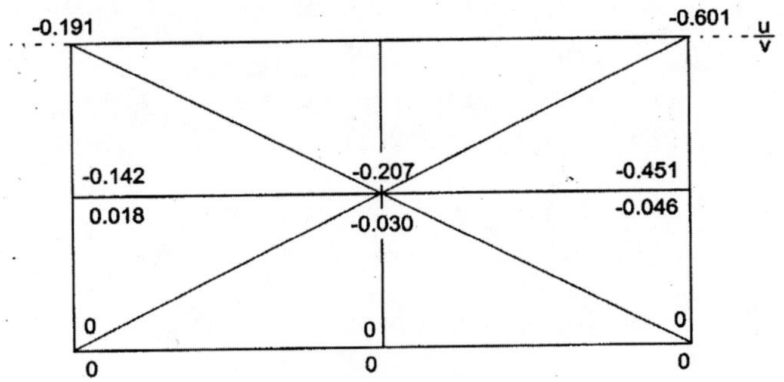

(c) Displacement

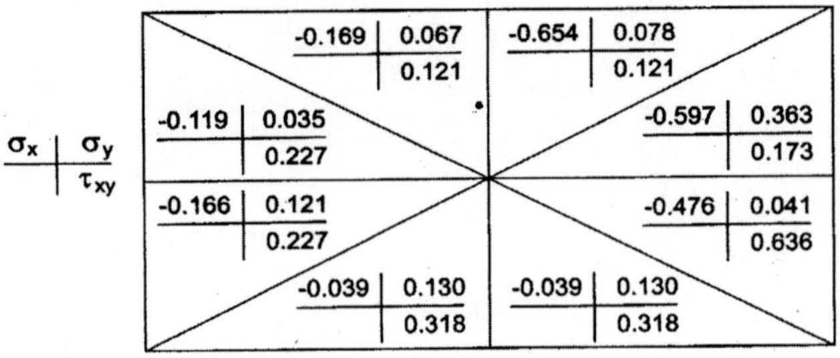

(d) Stresses

Fig. 9.6. The Plate example

Table 9.1. Input Data for the Plane Stress Problem Using Cst Element

```
1.0 0.3 1.0 0.5
3  2  18  10  1  6  8  2  2  2
0.0  0.5  0.5  0.0  0.0  0.25  1.0
1  2  5
0.0  0.5  0.0  0.0 0.25  0.25  1.0
1  5  4
0.0  0.5  0.0  0.0  0.0  0.25  1.0
4  5  7
0.0  0.0  -0.5  0.0  0.25  0.25  1.0
5  8  7
0.0  0.5  0.0  0.0  0.0  0.25  1.0
2  3  5
0.0  0.0  -0.5  0.0  0.25  0.25  1.0
3  6  5
0.0  0.5  0.5  0.0  0.0  0.25  1.0
5  6  9
0.0  0.5  0.0  0.0  0.25  0.25  1.0
 5  9  8
5 1  -0.125
11 1 -0.25
```

Table 9.2. Computer Output for the Plane Stress Problem Using Cst Element

```
PLANE STRESS PROBLEMS FOR TRIANGULAR ELEMENTS
D = 1.000000   PR = 0.300000  A = 1.000000   B = 0.500000
NN = 3  NFREE = 2      NT = 18NBH = 10        NLC = 1 NENP = 6
NELM = 8         NLDT = 2        NX = 2  NY = 2
Element-wise X coordinates, Y coordinates and thickness
0.0000  0.5000  0.5000  0.0000  0.0000  0.2500  1.0000
0.0000  0.5000  0.0000  0.0000  0.2500  0.2500  1.0000
0.0000  0.5000  0.0000  0.0000  0.0000  0.2500  1.0000
0.0000  0.0000 -0.5000  0.0000  0.2500  0.2500  1.0000
0.0000  0.5000  0.0000  0.0000  0.0000  0.2500  1.0000
0.0000  0.0000 -0.5000  0.0000  0.2500  0.2500  1.0000
0.0000  0.5000  0.5000  0.0000  0.0000  0.2500  1.0000
0.0000  0.5000  0.0000  0.0000  0.2500  0.2500  1.0000
Element-wise node numbers
1       1       2       5
2       1       5       4
3       4       5       7
```

```
4      5      8      7
5      2      3      5
6      3      6      5
7      5      6      9
8      5      9      8
The Loading Matrix
 0.00000e+000 0.00000e+000 0.00000e+000 0.00000e+000 -1.25000e-001
 0.00000e+000 0.00000e+000 0.00000e+000 0.00000e+000  0.00000e+000
-2.50000e-001 0.00000e+000 0.00000e+000 0.00000e+000  0.00000e+000
 0.00000e+000  0.00000e+000  0.00000e+000
The Displacement Matrix
-1.91045e-001 2.39539e-016 -2.85745e-001 -2.75714e-017 -6.01211e-001
-1.05971e-015 -1.41929e-001 1.77456e-002 -2.06899e-001  2.95080e-002
-4.51145e-001 -4.59991e-002 -6.17142e-016 6.99983e-016 -1.59153e-015
 6.48527e-016 -1.54133e-015 -5.00765e-016
The Stress Matrix
1 -1.69220e-001  6.72660e-002  1.21300e-001
2 -1.19390e-001  3.51654e-002  8.46100e-002
3 -1.66192e-001 -1.20840e-001  2.27401e-001
4 -3.89116e-002 -1.29705e-001  3.18307e-001
5 -6.54422e-001 -7.82946e-002  1.21300e-001
6 -5.97461e-001 -3.63235e-001  1.72789e-001
7 -4.76145e-001  4.11529e-002  6.35986e-001
8 -3.89116e-002 -1.29705e-001  3.18307e-001
```

REFERENCES AND SUGGESTED READING

9.1. K. H. Murray, Comments on the Convergence of the Finite Element Solutions, J. AIAA,v. 8, 1970, pp. 815–816.

9.2. O. C. Zienkiewicz and R. L. Taylor, The Finite Element Method, v. I and II, McGraw-Hill, 1989.

9.3. S. S. Rao, The Finite Element Method in Engineering, Second Edition, Pergamon Press, 1989.

9.4. R. D. Cook, D. S. Malkus and M. E. Plesha, Concepts and Applications of Finite Element Analysis, John Wiley & Sons, Third Edition, 1989.

EXERCISE 9

9.1. Deduce the stiffness matrix of a rectangle element in terms of normalized coordinates $\xi = x/a$ and $\eta = y/b$.

9.2. For the parallelogram plane stress element, derive the stiffness matrix in terms of ξ and η (Fig. 9.4)

9.3. For the parallelogram plane stress element, derive the stiffness matrix in terms of x and y.

9.4. The stiffness matrix of the parallelogram element is once formed in terms of ξ and η and then in terms of x and y. Do you think that the displacement function in both cases will satisfy the convergence criteria?

9.5. Write the computer program for the analysis of rectangular plates using rectangular plane stress elements.

9.6. What changes are to be incurred in the computer program in order to convert it for analysis of plates using rectangular elements? Make the necessary modifications.

9.7. Modify the computer program so that a plate can be used either triangular element or as a rectangular element or both.

9.8. Derive the stiffness matrix for the triangular element and verify the results from eqn. (9.25)

9.9. If a triangular element is subjected to distributed in plane forces X_1 and Y_1 per unit area in the direction of x and y axis, then by applying the principle of virtual work or by any other method, determine the equivalent nodal forces.

9.10. Construct stress matrix $\{\sigma\}$ for a triangular plane stress element for an isotropic plate.

9.11. Compute the plane strain stiffness matrix for the triangular element. The element has unit thickness. $\nu = 0.3$

❑❑❑

10

ISOPARAMETRIC AND OTHER ELEMENT REPRESENTATIONS AND NUMERICAL INTEGRATIONS

10.1 INTRODUCTION

When a structure subjected to plane stress is analysed by employing three-noded triangular or rectangular elements, two disadvantages may occur. *Firstly*, the structure may have irregular curved boundaries and as such more number of the straight edge elements will be needed to approximate the boundary. *Secondly*, a large number of these simple elements are needed to solve the problem to the desired degree of accuracy so as to require a considerable amount of computer time. Isoparametric element formulation that is introduced here is free from these drawbacks.

Isoparametric elements are capable of representing the boundary curvature. These elements are mapped into rectangular elements having specified number of nodes and then the necessary finite element formulation is made. Thus, isoparametric elements enable effective handling of structures with arbitrary shapes.

10.2 SHAPE FUNCTION OR INTERPOLATION FUNCTION

Let us have a relook into eqns. (9.10) and (9.11) and rewrite them as

$$\begin{Bmatrix} u \\ v \end{Bmatrix} = \begin{bmatrix} N_1 & 0 & N_2 & 0 & N_3 & 0 \\ 0 & N_1 & 0 & N_2 & 0 & N_3 \end{bmatrix} \begin{Bmatrix} u_1 \\ v_1 \\ u_2 \\ v_2 \\ u_3 \\ v_3 \end{Bmatrix} \tag{10.1}$$

or
$$\{f\} = [N]\{X\} \tag{10.2}$$

A careful look into eqn. (10.1) will reveal that $N_1 = 1$ at node 1 and zero at other nodes. Likewise $N_2 = 1$ at node 2 and zero at other nodes and so on. N_1, N_2 *etc.* are called shape function or interpolation function. It is termed as interpolation function because it is on the basis of this function that the field variable (displacement function in the present case) can be determined by interpolation at any point inside the element or at any point on the boundary. It is called the shape function because it is dependent on the shape of the element. There is one shape function corresponding to each degree of freedom of the element. The shape functions may be polynomials of any order, but the convergence criteria of Section 9.3 must be satisfied.

10.3 DETERMINATION OF SHAPE FUNCTIONS

Comparing eqn. (9.9) with eqn. (9.11), it is revealed that

$$[N] = [C]\,[A]^{-1} \tag{10.3}$$

The shape function matrix [*N*] can be derived from the displacement function as the starting point. But certain difficulties may be encountered in the process. It may be observed that for some elements, the inversion of [*A*] is considerably tedious and further in some cases [*A*] may become singular, *e.g.*, for triangular plate bending element where two sides are at right angles. The shape functions used in isoparametric elements remove these disadvantages, as they can be constructed directly, based on the basic property of shape function *i.e.*

$$\left.\begin{aligned} N_i &= 1 \text{ at } (\xi_i,\ \eta_i) \\ N_i &= 0 \text{ at } (\xi_i,\ \eta_j)\ (i \neq j) \end{aligned}\right\} \tag{10.4}$$

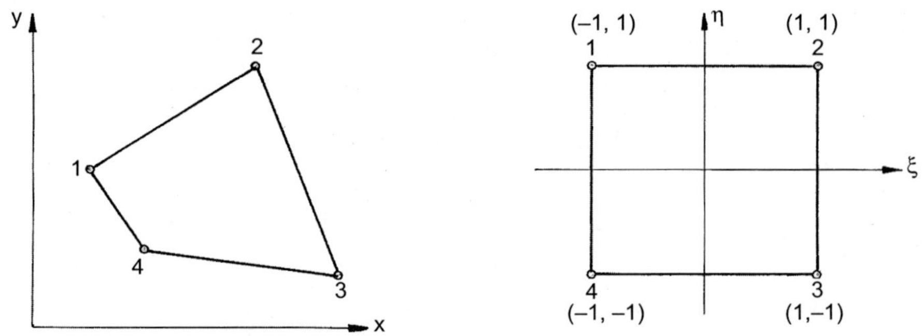

Fig. 10.1. Transformation from physical to isoparametric coordinates for the linear element

It is to be borne in mind that the edges of the element as it actually occurs may be irregular, but they are mapped into some regular shapes which facilities the analysis. In Fig. 10.1 an irregular shaped quadrilateral is mapped into a square. Similarly, in Fig. 10.2 the elements having highly curved edges are mapped into a square. The number of nodes in the isoparametric element is same as the number of nodes of the physical element and there is a node to node correspondence between the two elements. This aspect will be discussed later in this chapter. In the mean time shape functions for the square elements are discussed. The sides of the square elements are based on normalized coordinates such that their faces have values ±1.

10.3.1 Linear 2-D Element

The shape function for linear 2-D element of Fig. 10.1 (b) is given by

$$\left.\begin{aligned} N_1 &= \frac{1}{4}(1-\xi)(1+\eta) \\ N_2 &= \frac{1}{4}(1+\xi)(1+\eta) \\ N_3 &= \frac{1}{4}(1+\xi)(1-\eta) \\ N_4 &= \frac{1}{4}(1-\xi)(1-\eta) \end{aligned}\right\} \qquad (10.5)$$

It is seen that at node 1, $\xi = -1$ and $\eta = 1$ and when these values are substituted in N_1 of eqn. (10.5), N_1 becomes equal to 1. Similarly, if the coordinates of other nodes are substituted into N_1 of eqn. (10.5), N_1 becomes zero. Therefore, the basic requirement of shape function as given by eqn. (10.4) is satisfied. Other shape functions may be similarly verified.

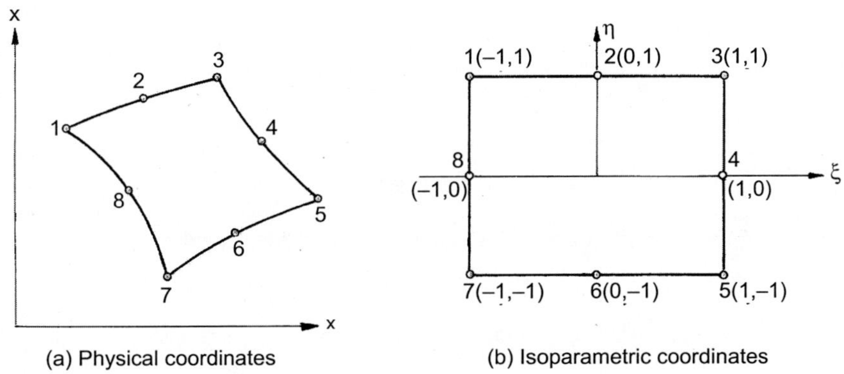

Fig. 10.2. Transformation from physical to isoparametric coordinates for quadratic element

For this four-noded element, the variation of the shape function is linear. Therefore, this is referred to as linear isoparametric element. As the shape function varies linearly on all sides, the continuity is satisfied.

The shape functions given by eqn. (10.5) can be written more compactly as given below:

$$N_r = \frac{1}{4}(1+\xi_0)(1+\eta_0) \qquad (10.6)$$

where $\xi_0 = \xi\,\xi_r$ and $\eta_0 = \eta\,\eta_r$

10.3.2 Quadratic 2-D Element

Referring to Fig. 10.3 (b), the shape function for the corner nodes is given by

$$N_r = \frac{1}{4}(1+\xi_0)(1+\eta_0)(\xi_0+\eta_0-1) \tag{10.7}$$

The shape function for mid-side nodes 2 and 6 (*i.e.*, where $\xi_r = 0$) are

$$N_r = \frac{1}{2}(1-\xi^2)(1+\eta_0) \tag{10.8}$$

and those for nodes 4 and 8 (*i.e.*, $\eta_r = 0$) are

$$N_r = \frac{1}{2}(1+\xi_0)(1-\eta^2) \tag{10.9}$$

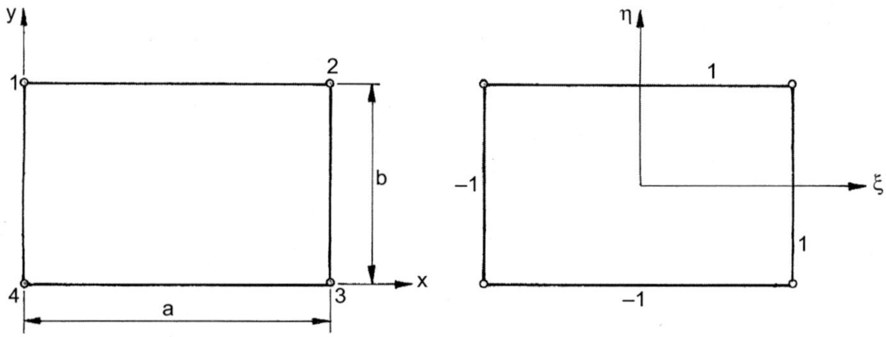

(a) Physical coordinates (b) Isoparametric coordinates

Fig. 10.3. Rectangular element and its isoparametric mapping

Shape functions for cubic and other higher order elements can be similarly written (10.1).

10.4 PLANE STRESS ISOPARAMETRIC LINEAR ELEMENT

A rectangular element with node numbers and actual coordinates has been indicated in Fig. 10.3. The element stiffness matrix of this element considering it as a linear isoparmetric element will be derived. The procedure to be presented is valid for any odd shaped quadrilateral element with nodes at four corners. The same analysis can be extended without any difficulty to quadratic or other higher order plane stress elements.

As indicated in the previous chapter, the displacement is specified in the two orthogonal directions in the plane of the plate. The displacement field is given by

$$\{f\} = \begin{Bmatrix} u \\ v \end{Bmatrix} \tag{10.10}$$

where u and v are the displacements parallel to x and y-directions. The nodal parameters are

$$\{X\}_e^T = \{u_1\ v_1 \,\vdots\, u_2\ v_2 \,\vdots\, u_3\ v_3 \,\vdots\, u_4\ v_4\} \tag{10.11}$$

10.4.1 Displacement Function in terms of Nodal Parameters

The shape function defines the geometry as well as the displacement field for the isoparametric element. Thus, at any point

$$\left.\begin{aligned} x &= N_1x_1 + N_2x_2 + N_3x_3 + N_4x_4 \\ y &= N_1y_1 + N_2y_2 + N_3y_3 + N_4y_4 \end{aligned}\right\} \tag{10.12}$$

and

$$\left.\begin{aligned} u &= N_1u_1 + N_2u_2 + N_3u_3 + N_4u_4 \\ v &= N_1v_1 + N_2v_2 + N_3v_3 + N_4v_4 \end{aligned}\right\} \tag{10.13}$$

For linear element, N_1 to N_4 are given by eqn. (10.5) in terms of ξ and η.

10.4.2 Strain-Nodal Parameter Relationship

The strain matrix associated with the plane stress problem is given by

$$\{\varepsilon\} = \begin{Bmatrix} \varepsilon_x \\ \varepsilon_y \\ \gamma_{xy} \end{Bmatrix} = \begin{Bmatrix} \dfrac{\partial u}{\partial x} \\ \dfrac{\partial v}{\partial y} \\ \dfrac{\partial u}{\partial y} + \dfrac{\partial v}{\partial x} \end{Bmatrix} = \sum_{r=1}^{4} \begin{bmatrix} \dfrac{\partial N_r}{\partial x} & 0 \\ 0 & \dfrac{\partial N_r}{\partial y} \\ \dfrac{\partial N_r}{\partial y} & \dfrac{\partial N_r}{\partial x} \end{bmatrix} \begin{Bmatrix} u_r \\ v_r \end{Bmatrix} \tag{10.14}$$

Thus, according to the previous notation

$$[B] = \sum_{r=1}^{4} [B]_r = \sum_{r=1}^{4} \begin{bmatrix} \dfrac{\partial N_r}{\partial x} & 0 \\ 0 & \dfrac{\partial N_r}{\partial y} \\ \dfrac{\partial N_r}{\partial y} & \dfrac{\partial N_r}{\partial x} \end{bmatrix} \tag{10.15}$$

The shape function derivatives may be calculated using the chain rule

$$\left.\begin{aligned} \frac{\partial N_r}{\partial x} &= \frac{\partial N_r}{\partial \xi}\cdot\frac{\partial \xi}{\partial x} + \frac{\partial N_r}{\partial \eta}\cdot\frac{\partial \eta}{\partial x} \\ \frac{\partial N_r}{\partial y} &= \frac{\partial N_r}{\partial \xi}\cdot\frac{\partial \xi}{\partial y} + \frac{\partial N_r}{\partial \eta}\cdot\frac{\partial \eta}{\partial y} \end{aligned}\right\} \tag{10.16}$$

Equation (10.16) can be written as

$$\begin{Bmatrix} \dfrac{\partial N_r}{\partial x} \\ \dfrac{\partial N_r}{\partial y} \end{Bmatrix} = \begin{bmatrix} \dfrac{\partial \xi}{\partial x} & \dfrac{\partial \eta}{\partial x} \\ \dfrac{\partial \xi}{\partial y} & \dfrac{\partial \eta}{\partial y} \end{bmatrix} \begin{Bmatrix} \dfrac{\partial N_r}{\partial \xi} \\ \dfrac{\partial N_r}{\partial \eta} \end{Bmatrix}$$

or

$$\begin{Bmatrix} \dfrac{\partial N_r}{\partial x} \\ \dfrac{\partial N_r}{\partial y} \end{Bmatrix} = [J]^{-1} \begin{Bmatrix} \dfrac{\partial N_r}{\partial \xi} \\ \dfrac{\partial N_r}{\partial \eta} \end{Bmatrix} \tag{10.17}$$

where $$[J] = \begin{pmatrix} \dfrac{\partial x}{\partial \xi} & \dfrac{\partial y}{\partial \xi} \\ \dfrac{\partial x}{\partial \eta} & \dfrac{\partial y}{\partial \eta} \end{pmatrix} \tag{10.18}$$

[J], the matrix of first derivatives, is referred to as the Jacobian matrix.

10.4.3 Evaluation of [B] Matrix

The nodes indicated as 1, 2, 3 and 4 in Fig. 10.3 have coordinates (0, b), (a, b) (a, 0) and (0, 0) respectively.

Substituting the values of N_1 to N_4 from eqn. (10.5) into eqn. (10.17) and then, differentiating x and y with respect to both ξ and η respectively,

$$\frac{\partial x}{\partial \xi} = \frac{\partial N_1}{\partial \xi}x_1 + \frac{\partial N_2}{\partial \xi}x_2 + \frac{\partial N_3}{\partial \xi}x_3 + \frac{\partial N_4}{\partial \xi}x_4$$

$$= 0 + \frac{a}{4}(1+\eta) + \frac{a}{4}(1-\eta) + 0 = \frac{a}{2}$$

Similarly,

$$\frac{\partial y}{\partial \xi} = -\frac{b}{4}(1+\eta) + \frac{b}{4}(1+\eta) + 0 + 0 = 0$$

$$\frac{\partial x}{\partial \eta} = 0 + \frac{a}{4}(1+\xi) - \frac{a}{4}(1+\xi) + 0 = 0$$

and $$\frac{\partial y}{\partial \eta} = \frac{b}{4}(1-\xi) + \frac{b}{4}(1+\xi) + 0 + 0 = \frac{b}{2}$$

Therefore, substituting the values of $\frac{\partial x}{\partial \xi}, \frac{\partial y}{\partial \xi}$ *etc.* into eqn. (10.18), yields

$$[J] = \begin{bmatrix} a/2 & 0 \\ 0 & b/2 \end{bmatrix} \tag{10.19}$$

Therefore,

$$[J]^{-1} = \begin{bmatrix} \dfrac{\partial \xi}{\partial x} & \dfrac{\partial \eta}{\partial x} \\ \dfrac{\partial \xi}{\partial y} & \dfrac{\partial \eta}{\partial y} \end{bmatrix} = \begin{bmatrix} 2/a & 0 \\ 0 & 2/b \end{bmatrix} \tag{10.20}$$

Thus, $$\frac{\partial \xi}{\partial x} = \frac{2}{a}, \frac{\partial \eta}{\partial y} = \frac{2}{b} \text{ and } \frac{\partial \eta}{\partial x} = \frac{\partial \xi}{\partial y} = 0$$

Now, eqn. (10.16) applied to node 1 yields

$$\frac{\partial N_1}{\partial x} = \frac{\partial N_1}{\partial \xi}\cdot\frac{\partial \xi}{\partial x} + \frac{\partial N_1}{\partial \eta}\cdot\frac{\partial \eta}{\partial x} = -\frac{(1+\eta)}{2a} \tag{10.20a}$$

and $$\frac{\partial N_1}{\partial y} = \frac{\partial N_1}{\partial \xi}.\frac{\partial \xi}{\partial y}+\frac{\partial N_1}{\partial \eta}.\frac{\partial \eta}{\partial y} = -\frac{(1-\xi)}{2b} \tag{10.21b}$$

Similarly, applying eqn. (10.16) to other nodes, yields

$$\begin{aligned}
\frac{\partial N_2}{\partial x} &= \frac{1+\eta}{2a}, \frac{\partial N_2}{\partial y} = \frac{1+\xi}{2b} \\
\frac{\partial N_3}{\partial x} &= \frac{1-\eta}{2a}, \frac{\partial N_3}{\partial y} = -\frac{(1+\xi)}{2b} \\
\frac{\partial N_4}{\partial x} &= -\frac{1-\eta}{2a}, \frac{\partial N_4}{\partial y} = -\frac{(1-\xi)}{2b}
\end{aligned} \tag{10.22}$$

Substituting the different values of the derivatives obtained in terms of ξ and η as given by eqns. (10.21) and (10.22) into eqn. (10.15) yield

$$[B]=\begin{bmatrix} -\frac{(1+\eta)}{2a} & 0 & \frac{(1+\eta)}{2a} & 0 & \frac{(1-\eta)}{2a} & 0 & -\frac{(1-\eta)}{2a} & 0 \\ 0 & \frac{1-\xi}{2b} & 0 & \frac{(1+\xi)}{2b} & 0 & -\frac{(1+\xi)}{2b} & 0 & -\frac{(1-\xi)}{2b} \\ \frac{1-\xi}{2b} & -\frac{(1+\eta)}{2a} & \frac{1+\xi}{2b} & \frac{1+\eta}{2a} & -\frac{(1+\xi)}{2b} & \frac{(1+\eta)}{2a} & -\frac{(1-\xi)}{2b} & -\frac{(1-\eta)}{2a} \end{bmatrix} \tag{10.23}$$

10.4.4 Element Stiffness Matrix

Following steps similar to those presented in the previous chapter, the element stiffness matrix can be shown as

$$[k]_e = t\int\int [B]^T[D][B]\,dx\,dy \tag{10.24}$$

As already seen $[B]$ is expressed in terms of ξ and η. The differential area $dx\,dy$ is to be replaced by

$$dx\,dy = |J|d\xi d\eta \tag{10.25}$$

$|J|$ is the determinant of Jacobian matrix $|J|$. From eqns. (10.24) and (10.25), $[k]_e$ can be written as

$$[k]_e = t\int\int [B]^T[D][B]\,|J|\,d\xi d\eta \tag{10.26}$$

Absolute value of $|J|$ is to be considered.

It is realized that explicit evaluation of $[k]_e$ is tedious even for this simplest case of the linear element. The practice that exists for the determining $[k]_e$ is through the use of numerical integration.

10.4.5 Convergence of Isoparametric Elements

In order that convergence criteria may be satisfied, the conditions of rigid body displacements, constant strain criteria and the inter-elemental compatibility condition must be fulfilled. In case of two nodes along the edge, the shape function is linear, for 3-noded edge it is quadratic, for 4-noded edge cubic. So the displacements are uniquely defined which ensures the continuity along the edge or rule out any discontinuity between adjacent elements common to the edge. The validity of other two criteria is shown below with respect to the plane stress element.

A linear plane stress field that satisfies both the conditions required for rigid body displacements and constant strain is given by

$$\left.\begin{aligned} u &= \alpha_1 + \alpha_2 x + \alpha_3 y \\ v &= \alpha_4 + \alpha_5 x + \alpha_6 y \end{aligned}\right\} \tag{10.27}$$

At node r, eqn. (10.27) assumes

$$\left.\begin{aligned} u_r &= \alpha_1 + \alpha_2 x_r + \alpha_3 y_4 \\ v_r &= \alpha_4 + \alpha_5 x_r + \alpha_6 y_r \end{aligned}\right\} \tag{10.28}$$

In isoparametric formulation

$$u = \sum_{r=1}^{n} N_r u_r \text{ and } v = \sum_{r=1}^{n} N_r v_r \tag{10.29}$$

where n is the number of nodes.

Substituting from eqn. (10.28) into eqn. (10.29) yields

$$\left.\begin{aligned} u &= \alpha_1 \sum_{r=1}^{n} N_r + \alpha_2 \sum_{r=1}^{n} N_r x_r + \alpha_3 \sum_{r=1}^{n} N_r y_r \\ v &= \alpha_4 \sum_{r=1}^{n} N_r + \alpha_5 \sum_{r=1}^{n} N_r x_r + \alpha_6 \sum_{r=1}^{n} N_r y_r \end{aligned}\right\} \tag{10.30}$$

Further noting that $x = \sum_{r=1}^{n} N_r x_r$ and $y = \sum_{r=1}^{n} N_r y_r$ eqn. (10.30) becomes

$$\left.\begin{aligned} u &= \alpha_1 \sum_{r=1}^{n} N_r + \alpha_2 x + \alpha_3 y \\ v &= \alpha_4 \sum_{r=1}^{n} N_r + \alpha_5 x + \alpha_6 y \end{aligned}\right\} \tag{10.31}$$

Displacements of eqn. (10.31) and eqn. (10.27) are same provided

$$\sum_{r=1}^{n} N_r = 1 \tag{10.32}$$

Although it may not look apparent, but eqn. (10.32) is valid for all types of isoparametric elements.

10.4.6 Concept of Isoparametric Element

The element mentioned above has been termed as isoparametric element, because the shape function parameter defining the displacements is same as that defining the geometry.

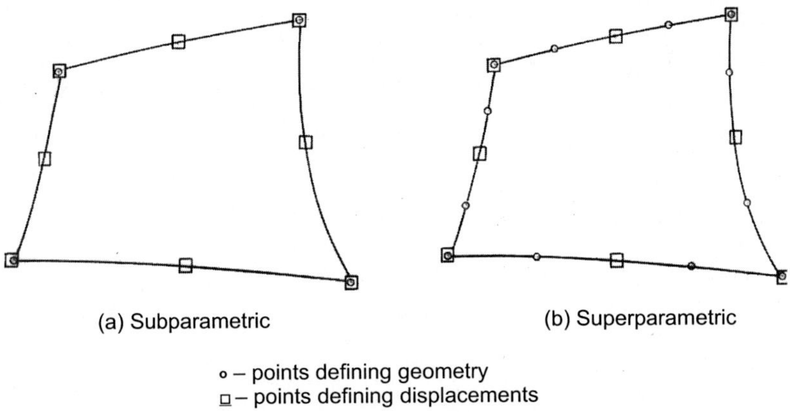

Fig. 10.4. Superparametric and subparametric elements

However, it is not necessary that the geometry and displacements may be defined by the same order model. The element in which geometry is represented by a lower order model than displacements is called subparametric (Fig. 10.4a). The element in which the converse is true is called superparametric element (Fig. 10.4b). For elements with straight edges, it is sufficient if the geometry is defined by 4-corner nodes whereas a higher order displacement field may be needed for the analysis. The choice of the type of element isoparametric, subparametric or superparamatric – will depend on the type of structure and the desired analysis. However, it is more important for the analysis of solids (10.1). For isoparametric, subparametric and some forms of superparametric elements, convergence criteria are satisfied.

It may be noted that the shape function which defines geometry is only reflected in the Jacobian matrix.

10.5 NUMERICAL INTEGRATION

Numerical integration is the method of evaluating the value of a definite integral based on certain numerical scheme. It is referred to as mechanical quadrature when applied to the integration of a single variable, and mechanical cubature when applied to the integration of two variables. However, in finite element literature, this distinction between quadrature and cubature is not made and all numerical integration schemes are referred to as quadrature.

Quadrature formulae may be divided into two main classes:

1. Newton-Cotes formula in which the absicca is usually divided into equal intervals.
2. Gaussian quadrature in which no restriction on equal spacing of absicca is made, although there may be other specific restrictions on the absicca and weight.

In Newton-Cotes formula, $(n + 1)$ equally spaced sampling points are used to integrate exactly a polynomial of maximum order n. On the other hand, in Gauss quadrature, for evaluation of a polynomial of maximum order $(2n - 1)$, n unequally spaced sampling points are needed. For performing the necessary integration to evaluate the element stiffness matrix in the finite element analysis, Gauss quadrature is employed, as it is usually most effective.

10.5.1 Gaussian Quadrature Formula

Gauss sought answer to the following question:

"If $\int_a^b f(x)\,dx$ is to be computed from a given number of values of $f(x)$, what selected values of abscissa in general give the most accurate result?"

He observed that points on the abscissa need not be equally spaced, but they should be symmetrical about the midpoint of the interval of integration. The detailed derivation has been given in the books of numerical analysis and as such it will not be discussed here. Only the procedure of evaluation will be indicated. For a function of one variable, the Gaussian quadrature rule is

$$I = \int_{-1}^{1} \phi(\xi)\,d\xi = \sum_{i=1}^{n} H_i\,\phi(\xi_i) \tag{10.33}$$

where n is the number of sampling points, ξ_i is the position corresponding to ith point, $\phi(\xi_i)$ and H_i are the values of the function and the weighting coefficients at that point (Fig. 10.5). The sampling points and weighting coefficients up to $n = 10$ is given in Table 10.1.

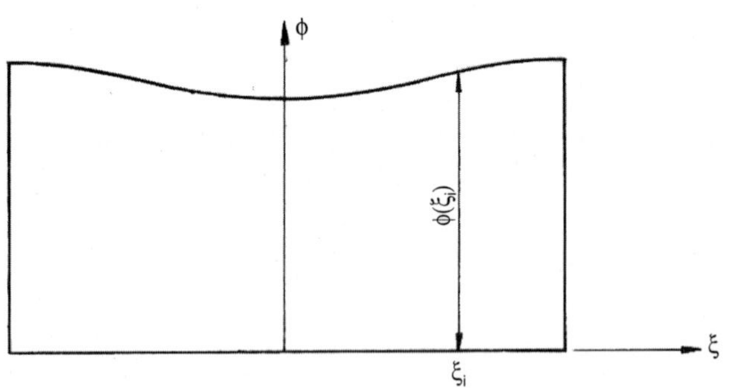

Fig. 10.5. Function ϕ (ξ)

10.5.2 Gaussian Integration of Two Variables

When a function having two variables are to be numerically integrated by Gaussian quadrature formula, it is done thus

$$I = \int_{-1}^{1} \int_{-1}^{1} \phi(\xi, \eta)\, d\xi\, d\eta$$

$$= \sum_{i=1}^{m} \sum_{j=1}^{n} H_i H_j\, \phi(\xi_i, \eta_j) \qquad (10.34)$$

In eqn. (10.34) the number of integrating points in both directions have been assumed to be different.

The above scheme can be extended to 3-dimensional integration.

Example: 10.1:

Evaluate the integral

$\int_{-1}^{1} \int_{-1}^{1} x^2 y^2\, dx\, dy$ using Gauss quadrature formula.

Let $$I = \int_{-1}^{1} \int_{-1}^{1} x^2 y^2\, dx\, dy$$

Both x and y in the above integral have quadratic terms. If $n = 1$, then polynomial of maximum order $(2 \times 1 - 1)$ or 1 can be evaluated exactly. So, $n = 1$ is insufficient in the present case. If $n = 2$, then the polynomial of maximum order $(2 \times 2 - 1)$ or 3 can be evaluated exactly. So, for performing the above integration, a 2-point Gaussian quadrature for both x and y is employed.

$$I = \sum_{i=1}^{m} \sum_{j=1}^{n} H_i H_j \phi(x_i, y_i)$$

Here $$m = n = 2$$

Therefore,

$$I = H_1 H_1 \phi(x_1, y_1) + H_1 H_2 \phi(x_1, y_2) + H_2 H_1 \phi(x_2, y_1)\ H_2 H_2 \phi(x_2, y_2)$$

$$= 1 \times 1 \times \left(-\frac{1}{\sqrt{3}}\right)^2 \times \left(-\frac{1}{\sqrt{3}}\right)^2 + 1 \times 1 \times \left(-\frac{1}{\sqrt{3}}\right)^2 \times \left(\frac{1}{\sqrt{3}}\right)^2$$

$$+ 1 \times 1 \times \left(\frac{1}{\sqrt{3}}\right)^2 \times \left(-\frac{1}{\sqrt{3}}\right)^2 + 1 \times 1 \times \left(\frac{1}{\sqrt{3}}\right)^2 \times \left(\frac{1}{\sqrt{3}}\right)^2$$

$$= \frac{4}{9}$$

Table 10.1. Abscissae and Weight Coefficients of the Gaussian Quadrature Formula

$$\int_{-1}^{1} f(x)\,dx = \sum_{j=1}^{n} H_i f(a_j)$$

$\pm a$				H		
			$n = 1$			
0.00000				2.00000		
			$n = 2$			
0.57735	02691	89626		1.00000	00000	00000
			$n = 3$			
0.77459	66692	41483		0.55655	55555	55556
0.00000	00000	00000		0.88888	88888	88889
			$n = 4$			
0.86113	63115	94053		0.34785	48451	37454
0.33998	10435	84856		0.65214	51214	62546
			$n = 5$			
0.90617	98459	38664		0.23692	68850	56189
0.53846	93101	05683		0.47862	86704	99366
0.00000	00000	00000		0.56888	88888	88889
			$n = 6$			
0.93246	95142	03152		0.17132	44923	79170
0.66120	93864	66265		0.36076	15730	48139
0.23861	91860	83197		0.46791	39345	72691
			$n = 7$			
0.94910	79123	42759		0.12948	49661	68870
0.74143	11855	99394		0.27970	53914	89277
0.40584	51513	77397		0.38183	00505	05119
0.00000	00000	00000		0.41795	91836	73469
			$n = 8$			
0.96028	98564	97536		0.10122	85362	90376
0.79666	64774	13627		0.22238	10344	53374
0.52553	24099	16329		0.31370	66458	77887
0.18343	46424	95650		0.36268	37833	78362
			$n = 9$			
0.96816	02395	07826		0.08127	43883	61574
0.83603	11073	26636		0.18064	81606	94857
0.61337	14327	00590		0.26061	06964	02935
0.32425	34234	03809		0.31234	70770	40003
0.00000	00000	00000		0.33023	93550	01260
			$n = 10$			
0.97390	65285	17172		0.06667	13443	08688
0.86506	33666	88985		0.14935	13491	50581
0.67940	95682	99024		0.21908	63625	15982
0.43339	53941	29247		0.26926	67193	09996
0.14887	43389	81631		0.29552	48247	14753

10.6 LAGRANGIAN INTERPOLATION

It is now realized that it is always preferable to express directly the assumed displacement function in terms of its nodal degrees of freedom. Lagrangian interpolation provides an easy means of generating shape functions.

Consider a line in Fig. 10.6 (c). It is divided into equally spaced points 1, 2, 3, ... *k*, ..., *n*. The coordinates of these points are $\xi_1, \xi_2, \xi_3, \ldots, \xi_k, \ldots, \xi_n$. The Lagrange polynomials satisfy the shape function properties given by eqn. (10.4). For different nodes, they are

$$N_1 = \frac{(\xi-\xi_2)(\xi-\xi_3)\cdots\cdots(\xi-\xi_n)}{(\xi_1-\xi_2)(\xi_1-\xi_3)\cdots\cdots(\xi_1-\xi_n)}$$

$$N_k = \frac{(\xi-\xi_1)(\xi-\xi_2)\cdots\cdots(\xi-\xi_{k-1})(\xi-\xi_{k+1})\cdots\cdots(\xi-\xi_n)}{(\xi_k-\xi_1)(\xi_k-\xi_2)\cdots\cdots(\xi_k-\xi_{k-1})(\xi_k-\xi_{k+1})\cdots\cdots(\xi_k-\xi_n)} \tag{10.35}$$

In two dimensions, the shape function is given by

$$N_{ij} = N_i N_j \tag{10.36}$$

where N_i *and* N_j are one-dimensional values along their respective axes as given by eqn. (10.35).

These shape functions are indeed very handy. But it suffers from two drawbacks. The presence of a number of internal nodes makes the subsequent analysis complicated. Secondly, polynomial terms contain some higher order terms, which remain devoid of lower order terms that affect the convergence of the problem.

10.7 NATURAL COORDINATES AND HIGHER ORDER TRIANGULAR ELEMENTS

A natural coordinate system is nothing but a local coordinate system which is dimensionless and whose maximum value can be unity.

10.7.1 One-Dimensional Element

For the line element of Fig. 10.7, the natural coordinates are L_1 and L_2 such that $L_1 = l_1/l$ and $L_2 = l_2/l$. For this line, following relationship holds good

$$\left.\begin{aligned} L_1 + L_2 &= 1 \\ x_1 L_1 + x_2 L_2 &= x \end{aligned}\right\} \tag{10.37}$$

It may be noticed that at point 1, $L_1 = 1$ and $L_2 = 0$ and at point 2, $L_1 = 0$ and $L_2 = 1$. From eqn. (10.37), the following equations are obtained

$$\begin{Bmatrix} L_1 \\ L_2 \end{Bmatrix} = \frac{1}{l}\begin{bmatrix} x_2 & -1 \\ -x_1 & 1 \end{bmatrix}\begin{Bmatrix} 1 \\ x \end{Bmatrix} \tag{10.38}$$

Differentiation of such polynomials with respect to *x* is given by the formula

$$\frac{d}{dx} = \sum_{i=1}^{2} \frac{\partial L_i}{\partial x} \cdot \frac{\partial}{\partial L_i} = \frac{1}{l}\left(\frac{\partial}{\partial L_2} - \frac{\partial}{\partial L_1}\right) \tag{10.39}$$

Integration of such polynomials with respect to x is given by

$$\int_l L_1^a L_2^b \, dx = \frac{a!b!}{(a+b+1)!} l \tag{10.40}$$

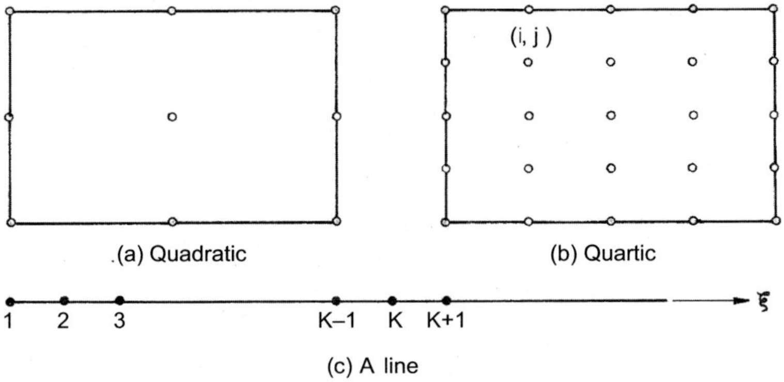

Fig. 10.6. Intervals for Lagrangian interpolation

10.7.2 Higher Order Triangular Elements

The natural coordinate system for triangular elements is known as area coordinates. If the total area of a triangle be A and a point P inside the triangle is such that the total triangle is divided into three areas A_1, A_2 and A_3 as shown in Fig. 10.8, then the area coordinates are given by

$$L_1 = A_1/A, \; L_2 = A_2/A, \; L_3 = A_3/A \tag{10.41}$$

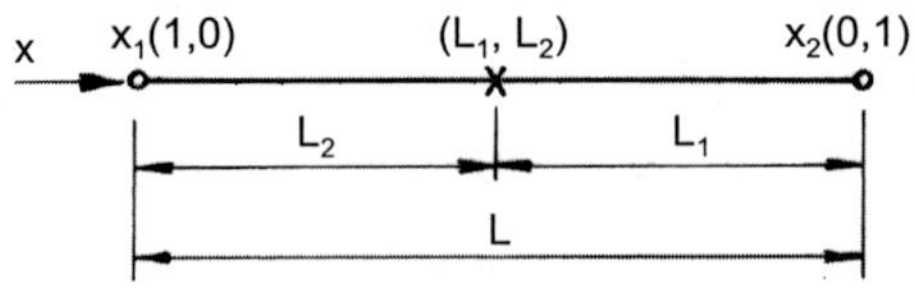

Fig. 10.7. Line element

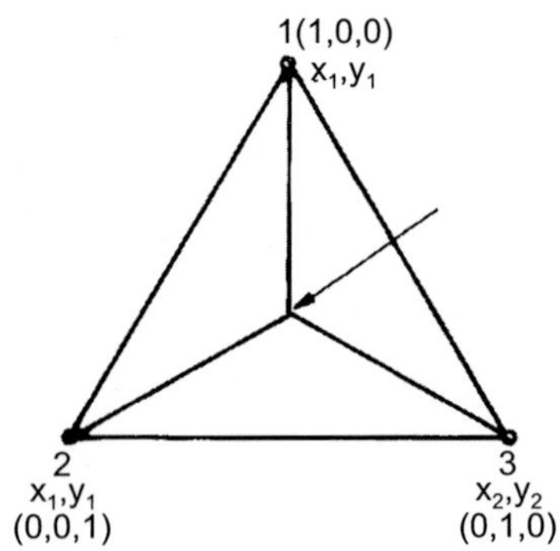

Fig. 10.8. Area coordinates

In Cartesian coordinate system, following relationship exists (Fig. 10.8),

$$\begin{Bmatrix} x \\ y \\ 1 \end{Bmatrix} = \begin{bmatrix} x_1 & x_2 & x_3 \\ y_1 & y_2 & y_3 \\ 1 & 1 & 1 \end{bmatrix} \begin{Bmatrix} L_1 \\ L_2 \\ L_3 \end{Bmatrix} \tag{10.42}$$

The last of the eqn. (10.42) indicates that two of the coordinates L_1, L_2 and L_3 are independent, Inverting eqn. (10.42), yields

$$\begin{Bmatrix} L_1 \\ L_2 \\ L_3 \end{Bmatrix} = \frac{1}{2A}\begin{bmatrix} b_1 & c_1 & a_1 \\ b_2 & c_2 & a_2 \\ b_3 & c_3 & a_3 \end{bmatrix}\begin{Bmatrix} x \\ y \\ 1 \end{Bmatrix} \tag{10.43}$$

where $b_1 = y_2 - y_3$, $c_1 = x_3 - x_2$ and $a_1 = x_2y_3 - x_3y_2$, a_2, b_2, c_2 *etc.* can be defined by chain rule.

The values of L_1, L_2 and L_3 are identical with those of N_1, N_2 and N_3 of eqn. (10.10) for 3-noded triangle. Therefore, for 3-noded triangle, the area coordinates are the shape functions.

Same concept can be extended to quadratic and other higher order triangles. The shape functions for a quadratic triangle of Fig. 10.9 are given as follows:

For corner nodes

$$N_1 = L_1(2L_1 - 1) \text{ and so on} \tag{10.44}$$

For mid-side nodes

$$N_4 = 4L_1L_2 \tag{10.45}$$

The differentiation rule to be applied is

$$\frac{\partial}{\partial x} = \sum_{i=1}^{8} \frac{\partial L_i}{\partial x}\cdot\frac{1}{\partial L_i} = \sum_{i=1}^{8} \frac{b_i}{2A}\cdot\frac{\partial}{\partial L_i} \tag{10.46}$$

$$\frac{\partial}{\partial y} = \sum_{i=1}^{8} \frac{\partial L_i}{\partial y}\cdot\frac{1}{\partial L_i} = \sum_{i=1}^{8} \frac{c_i}{2A}\cdot\frac{\partial}{\partial L_i}$$

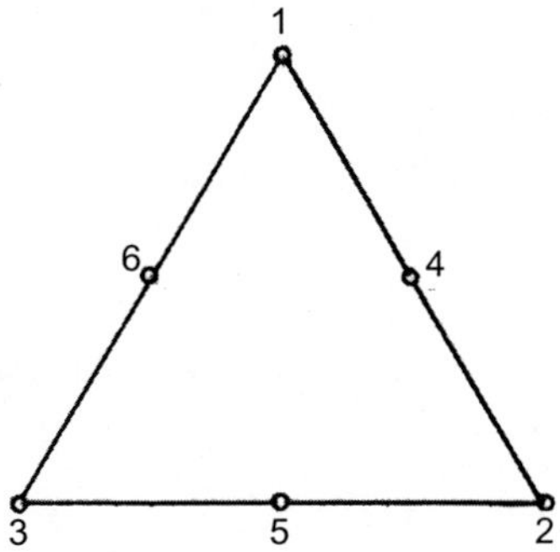

Fig. 10.9. Quadratic triangle

The integration rule to be applied is

$$\int_A L_1^r L_2^s L_3^t dx\,dy = \frac{r!s!t!}{(r+s+t+2)!}A \tag{10.47}$$

10.8 THE QUADRATIC TRIANGLE FOR THE PLANE STRESS PROBLEM

The quadratic triangle of Fig. 10.9 is considered. It has straight edges and mid-side nodes. The displacement vector is given by

$$\{X\}_e^T = \{u_1\ u_2\ u_3\ u_4\ u_5\ u_6\ v_1\ v_2\ v_3\ v_4\ v_5\ v_6\} \tag{10.48}$$

The displacement functions are given by

$$u = \sum_{i=1}^{6} N_i u_i,\ v = \sum_{i=1}^{6} N_i v_i \tag{10.49}$$

where the shape functions for the quadratic triangle (based on eqns. (10.44) and (10.45)) are

$$\left.\begin{matrix} N_1 = L_1(2L_1-1), & N_2 = L_2(2L_2-1), & N_3 = L_3(2L_3-1) \\ N_4 = 4L_1L_2, & N_5 = 4L_2L_3, & N_6 = 4L_3L_1 \end{matrix}\right\} \tag{10.50}$$

Next the strain-nodal displacement relationships are to be obtained.

$$\varepsilon_x = \frac{\partial u}{\partial x} = \sum_{i=1}^{6} \frac{\partial N_i}{\partial x} u_i \tag{10.51}$$

Again,

$$\frac{\partial N_i}{\partial x} = \sum_{j=1}^{3} \frac{\partial N_i}{\partial L_j} \frac{\partial L_j}{\partial x} \tag{10.52}$$

So,

$$\frac{\partial N_1}{\partial x} = \left(\frac{\partial N_1}{\partial L_1} \frac{\partial L_1}{\partial x} + \frac{\partial N_1}{\partial L_2} \frac{\partial L_2}{\partial x} + \frac{\partial N_1}{\partial L_3} \frac{\partial L_3}{\partial x} \right) \tag{10.53}$$

Using eqns. (10.43) and (10.50), eqn. (10.53) becomes

$$\frac{\partial N_1}{\partial x} = (4L_1 - 1) \frac{b_1}{2A} \tag{10.54}$$

Similarly,

$\frac{\partial N_2}{\partial x}, \frac{\partial N_3}{\partial x}$ and so on can be obtained

Equation (10.51) can be written as

$$\varepsilon_x = \frac{1}{2A}[(4L_1 - 1)b_1 \ (4L_2 - 1)b_2 \ (4L_3 - 1)b_3]$$

$$[4(L_2 b_1 + L_1 b_2)\ 4(L_3 b_2 + L_2 b_3)\ 4(L_1 b_3 + L_3 b_1)] \begin{Bmatrix} u_1 \\ u_2 \\ u_3 \\ u_4 \\ u_5 \\ u_6 \end{Bmatrix} \tag{10.55}$$

Similarly, ε_y and γ_{xy} can be expressed in terms of nodal displacements. Combining all three, the strain matrix can now be written as

$$\{\varepsilon\} = \begin{Bmatrix} \varepsilon_x \\ \varepsilon_y \\ \gamma_{xy} \end{Bmatrix} = [B]\{X\}_e \tag{10.56}$$

Equation (10.55) reveals that the strain is linearly varying within the element.

The formulation can be simplified on the following basis. The strain matrix $\{\varepsilon\}$ can interpolated from the matrix $\{\varepsilon_c\}$ indicating the strains at the corner nodes 1, 2 and 3.

$$\{\varepsilon\} = [Q]\{\varepsilon_c\} = [Q]\{\varepsilon_{x1}\ \varepsilon_{x2}\ \varepsilon_{x3}\ \varepsilon_{y1}\ \varepsilon_{y2}\ \varepsilon_{y3}\ \gamma_{xy1}\ \gamma_{xy2}\ \gamma_{xy3}\}^T \tag{10.57}$$

where ε_{x1} is ε_x for node 1 and so on.

$$[Q] = \begin{bmatrix} L_1 & L_2 & L_3 & 0 & 0 & 0 & 0 & 0 & 0 \\ 0 & 0 & 0 & L_1 & L_2 & L_3 & 0 & 0 & 0 \\ 0 & 0 & 0 & 0 & 0 & 0 & L_1 & L_2 & L_3 \end{bmatrix} \tag{10.58}$$

The strain-displacement relationship is now to be written for $[\varepsilon_c]$.

$$[\varepsilon_c] = [H]\{X\}_e \tag{10.59}$$

where $[H]$ is a matrix of order 9×12.

$H_{11} = 3b_1/2A$, $H_{12} = 3b_2/2A$ and so on. Therefore, $[H]$ is independent of x and y.

Combining eqns. (10.57) and (10.59), we get

$$\{\varepsilon\} = [Q]\,[H]\,\{X\}_e \tag{10.60}$$

Comparing eqn. (10.56) with eqn. (10.60), yields

$$[B] = [Q]\,[H] \tag{10.61}$$

The element stiffness matrix is given by

$$[k]_e = \int_V [B]^T[D][B]dV = [H]^T \int_V [Q]^T[D][Q]dV\,[H] \tag{10.62}$$

or,

$$[k]_e = t[H]^T \int_A [Q]^T[D][Q]dA\,[H] \tag{10.63}$$

For efficient evaluation of $[k]_e$ in eqn. (10.63), the integration rule given by eqn. (10.47) may be used. Numerical integration of area coordinates is given below.

10.9 NUMERICAL INTEGRATION OF AREA COORDINATES

If $f(L_1, L_2, L_3)$ is a function in terms of area coordinates L_1, L_2 and L_3, then its integral is of the following form.

$$I = \int_0^1 \int_1^{1-L_1} f(L_1, L_2, L_3)\,dL_2\,dL_1 \tag{10.64}$$

In order to numerically evaluate I, it is written as

$$I = \sum_{i=1}^{n} W_i f_i \tag{10.65}$$

where n is the number of sampling points f_i is the value of $f(L_1, L_2, L_3)$ at a specified point in the triangle, W_i is the weight appropriate to that point. A series of necessary sampling points and weights are given in Table 10.2 [10.8].

10.10 TRIANGULAR ISOPARAMETRIC ELEMENTS FOR THE ANALYSIS OF PLANE STRESS PROBLEMS

We take a quadratic triangle of arbitrary shape, shown in Fig. 10.10.

Like other isoparametric elements, we assume the following:

$$u = \sum N_i u_i,\ v = \sum N_i v_i,\ x = \sum N_i x_i,\ y = \sum N_i y_i \tag{10.66}$$

For quadratic triangle, the shape functions N_i are same as those given in eqn. (12.50). It has been shown in Art. 10.4 that one of the requirements for the computation of the stiffness matrix is the evaluation of the Jacobian matrix. It consists of terms such as

$$\frac{\partial x}{\partial \xi}, \frac{\partial y}{\partial \xi}, \frac{\partial x}{\partial \eta} \text{ and } \frac{\partial y}{\partial \eta}$$

$$\frac{\partial x}{\partial \xi} = \frac{\partial N_1}{\partial \xi} x_1 + \frac{\partial N_2}{\partial \xi} x_2 + \frac{\partial N_3}{\partial \xi} x_3 + \ldots \tag{10.67}$$

Therefore, $\frac{\partial N_i}{\partial \xi}$ is to be evaluated. In order to do that it may be noted that area coordinates are not independent. There are two independent quantities, as the constraint relation $L_1 + L_2 + L_3 = 1$ has to be satisfied.

The followings are assumed

$$\left.\begin{aligned} L_1 &= \xi \\ L_2 &= \eta \\ L_3 &= 1 - \xi - \eta \end{aligned}\right\} \tag{10.68}$$

Now,

$$\frac{\partial N_i}{\partial \xi} = \frac{\partial N_i}{\partial L_1} \cdot \frac{\partial L_1}{\partial \xi} + \frac{\partial N_i}{\partial L_2} \cdot \frac{\partial L_2}{\partial \xi} + \frac{\partial N_i}{\partial L_3} \cdot \frac{\partial L_3}{\partial \xi} = \frac{\partial N_i}{\partial L_1} - \frac{\partial N_i}{\partial L_3} \tag{10.69}$$

Similarly,

$$\frac{\partial N_i}{\partial \eta} = \frac{\partial N_i}{\partial L_2} - \frac{\partial N_i}{\partial L_3} \tag{10.70}$$

Rest of the procedure is similar to that given in Art. 10.4. The stiffness matrix of the triangular element is

$$[k]_e = \int_A [B]^T [D][B] t dA \tag{10.71}$$

where t is the element thickness.

The matrix $[B]$ consists of the terms expressed in area coordinates. Equation (10.71) is to be evaluated accordingly.

Table 10.2. Numerical Integration Formulae for Triangles

Order	Figure	Point	Triangular Coordinates	Weights
Linear	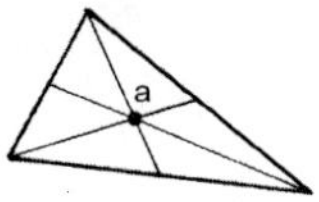	a	1/3, 1/3, 1/3	1
Quadratic	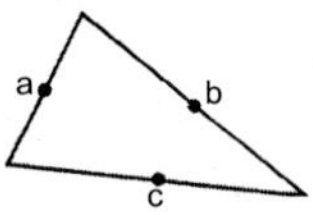	a	1/2, 1/2, 0	1/3
		b	0, 1/2, 1/2	1/3
		c	1/2, 0, 1/2	1/3
Cubic	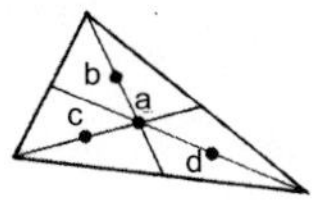	a	1/3, 1/3, 1/3	– 27/48
		b	0.6, 0.2, 0.2	25/48
		c	0.2 0.6, 0.2	
		d	0.2, 0.2, 0.6	
Quintic	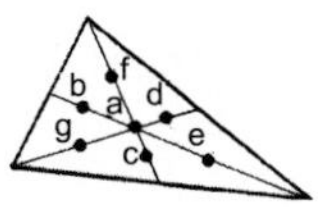	a	1/3, 1/3, 1/3	0.22500, 00000
		b	$\alpha_1, \beta_1, \beta_1$	0.13239, 41527
		c	$\beta_1, \alpha_1, \beta_1$	
		d	$\beta_1, \beta_1, \alpha_1$	
		e	$\alpha_2, \beta_2, \beta_2$	0.12593, 91805
		f	$\beta_2, \alpha_2, \beta_2$	
		g	$\beta_2, \beta_2, \alpha_2$	

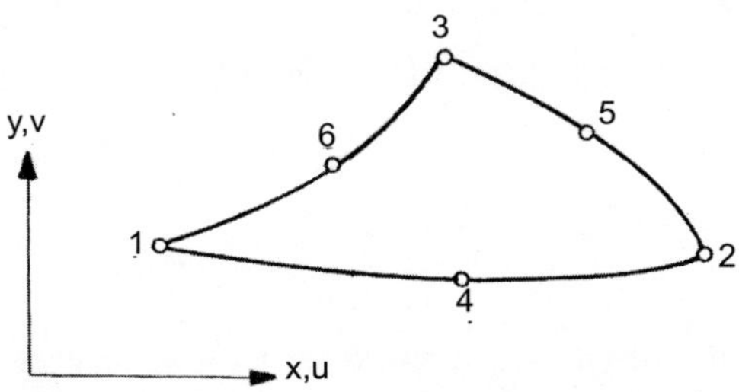

Fig. 10.10. Quadratic triangle of arbitrary shapes

10.11 ALLMAN'S TRIANGULAR PLANE STRESS ELEMENT

Allman's triangular finite element for the plane elasticity analysis has three degrees of freedom per node. With three corner nodes, total d.o.f per element is nine. Three d.o.f per node include 2 translations and a drilling d.o.f.

In the derivation of the stiffness matrix, it is assumed that the displaced shape of the element side is parabolic due to unequal drilling degrees of freedom at connecting nodes as shown in Fig. 10.11. The edge displacement of an element side consists of two parts: (*a*) a straight shape associated with the nodal d.o.f and (*b*) a parabolic shape associated with drilling d.o.f as shown in Fig. 10.11.

The displacement at any point on the side of the element is first to be found out. Let the edge normal displacement due to the unequal drilling d.o.f at the connecting nodes be δ (Fig. 10.1 (b)), then

$$\delta = a_0 + a_1 s + a_2 s^2 \tag{10.72}$$

Boundary conditions are

$$\left.\begin{array}{lll} \delta = 0 & \text{when} & s = 0 \\ \delta = 0 & \text{when} & s = L \\ \left.\dfrac{\partial \delta}{\partial s}\right|_{s=L} & -\left.\dfrac{\partial \delta}{\partial s}\right|_{s=0} & = -\, w_2 + w_1 \end{array}\right\} \tag{10.73}$$

Substituting the boundary conditions of eqn. (10.73) into eqn. (10.72) and then solving the resulting equation yields the values of the constants

$$a_0 = 1$$

$$a_1 = -\frac{w_1 - w_2}{2}, \quad a_2 = -\frac{w_1 - w_2}{2L} \tag{10.74}$$

Therefore, eqn. (10.72) takes the following form when the constants are substituted from eqn. (10.74),

$$\delta = \frac{(w_2 - w_1)\, s\,(L - s)}{2L} \tag{10.75}$$

Since the angle between normal to the side and x-axis is α, the components in the direction u and v are δ cos α and δ sin α respectively.

The displacement at any point on the side of the element connecting nodes i and j are

$$\begin{Bmatrix} u \\ v \end{Bmatrix} = \frac{L-s}{L}\begin{Bmatrix} u_i \\ v_i \end{Bmatrix} + \frac{s}{L}\begin{Bmatrix} u_j \\ v_j \end{Bmatrix} + \frac{s\,(L-s)}{2L}\left(w_j - w_i\right)\begin{Bmatrix} \cos\alpha \\ \sin\alpha \end{Bmatrix} \tag{10.76}$$

The displacements at mid-sides 4, 5 and 6 are as follows (Fig. 10.11 (*c*))

$$\begin{Bmatrix} u_4 \\ v_4 \end{Bmatrix} = \frac{1}{2}\begin{Bmatrix} u_1 \\ v_1 \end{Bmatrix} + \frac{1}{2}\begin{Bmatrix} u_2 \\ v_2 \end{Bmatrix} + \frac{w_2 - w_1}{8}\begin{Bmatrix} y_2 - y_1 \\ x_1 - x_2 \end{Bmatrix} \tag{10.77}$$

$$\begin{Bmatrix} u_5 \\ v_5 \end{Bmatrix} = \frac{1}{2}\begin{Bmatrix} u_2 \\ v_2 \end{Bmatrix} + \frac{1}{2}\begin{Bmatrix} u_3 \\ v_3 \end{Bmatrix} + \frac{w_3 - w_2}{8}\begin{Bmatrix} y_3 - y_2 \\ x_2 - x_3 \end{Bmatrix} \tag{10.78}$$

$$\begin{Bmatrix} u_6 \\ v_6 \end{Bmatrix} = \frac{1}{2}\begin{Bmatrix} u_1 \\ v_1 \end{Bmatrix} + \frac{1}{2}\begin{Bmatrix} u_3 \\ v_3 \end{Bmatrix} + \frac{w_1 - w_3}{8}\begin{Bmatrix} y_1 - y_3 \\ x_3 - x_1 \end{Bmatrix} \tag{10.79}$$

A relationship is now to be found out between 12 translational d.o.f of the original element having mid-side nodes to 9 degrees of freedom at the corner nodes.

They are

$$u_1 = u_1,\quad v_1 = v_1,\quad u_2 = u_2,\quad v_2 = v_2,\quad u_3 = u_3,\quad v_3 = v_3,$$

$$u_4 = \frac{u_1}{2}+\frac{u_2}{2}+\frac{y_2-y_1}{8}(w_2-w_1),\qquad v_4 = \frac{v_1}{2}+\frac{v_2}{2}+\frac{x_1-x_2}{8}(w_2-w_1)$$

$$u_5 = \frac{u_2}{2}+\frac{u_3}{2}+\frac{y_3-y_2}{8}(w_3-w_2),\qquad v_5 = \frac{v_2}{2}+\frac{v_3}{2}+\frac{x_2-x_3}{8}(w_3-w_2)$$

$$u_6 = \frac{u_1}{2}+\frac{u_3}{2}+\frac{y_1-y_3}{8}(w_1-w_3),\qquad v_6 = \frac{v_2}{2}+\frac{v_3}{2}+\frac{x_3-x_1}{8}(w_1-w_3) \tag{10.80}$$

The above equations in matrix form can be written as

$$\{u_1\ v_1\ u_2 \ldots v_6\}^T = [T]\{u_1\ v_1\ w_1 \ldots v_3\ w_3\} \tag{10.81}$$

$[T]$ is of size 12×9.

For the linear strain triangle (Fig. 10.11) (c)),

$$\begin{Bmatrix} u \\ v \end{Bmatrix} = \begin{bmatrix} N_1 & 0 & N_2 & 0 & N_3 & 0 & N_4 & 0 & N_5 & 0 & N_6 & 0 \\ 0 & N_1 & 0 & N_2 & 0 & N_3 & 0 & N_4 & 0 & N_5 & 0 & N_6 \end{bmatrix} \begin{Bmatrix} u_1 \\ v_1 \\ u_2 \\ \cdot \\ \cdot \\ \cdot \\ \cdot \\ v_6 \end{Bmatrix}$$

$$= [N][T]\begin{Bmatrix} u_1 \\ v_1 \\ w_1 \\ \cdot \\ \cdot \\ \cdot \\ v_3 \\ w_3 \end{Bmatrix} \tag{10.82}$$

Following appropriate steps, it can be shown that the stiffness matrix of the Allman's triangle is

$$[k]_e = \iint [T]^T [B]^T [D][B][T]\,dx\,dy \tag{10.83}$$

where

$$[B] = \begin{bmatrix} \frac{\partial N_1}{\partial x} & 0 & \frac{\partial N_2}{\partial x} & \cdots & \cdots & \cdots & \frac{\partial N_6}{\partial x} & 0 \\ 0 & \frac{\partial N_1}{\partial y} & 0 & \cdots & \cdots & \cdots & 0 & \frac{\partial N_6}{\partial y} \\ \frac{\partial N_1}{\partial y} & \frac{\partial N_1}{\partial x} & \frac{\partial N_2}{\partial y} & \cdots & \cdots & \cdots & \frac{\partial N_6}{\partial y} & \frac{\partial N_6}{\partial x} \end{bmatrix}_{3 \times 12} \tag{10.84}$$

and

$$[D] = \frac{Et}{1-\nu^2}\begin{bmatrix} 1 & \nu & 0 \\ \nu & 1 & 0 \\ 0 & 0 & \dfrac{1-\nu}{2} \end{bmatrix}$$

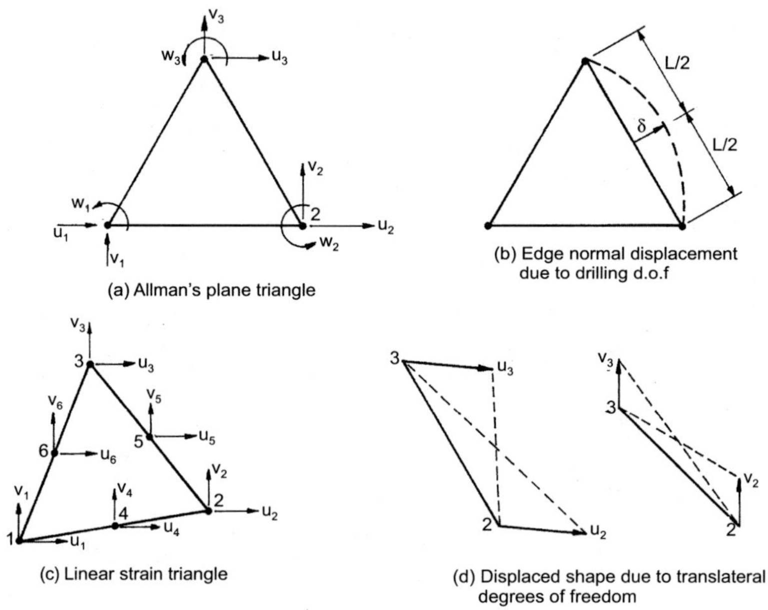

Fig. 10.11. Allman's triangle

10.12 COMPUTER PROGRAM FOR THE SOLUTION OF PLANE STRESS PROBLEM USING ISOPARAMETRIC ELEMENT

The computer program in FORTRAN for the solution of plane stress problems using isoparametric linear element has been given in Table 10.3 and is also given in the attached CD. The notations used have been explained at the beginning of the computer program and different steps have been indicated with 'comment' statements. The computer program as given solves only skew or rectangular orthotropic plates. However, with slight modification, the computer program can be used for the solution of plates of any general shape and material properties. The computer program in C for the same analysis has also been given in the CD.

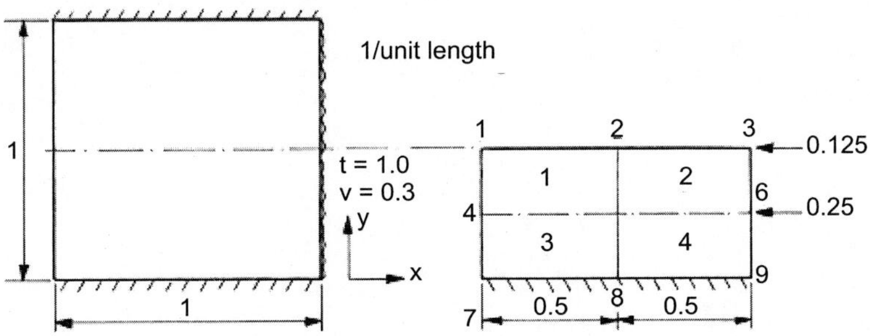

Fig. 10.12. A rectangular plate

Numerical integration has been performed by Gauss method. Subroutine GACUV supplied the necessary GAUSS-point values and their weighting function for two-point and three-point integration. GACUV, as given, is limited to 3 Gauss-point values.

The preparation of the input data of the computer program is pretty simple. The input data for the plate of Fig. 10.12 is shown in Table 10.4. The plate is analysed for bottom half and boundary conditions have been incorporated in the main computer program. The element and node numbers have been shown in Fig. 10.12 in which the plate is divided into 2 × 2 mesh. The computer output has been given in Table 10.5.

Table 10.3. Finite Element Analysis of Plane Stress Problems of Plates Using Isoparametric Linear Element

```
C       *************************************************************
**
C       *************************************************************
**
C       FINITE ELEMENT ANALYSIS OF PLATES USING ISOPARAMETRIC LINEAR
C                    PLANE STRESS ELEMENT
C                    ISOTROPIC PLATE CASE
C       *************************************************************
***
        DIMENSION STIFO(120,45),X(120,2),RGD(3,3),XG(4),YG(4),N
OD(100,4),
      *  ZX(4),ZY(4),SFD(4),SFDE(4),BB(3,8),ASAT(8,8),STIFF(8,8),
      *  PLOAD(120,2),DISP(8,2),STRESS(12,2),SFDZ(4)
        COMMON STIFO,PLOAD
        OPEN(7,FILE='isoin')
        OPEN(8,FILE='isout')
C       *************************************************************
*****
C       *************************************************************
*****
C       A       LENGTH ALONG X-AXIS
C       B       LENGTH ALONG Y-AXIS
C       THICK    THICKNESS OF THE PLATE
C       D       FLEXURAL RIGIDITY OF THE PLATE
```

```
C       PR      POISSON'S RATIO
C       EXL     SIZE OF THE ELEMENT IN THE X-DIRECTION
C       EYL     SIZE OF THE ELEMENT IN THE Y-DIRECTION
C       PHAI    ANGLE BETWEEN THE SIDES IN DEGREES (=90 FOR RECTANGULAR
)
C       NX,NY   MESH DIVISIONS IN THE X- AND Y-DIRECTIONS
C       NQ      NO. OF GAUSS POINTS FOR THE EVALUATION OF STIFF
C       NN      NO. OF NODES IN AN ELEMENT
C       NFREE   DEGREES OF FREEDOM OF A NODE
C       NT      TOTAL DEGREES OF FREEDOM OF THE ENTIRE PLATE
C       NBH     HALFBAND WIDTH OF THE STIFFNESS MATRIX
C       NQL     NO. OF GAUSS POINTS IN THE EVALUATION OF THE LOADING MA-
TRIX
C       NLC     NO. OF LOADING CONDITIONS
C       NELM    TOTAL NO. OF ELEMENTS
C       NENP    TOTAL DEGREES OF FREEDOM OF EACH ELEMENT
C       DXZ     PARTIAL DERIVATIVE OF X WITH RESPECT TO ZI
C       DXE     PARTIAL DERIVATIVE OF X WITH RESPECT TO ETA
C       DYZ     PARTIAL DERIVATIVE OF Y WITH RESPECT TO ZI
C       DYE     PARTIAL DERIVATIVE OF Y WITH RESPECT TO ETA
C       ZAC     VALUE OF THE JACOBIAN
C       DNX     DERIVATIVE OF THE SHAPE FUNCTION W R TO X
C       DNY     DERIVATIVE OF THE SHAPE FUNCTION W R TO Y
C       SFD     SHAPE FUNCTION
C       NOD     GLOBAL NO. OF NODES OF AN ELEMENT
C       ASAT    PART OF THE ELEMENT STIFFNESS MATRIX
C       RGD     RIGIDITY MATRIX
C       X       DISPLACEMENT MATRIX
C       PLOAD   OVERALL LOADING MATRIX
C       STRESS  STRESS VECTOR
C       STIFF   ELEMENT STIFFNESS MATRIX
C       STIFO   OVERAL STIFFNESS MATRIX
C       XG      X-COORDINATE OF THE ELEMENT
C       YG      Y-COORDINATE OF THE ELEMENT
C       ZX      NORMALISED X-COORDINATE
C       ZY      NORMALISED Y-COORDINATE
C       SFDE    PARTIAL DERIVATIVE OF SFD W R TO ETA
C       SFDZ    PARTIAL DERIVATIVE OF SFD W R TO ZI
C       ****************************************************************
****
C       ****************************************************************
****
C       ****************************************************************
*****

  100    FORMAT(2X,6G13.6)
  101    FORMAT(16I5)
C*****************************************************************
*******
C       INPUT OF MATERIAL AND GEOMETRIC PROPERTIES OF THE PLATE
C       ****************************************************************
*******
C       ****************************************************************
*******
  103   READ(7,*) PHAI,A,B
        READ(7,*) NN,NFREE,NQ,NLC,NX,NY
        READ(7,*) D,PR,THICK
```

```
C       ***************************************************************
***
C       OUTPUT OF THE MATERIAL PROPERTIES OF THE PLATE
C       ***************************************************************
****
C       ***************************************************************
****
        WRITE(8,*)'      ANALYSIS OF PLANE STRESS BY ISOPARAMETRIC
PLANE
     *  STRESS ELEMENT'
        EXL=A/NX
        EYL=B/NY
        NELM=NX*NY
        NX1=NX+1
        NY1=NY+1
        NT=(NX1*NY1)*NFREE
        NBH=(NX1+2)*NFREE
        NENP=NN*NFREE
        WRITE(8,*)'  EXL      EYL      PHAI      PR      D
THICK'
        WRITE(8,100)EXL,EYL,PHAI,PR,D,THICK
        WRITE(8,*)' NN NFREE   NT NELM NENP   NQ  NBH  NLC   NX   NY'
        WRITE(8,101)NN,NFREE,NT,NELM,NENP,NQ,NBH,NLC,NX,NY
        PI=3.1415927
        PHAI=PI*PHAI/180.0
C       ***************************************************************
***
C       ***************************************************************
***
C       INPUT AND OUTPUT OF PLATE PROPERTIES ARE OVER
C       ***************************************************************
***
C       NODE NUMBERS FOR DIFFERENT ELEMENTS ARE GENERATED
C       ***************************************************************
****
C       ***************************************************************
****
        NX1=NX+1
        DO 10 II=1,NELM
        NXI= (II-1)/NX
        DO 11 I=1,2
   11   NOD(II,I)=NXI*NX1+(II-NX*NXI)+I-1
        DO 14 I=3,4
   14   NOD(II,I)= NX1*(NXI+1)+(II-NX*NXI)+4-I
   10   CONTINUE
C       ***************************************************************
***
C       ***************************************************************
***
C       RIGIDITY MATRIX GENERATION STARTS
C       ***************************************************************
***
C       ***************************************************************
****
        DO 23 I=1,3
        DO 23 J=1,3
   23   RGD(I,J)=0.0
```

```
      PRR=PR*PR
      RGD(1,1)= D/(1.0-PR*PR)
      RGD(1,2)= PR*RGD(1,1)
      RGD(2,1)= RGD(1,2)
      RGD(2,2)= RGD(1,1)
      RGD(3,3)= (1.0-PR)*RGD(1,1)/2.0
      WRITE(8,*)'      RGD MATRIX'
      WRITE(8,100)((RGD(I,J),I=1,3),J=1,3)
C     ***************************************************************
****
C     ***************************************************************
*****
C     RIGIDITY MATRIX GENERATION OVER
C     ***************************************************************
*****
      DO 12 I=1,NENP
      DO 12 J=1,NENP
   12 STIFF(I,J)=0.0
C     ***************************************************************
*****
C     ***************************************************************
*****
C     ELEMENT COORDINATE INFORMATION IS PRESCRIBED
C     ***************************************************************
*****
C     ***************************************************************
****
      XG(1)= EYL*COS(PHAI)
      YG(1) = EYL*SIN(PHAI)
      XG(2)= EXL+XG(1)
      YG(2)= YG(1)
      XG(3)= EXL
      YG(3)= 0.0
      XG(4)= 0.0
      YG(4)= 0.0
      ZX(1)= -1.0
      ZY(1)= 1.0
      ZX(2)= 1.0
      ZY(2)= 1.0
      ZX(3)= 1.0
      ZY(3)= -1.0
      ZX(4)= -1.0
      ZY(4)= -1.0
C     ***************************************************************
******
C     ***************************************************************
*******
C     ELEMENT STIFFNESS MATRIX GENERATION STARTS
C     ***************************************************************
*****
C     ***************************************************************
*****
      DO 16 I=1,NQ
      DO 16 J=1,NQ
  790 CALL GACUV(I,J,ZI,ETA,HI,HJ,NQ)
  791 DO 19 L=1,4
      ZIZ= ZI*ZX(L)
```

```
      ETAZ= ETA*ZY(L)
      SFD(L)= 0.25*(1.0+ETAZ)*(1.0+ZIZ)
      SFDZ(L)= 0.25*(1.0+ETAZ)*ZX(L)
  19  SFDE(L)= 0.25*(1.0+ZIZ)*ZY(L)
      DO 20 II=1,3
      DO 20 JJ=1,NENP
  20  BB(II,JJ)= 0.0
      DXZ=0.0
      DYZ=0.0
      DXE=0.0
      DYE=0.0
      DO 21 KK=1,4
      DXZ= DXZ+SFDZ(KK)*XG(KK)
      DYZ= DYZ+SFDZ(KK)*YG(KK)
      DXE= DXE+SFDE(KK)*XG(KK)
      DYE= DYE+SFDE(KK)*YG(KK)
  21  CONTINUE
      ZAC= DXZ*DYE - DYZ*DXE
      DXZI= DYE/ZAC
      DYZI= -DYZ/ZAC
      DXEI= -DXE/ZAC
      DYEI= DXZ/ZAC
      DO 22 II=1,NN
      K=2*(II-1)
      DNX= SFDZ(II)*DXZI + SFDE(II)*DYZI
      DNY= SFDZ(II)*DXEI + SFDE(II)*DYEI
      BB(1,K+1)= DNX
      BB(2,K+2)= DNY
      BB(3,K+1)= DNY
      BB(3,K+2)= DNX
  22  CONTINUE
C     ***************************************************************
*****
C     STIFF MATRIX IS SYMMETRIC, ONLY UPPER HALF OF THE TRIANGLE IS
STORED
C     ***************************************************************
******
      DO 24 M=1,NENP
      DO 24 N=1,NENP
      IF(N-M) 24,611,611
 611  ASAT(M,N)= 0.0
      DO 24 L=1,3
      DO 24 KI=1,3
      ASAT(M,N)= ASAT(M,N)+ BB(L,M)*RGD(L,KI)*BB(KI,N)*ZAC*THICK
  24  CONTINUE
      DO 25 M=1,NENP
      DO 25 N=1,NENP
      IF (N-M) 25,615, 615
 615  STIFF(M,N)= STIFF(M,N)+ASAT(M,N)*HI*HJ
  25  CONTINUE
  16  CONTINUE
      DO 620 I=1,NENP
      DO 620 J=1,NENP
 620  STIFF(J,I)=STIFF(I,J)
      WRITE(8,*)'       THE STIFF MATRIX'
      WRITE(8,100)((STIFF(I,J),J=1,NENP),I=1,NENP)
C     ***************************************************************
********
```

```
C       ***************************************************************
********
C       ELEMENT STIFFNESS MATRIX GENERATION OVER
C       ***************************************************************
*******
C       ***************************************************************
********
C       ASSEMBLY OF OVERALL STIFFNESS MATRIX IN HALFBAND
C       ***************************************************************
*******
        DO 31 I=1,NT
        DO 31 J=1,NBH
   31    STIFO(I,J)= 0.0
        DO 26 II=1,NELM
        CALL ASSEM(NOD,NN,NFREE,II,STIFF)
   26    CONTINUE
C       ***************************************************************
*****
C       ***************************************************************
*****
C       PLOAD MATRIX IS INITIALIZED IN DO LOOP 32
C       ***************************************************************
******
C       ***************************************************************
*******
        DO 32 I=1,NT
        DO  32 J=1,NLC
   32    PLOAD(I,J)= 0.0
C       ***************************************************************
***
C       PLOAD MATRIX DATA ARE GIVEN AS THE INPUT
C       ***************************************************************
***
  312    READ(7,*)INP,JNP,VALUE
        IF(INP) 314,314,310
  310    PLOAD(INP,JNP)= PLOAD(INP,JNP)+VALUE
        GO TO 312
C       ***************************************************************
***
C       ***************************************************************
***
C       FITTING OF BOUNDARY CONDITIONS START
C       ***************************************************************
***
  314    NNO=NT/NFREE
        NX1= NX+1
        NBX= NNO-NX
        DO 37 MN= NBX,NNO
         NEWN= MN*NFREE-1
        NEW2= NEWN+1
        STIFO(NEWN,1)= 0.1E20
        STIFO(NEW2,1)=0.1E20
   37    CONTINUE
        DO 877  MN=1,NX1
        NEWN=MN*NFREE
  877    STIFO(NEWN,1)=0.1E20
C       **************************************************************
```

```
C       ***************************************************************
C       BOUNDARY CONDITION FITTING IS OVER
C       ***************************************************************
*
C       ***************************************************************
*
C       GAUSS IS A SUBROUTINE TO SOLVE SIMULTANEOUS EQUATIONS WHERE
C       THE COEFFICIENT MATRIX IS STORED IN HALFBAND FORM
C       ***************************************************************
***
        WRITE(8,*)'      THE STIFO MATRIX'
        WRITE(8,100)((STIFO(I,J),J=1,NBH),I=1,NT)
   320   CALL GAUSS(X,NBH,NT,NLC)
        WRITE(8,*)'                        THE DISPLACEMENT MATRIX'
        WRITE(8,100)((X(I,J),I=1,NT),J=1,NLC)
C       ***************************************************************
**
C       STRESS GENERATION STARTS
C       ***************************************************************
**
        WRITE(8,*)'               THE STRESS MATRIX'
        DO 41 II=1,NELM
        DO 518 LL=1,4
        DO 519 L=1,4
        ZIZ= ZX(LL)*ZX(L)
        ETAZ= ZY(LL)*ZY(L)
        SFD(L)= 0.25*(1.0+ETAZ)*(1.0+ZIZ)
        SFDZ(L)= 0.25*(1.0+ETAZ)*ZX(L)
   519   SFDE(L)= 0.25*(1.0+ZIZ)*ZY(L)
        DO 520 K=1,3
        DO 520 KK=1,NENP
   520   BB(K,KK)=0.0
        DXZ=0.0
        DYZ=0.0
        DXE=0.0
        DYE=0.0
        DO 521 KK=1,4
        DXZ=DXZ+SFDZ(KK)*XG(KK)
        DYZ=DYZ+SFDZ(KK)*YG(KK)
        DXE= DXE+SFDE(KK)*XG(KK)
   521   DYE= DYE+SFDE(KK)*YG(KK)
        DXZI= DYE/ZAC
        DYZI= -DYZ/ZAC
        DXEI=- DXE/ZAC
        DYEI= DXZ/ZAC
        DO 522 KK=1,NN
        K=2*(KK-1)
        DNX= SFDZ(KK)*DXZI + SFDE(KK)*DYZI
        DNY= SFDZ(KK)* DXEI +SFDE(KK)* DYEI
        BB(1,K+1)= DNX
        BB(2,K+2) = DNY
        BB(3,K+1) = DNY
   522   BB(3,K+2)= DNX
        DO 570 L=1,NLC
        DO 570 KK=1,NN
        DO 570 I=1,2
        K=(NOD(II,KK)-1)*NFREE+I
```

```
      J=(KK-1)*NFREE+I
  570  DISP(J,L)= X(K,L)
      DO 525 J=1,NLC
      DO 524 M=1,3
      STRESS(M,J)= 0.0
      DO 524 L=1,3
      DO 524 K=1,NENP
  524  STRESS(M,J)= STRESS(M,J)+ RGD(M,L)*BB(L,K)*DISP(K,J)
      WRITE(8,*)' ELE NO=  ',II,'   NODE NO=  ',LL,'     LOAD COND= '
     * ,J
      WRITE(8,100)(STRESS(M,J),M=1,3)
  525  CONTINUE
  518  CONTINUE
   41  CONTINUE
      GO TO 103
      END
C     ***********************************************************
C     ***********************************************************
      SUBROUTINE GACUV(I,J,ZI,ETA,HI,HJ,NQ)
      IF(NQ-2) 2,1,2
    1  IF(I-1) 4,3,4
    3 ZI= -0.577350269189626
      HI= 1.000000000000000
      GO TO 7
    4  ZI= 0.57735026918926
      HI= 1.000000000000000
    7  IF(J-1) 5,6,5
    6  ETA= -0.57735026918926
      HJ= 1.000000000000000
      GO TO 18
    5  ETA=  0.577350269189626
      HJ= 1.0000000000000000
      GO TO 18
    2  IF (I-1) 8,9,8
    9  ZI= -0.774596669241483
      HI=0.5555555555555556
      GO TO 12
    8  IF (I-2) 10,11,10
   11  ZI= 0.000000000000000
      HI= 0.8888888888888889
      GO TO 12
   10  ZI= 0.774596669241483
      HI= 0.555555555555556
   12  IF(J-1) 14,15,14
   15  ETA= -0.774596669241483
      HJ= 0.5555555555555556
      GO TO 18
   14  IF(J-2)  16,17,16
   17  ETA=0.0000000000000000
      HJ=0.8888888888888889
      GO TO 18
   16  ETA= 0.774596669241483
      HJ= 0.5555555555555556
   18  RETURN
      END
C     *******************************************************
```

Table 10.4. Input Data for the Plane Stress Problem

```
90.0  1.0  0.5
4  2   2   1   2  2
1.0 0.25 1.0
5  1  -0.125
11  1  -0.25
0  0  0.0
```

Table 10.5. Computer Output for the Plane Stress Problem Using Isoparametric Element

```
Analysis of Plane Stress by Isoparametric Plane Stress Element
Exl = 0.500    Eyl = 0.250    Phi(F) = 90.000        Pr = 0.250      D =
1.000   Thick = 1.000
NN = 4  NFREE = 2       NT = 18 NELM = 4        NENP = 8
NQ = 2  NBH = 10        NLC = 1 NX = 2  NY = 2
The RGD Matrix
  1.06667e+000  2.66667e-001  0.00000e+000
  2.66667e-001  1.06667e+000  0.00000e+000
  0.00000e+000  0.00000e+000  4.00000e-001
The Element Stiffness Matrix
  4.44445e-001 -1.66667e-001 -4.44444e-002  3.33333e-002
 -2.22222e-001  1.66667e-001 -1.77778e-001 -3.33333e-002

 -1.66667e-001  7.77778e-001 -3.33333e-002  2.88889e-001
  1.66667e-001 -3.88889e-001  3.33333e-002 -6.77778e-001

 -4.44444e-002 -3.33333e-002  4.44444e-001  1.66667e-001
 -1.77778e-001  3.33333e-002 -2.22222e-001 -1.66667e-001

  3.33333e-002  2.88889e-001  1.66667e-001  7.77778e-001
 -3.33333e-002 -6.77778e-001 -1.66667e-001 -3.88889e-001

 -2.22222e-001  1.66667e-001 -1.77778e-001 -3.33333e-002
  4.44444e-001 -1.66667e-001 -4.44444e-002  3.33333e-002

  1.66667e-001 -3.88889e-001  3.33333e-002 -6.77778e-001
 -1.66667e-001  7.77778e-001 -3.33333e-002  2.88889e-001

 -1.77778e-001  3.33333e-002 -2.22222e-001 -1.66667e-001
 -4.44444e-002 -3.33333e-002  4.44445e-001  1.66667e-001

 -3.33333e-002 -6.77778e-001 -1.66667e-001 -3.88889e-001
  3.33333e-002  2.88889e-001  1.66667e-001  7.77778e-001

The Overall Stiffness Matrix
  4.44445e-001 -1.66667e-001 -4.44444e-002  3.33333e-002  0.00000e+000
  0.00000e+000 -1.77778e-001 -3.33333e-002 -2.22222e-001  1.66667e-001

  1.00000e+019 -3.33333e-002  2.88889e-001  0.00000e+000  0.00000e+000
  3.33333e-002 -6.77778e-001  1.66667e-001 -3.88889e-001  0.00000e+000

  8.88889e-001  0.00000e+000 -4.44444e-002  3.33333e-002 -2.22222e-001
 -1.66667e-001 -3.55556e-001  0.00000e+000 -2.22222e-001  1.66667e-001

  1.00000e+019 -3.33333e-002  2.88889e-001 -1.66667e-001 -3.88889e-001
```

```
  0.00000e+000 -1.35556e+000  1.66667e-001 -3.88889e-001  0.00000e+000

  4.44444e-001  1.66667e-001  0.00000e+000  0.00000e+000 -2.22222e-001
 -1.66667e-001 -1.77778e-001  3.33333e-002  0.00000e+000  0.00000e+000

  1.00000e+019  0.00000e+000  0.00000e+000 -1.66667e-001 -3.88889e-001
 -3.33333e-002 -6.77778e-001  0.00000e+000  0.00000e+000  0.00000e+000

  8.88889e-001  0.00000e+000 -8.88889e-002  0.00000e+000  0.00000e+000
  0.00000e+000 -1.77778e-001 -3.33333e-002 -2.22222e-001  1.66667e-001

  1.55556e+000 -3.72529e-009  5.77778e-001  0.00000e+000  0.00000e+000
  3.33333e-002 -6.77778e-001  1.66667e-001 -3.88889e-001  0.00000e+000

  1.77778e+000  1.49012e-008 -8.88889e-002  0.00000e+000 -2.22222e-001
 -1.66667e-001 -3.55556e-001  0.00000e+000 -2.22222e-001  1.66667e-001

  3.11111e+000 -3.72529e-009  5.77778e-001 -1.66667e-001 -3.88889e-001
  0.00000e+000 -1.35556e+000  1.66667e-001 -3.88889e-001  0.00000e+000

  8.88889e-001  1.49012e-008  0.00000e+000  0.00000e+000 -2.22222e-001
 -1.66667e-001 -1.77778e-001  3.33333e-002  0.00000e+000  0.00000e+000

  1.55556e+000  0.00000e+000  0.00000e+000 -1.66667e-001 -3.88889e-001
 -3.33333e-002 -6.77778e-001  0.00000e+000  0.00000e+000  0.00000e+000

  1.00000e+019  1.66667e-001 -4.44444e-002 -3.33333e-002  0.00000e+000
  0.00000e+000  0.00000e+000  0.00000e+000  0.00000e+000  0.00000e+000

  1.00000e+019  3.33333e-002  2.88889e-001  0.00000e+000  0.00000e+000
  0.00000e+000  0.00000e+000  0.00000e+000  0.00000e+000  0.00000e+000

  1.00000e+019  1.49012e-008 -4.44444e-002 -3.33333e-002  0.00000e+000
  0.00000e+000  0.00000e+000  0.00000e+000  0.00000e+000  0.00000e+000

  1.00000e+019  3.33333e-002  2.88889e-001  0.00000e+000  0.00000e+000
  0.00000e+000  0.00000e+000  0.00000e+000  0.00000e+000  0.00000e+000

  1.00000e+019 -1.66667e-001  0.00000e+000  0.00000e+000  0.00000e+000
  0.00000e+000  0.00000e+000  0.00000e+000  0.00000e+000  0.00000e+000

  1.00000e+019  0.00000e+000  0.00000e+000  0.00000e+000  0.00000e+000
  0.00000e+000  0.00000e+000  0.00000e+000  0.00000e+000  0.00000e+000

The Displacement Matrix
 -1.56967e-001 -2.77353e-021 -2.78398e-001  1.82393e-021 -6.12379e-001
  8.95011e-021 -1.19232e-001 -2.18991e-002 -1.82388e-001 -3.04039e-002
 -4.91564e-001  5.42437e-002 -6.60647e-021 -6.10388e-021 -1.87891e-020
 -9.06911e-021 -1.21044e-020  7.17248e-021
The STRESS Matrix
Element No. = 1         Node No. = 1   Load Condition = 1
 -2.35693e-001  2.86729e-002 -6.03769e-002
Element No. = 1         Node No. = 2   Load Condition = 1
 -2.26622e-001  6.49602e-002 -1.53617e-001
Element No. = 1         Node No. = 3   Load Condition = 1
 -1.02302e-001  9.60402e-002 -1.60421e-001
Element No. = 1         Node No. = 4   Load Condition = 1
```

```
 -1.11374e-001  5.97529e-002 -6.71808e-002
Element No. = 2         Node No. = 1   Load Condition = 1
 -6.80062e-001 -4.83998e-002 -1.53617e-001
Element No. = 2         Node No. = 2   Load Condition = 1
 -7.70352e-001 -4.09563e-001 -1.93304e-001
Element No. = 2         Node No. = 3   Load Condition = 1
 -7.17436e-001 -3.96334e-001 -1.25586e-001
Element No. = 2         Node No. = 4   Load Condition = 1
 -6.27145e-001 -3.51707e-002 -8.58987e-002
Element No. = 3         Node No. = 1   Load Condition = 1
 -1.58092e-001 -1.27119e-001 -1.97575e-001
Element No. = 3         Node No. = 2   Load Condition = 1
 -1.67164e-001 -1.63406e-001 -2.98624e-001
Element No. = 3         Node No. = 3   Load Condition = 1
 -3.24308e-002 -1.29723e-001 -2.91820e-001
Element No. = 3         Node No. = 4   Load Condition = 1
 -2.33590e-002 -9.34360e-002 -1.90771e-001
Element No. = 4         Node No. = 1   Load Condition = 1
 -6.92007e-001 -2.94617e-001 -2.24102e-001
Element No. = 4         Node No. = 2   Load Condition = 1
 -6.01716e-001  6.65455e-002 -7.18784e-001
Element No. = 4         Node No. = 3   Load Condition = 1
  5.78599e-002  2.31440e-001 -7.86503e-001
Element No. = 4         Node No. = 4   Load Condition = 1
 -3.24308e-002 -1.29723e-001 -2.91820e-001
```

REFERENCES AND SUGGESTED READINGS

10.1. O. C. Zienkiewicz and R. L. Taylor, The Finite Element Method, v. 1 and v. 2, McGraw-Hill, 1989.

10.2. D. J. Allman, A Compatible Triangular Element Including Vertex Rotations for Plane Elasticity Analysis, Computers and Structures, v. 19, no. 1, 1984, pp. 1-8.

10.3. J. L. Batoz, K. J. Bathe and L. W. Ho, A Study of the Three-node Triangular Element, J. Num. Meth. Engg., v. 115, no. 12, 1980, pp. 1771–1812.

10.4. B. M. Irons, Engineering Application Numerical Integration in Stiffness Method, J. AIAA, v.14, 1966, pp. 2035–2037.

10.5. C. Maforano, S. Odorizzi and R. Vitaliani, Shortened Quadrature Rules for Finite Elements, Advances in Engineering Software, v. 4, no. 2, 1982.

10.6. J. E. Akin, Application and Implementation of Finite Element Methods, Academic Press, 1982.

10.7. M. J. Turner, R. W. Clough, H. C. Martin and L. P. Topp, Stiffness and Deflection Analysis of Complex Structures, J. Aero. Sc., v. 23, no. 9, 1956, pp. 805–823.

10.8. G. R. Cowper, Gaussian Quadrature Formulas for Triangles, International Journal for Numerical Methods in Engineering, v. 7, 1973, pp. 405–408.

10.9. G. R. Cowper, G. M. Lindberg and M. B. Olson, A Shallow Shell Finite Element of Triangular Shape, International Journal of Solids and Structures, v. 6, 1970, pp. 1133–1156.

EXERCISE 10

10.1. Write the shape functions of the quadratic triangle and verify the shape function properties.

10.2. Write the interpolation function for a line segment having 3 nodes in terms of natural coordinates.

10.3. For a cubic triangle, write the shape functions in terms of area coordinates and verify the properties of the shape function.

10.4. Verify the convergence criteria for the 3-noded triangle where shape functions have been expressed in terms of area coordinates.

10.5. For the quadratic triangle, derive the stiffness matrix of the element.

10.6. Determine the shape functions for isoparametric cubic element.

10.7. For the derivation of the stiffness matrix of the rectangular element presented in the chapter by using isoparametric element, evaluate explicitly the stiffness matrix of the element.

10.8. For the computer program for the solution of plane stress problems by the use of isoparametric linear element presented in this chapter, what changes in the computer program will be needed if a plane strain problem is to be solved by the use of isoparametric quadratic element?

10.9. For isoparametric quadratic element, develop expressions for the nodal forces due to the presence of body forces.

10.10. Construct a displacement field parallel to x-axis (*i.e.* u) for the element.

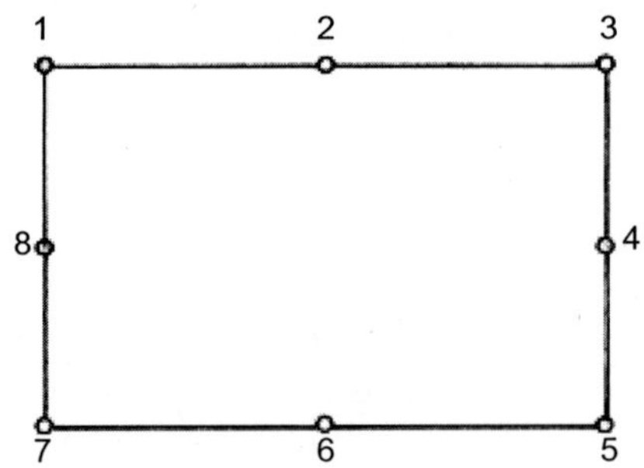

Prob. 10.10

❑❑❑

11

FINITE ELEMENT ANALYSIS OF PLATE BENDING PROBLEMS

11.1 INTRODUCTION

Bending of plates forms another important class of problems. Solutions obtained by the application of classical theory of plate flexure are limited mainly to simple types of plates with simple loading and boundary conditions. With the advent of the finite element method, the plate bending problems have received considerable attention. As a result of which, a large number of different plate-bending element formulations have been made (11.1). This chapter is limited to the discussion of some of these elements.

For most of the elements, the starting point is the choice of the displacement function. However, attempts have been made to directly formulate the shape functions. The determination of the shape functions now becomes more difficult. The continuity of displacements along the edges now implies continuity of slope. The elements, which are formulated by satisfying both the deflection and slope continuity along the edges, are referred to as conforming elements. Elements, which preserve the continuity of deflection, but violate the condition of slope compatibility is referred to as 'non-conforming elements'. These elements, however, do converge.

For conforming elements, the formulation may be quite tedious. Though these elements gradually converge to the correct solution, they may yield inferior accuracy in some cases. Hence, non-conforming elements are of much greater practical usage.

11.2 BEAM ELEMENT

A uniform beam element is considered and is shown in Fig. 11.1. Axial and torsional deformations have not been considered. At each node at two ends of the member, there are two degrees of freedom – deflection w and rotation dw/dx. The steps for deriving the stiffness matrix of the beam element are same as those for plane elasticity problems. The derivation of the stiffness matrix of the beam element is presented in the following. The displacement and the force vectors for the beam element are

$$\{X\}_e^T = \{X_1 \quad X_2 \quad X_3 \quad X_4\} \qquad (11.1)$$

$$\{P\}_e^T = \{P_1 \quad P_2 \quad P_3 \quad P_4\} \qquad (11.2)$$

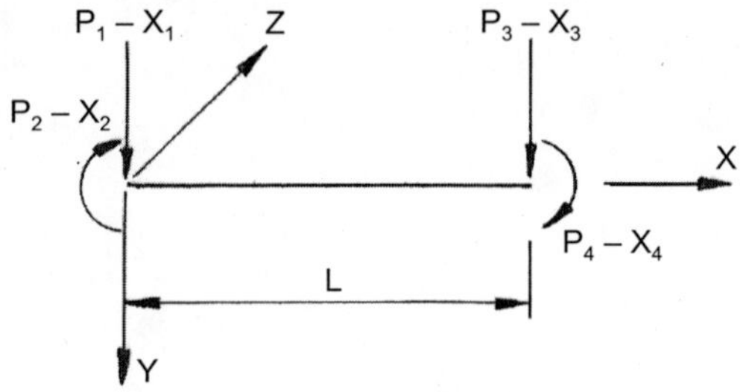

Fig. 11.1. Beam element

11.2.1 Displacement Function

Total degrees of freedom for this element are four. Therefore, four constants are to be associated with the displacement function. Unlike plane elasticity problems, where u and v have been independently prescribed, once the deflection is written, the slope can be obtained by its differentiation. Therefore, all the four unknown coefficients are reflected in the expression for deflection. The expression for deflection is given by the following polynomial expression.

$$\{f\} = w = \alpha_1 + \alpha_2 x + \alpha_3 x^2 + \alpha_4 x^3 \qquad (11.3)$$

or

$$\{f\} = [1 \quad x \quad x^2 \quad x^3] \begin{Bmatrix} \alpha_1 \\ \alpha_2 \\ \alpha_3 \\ \alpha_4 \end{Bmatrix}$$

or

$$\{f\} = [C]\{\alpha\} \qquad (11.4)$$

Though f indicates a single quantity for this element, matrix notation of f is given for the sake of generality.

11.2.2 Displacement Function in Terms of Nodal Displacements

Differentiating w with respect to x in eqn. (11.3) yields

$$\frac{dw}{dx} = \alpha_2 + 2\alpha_3 + 3\alpha_4 x^2 \qquad (11.5)$$

Node i is chosen as the origin.

At node i, $x = 0$ and at node j, $x = L$. Substituting the respective values of x in eqns. (11.3) and (11.5).

$$\left.\begin{aligned} X_1 &= (w)_{x=0} \\ X_2 &= \left(\frac{dw}{dx}\right)_{x=0} \\ X_3 &= (w)_{x=L} = \alpha_1 + \alpha_2 L + \alpha_3 L^2 + \alpha_4 L^3 \\ X_4 &= \left(\frac{dw}{dx}\right)_{x=L} = \alpha_2 + 2\alpha_3 L + 3\alpha_4 L^2 \end{aligned}\right\} \qquad (11.6)$$

Equation (11.6) in matrix form becomes

$$\begin{Bmatrix} X_1 \\ X_2 \\ X_3 \\ X_4 \end{Bmatrix} = \begin{bmatrix} 1 & 0 & 0 & 0 \\ 0 & 1 & 0 & 0 \\ 1 & L & L^2 & L^3 \\ 0 & 1 & 2L & 3L^2 \end{bmatrix} \begin{Bmatrix} \alpha_1 \\ \alpha_2 \\ \alpha_3 \\ \alpha_4 \end{Bmatrix} \tag{11.7}$$

or

$$\{X\}_e = [A]\{\alpha\} \tag{11.8}$$

Inverting eqn. (11.8) becomes

$$\{\alpha\} = [A]^{-1}\{X\}_e \tag{11.9}$$

Substituting $\{\alpha\}$ from eqn. (11.9) into eqn. (11.4) yields

$$\{f\} = [C][A]^{-1}\{X\}_e$$

$$\{f\} = w = [N]\{X\}_e \tag{11.10}$$

For the beam element,

$$[A]^{-1} = \begin{bmatrix} 1 & 0 & 0 & 0 \\ 0 & 1 & 0 & 0 \\ \dfrac{-3}{L^2} & \dfrac{-2}{L} & \dfrac{3}{L^2} & \dfrac{-1}{L} \\ \dfrac{2}{L^3} & \dfrac{1}{L^2} & -\dfrac{2}{L^3} & \dfrac{1}{L^2} \end{bmatrix} \tag{11.11}$$

$$[N] = \left[\left(1-\frac{3x^2}{L^2}+\frac{2x^3}{L^3}\right)\left(x-\frac{2x^2}{L}+\frac{x^3}{L^2}\right)\left(\frac{3x^2}{L^2}-\frac{2x^2}{L^3}\right)\left(\frac{-x^2}{L}+\frac{x^3}{L^2}\right)\right] \tag{11.12}$$

11.2.3 Strain (Curvature) : Nodal Parameter Relationship

For the beam element, the only 'strain', component present is the curvature where the shear deformation is not considered. The 'strain' is then

$$\{\varepsilon\} = -\frac{d^2w}{dx^2} \tag{11.13}$$

Substituting $[N]$ from eqn. (11.12) into eqn. (11.10) and then differentiating twice, it gives eqn. (11.13) as

$$[\varepsilon] = \left[\left(\frac{6}{L^2}-\frac{12x}{L^3}\right)\left(\frac{4}{L}-\frac{6x}{L^2}\right)\left(-\frac{6x}{L^2}+\frac{12}{L^3}\right)\left(\frac{2}{L}-\frac{6x}{L^2}\right)\right]\{X\}_e \tag{11.14}$$

or

$$\{\varepsilon\} = [B]\{X\}_e \tag{11.15}$$

11.2.4 Stress (Moment) : Strain (Curvature) Relationship

The 'stress' referred to in the beam element is the bending moment. The following is the moment-curvature relationship

$$M = -EI\frac{d^2w}{dx^2} \tag{11.16}$$

In conformity with the standard notations used, eqn. (11.16)

$$\{\sigma\} = [D]\{\varepsilon\} \tag{11.17}$$

$$\{\sigma\} = [M] \text{ and } [D] = [EI]$$

Combining eqns. (11.15) and (11.17), yields

$$\{\sigma\} = [D][B]\{X\}_e \tag{11.18}$$

11.2.5 Derivation of the Element Stiffness Matrix

The element stiffness matrix of the beams is derived here using the principle of virtual work.

Let a virtual displacement $\delta\{X\}_e$ be imposed at the nodes. The work done by the external forces due to this virtual displacement is given by

$$\delta\{X\}_e^T\{P\}_e$$

i.e., they constitute the product of the individual nodal forces and their corresponding virtual displacements.

Due to the virtual displacements, the displacements and the strains will undergo virtual changes and they become (from eqns. (11.10) and (11.15))

$$\delta\{f\} = \delta w = [N]\,\delta\{X\}_e \tag{11.19}$$

$$\delta\{\varepsilon\} = [B]\,\delta\{X\}_e \tag{11.20}$$

The internal work done per unit length by the stresses (moment) is

$$\delta\{\varepsilon\}^T\{\sigma\}$$

Substituting respective values from eqns. (11.18) and (11.20), the internal work per unit length = $\delta\{X\}_e^T[B]^T[D]\{X\}_e$.

Total internal work over the entire element is

$$\delta\{X\}_e^T\int_0^L[B]^T[D][B]\,dx\,\{X\}_e$$

Equating the external work to the internal work

$$\delta\{X\}_e^T\{P\}_e = \delta\{X\}_e^T\int_0^L[B]^T[D][B]\,dx\,\{X\}_e$$

or,
$$\{P\}_e = \int_0^L[B]^T[D][B]dx\,\{X\}_e$$

or,
$$\{P\}_e = \{K\}_e\,\{X\}_e \tag{11.21}$$

where

$$[k]_e = \int_0^L[B]^T[D][B]dx \tag{11.22}$$

$[k]_e$ can be explicitly evaluated. After performing the matrix multiplication, eqn. (11.22) becomes

$$[k]_e = EI\int_0^L \begin{bmatrix} \frac{36}{L^4}-\frac{144x}{L^5}+\frac{144x^2}{L^6} & & symmetrical & \\ \frac{24}{L^3}-\frac{84x}{L^4}+\frac{72x^2}{L^5} & \frac{16}{L^2}-\frac{48x}{L^3}+\frac{36x^2}{L^4} & & \\ -\frac{36}{L^4}+\frac{144x}{L^5}-\frac{144x^2}{L^6} & -\frac{24}{L^3}+\frac{84x}{L^4}-\frac{72x^2}{L^5} & \frac{36}{L^4}-\frac{144x}{L^5}+\frac{144x^2}{L^6} & \\ \frac{12}{L^3}-\frac{60x}{L^4}+\frac{72x^2}{L^5} & \frac{8}{L^2}-\frac{36x}{L^3}+\frac{36x^2}{L^4} & -\frac{12}{L^3}+\frac{60x}{L^4}-\frac{72x^2}{L^5} & \frac{4}{L^2}-\frac{24x}{L^3}+\frac{36x^2}{L^4} \end{bmatrix} dx$$

$$[k]_e = \frac{EI}{L^3}\begin{bmatrix} 12 & symmetrical & & \\ 6L & 4L^2 & & \\ -12 & -6L & 12 & \\ 6L & 2L^2 & -6L & 4L^2 \end{bmatrix} \tag{11.23}$$

11.2.6 Determination of Equivalent Loading on the Beam

If the loads applied are such that they correspond to the nodal displacements, then the loading vector can be formed directly by intuitive reasoning. But in many cases, loads act inside the element or the nodal parameters may be such that there may not exist any corresponding 'physical quantity'. For such cases consistent loading vector or equivalent loading at the nodes are to be obtained by the application of variational principles.

Let $\{P\}_e$ be the equivalent nodal forces. Due to the virtual displacement $\delta\{X\}_e^T$, the work done by the nodal forces is given by

$$\delta\{X\}_e^T\{P\}_e$$

If $\{p\}$ is the loading within the element - it may be distributed or it may have a discrete number of concentrated loads or it may have any other variation within the element, then the work done by this loading is

$$\int_0^L \delta\{f\}^T\{p\}\,dx$$

From eqn. (11.19), the work done by the load inside the element is

$$\int_0^L \delta\{X\}_e^T[n]^T\{p\}\,dx$$

Equating both the work,

$$\delta\{X\}_e^T\{P\}_e = \{X\}_e^T\int_0^L \delta[N]^T\{p\}\,dx$$

or

$$\{P\}_e = \int_0^L [N]^T\{p\}\,dx \tag{11.24}$$

Equation (11.24) is the next general expression for determining equivalent nodal forces. If $\{p\}$ is uniformly distributed load, *i.e.* $\{P\} = q$, then eqn. (11.24) becomes

$$\{P\}_e = q\int_0^L [N]^T\, dx \tag{11.25}$$

For the beam element, substituting the value of $[N]$ from eqn. (11.12] into eqn. (11.25) and performing the necessary integration, yields

$$\{P\}_e = \begin{Bmatrix} qL/2 \\ qL^2/12 \\ qL/2 \\ -qL^2/12 \end{Bmatrix} \tag{11.26}$$

11.3 RECTANGULAR PLATE BENDING ELEMENT

The rectangular elements can be effectively used for plates having rectangular edges. Rectangular elements can also be employed for irregular plates in conjunction with other types of elements (*e.g.* triangular elements). A node of the plate bending element will have three degrees of freedom – the transverse deflection and two orthogonal rotations.

The rectangular plate bending element along with their dimensions, coordinate system and node numbering as shown in Fig. 11.2(a). The positive directions of all the displacements are shown in Fig 11.2(b). The positive directions of rotations are indicated by right hand screw rule. The displacements at node 1 are $\{w_1\ \theta_{x1}\ \theta_{y1}\}$ and the corresponding forces are $\{P_1\ M_{x1}\ M_{y1}\}$. Therefore, complete displacement vectors for this element are

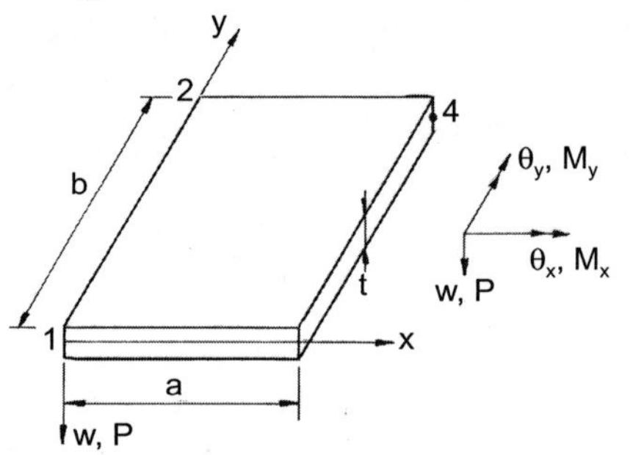

Fig. 11.2. Rectangular plate bending element

$$\{X\}_e^T = \{w_1\ \theta_{x1}\ \theta_{y1}\ w_2\ \theta_{x2}\ \theta_{y2}\ w_3\ \theta_{x3}\ \theta_{y3}\ w_4\ \theta_{x4}\ \theta_{y4}\} \tag{11.27}$$

$$\{P\}_e^T = \{P_1\ M_{x1}\ M_{y1}\ P_2\ M_{x2}\ M_{y2}\ P_3\ M_{x3}\ M_{y3}\ P_4\ M_{x4}\ M_{y4}\} \tag{11.28}$$

11.3.1 Displacement Function

There are three degrees of freedom associated with each node. So for a four-noded rectangle, there are in all twelve degrees of freedom. The polynomial expression to be chosen should have twelve terms. A complete cubical polynomial (Fig. 11.3) contains 10 terms of quartic polynomial are to be eliminated. A suitable function is given by

$$\{f\} = w = \alpha_1 + \alpha_2 x + \alpha_3 y + \alpha_4 x^2 + \alpha_5 xy + \alpha_6 y^2 + \alpha_7 x^3 + \alpha_8 x^2 y + \alpha_9 xy^2 + \alpha_{10} y^3 + \alpha_{11} x^3 y + \alpha_{12} xy^3 \tag{11.29}$$

or,
$$\{f\} = [1\ x\ y\ x^2\ xy\ y^2\ x^3\ x^2y\ xy^2\ y^3\ x^3y\ xy^3]\begin{Bmatrix}\alpha_1\\ \alpha_2\\ \vdots\\ \alpha_{12}\end{Bmatrix} \tag{11.30}$$

or,
$$\{f\} = [C]\{\alpha\} \tag{11.31}$$

The displacement function of eqn. (11.29) gives the following expression for rotations.

$$\theta_x = -\frac{\partial w}{\partial y} = -(\alpha_3 + \alpha_5 x + 2\alpha_6 y + \alpha_3 x^2 + 2\alpha_9 xy + 3\alpha_{10} y^2 + \alpha_{11} x^3 + 3\alpha_{12} xy^2) \tag{11.32}$$

and

$$\theta_y = \frac{\partial w}{\partial x} = \alpha_2 + 2\alpha_4 x + \alpha_5 y + 3\alpha_7 x^2 + 2\alpha_8 xy + \alpha_9 y^2 + 3\alpha_{11} x^2 y + \alpha_{12} y^3 \tag{11.33}$$

11.3.1.1 Checking the continuity requirements of the displacement function

Let us verify whether the displacement function chosen satisfies the continuity of the slopes along the edges. Let it be checked with respect to the edge $x = a$.

$$w = \alpha_1 + \alpha_2.a + \alpha_3 y + \alpha_4.a^2 + \alpha_5 ay + \alpha_6 y^2 + \alpha_7 a^3 + \alpha_8 a^2 y + \alpha_9 ay^2 + \alpha_{10} y^3 + \alpha_{11} a^3 y + \alpha_{12} ay^3$$

$$w = (\alpha_1 + \alpha_2 a + \alpha_4 a^2 + \alpha_7 a^3) + (\alpha_3 + \alpha_5 a + \alpha_8 \cdot a^2 + \alpha_{11} a^3)y + (\alpha_6 + \alpha_3 a)y^2 + (\alpha_{10} + \alpha_{12} a)y^3 \tag{11.34}$$

The expression given by (11.34) is a cubic polynomial in y. If the flexural action of the edge in the direction of y is considered (treating the edge as a beam), a cubic polynomial uniquely defines the end deflections and end slopes. Thus the compatibility of the deflections among the common boundary between the adjacent elements is ensured.

At $x = a$, θ_y is given by

$$(\theta_y)_{x=a} = \left(\frac{\partial w}{\partial x}\right)_{x=a} = (\alpha_2 + 2\alpha_4 a + 3\alpha_7 a^2) + (\alpha_5 + 2\alpha_8 a + 3\alpha_{11} a^2) + (\alpha_5 + 2\alpha_8 a + 3\alpha_{11} a^2)y + \alpha_9 y^2 + \alpha_{12} y^3 \tag{11.35}$$

This cross-rotation term θ_y along the edge is similar to the angle of twist in a bar subjected to torsion. For this bar, a two-term polynomial uniquely determines the displacement field. Equation (11.35) represents a cubic function, which does not uniquely define the cross-rotation values along the edge. θ_y is, therefore, discontinuous along this edge. Similar conclusions can be arrived at the other edges of the element.

This element therefore, does not satisfy all slope continuity requirements and is termed as 'non-conforming'. This aspect has been discussed at the end of the chapter.

11.3.2 Displacement Function Expressed in Terms of Nodal Displacements

The coordinates of nodes 1, 2, 3 and 4 are (0, 0), (0, b), (a, 0) and (a, b) respectively. Substituting the values of the nodal coordinates in eqns. (11.29), (11.32) and (11.33) respectively, the following equations result.

Inverting eqn. (11.37)

$$\{\alpha\} = [A]^{-1}\{X\}_e \tag{11.38}$$

Combining eqns. (11.31) and (11.38) yields

$$\{f\} = w = [C][A]^{-1}\{X\}_e$$

or

$$\{f\} = w = [N]\{X\}_e \tag{11.39a}$$

where

$$[N] = [C][A]^{-1} \tag{11.39b}$$

$$\begin{Bmatrix} w_1 \\ \theta_{x1} \\ \theta_{y1} \\ w_2 \\ \theta_{x2} \\ \theta_{y2} \\ w_3 \\ \theta_{x3} \\ \theta_{y3} \\ w_4 \\ \theta_{x4} \\ \theta_{y4} \end{Bmatrix} = \begin{bmatrix} 1 & 0 & 0 & 0 & 0 & 0 & 0 & 0 & 0 & 0 & 0 & 0 \\ 0 & 0 & -1 & 0 & 0 & 0 & 0 & 0 & 0 & 0 & 0 & 0 \\ 0 & 1 & 0 & 0 & 0 & 0 & 0 & 0 & 0 & 0 & 0 & 0 \\ 1 & 0 & b & 0 & 0 & b^2 & 0 & 0 & 0 & b^3 & 0 & 0 \\ 0 & 0 & -1 & 0 & 0 & -2b & 0 & 0 & 0 & -3b^2 & 0 & 0 \\ 0 & 1 & 0 & 0 & b & 0 & 0 & 0 & b^2 & 0 & 0 & b^3 \\ 1 & a & 0 & a^2 & 0 & 0 & a^3 & 0 & 0 & 0 & 0 & 0 \\ 0 & 0 & -1 & 0 & -a & 0 & 0 & -a^2 & 0 & 0 & -a^3 & 0 \\ 0 & 1 & 0 & 2a & 0 & 0 & 3a^2 & 0 & 0 & 0 & 0 & 0 \\ 1 & a & b & a^2 & ab & b^2 & a^3 & a^2b & ab^2 & b^3 & a^3b & ab^3 \\ 0 & 0 & -1 & 0 & -a & -2b & 0 & -a^2 & -2ab & -3b^2 & -a^3 & -3ab^2 \\ 0 & 1 & 0 & 2a & b & 0 & 3a^2 & 2ab & b^2 & 0 & 3a^2b & b^3 \end{bmatrix} \begin{Bmatrix} \alpha_1 \\ \alpha_2 \\ \alpha_3 \\ \alpha_4 \\ \alpha_5 \\ \alpha_6 \\ \alpha_7 \\ \alpha_8 \\ \alpha_9 \\ \alpha_{10} \\ \alpha_{11} \\ \alpha_{12} \end{Bmatrix} \tag{11.36}$$

or

$$\{X\}_e = [A]\{\alpha\} \tag{11.37}$$

11.3.3 Strain-Nodal Parameter Relationship

The 'strains' in plate bending problem are the curvatures. The strain matrix is given by

$$\{\varepsilon\} = \begin{Bmatrix} -\dfrac{\partial^2 w}{\partial x^2} \\ -\dfrac{\partial^2 w}{\partial y^2} \\ 2\dfrac{\partial^2 w}{\partial x \partial y} \end{Bmatrix} \tag{11.40}$$

By directly differentiating w given in eqn. (11.29) with respect to the quantities indicated in eqn. (11.40), yields

$$\{\varepsilon\} = \begin{Bmatrix} -(2\alpha_4 + 6\alpha_7 x + 2\alpha_8 y + 6\alpha_{11} xy) \\ -(2\alpha_6 + 2\alpha_9 x + 2\alpha_{10} y + 6\alpha_{12} xy) \\ 2\alpha_5 + 4\alpha_8 x + 4\alpha_9 y + 6\alpha_{11} x^2 + 6\alpha_{12} y^2 \end{Bmatrix} \tag{11.41}$$

or

$$\{\varepsilon\} = [Q]\{\alpha\} \tag{11.42}$$

where

$$[Q] = \begin{bmatrix} 0 & 0 & 0 & -2 & 0 & 0 & -6x & -2y & 0 & 0 & -6xy & 0 \\ 0 & 0 & 0 & 0 & 0 & -2 & 0 & 0 & -2x & -6y & 0 & -6xy \\ 0 & 0 & 0 & 0 & 2 & 0 & 0 & 4x & 4y & 0 & 6x^2 & 6y^2 \end{bmatrix} \tag{11.43}$$

Substituting $\{\alpha\}$ from eqn. (11.38) into eqn. (11.42), yields

$$\{\varepsilon\} = [Q][A]^{-1}\{X\}_e$$

or

$$\{\varepsilon\} = [B]\{X\}_e \tag{11.44}$$

Equations (11.42) and (11.43) reveal that the displacement function of eqn. (11.29) satisfies one of the requirements of convergence, as it contains constant strain (curvature) terms.

11.3.4 Stress (moment) – Strain (curvature) Relationship

The moment-curvature relationship for the orthotropic plate has been deduced in Chapter 6 and they are reproduced below for orthotropic plate

$$\{\sigma\} = \begin{Bmatrix} M_x \\ M_y \\ M_{xy} \end{Bmatrix} = \begin{bmatrix} D_x & D_1 & 0 \\ D_1 & D_y & 0 \\ 0 & 0 & D_{xy} \end{bmatrix} = \begin{Bmatrix} -\dfrac{\partial^2 w}{\partial x^2} \\ -\dfrac{\partial^2 w}{\partial y^2} \\ 2\dfrac{\partial^2 w}{\partial x \partial y} \end{Bmatrix} \tag{11.45}$$

The orthotropic constants are given by

$$D_x = \frac{(EI)_x}{1-\nu_x\nu_y},\ D_y = \frac{(EI)_y}{1-\nu_x\nu_y},\ D_1 = \nu_y D_x = D_y\,\nu_x \text{ and } D_{xy} = \frac{\left(1-\sqrt{\nu_x\nu_y}\right)}{2}\sqrt{D_x D_y}$$

For an isotropic plate the constants of eqn. (11.45) will be

$$D_x = D_y = D = \frac{Et^3}{12\left(1-\nu^2\right)}$$

$$D_1 = \nu D \tag{11.46}$$

$$D_{xy} = \frac{1-\nu}{2}D$$

Equation (11.45) in compact form, becomes

$$\{\sigma\} = [D]\{\varepsilon\} \tag{11.47}$$

Substituting $\{\varepsilon\}$ from eqn. (11.44) into eqn. (11.47), yields

$$\{\sigma\} = [D][B]\{X\}_e \tag{11.48}$$

$[D][B] =$

$$\frac{1}{ab}\begin{bmatrix} ZA \\ ZB & -ZC \\ ZD & ZE & ZF \\ ZG & ZH & 0 & ZA \\ -ZI & ZJ & 0 & ZB & ZC \\ -ZD & 0 & -ZF & ZD & ZE & ZF \\ -ZK & 0 & -ZM & 0 & 0 & 0 & ZA \\ -ZL & 0 & -ZN & 0 & 0 & 0 & ZB & -ZC \\ -ZD & ZE & 0 & ZD & 0 & 0 & ZD & -ZE & -ZF \\ 0 & 0 & 0 & -ZK & 0 & -ZM & -ZG & ZH & 0 & ZA \\ 0 & 0 & 0 & -ZL & 0 & -ZN & ZI & ZJ & 0 & ZB & ZC \\ -ZD & 0 & 0 & ZD & ZE & 0 & ZD & 0 & -ZF & -ZD & -ZE & ZF \end{bmatrix} \tag{11.49}$$

The product $[D]$ $[B]$ has been explicitly evaluated and it is given by eqn. (11.49), where

$$p = \frac{a}{b} \qquad ZG = 6D_1 p$$

$$ZA = 6D_x p^{-1} + 6D_1 p \qquad ZH = 2MD_1 a$$

$$ZB = 6D_y p + 6D_1 p^{-1} \qquad ZI = 6D_y p$$

$$ZC = 4D_y a \qquad ZJ = 2D_y a$$

$$ZD = 2D_{xy} \qquad ZK = 6D_x p^{-1}$$

$$ZE = 2bD_{xy} \qquad ZL = 6D_1 p^{-1}$$

$$ZF = 2aD_{xy} \qquad ZM = 2D_x b$$

$$ZN = 2D_1 b$$

11.3.5 Derivation of the Element Stiffness Matrix

The element stiffness matrix is derived here by applying the principle of minimum potential energy. The potential energy of the plate element is given by (11.2)

$$\Phi = \frac{1}{2}\int_0^a\int_0^b\left(-M_x\frac{\partial^2 w}{\partial x^2} - M_y\frac{\partial^2 w}{\partial y^2} + 2M_{xy}\frac{\partial^2 w}{\partial x\,\partial y}\right)dx\,dy$$

$$-\int_0^a\int_0^b \{f\}^T q\;dx\,dy \tag{11.50}$$

where q is any discrete loading inside the element. Based on the notations used so far for the rectangular plate-bending element, eqn. (11.50) can be written as

$$\Phi = \frac{1}{2}\int_0^a \int_0^b \{\varepsilon\}^T \{\sigma\}\, dx\, dy - \int_0^a \int_0^b w^T q\, dx\, dy \tag{11.51}$$

Combining eqns. (11.39), (11.44), (11.48) and (11.51), yields

$$\Phi = \frac{1}{2}\int_0^a \int_0^b \{X\}_e^T [B]^T [D][B]\{X\}_e\, dx\, dy - \int_0^a \int_0^b \{X\}_e^T [N]^T q\, dx\, dy \tag{11.52}$$

The principle of minimum potential energy requires

$$\left\{\frac{\partial \Phi}{\partial \{X\}_e}\right\} = \{0\} \tag{11.53}$$

Therefore, performing the partial differentiation of eqn. (11.52), yields

$$\left\{\frac{\partial \Phi}{\partial \{X\}_e}\right\} = \int_0^a \int_0^b [B]^T [D][B]\, dx\, dy\, \{X\}_e - \int_0^a \left| \int_0^a \right| [N]^T q\, dx\, dy = \{0\}$$

or,

$$[k]_e \{X\}_e = \{P\}_e \tag{11.54}$$

where

$$[k]_e = \int_0^a \int_0^b [B]^T [D][B]\, dx\, dy \tag{11.55}$$

$$\{P\}_e = \int_0^a \int_0^b [N]^T q\, dx\, dy \tag{11.56}$$

$[k]_e$ can be explicitly evaluated, but has not been presented here. The complete matrix $[k]_e$ for the skew element is given in the next section. By putting angle $\phi = 90^0$ for the skew element, the element stiffness matrix $[k]_e$ for the rectangular element can be obtained (Table 11.1).

If uniform load is present in the plate element *i.e.*, q is constant, then the element loading matrix is given by

$$\{P\}_e = q\, ab \begin{Bmatrix} 1/4 \\ -b/24 \\ \underline{a/24} \\ 1/4 \\ b/24 \\ \underline{a/24} \\ 1/4 \\ -b/24 \\ \underline{-a/24} \\ 1/4 \\ b/24 \\ -a/24 \end{Bmatrix} \tag{11.57}$$

11.4 PARALLELOGRAM ELEMENT OF PLATE BENDING

Many practical structures have intersecting edges, which are not orthogonal; some of them may have highly irregular boundaries. Out of them, structures like swept wings of aircrafts or skew bridges, and some others can be idealized as a plate having the shape of a parallelogram. Rectangular elements will not be able to fit into their boundary. If the rectangular elements are generalized into the shape of a quadrilateral, then the criterion of constant curvature cannot be satisfied. Parallelogram elements may be effectively used for the analysis of these plates.

A parallelogram element with dimensions and the coordinate system has been shown in Fig. 11.3. There are three degrees of freedom at each node – the unknown rotations are assumed along skew coordinate system. As such the displacements at node 1 are $\{w_1 \ \theta_{\xi 1} \ \theta_{\eta 1}\}$. The complete displacement and the force vectors for this element are

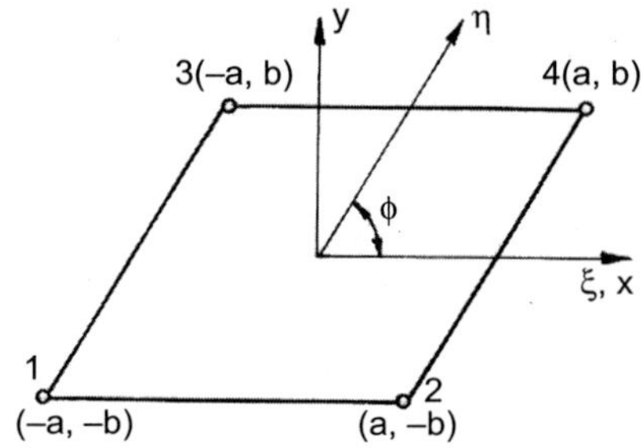

Fig. 11.3. Parallelogram element

$$\{X\}_e^T = \{w_1 \ \theta_{\xi 1} \ \theta_{\eta 1} \ w_2 \ \theta_{\xi 2} \ \theta_{\eta 2} \ w_3 \ \theta_{\xi 3} \ \theta_{\eta 3} \ w_4 \ \theta_{\xi 4} \ \theta_{\eta 4}\}$$

$$\{P\}_e^T = \{P_1 \ M_{\xi 1} \ M_{\eta 1} \ P_2 \ M_{\xi 2} \ M_{\eta 2} \ P_3 \ M_{\xi 3} \ M_{\eta 3} \ P_4 \ M_{\xi 4} \ M_{\eta 4}\} \quad (11.58)$$

In Fig. 11.3, two coordinates systems have been shown

$$\begin{Bmatrix} x \\ y \end{Bmatrix} = \begin{bmatrix} 1 & \cos\phi \\ 0 & \sin\phi \end{bmatrix} \begin{Bmatrix} \xi \\ \eta \end{Bmatrix} \quad (11.59)$$

Inverting eqn. (11.59)

$$\begin{Bmatrix} \xi \\ \eta \end{Bmatrix} = \begin{bmatrix} 1 & -\cot\phi \\ 0 & \operatorname{cosec}\phi \end{bmatrix} \begin{Bmatrix} x \\ y \end{Bmatrix} \quad (11.60)$$

11.4.1 Displacement Function

The displacement function for the plate is

$$\{f\}^T = w = \alpha_1 + \alpha_2\xi + \alpha_3\eta + \alpha_4\xi^2 + \alpha_6\eta^2 + \alpha_7\xi^3 + \alpha_8\xi^2\eta + \alpha_9\xi\eta^2 + \alpha_{10}\eta^3 + \alpha_{11}\xi^3\eta + \alpha_{12}\xi\eta^3 \quad (11.61)$$

The rotations are given by

$$\theta_\xi = -\frac{\partial w}{\partial \eta} = -(\alpha_3 + \alpha_5\xi + 2\alpha_6\eta + \alpha_8\xi^2 + 2\alpha_9\xi\eta + 3\alpha_{10}\eta^2 + \alpha_{11}\xi^3 + 3\alpha_{12}\xi\eta^2) \quad (11.62)$$

$$\theta_\eta = \frac{\partial w}{\partial \xi} = \alpha_2 + 2\alpha_4\xi + \alpha_5\eta + 3\alpha_7\xi^2 + 2\alpha_8\xi\eta + \alpha_9\eta^2 + 3\alpha_{11}\xi^2\eta + \alpha_{12}\eta^3 \quad (11.63)$$

11.4.2 Displacement Function in Terms of Nodal Displacements

Following exactly the same procedure by which eqn. (11.36) is obtained with the exception that the coordinates of the nodes are in terms of skew coordinates system, the following familiar equations can be obtained.

$$\{\alpha\} = [A]^{-1}\{X\}_e \tag{11.64}$$

$$\{f\} = w = [N]\{X\}_e \tag{11.65}$$

11.4.3 Curvature-Nodal Parameter Relationship

The curvatures in cartesian coordinate system are given by eqn. (11.40). As the displacement function w is a function of x and y, the curvatures are to be expressed in terms of skew coordinates, that is, a relationship of curvatures between the two axis systems is to be obtained.

Performing the necessary differentiation in eqn. (11.60), following results are obtained.

$$\frac{\partial \xi}{\partial x} = 1, \quad \frac{\partial \xi}{\partial y} = -\cot\phi, \quad \frac{\partial \eta}{\partial x} = 0, \quad \frac{\partial \eta}{\partial y} = \operatorname{cosec}\phi \tag{11.66}$$

By chain rule,

$$\left.\begin{aligned} \frac{\partial w}{\partial x} &= \frac{\partial w}{\partial \xi}\cdot\frac{\partial \xi}{\partial x} + \frac{\partial w}{\partial \eta}\cdot\frac{\partial \eta}{\partial x} \\ \frac{\partial w}{\partial y} &= \frac{\partial w}{\partial \xi}\cdot\frac{\partial \xi}{\partial y} + \frac{\partial w}{\partial \eta}\cdot\frac{\partial \eta}{\partial y} \end{aligned}\right\} \tag{11.67}$$

Substituting the values of the differentiated quantities of eqn. (11.66) into eqn. (11.67), yields

$$\left.\begin{aligned} \frac{\partial w}{\partial x} &= \frac{\partial w}{\partial \xi} \\ \frac{\partial w}{\partial y} &= -\cot\phi\frac{\partial w}{\partial \xi} + \operatorname{cosec}\phi\frac{\partial w}{\partial \eta} \end{aligned}\right\} \tag{11.68}$$

Equation (11.68) reveals

$$\frac{\partial}{\partial x} = \frac{\partial}{\partial \xi}$$

Hence,

$$\frac{\partial^2}{\partial x^2} = \frac{\partial^2}{\partial \xi^2} \tag{11.69}$$

Similarly, second equation of eqn.(11.68) yields in terms of the operator

$$\frac{\partial}{\partial y} = -\cot\phi\frac{\partial}{\partial \xi} + \operatorname{cosec}\phi\frac{\partial}{\partial \eta} \tag{11.70}$$

Hence,

$$\frac{\partial^2}{\partial y^2} = \cot^2\phi\frac{\partial^2}{\partial \xi^2} + \text{cosec}^2\phi\frac{\partial^2}{\partial \eta^2} - 2\cot\phi\,\text{cosec}\,\phi\frac{\partial^2}{\partial \xi\partial \eta} \qquad (11.71)$$

Similarly,

$$\frac{\partial^2}{\partial x\partial y} = \cot\phi\frac{\partial^2}{\partial \xi^2} + \text{cosec}\,\phi\frac{\partial^2}{\partial \xi\partial \eta} \qquad (11.72)$$

Relations (11.69), (11.71) and (11.72) when written in matrix form become

$$\begin{Bmatrix} -\dfrac{\partial^2 w}{\partial x^2} \\ -\dfrac{\partial^2 w}{\partial y^2} \\ 2\dfrac{\partial^2 w}{\partial x\partial y} \end{Bmatrix} = \begin{bmatrix} 1 & 0 & 0 \\ \cot^2\phi & \text{cosec}^2\phi & \cot\phi\,\text{cosec}\,\phi \\ 2\cot\phi & 0 & \text{cosec}\,\phi \end{bmatrix} \begin{Bmatrix} -\dfrac{\partial^2 w}{\partial \xi^2} \\ -\dfrac{\partial^2 w}{\partial \eta^2} \\ 2\dfrac{\partial^2 w}{\partial \xi\partial \eta} \end{Bmatrix} \qquad (11.73)$$

Equation (11.73) in compact form is written as

$$\{\varepsilon\}xy = [H]\{\varepsilon\}_{\xi\eta} \qquad (11.74)$$

By differentiation of eqn. (11.65), the curvatures in skew coordinates can be expressed as

$$\{\varepsilon\}_{\xi\eta} = [B]\{X\}_e \qquad (11.75)$$

Combining eqns. (11.74) and (11.75), yields

$$\{\varepsilon\}_{xy} = [H][B]\{X\}_e \qquad (11.76)$$

11.4.4 NMoment-Curvature Relationship

In cartesian system, moment-curvature relationship is given by

$$\{\sigma\}_{xy} = [D]\ \{\varepsilon\}_{xy} \qquad (11.77)$$

where $\{\sigma\}_{xy}^T = \{M_x\ M_y\ M_{xy}\}$

In eqn. (11.77), substitution of $\{\varepsilon_{xy}\}$ from eqn. (11.76) yields

$$\{\sigma\}_{xy} = [D][H][B]\{X\}_e \qquad (11.78)$$

11.4.5 Element Stiffness Matrix

Total potential energy is given by

$$\Phi = \frac{1}{2}\iint \{\varepsilon\}_{xy}^T\{\sigma\}_{xy}\ dx\,dy - \iint w^T q\ dx\ dy \qquad (11.79)$$

Substituting $\{\varepsilon\}_{xy}$, $\{\sigma\}_{xy}$ and w from eqn. (11.76), (11.78) and (11.65) into eqn. (11.79), yields

$$\Phi = \frac{1}{2}\iint \{X\}_e^T[B]^T[H]^T[D][H][B]\ dx\,dy - \iint [N]^T q\ dx\,dy \qquad (11.80)$$

Applying the principle of minimum potential energy, eqn. (11.80) reduces to

$$[k]_e\{X\}_e = \{P\}_e \tag{11.81}$$

where
$$[k]_e = \iint [B]^T [H]^T [D][H][B]\, dx\, dy \tag{11.82}$$

$$\{P\}_e = = \iint [N]^T q\, dx\, dy \tag{11.83}$$

$[B]$ matrix of eqn. (11.82) has been expressed in terms of ξ and η. Therefore, the integration is to be performed in terms of ξ and η

It has been shown in eqn. (11.25) that

$$dx\, dy = |J|\, d\xi\, d\eta \tag{11.84}$$

where

$$|J| = \begin{vmatrix} \dfrac{\partial x}{\partial \xi} & \dfrac{\partial y}{\partial \xi} \\ \dfrac{\partial x}{\partial \eta} & \dfrac{\partial y}{\partial \eta} \end{vmatrix} = \begin{vmatrix} 1 & 0 \\ \cot\phi & \sin\phi \end{vmatrix} = \sin\phi \tag{11.85}$$

The element stiffness matrix of eqn. (11.82) then finally becomes

$$[k]_e = \sin\phi \iint [B]^T [H]^T [D][H][B]\, d\xi\, d\eta \tag{11.86}$$

The explicit value of $[k]_e$ is given in Table 11.1.

11.5 HERMITIAN POLYNOMIAL INTERPOLATION

Hermitian polynomials have been used to write down directly the shape functions for a conforming plate bending element. Hermitian polynomial interpolation is useful in cases where higher derivatives are introduced as degrees of freedom of the plate.

Slopes are the derivatives of the deflection function. So, when the formulation is thought of in terms of directly choosing the shape function (eqn. (11.65)), the function should be such that if it is one, its derivative/derivatives are zero at a particular node. The choice of such a function can be conveniently done by using Hermitian polynomials. Mathematically, a Hermitian polynomial is given as $N_{ri}^{m}(x)$. Here the polynomial is of order $2m + 1$ and is such that at $x = x_i$

$$\left.\begin{aligned} &\frac{d^s N}{dx^s} = 1 \text{ when } s = r \text{ for } r = 0 \text{ to } m \\ \text{and} \quad &\frac{d^s N}{dx^s} = 0 \text{ when } s \neq r \text{ when } x = x_j \end{aligned}\right\} \tag{11.87}$$

This is illustrated with an example.

A flexural line element having deflection and rotation at its two ends is shown in Fig. 11.4.

Table 11.1. Element Stiffness Matrix of the Skew Element

$$
[K]_S = \frac{D}{4Sin^3\phi}\left[\begin{array}{cccccccccccc}
16SA+8SB+16SD+\frac{28}{5}SF & & & & & & & & & & & \\
8SA+4SB+\frac{2}{5}SF & \frac{16}{3}SA+2SC+\frac{8}{15}SF & & & & & & & & & & \\
4KB+8KD+\frac{2}{5}KF & 4KB+\frac{1}{3}KC+\frac{1}{3}KE & \frac{16}{3}WD+2WE+\frac{8}{15}WF & & & & & & & & & \\
-16SA-8SB+8SD-\frac{28}{5}SF & -8SA-\frac{2}{5}SF & -4KB+4KD+4KE-\frac{2}{5}KF & 16SA+8SB+16SD+\frac{28}{5}SF & & & & & & & & \\
8SA+\frac{2}{5}SF & \frac{8}{3}SA-\frac{2}{15}SF & -\frac{1}{3}KC-\frac{1}{3}KE & -8SA-4SB-\frac{2}{5}SF & \frac{16}{3}SA-2SC+\frac{8}{15}SF & & & & & & & \\
-4KB+4KD-4KE-\frac{2}{5}KF & -\frac{1}{3}KC-\frac{1}{3}KE & \frac{8}{3}WD-\frac{8}{15}WF & 4KB+8KD+\frac{2}{5}KF & -4KB+\frac{1}{3}KC+\frac{1}{3}KE & \frac{16}{3}WD-2WE+\frac{8}{15}WF & & & & & & \\
8SA-8SB-16SD-\frac{28}{5}SF & 4SA+4SC-4SB-\frac{2}{5}SF & -8KD-\frac{2}{5}KF & -8SA+8SB-8SD+\frac{28}{5}SF & 4SA-4SC-\frac{2}{5}SF & -4KB+4KE+\frac{2}{5}KF & 16SA+8SB+16SD+\frac{28}{5}SF & & & & & \\
4SA-4SB-4SC-\frac{2}{5}SF & \frac{8}{3}SA-\frac{8}{15}SF & -\frac{1}{3}KC-\frac{1}{3}KE & -4SA+4SC+\frac{2}{5}SF & \frac{4}{3}SA-2SC+\frac{2}{15}SF & \frac{1}{3}KC+\frac{1}{3}KE & 8SA+4SB+\frac{2}{5}SF & \frac{16}{3}SA-2SC+\frac{8}{15}SF & & & & \\
8KD+\frac{2}{5}KF & -\frac{1}{3}KC-\frac{1}{3}KE & \frac{8}{3}WD-\frac{2}{15}WF & 4KD-4KE-\frac{2}{5}KF & \frac{1}{3}KC+\frac{1}{3}KE & \frac{4}{3}WD-2WE+\frac{2}{15}WF & -4KB-8KD-\frac{2}{15}KF & -4KB+\frac{1}{3}KC+\frac{1}{3}KE & \frac{16}{3}WD-2WE+\frac{8}{15}WF & & & \\
-8SA+8SB-8SD+\frac{28}{5}SF & -4SA-4SC+\frac{2}{5}SF & -4KD-4KE+\frac{2}{5}KF & 8SA-8SB-16SD-\frac{28}{5}SF & -4SA+4SB+4SC+\frac{2}{5}SF & -8KD-\frac{2}{5}KF & -16SA-8SB-\frac{28}{5}SF+8SD & -8SA-\frac{2}{5}SF & 4KB-4KD+4KE+\frac{2}{5}KF & 16SA+8SB+\frac{28}{5}SF+16SD & & \\
4SA+4SC-\frac{2}{5}SF & \frac{4}{3}SA+2SC+\frac{2}{15}SF & \frac{1}{3}KC+\frac{1}{3}KE & -4SA+4SB-4SC+\frac{2}{5}SF & \frac{8}{3}SA-\frac{8}{15}SF & -\frac{1}{3}KC-\frac{1}{3}KE & 8SA+\frac{2}{5}SF & \frac{8}{3}SA-\frac{2}{15}SF & -\frac{1}{3}KC-\frac{1}{3}KE & -8SA-4SB-\frac{2}{5}SF & \frac{16}{3}SA+2SC+\frac{8}{15}SF & \\
4KD+4KE-\frac{2}{5}KF & \frac{1}{3}KC+\frac{1}{3}KE & \frac{4}{3}WD+2WE+\frac{2}{15}WF & 8KD+\frac{2}{5}KF & -\frac{1}{3}KC-\frac{1}{3}KE & \frac{8}{3}WD-\frac{2}{15}WF & 4KB-4KD-4KE+\frac{2}{5}KF & -\frac{1}{3}KC-\frac{1}{3}KE & \frac{8}{3}WD-\frac{8}{15}WF & -4KD-8KD-\frac{2}{5}KF & 4KB+\frac{1}{3}KC+\frac{1}{3}KE & \frac{16}{3}KD+2KE+\frac{8}{15}KF
\end{array}\right]
$$

$\frac{b}{a}=p$ $\quad A=1$ $\quad B=\upsilon Sin^2\phi+\upsilon Cos^2\phi$

$C=-2Cos\phi$ $\quad D=1$ $\quad E=-2Cos\phi$

$F=2(1-\upsilon)+2(1+\upsilon)Cos^2\phi$

$SA=p\cdot A$ $\quad KA=p^2\cdot A$ $\quad SD=\frac{B}{P}$ $\quad WB=p\cdot B$

$SC=C$ $\quad KC=p\cdot c$ $\quad SD=\frac{D}{p^2}$ $\quad WD=\frac{D}{p}$

$SE=\frac{E}{p^2}$ $\quad KE=\frac{E}{P}$ $\quad WE=E$

$SF=\frac{F}{p}$ $\quad KF=F$ $\quad WF=p.F$

$KB=B$ $\quad KD=\frac{D}{p^2}$

The function is given as

$$f = N_{01}\, w_1 + N_{11}\, \theta x_1 + N_{02}\, w_2 + N_{12}\, \theta x_2 \tag{11.88}$$

Now, there exist two conditions per node *i.e.* in all four conditions are available for the construction of N_{01}, N_{11} *etc.* $N_{01} = 1$ when associated with w_1 and must be zero when the evaluation is performed for the remaining degrees of freedom. Therefore, in the present case, a cubic polynomial can be chosen, because the four constants associated with it can be determined on the basis of four conditions mentioned above.

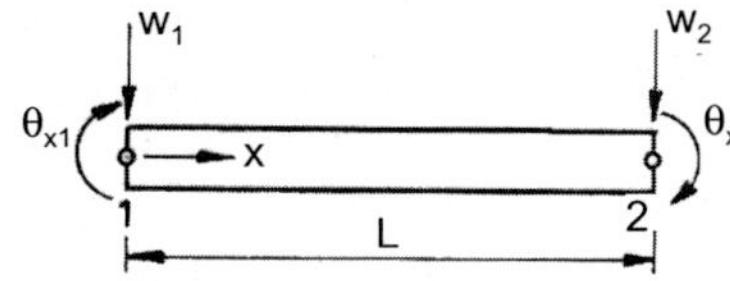

Fig. 11.4. Rectangular plate bending element

The polynomial then is

$$N_i = \alpha_1 + \alpha_2 x + \alpha_3 x^2 + \alpha_4 x^3 \tag{11.89}$$

Let us look into the procedure for determination of N_{11}, *i.e.* the shape function corresponding to θ_{x1} term. Substituting necessary conditions, the following four equations result

$$\left.\begin{aligned} &\text{At } x = 0,\ N_i = 0 \text{ gives } \alpha_1 = 0 \\ &\text{At } x = 0,\ \frac{dN_i}{dx} = 1 \text{ gives } \alpha_2 = 1 \\ &\text{At } x = L,\ N_i = 0 \text{ gives } \alpha_2 L + \alpha_3 L^2 + \alpha_4 L^3 = 0 \\ &\text{At } x = L,\ \frac{dN_i}{dx} = 0 \text{ gives } \alpha_2 + 2\alpha_3 L + 3\alpha_4 L^2 = 0 \end{aligned}\right\} \tag{11.90}$$

By the solution of the four equations of eqn. (11.90), results

$$N_{11} = x\,(1 - \xi)^2 \quad \text{where } \xi = \frac{x}{L} \tag{11.91}$$

Other functions can be similarly obtained.

$$N_{01} = 1 - 3\xi^2 + 2\xi^3,\ N_{02} = 3\xi^2 - 2\xi^3,\ N_{12} = x(\xi^2 + \xi) \tag{11.92}$$

These Hermitian interpolation functions can be extended into two dimensions. This is illustrated in the conforming plate bending element discussed in the next section.

11.6 A CONFORMING PLATE BENDING ELEMENT

Based on the concept of Hermitian polynomial introduced, it can be verified that at node i, the shape function is given by

$$N_i = [N_{0i}(x)\, N_{0i}(y),\ N_{1i}(x)\, N_{0i}(y),\ N_{0i}(x)\, N_{1i}(y),\ N_{1i}(x)\, N_{1i}(y)] \tag{11.93}$$

which correspond to w, $\dfrac{\partial w}{\partial x}$, $\dfrac{\partial w}{\partial y}$ and $\dfrac{\partial^2 w}{\partial x\, \partial y}$

Introducing Hermitian polynomials, the continuity of slopes along edges is satisfied and hence the element analyzed on this basis can be termed as conforming element.

For the element of Fig. 11.2, the degrees of freedom per node are

$$w, \theta_x = \frac{\partial w}{\partial x}, \theta_y = \frac{\partial w}{\partial y} \text{ and } \theta_{xy} = \frac{\partial w^2}{\partial x \, \partial y}$$

There are, therefore, 16 degrees of freedom for this element.

The displacement function for this element as chosen is given below.

$$\left.\begin{aligned} w = {} & N_{01}(x)\, N_{01}(y) w_1 + N_{11}(x)\, N_{01}(y)\, \theta_{x1} + N_{01}(x)\, N_{11}(y) \theta_{y1} + N_{11}(x)\, N_{11}(y) \theta_{xy1} \\ & + N_{02}(x)\, N_{02}(y) w_2 + N_{12}(x)\, N_{02}(y) \theta_{x2} + N_{02}(x)\, N_{12}(y) \theta_{y2} + N_{12}(x)\, N_{12}(y) \theta_{xy2} \\ & + N_{03}(x)\, N_{03}(y) w_3 + N_{13}(x)\, N_{03}(y) \theta_{x3} + N_{03}(x)\, N_{13}(y) \theta_{y3} + N_{13}(x)\, N_{13}(y) \theta_{xy3} \\ & + N_{04}(x)\, N_{04}(y) w_4 + N_{14}(x)\, N_{04}(y) \theta_{x4} + N_{04}(x)\, N_{14}(y) \theta_{y4} + N_{14}(x)\, N_{14}(y) \theta_{xy4} \end{aligned}\right\} \tag{11.94}$$

Once, the displacement function has been defined, the rest of the derivation follows similar lines to those already presented.

11.7 ISOPARAMETRIC PLATE BENDING ELEMENT

Isoparametric bending element can be used for the analysis of both thick and thin plates. So this element presents an advantage over the other plate bending elements discussed so far – it can take into account the shear deformation of the plate. The formulation of this element is more elegant than that of other plate flexure elements, which incorporate shear. This element has been effectively used for the analysis of sandwich and cellular plates (11.11, 11.12).

In the theory of these plates it has been assumed that normal to the middle surface before deformation remains normal after deformation of the plate. When shearing deformation is considered this assumption does not remain valid and it is to be modified. The derivation of the element is based on the suggestion by Mindlin that normals to the middle plane of the plate before bending remain straight, but not necessarily normal to the middle surface after bending. Therefore, the rotations θ_x and θ_y of the plate shown in Fig. 11.5 are given by

$$\begin{Bmatrix} \theta_x \\ \theta_y \end{Bmatrix} = \begin{Bmatrix} \dfrac{\partial w}{\partial x} + \phi_x \\ \dfrac{\partial w}{\partial y} + \phi y \end{Bmatrix} \tag{11.95}$$

where ϕ_x and ϕ_y are average shear rotation over the thickness of the plate.

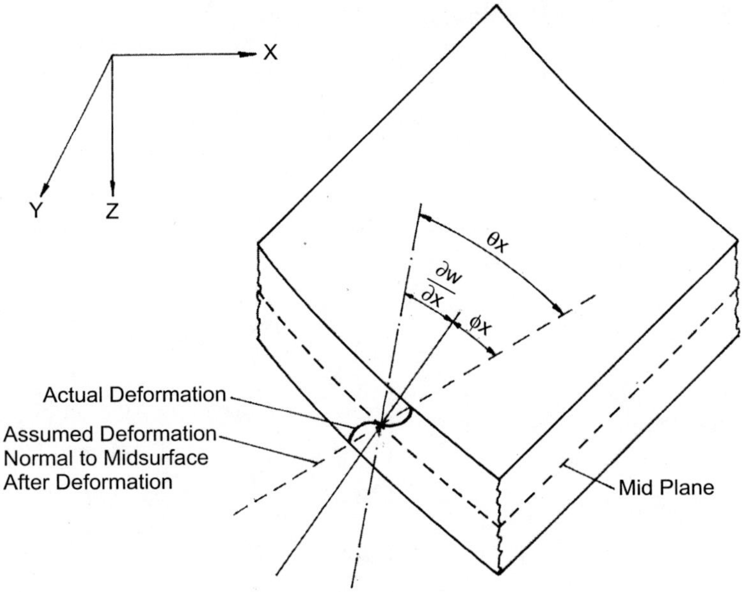

Fig. 11.5. Deformation of the plate in XZ plane

The displacement of the plate may be fully described by

$$\{f\} = \begin{Bmatrix} w \\ \theta_x \\ \theta_y \end{Bmatrix} \tag{11.96}$$

11.7.1 Displacement Function

A general formulation is presented so that the derivation is valid for all types of isoparametric elements – linear, quadratic, cubic *etc.*, only the appropriate shape function is to be substituted.

The coordinates within an element are

$$\begin{Bmatrix} x \\ y \end{Bmatrix} = \sum_{r=1}^{n} \begin{bmatrix} 1 & 0 \\ 0 & 1 \end{bmatrix} N_r \begin{Bmatrix} x_r \\ y_r \end{Bmatrix} \tag{11.97}$$

where N_r is the shape function as discussed in Sec. 10.3, x_r, y_r are the nodal coordinates and n is the number of nodes. A quadratic element after mapping is shown in Fig. 11.6.

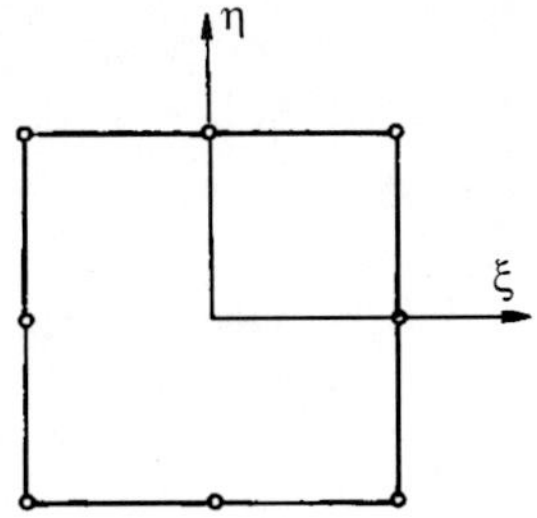

Fig. 11.6. A non-dimensional mapped element

Similarly, the displacement function is given by

$$\begin{Bmatrix} w \\ \theta_x \\ \theta_y \end{Bmatrix} = \sum_{r=1}^{n} N_r \begin{bmatrix} 1 & 0 & 0 \\ 0 & 1 & 0 \\ 0 & 0 & 1 \end{bmatrix} \begin{Bmatrix} w_r \\ \theta_{xr} \\ \theta_{yr} \end{Bmatrix} \tag{11.98}$$

or, $$\{f\} = [N]\,\{X\}_e \tag{11.99}$$

11.7.2 Strain-Nodal Displacement Relationship

The generalized strains are

$$\{\varepsilon\}^T = \left\{ -\frac{\partial \theta_x}{\partial x} \; -\frac{\partial \theta_y}{\partial y} \; -\left(\frac{\partial \theta_x}{\partial y} + \frac{\partial \theta_y}{\partial x}\right) \; -\phi_x \; -\phi_y \right\} \tag{11.100}$$

Therefore, performing the necessary differentiation with respect to the different quantities given in eqn. (11.98), yields

$$\begin{Bmatrix} -\dfrac{\partial \theta_x}{\partial x} \\ -\dfrac{\partial \theta_y}{\partial y} \\ -\left(\dfrac{\partial \theta_x}{\partial y} + \dfrac{\partial \theta_y}{\partial x}\right) \\ -\phi_x \\ -\phi_y \end{Bmatrix} = \sum_{r=1}^{n} \begin{bmatrix} 0 & -\dfrac{\partial N_r}{\partial x} & 0 \\ 0 & 0 & -\dfrac{\partial N_r}{\partial y} \\ 0 & -\dfrac{\partial N_r}{\partial y} & -\dfrac{\partial N_r}{\partial x} \\ \dfrac{\partial N_r}{\partial x} & -N_r & 0 \\ \dfrac{\partial N_r}{\partial y} & 0 & -N_r \end{bmatrix} \begin{Bmatrix} w_r \\ \theta_{xr} \\ \theta_{yr} \end{Bmatrix} \tag{11.101}$$

While deriving eqn. (11.101), the following assumption has been made

$$\left.\begin{aligned} -\phi_x &= \frac{\partial w}{\partial x} - \theta_x \\ -\phi_y &= \frac{\partial w}{\partial y} - \theta_y \end{aligned}\right\} \tag{11.102}$$

Equation. (11.101) written in compact form becomes

$$\{\varepsilon\} = \sum_{r=1}^{n} [B]_r \{X_r\}_e \tag{11.103}$$

or,

$$\{\varepsilon\} = [B]\{X\}_e \tag{11.104}$$

The shape function derivatives are to be obtained by chain rules as given in Sec. 10.3.

11.7.3 Stress-Strain Relationship

The stress components to be considered in this case are

$$\{\sigma\}^T = \{M_x \; M_y \; M_{xy} \; V_x \; V_y\} \tag{11.105}$$

The stress-strain relationship for an orthotropic plate is

$$\begin{Bmatrix} M_x \\ M_y \\ M_{xy} \\ V_x \\ V_y \end{Bmatrix} = \begin{bmatrix} D_x & D_1 & 0 & 0 & 0 \\ D_1 & D_y & 0 & 0 & 0 \\ 0 & 0 & D_{xy} & 0 & 0 \\ 0 & 0 & 0 & S_x & 0 \\ 0 & 0 & 0 & 0 & S_y \end{bmatrix} \begin{Bmatrix} -\dfrac{\partial \theta_x}{\partial x} \\ -\dfrac{\partial \theta_y}{\partial y} \\ -\left(\dfrac{\partial \theta_x}{\partial y} + \dfrac{\partial \theta_y}{\partial x}\right) \\ -\phi_x \\ -\phi_y \end{Bmatrix} \tag{11.106}$$

where S_x and S_y are shear rigidities. For an isotropic plate they are

$$S_x = S_y = \frac{Gt}{1.2} \tag{11.107}$$

In compact form, eqn. (11.106) becomes

$$\{\sigma\} = [D]\{\varepsilon\} \tag{11.108}$$

11.7.4 Element Stiffness Matrix

The element stiffness matrix defined in the usual way is given by

$$[k]_e = \int_{-1}^{1} \left| \int_{-1}^{1} \right| [B]^T [D][B][J]\, d\varepsilon\, d\eta \tag{11.109}$$

The integration is to be carried out numerically by employing Gaussian Quadrature. For this isoparametric plate-bending element, it has been found that the best results of the integration can be obtained by using 'reduced integration technique'. This is explained below.

11.7.5 Reduced Integration Technique

It has already been stated that Gaussian integration of n points gives exact values for the integral of polynomials of degree $(2n - 1)$ in x. In order to evaluate the final values of element stiffness matrix, elements of $[B]^T$ $[D]$ $[B]$ are to be studied in details. The shear strain components are functions of N_r and its derivative whereas flexural strain components are functions of derivatives of N_r only. Therefore, for all the cases of linear, quadratic and other isoparametric elements, shear strain components are one order higher than flexural strain components, which introduces an inconsistency in the analysis that is responsible for the poor behaviour of these elements in shear. This requires more detailed discussion. Isoparametric quadratic element is taken up first.

11.7.5.1 Isoparametric quadratic element

For an isoparametric quadratic element, the shape functions are quadratic in ξ and η and their derivatives are linear. It can be seen in $[B]$ matrix of eqn. (11.101) that flexural strain terms are derivatives of shape function and hence vary linearly whereas shear strain terms are functions of both the shape functions and its derivatives and have a quadratic variation. Thus, shear strain terms have a higher order variation, which introduces inconsistency. Thus, for the exact evaluation of flexural components in eqn. (11.109), a 2 × 2 Gaussian quadrature rule is to be employed, whereas exact evaluation of the shear contribution is to be done by 3 × 3 Gaussian quadrature. The results of the shear forces obtained by 3 × 3 Gaussian integration for the thin plate have been found to be erratic. On the other hand, it has been found that a reduced integration of 2 × 2 (which also is economic) gives vastly superior results.

It is preferable to have all terms for moments and shears having the same bilinear variation. Figure 11.7 shows a least square straight line fit to a parabola and it has been shown that they intersect at Gauss point values. Thus, if a 2 × 2 Gaussian integration is performed, the quadratic function is sampled at points such that the result will be identical to that obtained by use of 'smoothed' (linear) shape functions. So both the functions are reduced to the same order by the least square approximation. Thus, quadratic function is replaced by a linear function.

The argument presented above is no doubt approximate, but it has been found that the use of 2 × 2 integration has provided good convergence characteristics.

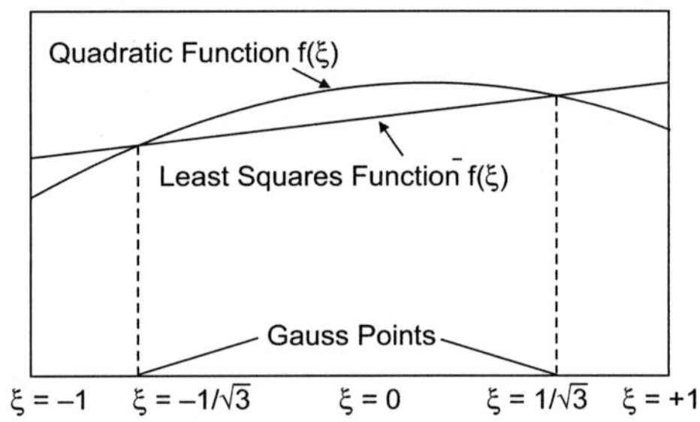

Fig. 11.7. A least square straight line fitting the quadratic function

11.7.5.2 Isoparametric linear element

For an isoparametric linear element, shape functions are linear in ξ and η and their derivatives are constants (11.4). $[B]$ matrix in eqn. (11.101] reveals that moment terms are constant whereas shear terms are linear functions of ξ and η.

The elements of $[B]^T$ $[D]$ $[B]$ in the element stiffness matrix can have a maximum of second order variation and accordingly a 2 × 2 Gauss integration is to be employed for exact evaluation. A reduced one-point integration is to be performed for removing the inherent inconsistency.

'If an n th order function $f(\xi)$ (where $-1 \leq \xi \leq 1$) is sampled at n points of an n-point Gauss rule, then the n values of the functions $f(\xi)$ at these points uniquely define a function of order $(n-1)$ which is a least square fit of the function $f(\xi)$. The interpretation of the above statement in the case of a linear function is represented in Fig. 11.8 (11.4). The original trapezium is substituted by a rectangle – the inclined side of the trapezium and the horizontal

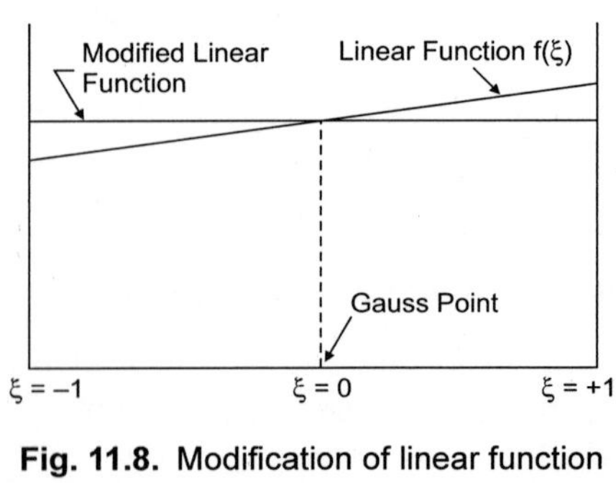

Fig. 11.8. Modification of linear function

top side of the rectangle intersects at the centre *i.e.,* at the Gauss point for one-point integration. The linear terms of these are thus replaced by the their Gauss point values which are zero in the present case, and so will yield the exact value of the function at that point. This reduced one-point integration not only economises the solution, but also improves the analysis to a great extent.

11.8 SMOOTHED STRESSES

It has been found that the values of the moments acting at the nodes are reasonably accurate, whereas the nodes are the worst sampling points for shear. As reduced integration is equivalent to the use of substitute lower order variation it is suggested that the values of the stresses be determined at Gauss points and then extrapolated to get the nodal values. These stresses are referred to as smoothed stresses.

Stress can be smoothed in another way (11.3). In reduced integration technique, the element stiffness matrix has been manipulated whereas $[B]$ matrix remains unchanged. If $[B]$ matrix is modified by least square smoothing, the variation of both the stress and strain field within the element is improved. This is done in the quadratic element by substituting the quadratic terms in $[B]$ matrix by their least square linear fit, *i.e.*, ξ^2 and η^2 are replaced by 1/3. In this way, nodes can be taken as the sampling points and the values will be identical to those obtained from bilinear extrapolation of Gauss point values.

Similar smoothing can also be done in the linear element. The linear terms of ξ, η in $[B]$ matrix are replaced by their Gauss point values *i.e.*, zero in this case. The spurious shear effect is thus removed. This reduces isoparametric linear element to a constant moment, constant shear element.

11.9 TRIANGULAR PLATE BENDING ELEMENTS

Considerable attempt has been made to develop triangular plate bending elements. In the initial stages difficulties have been encountered in the formulation. A three-noded triangle will have in all 9 degrees of freedom. To choose satisfactory displacement function based on polynomial expansion posed problems. A complete quadratic polynomial will contain 6 terms and a complete cubic will contain 10 terms. To maintain geometric isotropy and then the choice of 9 terms, led into serious difficulties. Two suggested displacement functions are given below.

$$w = \alpha_1 + \alpha_2 x + \alpha_3 y + \alpha_4 x^2 + \alpha_5 y^2 + \alpha_6 x^3 + \alpha_7 x^2 y + \alpha_8 xy^2 + \alpha_9 y^3 \tag{11.110}$$

and

$$w = \alpha_1 + \alpha_2 x + \alpha_3 y + \alpha_4 x^2 + \alpha_5 xy + \alpha_6 y^2 + \alpha_7 x^3 + \alpha_8 (x^2 y + xy^2) + \alpha_9 y^3 \tag{11.111}$$

In eqn. (11.110) omitting the '*xy*' term resulted in the assumed displacement function incapable of representing constant twist term within the element. This made the element far too stiff. Though the formulation based on the displacement function (11.111) gave reasonably good results, yet it was observed that the matrix $[A]$ relating the generalized displacements to the nodal displacements was becoming singular for certain orientation of the element.

Attempts to develop a satisfactory element may fall under one of the following classes:

(*a*) The element may be defined with 6 degrees of freedom – three vertical displacements at the three corner nodes and three slopes at the midside nodes. The displacement function can then be represented by a complete quadratic polynomial. Inter-element displacement continuity conditions however could not be satisfied.

(*b*) In order to obtain a displacement function, which can be represented by a complete cubic polynomial of 10 terms, insert an unknown displacement at an internal node. Then the effect of the internal node is eliminated by static condensation. In this case also the inter-elemental continuity conditions cannot be satisfied.

(*c*) In order to satisfy inter-elemental continuity conditions, a complete polynomial requires 21 terms so that the total degrees of freedom are 21.

(*d*) Satisfactory stiffness matrix having 9 degrees of freedom has been achieved by using area coordinates. Continuous displacement field has been obtained by superimposition of proper combination of shape functions.

For further details of triangular plate bending elements, Ref. 11.5 may be consulted.

11.10 DKT ELEMENT

An efficient triangular plate bending element has been proposed. It is known as discrete Kirchhoff triangle or DKT element [11.6,11.7]. The element is developed on the basis of imposition of zero transverse shear strains at specific points. The basic steps for the derivation of stiffness matrix are given below.

The initial element with degrees of freedom is shown in Fig. 11.9(a). It has 6 nodes and 12 d.o.f and involves the area coordinates in the formulation. The final 9 d.o.f element is shown in Fig. 11.9 (b).

The starting point is a straight-sided element with corner and midside nodes as shown in Fig. 11.9(a). The slopes θ_x and θ_y at the plate middle surface are expressed in terms of nodal rotations as follows:

$$\theta_x = \sum_{i=1}^{6} N_i\,\theta_{xi},\ \theta_y = \sum_{i=1}^{6} N_i\,\theta_{yi} \tag{11.112}$$

where N_i are the shape functions expressed in area coordinates [eqn. (11.50)].

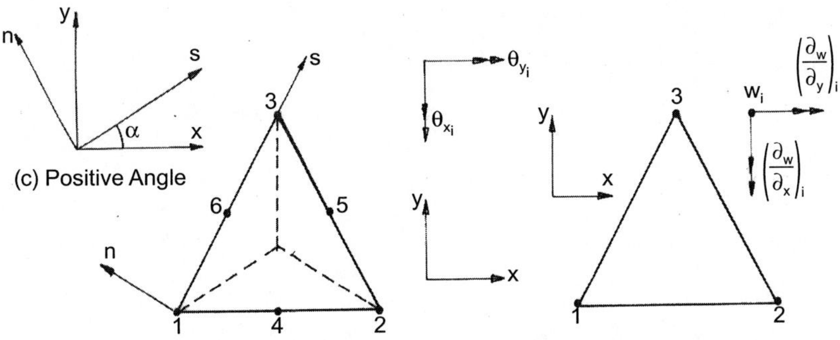

Fig. 11.9. DKT Element

A nine d.o.f element is sought that has nodal d.o.f as shown in Fig. 11.9(b). Accordingly, 12 d.o.f θ_{xi} and θ_{yi} at nodes 1 through 6 must be expressed in terms of w_i, θ_{xi} and θ_{yi} at only corner nodes.

The lateral deflection is assumed to vary cubically in an edge-tangent coordinate system. The value of s is equal to zero at one node and equal to the length of the side at the other node.

The lateral deflection w along the edge, say 13, can be expressed as

$$w = a_0 + a_1 s + a_2 s^2 + a_3 s^3 \tag{11.113}$$

where a_0, ..., a_4 are constants.

The rotations at any midside node, say 4 are to be obtained. This can be done by differentiating w given by eqn. (11.10) after substituting $[N]$ from eqn. (11.12) and it becomes

$$\left(\frac{\partial w}{\partial s}\right)_4 = -\frac{3}{2L_{12}} w_1 - \frac{1}{4}\left(\frac{\partial w}{\partial s}\right)_1 + \frac{3}{2L_{12}} w_2 - \frac{1}{4}\left(\frac{\partial w}{\partial s}\right)_2 \tag{11.114}$$

where L_{12}, is the length of the edge 12.

The detailed derivation of it is shown later.

Similarly,

$$\left(\frac{\partial w}{\partial s}\right)_5 = -\frac{3}{2L_{23}} w_2 - \frac{1}{4}\left(\frac{\partial w}{\partial s}\right)_2 + \frac{3}{2L_{23}} w_3 - \frac{1}{4}\left(\frac{\partial w}{\partial s}\right)_3 \tag{11.115}$$

$$\left(\frac{\partial w}{\partial s}\right)_6 = -\frac{3}{2L_{13}} w_3 - \frac{1}{4}\left(\frac{\partial w}{\partial s}\right)_3 + \frac{3}{2L_{13}} w_1 - \frac{1}{4}\left(\frac{\partial w}{\partial s}\right)_1 \tag{11.116}$$

The slopes $\frac{\partial w}{\partial y}$ in eqns. (11.114) to (11.116) can be expressed in terms of $\frac{\partial w}{\partial x}$ and $\frac{\partial w}{\partial y}$ by transforming the coordinates.

11.10.1 Constraint Equations

The following constraint equations are used.

1. Transverse shear strains γ_{yz} and γ_{zx} vanish at corners 1, 2 and 3. Thus,

$$\theta_{xi} = \left(\frac{\partial w}{\partial x}\right)_i, \theta_{yi} = \left(\frac{\partial w}{\partial x}\right)_i, \quad \text{for } i = 1, 2, 3 \tag{11.117}$$

2. Transverse shear strain γ_{sz} vanishes at nodes 4, 5 and 6.
   Thus,

$$\theta_{si} = \left(\frac{\partial w}{\partial s}\right)_i, \quad \text{for } i = 4, 5, 6 \tag{11.118}$$

$\left(\frac{\partial w}{\partial s}\right)_i$ is expressed in terms of $\left(\frac{\partial w}{\partial x}\right)_i$ and $\left(\frac{\partial w}{\partial y}\right)_i$ by coordinate transformation.

3. Normal slopes are assumed to vary linearly along the edge. Thus

$$\left.\begin{aligned}
\theta_{n4} &= \frac{1}{2}\left[\left(\frac{\partial w}{\partial n}\right)_1 + \left(\frac{\partial w}{\partial n}\right)_2\right], \\
\theta_{n5} &= \frac{1}{2}\left[\left(\frac{\partial w}{\partial n}\right)_2 + \left(\frac{\partial w}{\partial n}\right)_3\right], \\
\theta_{n6} &= \frac{1}{2}\left[\left(\frac{\partial w}{\partial n}\right)_3 + \left(\frac{\partial w}{\partial n}\right)_1\right],
\end{aligned}\right\} \tag{11.119}$$

Applying the constraint equations mentioned above, 12 d.o.f of the initial element an expressed in terms of 9 d.o.f at the corner nodes. Mathematically,

$$\{\theta_{x1}\ \theta_{y1} \ldots\ldots\ldots\ldots \theta_{y6}\}^T = [T]\left\{w_1 \left(\frac{\partial w}{\partial x}\right)_1 \left(\frac{\partial w}{\partial x}\right)_1 \ldots\ldots\ldots\ldots \left(\frac{\partial w}{\partial y}\right)_3\right\}^T \tag{11.120}$$

where $[T]$ is the transformation matrix of size 12 × 9

11.10.2 Transformation Matrix

α is the angle between the edge tangent axis s and Cartesian coordinate axis x as shown in Fig. 11.9 (c).

Applying chain rule

$$\left.\begin{aligned}
\frac{\partial w}{\partial s} &= \frac{\partial w}{\partial x}\frac{\partial x}{\partial s} + \frac{\partial w}{\partial y}\frac{\partial y}{\partial s} \\
\frac{\partial w}{\partial n} &= \frac{\partial w}{\partial x}\frac{\partial x}{\partial n} + \frac{\partial w}{\partial y}\frac{\partial y}{\partial n}
\end{aligned}\right\} \tag{11.121}$$

or,

$$\left.\begin{aligned} w_{,s} &= \cos\alpha\, w_{,x} + \sin\alpha\, w_{,y} \\ w_{,s} &= \sin\alpha\, w_{,x} + \cos\alpha\, w_{,y} \end{aligned}\right\} \tag{11.122}$$

where $w,_s = \dfrac{\partial w}{\partial s}$, $\quad w,_n = \dfrac{\partial w}{\partial n}$ *etc.*

Similarly,

$$\left.\begin{aligned} w_{,x} &= \cos\alpha\, w_{,s} + \sin\alpha\, w_{,n} \\ w_{,y} &= \sin\alpha\, w_{,s} + \cos\alpha\, w_{,n} \end{aligned}\right\} \tag{11.123}$$

Let us proceed to express θ_x, θ_y at the midside nodes in terms of the d.o.f of the corner nodes. Midside node is 4 along the edge 12. For this node 4, applying the second constraint

$$\theta_{s4} = w,_{s4} \tag{11.124}$$

Applying the third constraint

$$\theta_{n4} = 0.5(w,_{nl} + w,_{n2}) \tag{11.125}$$

From eqn. (11.122), we can write

$$\begin{Bmatrix} w,_{s1} \\ w,_{n1} \end{Bmatrix} = \begin{bmatrix} \cos\alpha & \sin\alpha \\ -\sin\alpha & \cos\alpha \end{bmatrix} \begin{Bmatrix} w,_{x1} \\ w,_{y1} \end{Bmatrix} \tag{11.126}$$

and

$$\begin{Bmatrix} w,_{s2} \\ w,_{n2} \end{Bmatrix} = \begin{bmatrix} \cos\alpha & \sin\alpha \\ -\sin\alpha & \cos\alpha \end{bmatrix} \begin{Bmatrix} w,_{x2} \\ w,_{y2} \end{Bmatrix} \tag{11.127}$$

From eqns. (11.119) and making proper substitutions from eqns. (11.126) and (11.127). we get

$$\theta_{n4} = \theta_{n4} = \frac{1}{2}[(w,_{x1} + w,_{x2})(-\sin\alpha) + (w,_{y1} + w,_{y2})(\cos\alpha)] \tag{11.128}$$

In order to determine θ_{s4}, a_0, …, a_4 in eqn. (11.113) are to be evaluated.

$$\left.\begin{aligned} w\Big|_{s=0} &= w_1 \\ w\Big|_{s=L_{12}} &= w_2 \\ \frac{\partial w}{\partial s}\Big|_{s=0} &= w,_{s1} \\ \frac{\partial w}{\partial s}\Big|_{s=L_{12}} &= w,_{s2} \end{aligned}\right\} \tag{11.129}$$

Substituting the 4 conditions of eqn. (11.129) for the edge 12 into eqn. (11.113) yields

$$\left.\begin{aligned} a_0 &= w_1 \\ a_0 + a_1 L_{12} + a_2 L_{12}^2 + a_3 L_{12}^3 &= w_2 \\ a_1 + 2a_2.0 + 3a_3.0 &= w,_{s1} \\ a_1 + 2a_2 L_{12} + 3a_3 L_{12}^2 &= w,_{s2} \end{aligned}\right\} \tag{11.130}$$

Solving 4 equations given by eqn. (11.130) yields

$$\left.\begin{aligned} a_0 &= w_1 \\ a_1 &= w,_{sl} \\ a_2 &= \frac{3}{L_{12}^2}(w_2 - w_1) - \frac{w,_{s2} + 2w,_{s1}}{L_{12}} \\ a_3 &= \frac{w,_{sl} + w,_{s2}}{L_{12}^2} - \frac{2}{L_{12}^2}(w_2 - w_1) \end{aligned}\right\} \tag{11.131a}$$

Substituting the constants from eqn. (11.131) into eqn. (11.113) yields

$$w,_{s4} = -\frac{3}{2L_{12}} w_1 - \frac{1}{4} w,_{s1} + \frac{3}{2L_{12}} w_2 - \frac{1}{4} w,_{s2} \tag{11.131b}$$

Substituting $w,_{s1}$ and $w,_{s2}$ from eqn. (11.122) into eqn. (11.131a) yields

$$\begin{aligned} \theta_{s4} = w,_{s4} = & -\frac{3}{2L_{12}} w_1 - \frac{1}{4}(w,_{x1} \cos\alpha + w,_{y1} \sin\alpha) \\ & + \frac{3}{2L_{12}} w_2 - \frac{1}{4}(w,_{x2} \cos\alpha + w,_{y2} \sin\alpha) \end{aligned} \tag{11.132}$$

Now,

$$\begin{aligned} \theta_{x4} &= w,_{x4} \cos\alpha - w,_{n4} \sin\alpha \\ &= \cos\alpha\left[-\frac{3}{2L_{12}} w_1 - \frac{1}{4}(w,_{x1} \cos\alpha + w,_{y1} \sin\alpha)\right] \\ & \quad + \frac{3}{2L_{12}} w_1 - \frac{1}{4}(w,_{x2} \cos\alpha + w,_{y2} \sin\alpha) \\ & \quad - \left[\frac{1}{2}(w,_{x1} + w,_{x2})(-\sin\alpha) + \frac{1}{2}(w,_{y1} + w,_{y2})(\cos\alpha)\right] \sin\alpha \end{aligned} \tag{11.133}$$

Equation (11.133) has been obtained after combining eqns. (11.123), (11.119) and (11.125).

Rearranging the terms of eqn. (11.133), yields

$$\begin{aligned} \theta_{x4} = & \left(-\frac{3}{2L_{12}} \cos\alpha\right) w_1 + \left(\frac{1}{2} - \frac{3}{4}\cos^2\alpha\right) w,_{x1} + \left(-\frac{3}{4} \sin\alpha \cos\alpha\right) w,_{y1} \\ & + \left(\frac{3}{2L_{12}} \cos\alpha\right) w_2 + \left(\frac{1}{2} - \frac{3}{4}\cos^2\alpha\right) w,_{x2} + \left(-\frac{3}{4} \sin\alpha \cos\alpha\right) w,_{y2} \end{aligned} \tag{11.134}$$

Similarly,

$$\begin{aligned} \theta_{x4} = & \left(-\frac{3}{2L_{12}} \sin\alpha\right) w_1 + \left(-\frac{3}{4}\cos\alpha \sin\alpha\right) w,_{x1} + \left(\frac{1}{2} - \frac{3}{4}\sin^2\alpha\right) w,_{y1} + \left(\frac{3}{2L_{12}} \sin\alpha\right) w_2 \\ & + \left(-\frac{3}{4}\cos\alpha \sin\alpha\right) w_2 + \left(-\frac{3}{4} \cos\alpha \sin\alpha\right) w,_{x2} + \left(\frac{1}{2} - \frac{3}{4}\sin^2\alpha\right) w,_{y2} \end{aligned} \tag{11.135}$$

Following the procedure mentioned above, θ_{x5}, θ_{y5}, θ_{x6}, and θ_{y6} can be expressed in terms of d.o.f. of the corner nodes.

The transformation matrix is given by eqn. (11.136)

$$[T] = \begin{bmatrix} 0 & 1 & 0 & 0 & 0 & 0 & 0 & 0 & 0 \\ 0 & 0 & 1 & 0 & 0 & 0 & 0 & 0 & 0 \\ 0 & 0 & 0 & 0 & 1 & 0 & 0 & 0 & 0 \\ 0 & 0 & 0 & 0 & 0 & 1 & 0 & 0 & 0 \\ 0 & 0 & 0 & 0 & 0 & 0 & 0 & 1 & 0 \\ 0 & 0 & 0 & 0 & 0 & 0 & 0 & 0 & 1 \\ -\frac{3C}{2L_{12}} & \frac{1}{2}-\frac{3C^2}{4} & -\frac{3SC}{4} & \frac{3C}{2L_{12}} & \frac{1}{2}-\frac{3C^2}{4} & -\frac{3SC}{4} & 0 & 0 & 0 \\ -\frac{3S}{2L_{12}} & -\frac{3CS}{4} & \frac{1}{2}-\frac{3S^2}{4} & \frac{3S}{2L_{12}} & -\frac{3SC}{4} & \frac{1}{2}-\frac{3S^2}{4} & 0 & 0 & 0 \\ 0 & 0 & 0 & -\frac{3C}{2L_{23}} & \frac{1}{2}-\frac{3C^2}{4} & -\frac{3SC}{4} & \frac{3C}{2L_{23}} & \frac{1}{2}-\frac{3C^2}{4} & -\frac{3SC}{4} \\ 0 & 0 & 0 & -\frac{3S}{2L_{23}} & -\frac{3SC}{4} & \frac{1}{2}-\frac{3S^2}{4} & \frac{3S}{2L_{23}} & -\frac{3SC}{4} & \frac{1}{2}-\frac{3S^2}{4} \\ \frac{3C}{2L_{31}} & \frac{1}{2}-\frac{3C^2}{4} & -\frac{3SC}{4} & 0 & 0 & 0 & -\frac{3C}{2L_{31}} & \frac{1}{2}-\frac{3C^2}{4} & -\frac{3SC}{4} \\ \frac{3S}{2L_{31}} & -\frac{3SC}{4} & \frac{1}{2}-\frac{3S^2}{4} & 0 & 0 & 0 & \frac{3S}{2L_{31}} & -\frac{3SC}{4} & \frac{1}{2}-\frac{3S^2}{4} \end{bmatrix} \tag{11.136}$$

$$C = \cos\alpha \quad ; \quad S = \sin\alpha$$

11.10.3 Element Stiffness Matrix

Equation (11.120) is written in more compact form as

$$\{X_\theta\} = [T]\{X\}_e \tag{11.137}$$

Since θ_x and θ_y vary quadratically, these can be related to nodal degrees of freedom as

$$\begin{Bmatrix} \theta_x \\ \theta_y \end{Bmatrix} = \begin{bmatrix} N_1 & 0 & N_2 & 0 & \cdots & N_6 & 0 \\ 0 & N_1 & 0 & N_2 & \cdots & 0 & N_6 \end{bmatrix} \{\theta_{x1} \cdots\cdots \theta_{x6}\}^T$$

$$= [N]\{X_\theta\} \tag{11.138}$$

where N_1, N_2 *etc.* are defined in eqn. (10.50) and the relation between area coordinate and the Cartesian coordinate is given by eqn. (10.43). The strain vector is given by

$$\{\varepsilon_x \; \varepsilon_y \; \gamma_{xy}\}^T = \left[-z\frac{\partial\theta_x}{\partial x} \; -z\frac{\partial\theta_y}{\partial y} \; -z\left(\frac{\partial\theta_y}{\partial y} + \frac{\partial\theta_x}{\partial y}\right) \right]$$

or
$$\{\varepsilon\} = -z[\partial]\begin{Bmatrix}\theta_x \\ \theta_y\end{Bmatrix} \tag{11.139}$$

where
$$[\partial] = \begin{bmatrix}\frac{\partial}{\partial x} & 0 \\ 0 & \frac{\partial}{\partial y} \\ \frac{\partial}{\partial y} & \frac{\partial}{\partial x}\end{bmatrix}$$

Combining eqns. (11.137), (11.138) and (11.139) yields,
$$\{\varepsilon\} = -z[B_\theta][T]\{X\}_e \tag{11.140}$$

where

$$[B_\theta] = \begin{bmatrix}\frac{\partial N_1}{\partial x} & 0 & \frac{\partial N_2}{\partial x} & 0 & \cdots & \frac{\partial N_6}{\partial x} & 0 \\ 0 & \frac{\partial N_1}{\partial y} & 0 & \frac{\partial N_2}{\partial y} & \cdots & 0 & \frac{\partial N_6}{\partial y} \\ \frac{\partial N_1}{\partial y} & \frac{\partial N_1}{\partial x} & \frac{\partial N_2}{\partial y} & \frac{\partial N_2}{\partial x} & \cdots & \frac{\partial N_6}{\partial y} & \frac{\partial N_6}{\partial x}\end{bmatrix} \tag{11.141}$$

The stress-strain relationship for an isotropic plate is given by

$$\begin{Bmatrix}\sigma_x \\ \sigma_y \\ \sigma_{xy}\end{Bmatrix} = \frac{E}{1-\nu^2}\begin{bmatrix}1 & \nu & 0 \\ \nu & 1 & 0 \\ 0 & \nu & \frac{1-\nu}{2}\end{bmatrix}\begin{Bmatrix}\varepsilon_x \\ \varepsilon_y \\ \gamma_{xy}\end{Bmatrix} \tag{11.142}$$

or
$$\{\sigma\} = [D]\{\varepsilon\} \tag{11.143}$$

From eqn. (11.140), increment in strain is given by
$$\partial\{\varepsilon\} = -z[B_\theta][T]\,\partial\{X\}_e \tag{11.144}$$

Strain energy stored in the element becomes
$$U = \iiint \partial\{\varepsilon\}^T\{\sigma\}\,dv \tag{11.145}$$

Combining eqns. (11.145), (11.143) and (11.144) yields,
$$U = \iiint (-z)\partial\{X\}_e^T[T]^T[B_\theta]^T[D](-z)[B_\theta][T]\{X\}_e\,dx\,dy\,dz$$

or
$$U = \delta\{X\}_e^T \iint [T]^T[B_\theta]^T[D][B_\theta][T]\,dx\,dy \tag{11.146}$$

where

$$[D] = \frac{Et^3}{12(1-\nu^2)}\begin{bmatrix} 1 & \nu & 0 \\ \nu & 1 & 0 \\ 0 & 0 & \frac{1-\nu}{2} \end{bmatrix}$$

It can be shown that the element stiffness matrix is given by

$$[K]_e = \iint [T]^T [B_\theta][D][B_\theta][T]\, dx\, dy \tag{11.147}$$

11.11 THE PATCH TEST

It has been shown in Sec. 11.3 that the four-noded rectangular plate bending element is a non-conforming element. Despite its non-conformity, this element is a useful one and results obtained by using this element do converge to the correct solution. There are many other non-conforming elements which have been successfully used in the stress analysis. Non-conformity should not be treated as a numerical disadvantage; the presence of kinks at the inter-elemental boundaries of the structure makes it more flexible and thus tends to compensate for whatever over-stiff representation made within the individual element.

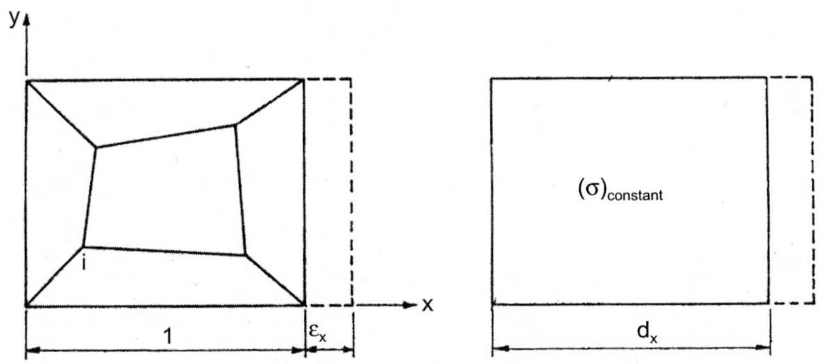

Fig. 11.10. Patch test for plane stress element

Are all non-conforming elements acceptable? What is its test of acceptability? The basic requirement of the element is that it should converge, that is, the results tend towards the correct solution with the refinement of the mesh. The patch test has been devised to answer to these queries [11.7]. It is a simple test that can be performed numerically to check the validity of an element formulation and its programmed implementation.

In the test, one assembles a small number of elements into a patch which should have at least one nodal point completely surrounded by elements. The test is now performed to see whether a patch of elements subjected to a state of constant strain can reproduce the exact constitutive behaviour of the material and will result in correct stresses when it becomes infinitesimally small. Though there is a requirement that the patch size be infinitesimal, but for elements where polynomial displacement functions are used, the size of the element may be assumed to possess any value [11.8, 11.9].

A rigid body displacement of the patch will cause no strain, and if proper stress-strain relationships are used, no changes in stress will result.

A physical problem is based on differential equations. For many such problems, exact solution exists. The same problem when attempted by the finite element method results in a set of algebraic equations of the following form:

$$[K]\{X\} = \{P\} \tag{11.148}$$

which when solved given an approximation to the differential equation and its boundary conditions.

The test can be carried out in one of the two ways. A patch of elements is made and a node i inside the patch is selected which is connected to the elements. It is to be remembered that inside nodes should not be fixed. The differential equation of the arbitrary system is solved and the parameters X at all nodes of a patch, which assembles completely the nodal variable, X_i are determined.

In the first test, the exact values of the parameters X substituted into the ith equation and then the following equation is verified:

$$K_{ij}X_j - P_i = 0 \tag{11.149}$$

In the other test, only the values of X from the exact solution corresponding to the boundaries of the patch are inserted and X_i is then formed as

$$X_i = K_{ii}^{-1}(P_i - K_{ij}X_j) \quad j \neq i \tag{11.150}$$

and then compare with the value.

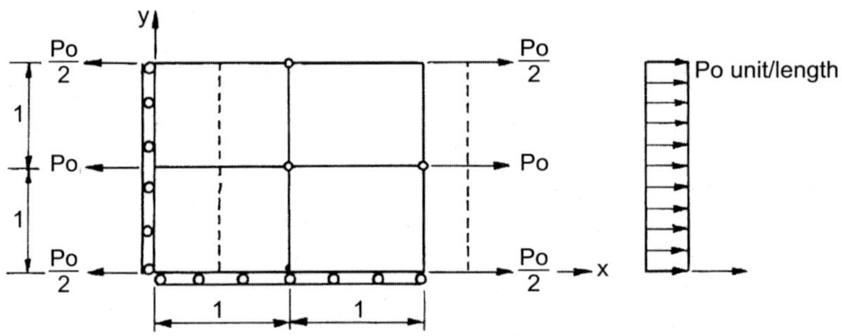

Fig. 11.11. Patch test for plane stress

11.11.1 The Patch Test for the Plane Stress Element

A small number of elements are assembled into a patch. At least one node is placed within the patch (Fig. 11.11), built of four-noded elements. Nodes at the boundary are loaded by consistent loads appropriate to a state of constant stress. Supports are to be placed in such a manner that rigid body motions are to be prevented. In this example, if instead of roller on the left edges, it is considered to be clamped, then it would have been fatal for the patch test, as the effect of Poisson's ratio in the y-direction would have been prevented. By applying the theory of elasticity approach, the stresses at the nodes are to be computed. Then the patch is to be studied by one of the above approaches (either eqn. (11.149) or eqn. (11.150).

If the agreement is good, then the patch is passed. The patch test is then to be repeated for constant σ_y and constant τ_{xy}, as the test must be repeated for all other constant stress states demanded by the element.

11.11.2 The Patch Test for Plate Bending Elements

In plate bending problems, the terms corresponding to 'strains' are curvatures. Therefore, constant strain terms in patch test are of constant curvatures. For performing the patch test, the patch is to be subjected either a set of prescribed displacements or boundary loads which correspond to a set of constant curvature.

11.11.2.1 Prescribed displacement approach

Suggestions have been made as to the form of deflection to be assessed. They are as follows:

$$w = \frac{C\left(x^2 + xy + y^2\right)}{2} \tag{11.151}$$

$$\frac{\partial w}{\partial x} = C\left(x + \frac{y}{2}\right) \tag{11.152}$$

$$\frac{\partial w}{\partial y} = C\left(\frac{x}{2} + y\right) \tag{11.153}$$

$$\frac{\partial^2 w}{\partial x \partial y} = \frac{C}{2} \tag{11.154}$$

where C is assumed to be the value of the deflection that it is much smaller than the thickness, One of the above two approaches can be adopted. For the case of plate bending with prescribed displacements, they are restated as follows:

1. The values of w, $\partial w/\partial x$ and $\partial w/\partial y$ at the boundary nodes 1, 2, 3 and 4 as prescribed [Fig. 11.12]. Based on these values, the assembled stiffness equations corresponding to the degrees of freedom at the inner nodes, 5, 6, 7 and 8 are obtained. Then determine the curvature values of each element and check whether they are constants.
2. The values of w, $\partial w/\partial x$ and $\partial w/\partial y$ at all eight nodes are prescribed. These values impose a state of constant curvature in each element. Then compute total lateral forces and bending moments at each of the four inner nodes using the assembled stiffness equations. The patch test is passed if the lateral forces and moments at each inner node obtained as a summation of contributions from all elements surrounding the node, vanish.

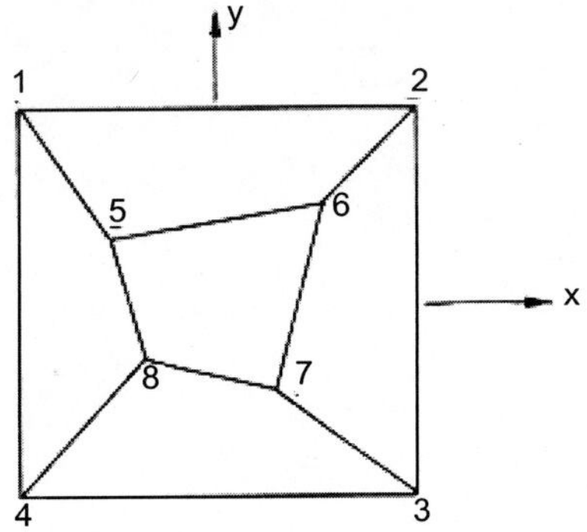

Fig. 11.12. Patch test for plate bending

11.11.2.2 Prescribed boundary load approach

The boundary loads to be applied should induce constant curvature. The appropriate loading conditions for these cases are shown in Fig. 11.13. For both *y* constant curvature and constant twist tests, only a quadrant of a plate need be analysed taking advantage of the symmetry and antisymmetry conditions.

When the constant curvature test is to be performed, a constant moment M_0 is applied in the *x*-direction while M_y and M_{xy} are kept zero. Then using moment-curvature relationship given by eqn. (7.27) yields

$$\frac{\partial^2 w}{\partial y^2} = -\nu\frac{\partial^2 w}{\partial x^2} = -\frac{12\nu M_0}{Eh^3} \text{ and } \frac{\partial^2 w}{\partial x \partial y} = 0 \qquad (11.155)$$

Similarly, the constant curvature test can be performed in the *y*-direction.

For the constant twist test, the centre and the four midside points of the deflected plate lie in the same plane with no deflection. Attention is focused of a quadrant. Any of the quadrants is balanced by four corner loads in the same manner as for the whole plate based on equilibrium. The centre and four midsides of the quadrant again lie in the same plane, but not parallel to *xy* plane. On further division of this quadrant into four subquadrants reveals that each subquadrant is under same type of corner load. Thus, it can be divided into infinite number of quadrants. Taking an infinitesimal quadrant, each of the four corner loads is split into two halves to produce two twisting moments $\frac{P}{2}dx$ and $\frac{P}{2}dy$. Thus, the plate is subjected to constant twisting moment $(Pdx/2)/dx$ and $(Pdy/2)/dy$ at all places and remains anticlastic, that is after deflection. Thus

$$\text{for} \qquad M_x = M_y = 0, \frac{\partial^2 w}{\partial x^2} = \frac{\partial^2 w}{\partial y^2} = 0 \qquad (11.156a)$$

$$\text{for} \qquad M_{xy} = -M_{yz} = \frac{P}{2}, \frac{\partial^2 w}{\partial x \partial y} = \frac{6P(1+\nu)}{Eh^3} \qquad (11.156b)$$

The patch test can now be performed in one of the two ways mentioned earlier.

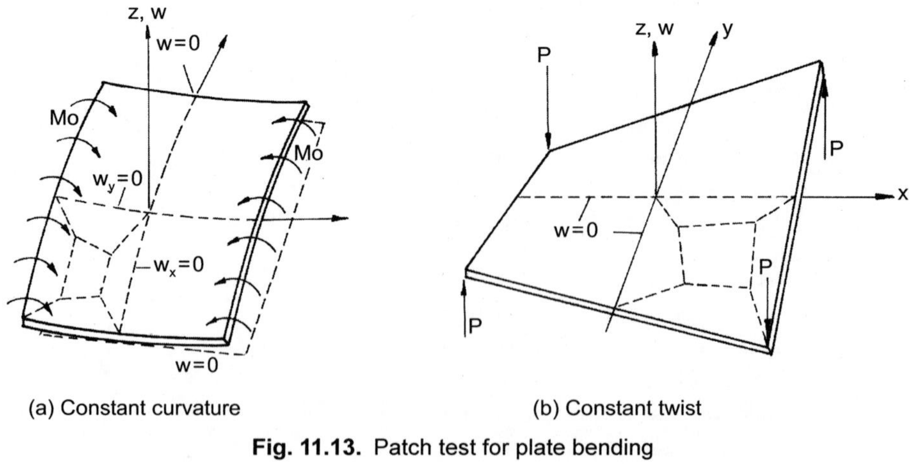

(a) Constant curvature (b) Constant twist

Fig. 11.13. Patch test for plate bending

The outside boundary of the patch is chosen as rectangular so as to facilitate the enforcement of constant curvature of the plate. For quadrilateral types of elements, the mesh sizes should be as irregular as possible.

11.12 HORIZONTALLY CURVED ISOPARAMETRIC BEAM

The horizontally curved isoparametric beam formulation given below follows from Ref. 11.10.

11.12.1 Displacement Function in Terms of Nodal Parameters

A one-dimensional parabolic isoparametric element with three nodes is chosen (Fig. 11.14). Shape functions in the three nodes in terms of the natural coordinate ξ are defined by

$$\left.\begin{aligned} N_i &= \frac{1}{2}\xi_0(1+\xi_0) \text{ for } i = 1 \text{ or } 3 \text{ and } \xi_0 = \xi\xi_i \\ N_i &= 1-\xi^2 \end{aligned}\right\} \tag{11.157}$$

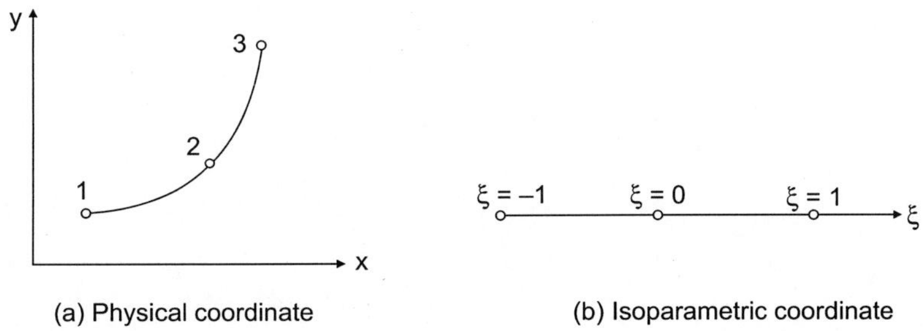

Fig. 11.14. Horizontally curved beam and its isoparametric mapping

Thus, at any point on the beam axis

$$\left.\begin{aligned} x &= \sum_{i=1}^{3} N_i x_i \\ y &= \sum_{i=1}^{3} N_i y_i \end{aligned}\right\} \tag{11.158}$$

Nodal displacements at the node i are taken as

$$\{X\}_i = \begin{Bmatrix} w_i \\ \theta_{si} \\ \theta_{ti} \end{Bmatrix} \tag{11.159}$$

where w_i the displacement of the node in the vertical z-direction, θ_{si} is the total rotation of the cross-section due to bending and shear strains and θ_{ti} is the angle of twist at the node (*see* Fig. 11.15).

The following relationships are well known in the isoparametric formulation.

$$\theta_s = \frac{dw}{ds} + \phi \tag{11.160}$$

where ϕ is the rotation due to transverse shear.

$$\left.\begin{aligned} w &= \sum_{i=1}^{3} N_i w_i \\ \theta_s &= \sum_{i=1}^{3} N_i \theta_{si} \\ \theta_t &= \sum_{i=1}^{3} N_i \theta_{ti} \end{aligned}\right\} \tag{11.161}$$

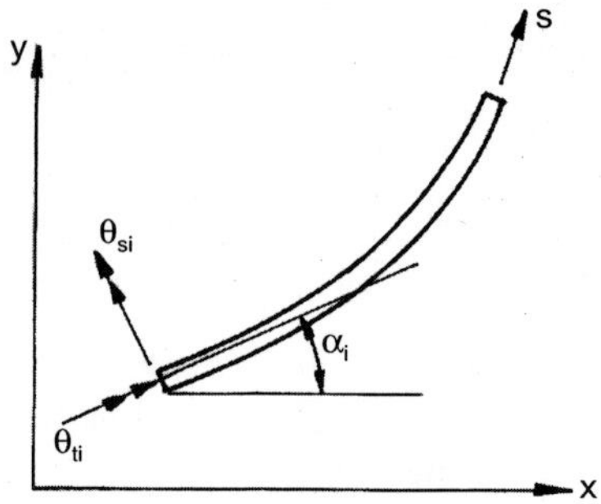

Fig. 11.15. Positive directions of model rotations

The displacement matrix at any point is given by

$$f = \begin{Bmatrix} w \\ \theta_s \\ \theta_t \end{Bmatrix} = [N]\{X\}_e = \{[n_1][n_2][n_3]\} \begin{Bmatrix} \{x\}_1 \\ \{x\}_2 \\ \{x\}_3 \end{Bmatrix} \tag{11.162}$$

where
$$[n_1] = \begin{bmatrix} N_1 & & 0 \\ & N_1 & \\ 0 & & N_1 \end{bmatrix}$$

11.12.2 Stress-Strain Relations

For a beam horizontally curved in plan, the bending moment and the torque are given by

$$T = GJ\left(\frac{d\theta_t}{ds} - \frac{\theta_s}{R}\right) \tag{11.163}$$

$$M = -EI\left(\frac{d\theta_s}{ds} + \frac{\theta_t}{R}\right) \tag{11.164}$$

where R is the radius of curvature of the beam and EI and GJ are the flexural rigidity and torsional rigidity respectively. Stress-resultants for the horizontally curved beam is related as follows:

$$\begin{Bmatrix} M \\ V \\ T \end{Bmatrix} = \begin{bmatrix} EI & 0 & 0 \\ 0 & GA_s & 0 \\ 0 & 0 & GJ \end{bmatrix} \begin{Bmatrix} -\left(\frac{d\theta_s}{ds} - \frac{\theta_t}{R}\right) \\ \frac{dw}{ds} - \theta_s \\ \frac{d\theta_t}{ds} - \frac{\theta_s}{R} \end{Bmatrix} \tag{11.165}$$

where A_s is the shear area of the plate

or, $$\{\sigma\} = [D]\{\varepsilon\} \tag{11.166}$$

11.12.3 Strain-Displacement Relationship

From eqn. (11.148), the strain components are given by

$$\{\varepsilon\} = \begin{Bmatrix} -\left(\dfrac{d\theta_s}{ds}+\dfrac{\theta_t}{R}\right) \\ \dfrac{dw}{ds}-\theta_s \\ \dfrac{d\theta_t}{ds}-\dfrac{\theta_s}{R} \end{Bmatrix} \tag{11.167}$$

Substituting relations given by eqn. (11.161) into eqn. (11.167), yields

$$\{\varepsilon\} = [B]\{X\}_e \tag{11.168}$$

where $$[B] = \sum_{i=1}^{3}[B]_i = \sum_{i=1}^{3}\begin{bmatrix} 0 & -\dfrac{dN_i}{ds} & -\dfrac{N_i}{R} \\ \dfrac{dN_i}{ds} & -N_i & 0 \\ 0 & -\dfrac{N_i}{R} & \dfrac{dN_i}{ds} \end{bmatrix} \tag{11.169}$$

Equation (11.169) contains a term $\dfrac{dN_i}{ds} = \dfrac{dN_i}{d\xi}.\dfrac{d\xi}{ds} = \dfrac{1}{J}N_i'$ (11.170)

where $$J = \sqrt{\left[\sum(N_i'x_i)^2+\sum(N_i'y_i)^2\right]} \tag{11.171}$$

The curvature in cartesian coordinate is given by

$$\frac{1}{R} = \frac{\dfrac{d^2y}{dx^2}}{\left[1+\left(\dfrac{dy}{dx}\right)^2\right]^{1/2}} \tag{11.172}$$

However,

$$\frac{dy}{dx} = \frac{\sum_{i=1}^{3}N_i'y_i}{\sum_{i=1}^{3}N_i'x_i} \tag{11.173}$$

Also $$\frac{d^2y}{dx^2} = \frac{d}{d\xi}\left(\frac{dy}{dx}\right).\frac{d\xi}{dx} = \frac{\sum N_i'x_i\sum N_i''y_i-\sum N_i''x_i\sum N_i'y_i}{\left(\sum N_i'x_i\right)^3} \tag{11.174}$$

The curvature in eqn. (11.172) can be obtained from eqns. (11.173) and (11.174). Combining eqns. (11.169) and (11.170), yields

$$[B]_i = \begin{bmatrix} 0 & -\dfrac{N_i'}{J} & -\dfrac{N_i}{R} \\ \dfrac{N_i'}{J} & -N_i & 0 \\ 0 & -\dfrac{N_i}{R} & \dfrac{N_i'}{J} \end{bmatrix} \tag{11.175}$$

11.12.4 Element Stiffness Matrix

The element stiffness matrix of the order 9 × 9 is given by

$$[k]_e = \int [B]^T [D][B]\, ds \tag{11.176}$$

The rotations tangential and normal to the curve have been considered as unknowns in the local coordinate system. They are required to be transformed in the global degrees of freedom through the transformation matrix.

11.13 NON-UNIFORM STRAIGHT BEAM ELEMENT

In many practical structures, beams with tapered cross-sections are provided. It is rather common to divide the tapered beam for the analysis into a number of elements and consider each element as uniform. This stepped beam approach may not yield accurate results unless the beam is divided into a sufficiently large number of elements. In the following, the stiffness matrix of the beam element is formulated by considering the variation of the depth within the element.

In Fig. 11.16, a non-uniform beam has been shown. I is the second moment of area at node 1, the length of the element is L and the modulus of elasticity is E. The second moment of area at any distance x from 1 is given by

$$I(x) = I_0\left[1 + r\left(\frac{x}{L}\right)^{\alpha}\right] \tag{11.177}$$

where r and α are parameters associated with the variation of the geometry along the length. For example, for the tapered beam element of Fig. 11.16 (b),

$$r = (m-1) \text{ and } \alpha = 1$$

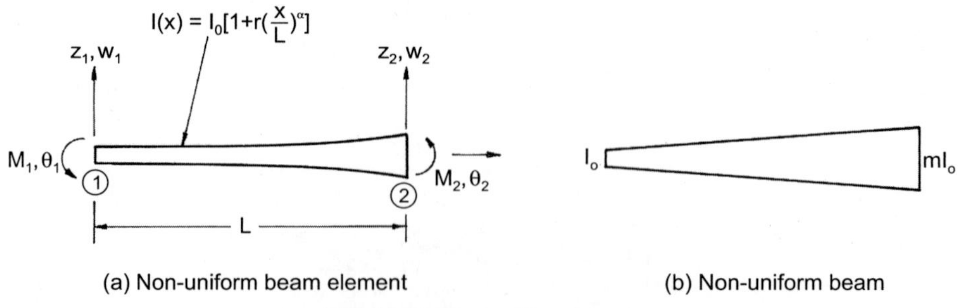

(a) Non-uniform beam element (b) Non-uniform beam

Fig. 11.16. Non-uniform beam

The strain energy for the non-uniform beam element is

$$U = \frac{1}{2}\int_0^L EI(x)\left(\frac{\partial^2 w}{\partial x^2}\right)dx \tag{11.178}$$

The displacement function for a non-uniform beam remains same as that for the uniform beam (*see* eqn. (11.10)).

The element stiffness matrix for the non-uniform beam is given by Yang [11.11].

$$[k]_e = \frac{EI_0}{L}\begin{bmatrix} \frac{12}{L^2}C_{11} & & & \\ \frac{6}{L}C_{21} & 4C_{22} & & \\ -\frac{12}{L^2}C_{11} & -\frac{6}{L}C_{21} & \frac{12}{L^2}C_{11} & \\ \frac{6}{L}C_{41} & 2C_{42} & -\frac{6}{L}C_{41} & 4C_{44} \end{bmatrix} \tag{11.179}$$

where

$$C_{11} = 1+3r\left(\frac{1}{\alpha+1}-\frac{4}{\alpha+2}+\frac{4}{\alpha+3}\right)$$

$$C_{21} = 1+2r\left(\frac{2}{\alpha+1}-\frac{7}{\alpha+2}+\frac{6}{\alpha+3}\right)$$

$$C_{22} = 1+r\left(\frac{4}{\alpha+1}-\frac{12}{\alpha+2}+\frac{9}{\alpha+3}\right)$$

$$C_{41} = 1+2r\left(\frac{1}{\alpha+1}-\frac{5}{\alpha+2}+\frac{6}{\alpha+3}\right)$$

$$C_{42} = 1+2r\left(\frac{2}{\alpha+1}-\frac{9}{\alpha+2}+\frac{9}{\alpha+3}\right)$$

$$C_{44} = 1+r\left(\frac{2}{\alpha+1}-\frac{6}{\alpha+2}+\frac{9}{\alpha+3}\right)$$

11.14 COMPUTER PROGRAM FOR ISOPARAMETRIC QUADRATIC BENDING ELEMENT

The computer programs in *C* and FORTRAN for the solution of plate bending problems using isoparametric plate bending element have been given in the attached *CD*. The given computer programs consider two load cases – uniform load throughout the plate and concentrated load at mid-span. It has been assumed in the computer program that all elements have identical dimensions so that the element stiffness matrix and the element loading matrix have been computed for one typical element and then assembled for all the elements to form the overall stiffness and overall loading matrices.

Gaussian numerical integration has been done for the evaluation of the element stiffness matrix and the element loading matrix for the uniform load. The overall stiffness matrix is stored in half band form and simultaneous equations are solved with the help of subroutine GAUSS. For calculating STIFF, 2-point integration is to be performed.

For obtaining shear forces at the nodes, smoothed stresses have been used, *i.e.* ξ^2 and η^2 values in SFD, SFDE, SFDZ *etc.* are reduced to 1/3. This helps in directly obtaining the nodal shear forces (11.3) and interpolation of the nodal values from those at the GAUSS points may be avoided. Another elegant feature of the computer program is that the mesh generation of the rectangular plate is automatically done.

The boundary conditions specified in the computer program are also general. It can take up 3 support conditions – simple, fixed and free as also the symmetry about the plate centre lines. NBON1, NBON2, NBON3 and NBON4 refer to the four edges in the order 1, 2, 3 and 4 specified in the Fig. 11.17. Edge numbers 2 and 3 can have symmetry depending on the plate and the loading. They will have values 1 if free, 2 if simply supported and 3 if fixed depending upon the edge conditions.

The items of the computer output which will be of interest to the analyst are the nodal displacements and stress resultants at nodes of interest.

A square plate having all edges simply supported has been analysed on the basis of a 4 × 4 mesh division. In order to treat it as a thin plate $\frac{S_x a^2}{D} = \frac{S_y b^2}{D} = 35000$ has been assumed. The input data of the plate is given in Table 11.2. The computer output of the plate is shown in Table 11.3.

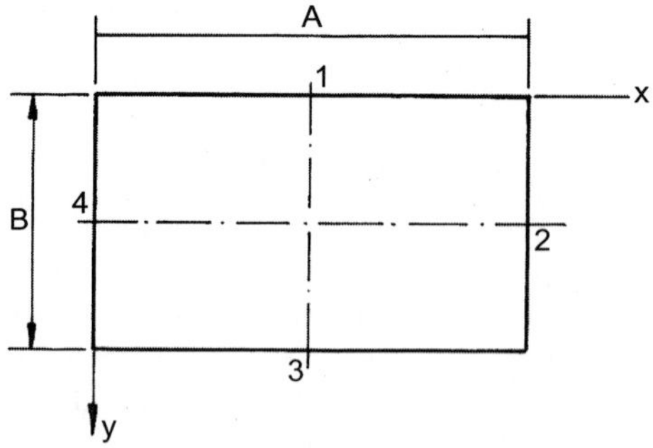

Fig. 11.17. Edge Coding

Table 11.2. Input Data for Plate Bending Problem

```
1.0 1.0 90.0 1.0 1.0
4  4  8 3 195 51 2 2 2
2 2 2 2 0 0
1.0 1.0 0.3 0.35 35000.0  35000.0 0.3
6  1.0   -1.0
7  -1.0   -1.0
9  1.0  1.0
10  -1.0  1.0
0   0  0
```

Table 11.3. Computer Output of the Plate Bending Problem

```
                A              B          PHAI               QL             QLC
          1.00000        1.00000         90.0000        1.00000        1.00000
     NX    NY    NN NFREE    NT   NBH     NQ  NQL NELM    NENP        NLC
      4     4     8     3  195    51      2    2   16      24          2
        NBON1      NBON2      NBON3       NBON4        NSYMX        NSYMY
          2          2          2           2            0            0
     DX           DY              D1              DXY           SX            SY
 1.00000       1.00000        .300000         .350000       35000.0       35000.0
          NODE   NUMBERS
      THE ELEMENT NO.1
  1    2    3   11   17   16   15   10
      THE ELEMENT NO.2
  3    4    5   12   19   18   17   11
      THE ELEMENT NO.3
  5    6    7   13   21   20   19   12
      THE ELEMENT NO.4
  7    8    9   14   23   22   21   13
      THE ELEMENT NO.5
 15   16   17   25   31   30   29   24
      THE ELEMENT NO.6
 17   18   19   26   33   32   31   25
      THE ELEMENT NO.7
 19   20   21   27   35   34   33   26
      THE ELEMENT NO.8
 21   22   23   28   37   36   35   27
      THE ELEMENT NO.9
 29   30   31   39   45   44   43   38
      THE ELEMENT NO.10
 31   32   33   40   47   46   45   39
      THE ELEMENT NO.11
 33   34   35   41   49   48   47   40
      THE ELEMENT NO.12
 35   36   37   42   51   50   49   41
      THE ELEMENT NO.13
 43   44   45   53   59   58   57   52
      THE ELEMENT NO.14
 45   46   47   54   61   60   59   53
      THE ELEMENT NO.15
 47   48   49   55   63   62   61   54
      THE ELEMENT NO.16
 49   50   51   56   65   64   63   55
            THE OVERALL STIFFNESS MATRIX
  THE PLOAD MATRIX
```

```
                      THE DISPLACEMENT MATRIX
-.629004E-20  .178275E-19 -.221340E-19  .337624E-19 -.261318E-19  .337624E-19
-.221340E-19  .178275E-19 -.629004E-20  .125736E-19  .115841E-02  .159925E-02
 .115841E-02  .125736E-19 -.143785E-19  .116400E-02  .210058E-02  .270585E-02
 .290003E-02  .270585E-02  .210058E-02  .116400E-02 -.143785E-19  .281087E-19
 .270839E-02  .374553E-02  .270839E-02  .281087E-19 -.240665E-19  .159967E-02
 .290129E-02  .374527E-02  .401899E-02  .374527E-02  .290129E-02  .159967E-02
-.240665E-19  .328166E-19  .270549E-02  .374454E-02  .270549E-02  .328166E-19
-.214072E-19  .115510E-02  .209631E-02  .270514E-02  .289913E-02  .270514E-02
 .209631E-02  .115510E-02 -.214072E-19  .178260E-19  .115561E-02  .159838E-02
 .115561E-02  .178260E-19 -.819565E-20  .181994E-19 -.222819E-19  .340273E-19
-.266436E-19  .340273E-19 -.222819E-19  .181994E-19 -.819565E-20  .398811E-21
-.167851E-20  .191891E-20  .170956E-19 -.128952E-19  .170956E-19  .191889E-20
-.167849E-20  .398811E-21  .828165E-20  .249098E-02  .367225E-02  .249098E-02
 .828168E-20 -.127838E-19  .248039E-02  .473497E-02  .639662E-02  .697028E-02
 .639662E-02  .473497E-02  .248039E-02 -.127838E-19  .278137E-19  .639180E-02
 .993486E-02  .639180E-02  .278138E-19 -.168106E-19  .367146E-02  .696790E-02
 .993535E-02  .113114E-01  .993535E-02  .696790E-02  .367146E-02 -.168106E-19
 .188887E-19  .639731E-02  .993674E-02  .639731E-02  .188886E-19  .541049E-21
 .249726E-02  .474306E-02  .639797E-02  .697199E-02  .639797E-02  .474306E-02
 .249726E-02  .541054E-21 -.167564E-20  .249629E-02  .367390E-02  .249629E-02
-.167567E-20  .401141E-20 -.238363E-20  .219916E-20  .165933E-19 -.119250E-19
 .165933E-19  .219918E-20 -.238364E-20  .401141E-20
              CENTRAL DEFLECTION
   4.018994436947302E-003   1.131136125263715E-002
                         THE VALUES OF MOMENTS AND SHEARS
ZI=  1.00000000000000 ETA=  -1.00000000000000 ELEMENT NO.=  6 LOAD NO=1
 .493820E-01  .497628E-01 -.134819E-03 -3.19236      2.55931
ZI=  1.00000000000000 ETA=  -1.00000000000000 ELEMENT NO.=  6 LOAD NO=2
 .292060      .291338      .255590E-03  162.483     -161.283
ZI=  -1.00000000000000 ETA=  -1.00000000000000 ELEMENT NO.=  7 LOAD NO=1
 .493820E-01  .497628E-01  .134820E-03  3.19236      2.55931
ZI=  -1.00000000000000 ETA=  -1.00000000000000 ELEMENT NO.=  7 LOAD NO=2
 .292060      .291338     -.255583E-03 -162.483     -161.283
ZI=  1.00000000000000 ETA=  1.00000000000000 ELEMENT NO.=  9 LOAD NO=1
 .426605E-01  .373798E-01  .122599E-03  18.9528     -3.59912
ZI=  1.00000000000000 ETA=  1.00000000000000 ELEMENT NO.=  9 LOAD NO=2
 .715450E-01  .970611E-01 -.547278E-03  26.0118      9.10107
ZI=  -1.00000000000000 ETA=  1.00000000000000 ELEMENT NO.=  10 LOAD NO=1
 .396778E-01  .364850E-01  .223782E-03  12.3734     -3.59912
ZI=  -1.00000000000000 ETA=  1.00000000000000 ELEMENT NO.=  10 LOAD NO=2
-.695150E-02  .735121E-01 -.739110E-03  209.199      9.10107
```

REFERENCES AND SUGGESTED READINGS

11.1. M. M. Hrabok and T. M. Hradey, A Review and Catalog of Plate Bending Finite Elements, Computers and Structures, v. 19, no. 3, 1984, pp. 479-495.

11.2. S. Timoshenko and S. Woinsky-Krieger, Theory of Plates and Shells, 2nd Edition, McGraw-Hill Book Co., 1959.

11.3. A. S. Mawaneya, Shear in Isoparametric Beam and Plate Bending Elements, Proceedings of the Institution of Civil Engineers (London), v. 61, Part 2, March 1976, pp.197–204.

11.4. M. Mukhopadhyay and D. K. Dinker, Isoparametric Linear Bending Element and One-point Integration, Computers and Structures, v. 9, 1978, pp. 365–369.

11.5. E. B. Becker, G. F. Carey and J. T. Oden, Finite Element Method, v1 – v. 5, Prentice Hall 1981.

11.6. J. L. Batoz, An Explicit Formulation for an Efficient Triangular Plate Bending Element, Int. J. for Numerical Methods in Enggr., v. 18, 1982, pp. 1077–1089.

11.7. R. D. Cook, D. S. Malkus and M. E. Plesha, Concepts and Applications of the Finite Element Analysis, 3rd Edition, John Wiley and Sons, 1989.

11.8. G. P. Bazeley, Y. K. Cheung, B. M. Irons and O. C. Zienkiewicz, Triangular Elements in Bending - Conforming and Non-conforming Solutions, Proc. Conf. on Matrix Methods of Structural Mechanics, Air Force Institute of Technology, Wright-Patterson Air Force Base, Ohio, USA, 1965.

11.9. O. C. Zienkiewicz and R. L. Taylor, The Finite Element Method, v. 1, 4th Edition, McGraw-Hill, 1989.

11.10. S. Dasgupta and D. Sengupta, Horizontally Curved Isoparametric Beam Element with or without Elastic Foundation Including Effect of Shear Deformation, Computers and Structures, v. 29, no. 6, 1988, pp. 967–973.

11.11. T. Y. Yang, Finite Element Structural Analysis, Prentice Hall Inc., USA, 1986.

11.12. E. M. Hinton, A. Razzaque, O. C. Zienkiewicz and J. D. Davies, A Simple Finite Element Solution for Plates of Homogeneous, Sandwich and Cellular Construction, Proc. Inst. of Civil Engineers (London), Part 2, v. 59, March 1975, pp. 43–65.

EXERCISE 11

11.1. Based on Table 11.1, calculate central deflection for a clamped square plate subjected to a concentrated load at the centre. Divide the plate into four quadrants and do not take advantages of the symmetry.

11.2. Derive the stiffness matrix for the 4-noded quadrilateral plate bending element.

11.3. Obtain a quadrilateral element from four complete triangular elements. State the procedure for deriving the stiffness matrix of the four-noded quadrilateral from the assembly of four triangular elements.

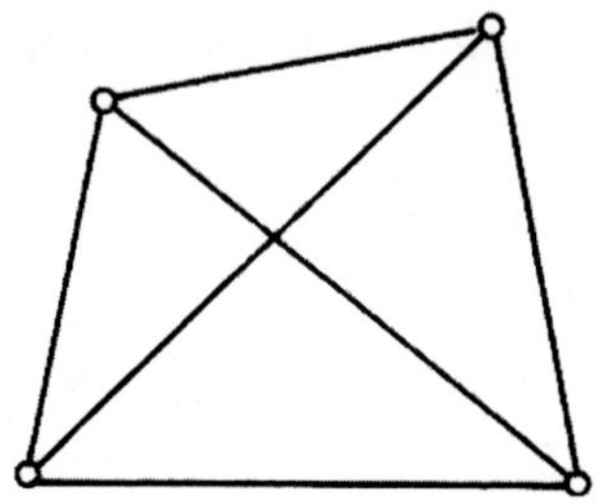

Prob. 11.3.

11.4. To incorporate the transverse shear effect, five degrees of freedom are assumed, they are, $w, \theta_x, \theta_y, \gamma_x$ and γ_y. Given $\theta_x = \dfrac{\partial w}{\partial x} + \gamma_x$ and $\theta_y = \dfrac{\partial w}{\partial y} + \gamma_y$.

Consider a four-noded rectangular element having twenty degrees of freedom, derive the element stiffness matrix starting with a displacement function of 20 terms.

11.5. Derive the element stiffness for an isoparametric beam element.

11.6. Extend the derivation of Prob. 11.5 to incorporate the twisting effect of the bar.

11.7. (*a*) What would be the effect on the results if two-point integration is performed on isoparametric linear plate bending element?

(*b*) Discuss the reduced integration technique with respect to isoparametric cubic plate bending element.

❑❑❑

12

FINITE ELEMENT ANALYSIS OF SHELLS

12.1 INTRODUCTION

A shell is a curved surface in space. Usually, it is thin in comparison to its span. There exists very limited scope for obtaining analytical solution for shell problems. For shells having arbitrary shape and loading conditions, irregular stiffening and support conditions, cut-outs and many other aspects of practical design, a closed-bound solution is practically impossible. As such the finite element method is well suited more to the structural analysis of shells. The development of the finite element method for the shell analysis is not without formidable difficulties and pitfalls.

It is seen that there are three distinct approaches to the representation of the shell structures. They are (1) flat elements, (2) curved elements and (3) three-dimensional elements. The finite element analysis of these elements will be presented one after another.

12.2 FLAT SHELL ELEMENT

A curved shell may be considered to consist of flat elements. Only triangular elements are used. For many practical purposes, finite element solutions obtained on the basis of flat elements give acceptable results.

A triangular element subjected simultaneously to membrane and bending actions are shown in Fig. 12.1.

The element is lying in the $x'y'$ plane, and the d.o.f corresponding to the membrane action and the bending action are shown in Fig. 12.1. The displacement vector is given by

$$\{X\}_e^T = \{u_1\ u_2\ u_3\ v_1\ v_2\ v_3\ w_1\ w_2\ w_3\ \theta_{x1}\ \theta_{x2}\ \theta_{x3}\ \theta_{y1}\ \theta_{y2}\ \theta_{y3}\} \quad (12.1)$$

The stiffness matrix can be formed by combining the membrane stiffness with the bending stiffness. The membrane stiffness part is same as a CST element. To this, the triangular plate bending stiffness such as obtained from DKT element may be added. While this is being done, let it be borne in mind that $\theta_{z'}$ component has not been considered in the displacement vector. It will be convenient in the assembly stage if θ_z is considered. This can be incorporated in the displacement vector and the stiffness matrix may be modified by incorporating an appropriate number of zeros. For the flat shell element in local coordinates, the stiffness matrix is given by

$$[K]_e\,\{X\}_e = \begin{bmatrix} \underset{6\times 6}{[K_m]} & \underset{6\times 9}{[0]} & \underset{6\times 3}{[0]} \\ \underset{9\times 6}{[0]} & \underset{9\times 9}{[K_p]} & \underset{9\times 3}{[0]} \\ \underset{3\times 6}{[0]} & \underset{3\times 9}{[0]} & \underset{3\times 3}{[0]} \end{bmatrix} \begin{Bmatrix} u_1 \\ u_2 \\ \vdots \\ v_2 \\ \hline w_1 \\ \\ \theta_{x1} \\ \\ \theta_{y3} \\ \hline \theta_{z_1} \\ \theta_{z2} \\ \theta_{z3} \end{Bmatrix} \tag{12.2}$$

Instead of arranging the displacement vector as per eqn. (12.1) it can be arranged node-wise.

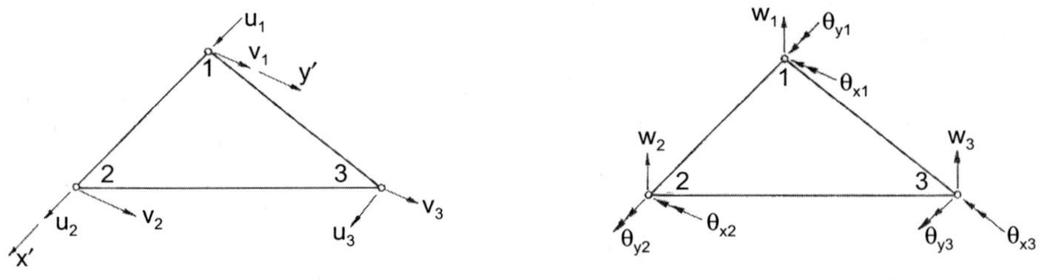

(a) Membrane action

(b) Bending action

Fig. 12.1. Flat triangular element

The displacement vector for a typical node i is then given by

$$\{X_i\}^T = \{u_i\, v_i\, w_i\, \theta_{xi}\, \theta_{yi}\, \theta_{zi}\} \tag{12.3}$$

The element stiffness matrix can then be accordingly cast. In that case, the part of the stiffness matrix involving the displacement at node i may be written as

$$[K'_{rs}] = \begin{bmatrix} [K^m_{rs}] & 0 & 0 & 0 & 0 \\ 2\times 2 & 0 & 0 & 0 & 0 \\ 0 \quad 0 & & [K^b_{rs}] & & 0 \\ 0 \quad 0 & & & & 0 \\ 0 \quad 0 & & 3\times 3 & & 0 \\ 0 \quad 0 & 0 & 0 & 0 & 0 \end{bmatrix} \tag{12.4}$$

Formulating the stiffness matrix in the manner of eqn. (12.4) has some advantages when it is to be converted into global axes-system.

12.2.1 Transformation of the Stiffness Matrix and Assembly

The stiffness matrix in the local coordinates is to be transformed along a common global system. The direction cosines between local and global axes are to be obtained.

An arbitrary shell divided into triangular elements is shown in Fig. 12.2. The local axis system is chosen for a typical element. Suppose local x'-axis is directed along the side 12. The coordinates of node 1 and 2 are (x_1, y_1, z_1) and (x_2, y_2, z_2) with respect to global axis system. Thus, the vector V_{12} in terms of global coordinates are

$$V_{12} = \begin{Bmatrix} x_2 - x_1 \\ y_2 - y_1 \\ z_2 - z_1 \end{Bmatrix} = \begin{Bmatrix} x_{21} \\ y_{21} \\ z_{21} \end{Bmatrix} \tag{12.5}$$

The direction cosines are given by the components of this vector by its length *i.e.*, defining a unit vector

$$v_{x'} = \begin{Bmatrix} \lambda_{x'x} \\ \lambda_{x'y} \\ \lambda_{x'z} \end{Bmatrix} = \frac{1}{l_{ij}} \begin{Bmatrix} x_{21} \\ y_{21} \\ z_{21} \end{Bmatrix} \tag{12.6}$$

where $\quad l_{ij} = \sqrt{x_{21}^2 + y_{21}^2 + z_{21}^2}$

The direction cosines in the other two directions are also to be obtained. z' direction is normal to the plane of the triangle. By forming cross products of two vectors V_{12} and V_{13}, we obtain a vector parallel to z' direction.

$$V_{z'} = V_{12} \times V_{13} = \begin{Bmatrix} y_{21} z_{31} - z_{21} y_{31} \\ z_{21} x_{31} - x_{21} z_{31} \\ x_{21} y_{31} - y_{21} x_{31} \end{Bmatrix} \tag{12.7}$$

Length of this vector is

$$l_{z'} = (y_{21} z_{31} - z_{21} y_{31})^2 + (z_{21} x_{31} - z_{31} x_{21})^2 + (x_{21} y_{31} - y_{21} x_{31})^2$$
$$= 2\Delta = 2 \text{ x area of the triangle}$$

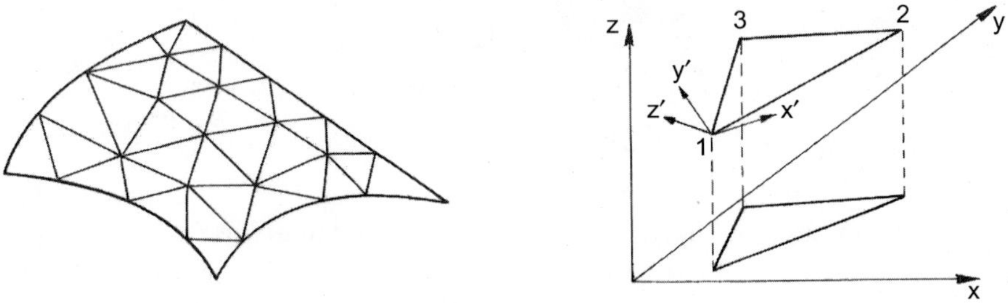

(a) Arbitrary shell

(b) A typical triangular element with local and global coordinates

Fig. 12.2. An arbitrary shell with different axis system

Therefore, unit vector of V_z' is

$$v_{z'} = \begin{Bmatrix} \lambda_{z'x} \\ \lambda_{z'y} \\ \lambda_{z'z} \end{Bmatrix} = \frac{1}{2\Delta} \begin{Bmatrix} y_{21}z_{31} - z_{21}y_{31} \\ z_{21}x_{31} - x_{21}z_{31} \\ x_{21}y_{31} - y_{21}x_{31} \end{Bmatrix} \tag{12.8}$$

Finally, the direction cosines of y'-axis can be obtained in a similar manner to the direction cosines of a vector normal to both x' and z' directions. It is

$$v_{y'} = \begin{Bmatrix} \lambda_{y'x} \\ \lambda_{y'y} \\ \lambda_{y'z} \end{Bmatrix} = v_{z'} \times v_{x'} = \begin{Bmatrix} \lambda_{z'y}\lambda_{x'z} - \lambda_{z'z}\lambda_{x'y} \\ \lambda_{z'z}\lambda_{x'x} - \lambda_{z'x}\lambda_{x'z} \\ \lambda_{z'x}\lambda_{x'y} - \lambda_{z'y}\lambda_{x'x} \end{Bmatrix} \tag{12.9}$$

It may be noted that as the cross product of vectors in eqn. (12.7) involves only unit vectors, so for the determination of direction cosines of z'-axis will not entail any division by length.

Therefore, the rotational matrix between two sets of axis system is given by

$$[\lambda] = \begin{Bmatrix} \lambda_{x'x} & \lambda_{x'y} & \lambda_{x'z} \\ \lambda_{y'x} & \lambda_{y'y} & \lambda_{y'z} \\ \lambda_{z'x} & \lambda_{z'y} & \lambda_{z'z} \end{Bmatrix} \tag{12.10}$$

The transformation matrix $[T]$ for the total shell element can now be built up. The stiffness matrix of the element in global coordinate can be expressed as

$$[K]_e = [T]^T [K'] [T] \tag{12.11}$$

If the stiffness matrix is formed in the manner given in eqn. (12.4), then the nodal transformation matrix is

$$[L] = \begin{bmatrix} [\lambda] & [0] \\ [0] & [\lambda] \end{bmatrix} \tag{12.12}$$

and the transformation matrix $[T]$ will be

$$[T] = \begin{bmatrix} [L] & [0] & [0] \\ [0] & [L] & [0] \\ [0] & [0] & [L] \end{bmatrix} \tag{12.13}$$

Typical stiffness submatrix in global coordinates will now become

$$[K_{rs}]_e = [L]^T [K'_{rs}]_e [L] \tag{12.14}$$

where $[K'_{rs}]_e$ is given by eqn. (12.4).

It may be noted that because of 0 in some diagonal elements of the stiffness matrix (eqn. (12.2)), the stiffness matrix will be singular if all the elements assembled are coplanar. One approach of removing singularity is to eliminate elements of stiffness matrix related to θ_z terms, thus reducing the size of element stiffness matrix from 18 × 18 to 15 × 15. In another approach, in order to obviate the difficulty, the following modification has been suggested. The 3 × 3 on diagonal null matrix shown in eqn. (12.2) is replaced by the following equation:

$$\begin{Bmatrix} M_{21} \\ M_{22} \\ M_{23} \end{Bmatrix} = \alpha EV \begin{bmatrix} 1.0 & -0.5 & -0.5 \\ -0.5 & 1.0 & -0.5 \\ -0.5 & -0.5 & 1.0 \end{bmatrix} \begin{Bmatrix} \theta_{z1} \\ \theta_{z2} \\ \theta_{z3} \end{Bmatrix} \tag{12.15}$$

where E is the elastic modulus, V is the element volume and α is a number such as 0.3 or less. So, some fictitious stiffness has been introduced corresponding to θ_z terms, but it offers no resistance to the mode $\theta_{z1} = \theta_{z2} = \theta_{z3}$ or to any other rigid body motion.

12.3 SHELL OF REVOLUTION

Both bending and membrane forces that occur in shells have been considered for the shells of revolution. The structure is axi-symmetric. In addition, the case of axi-symmetric loading only is dealt with.

A shell is divided into a series of frustrum. Three displacements u, v and w in the middle surface of the shell have been shown in Fig. 12.3. Due to axi-symmetric loading, the v component will be zero, u and w are the tangential and normal directions respectively.

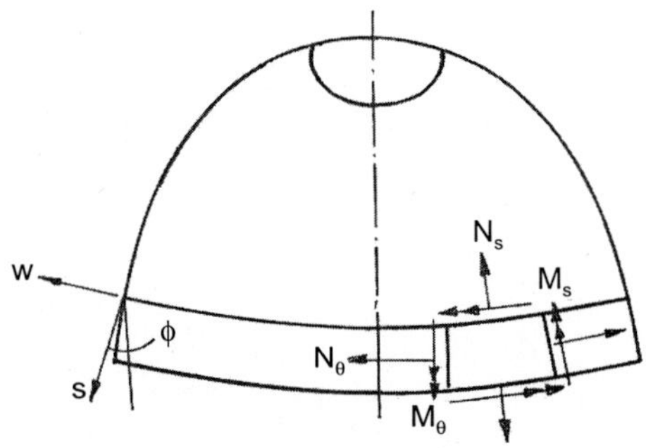

Fig. 12.3. Shell of revolution

To simplify the analysis, the frustrum is assumed to be of conical shape as shown in Fig. 12.4. The local axis system corresponds to u, w, $\frac{dw}{ds}$. The global axis system corresponds to $\bar{u}$, $\bar{w}$, and $\frac{d\bar{w}}{ds}$. Their relation at node i is given by

$$\begin{Bmatrix} u_i \\ w_i \\ \left(\dfrac{dw}{ds}\right)_i \end{Bmatrix} = \begin{bmatrix} \cos\phi & \sin\phi & 0 \\ -\sin\phi & \cos\phi & 0 \\ 0 & 0 & 1 \end{bmatrix} \begin{Bmatrix} \bar{u}_i \\ \bar{w}_i \\ \beta_i \end{Bmatrix} \tag{12.16}$$

or $$\{X\}_i = [\lambda]\{\bar{X}\}_t \tag{12.17}$$

u is assumed to vary linearly with s and with w as cubic in s. Thus

$$u = \alpha_1 + \alpha_2 s \tag{12.18}$$

$$w = \alpha_3 + \alpha_4 s + \alpha_5 s^2 + \alpha_6 s^3$$

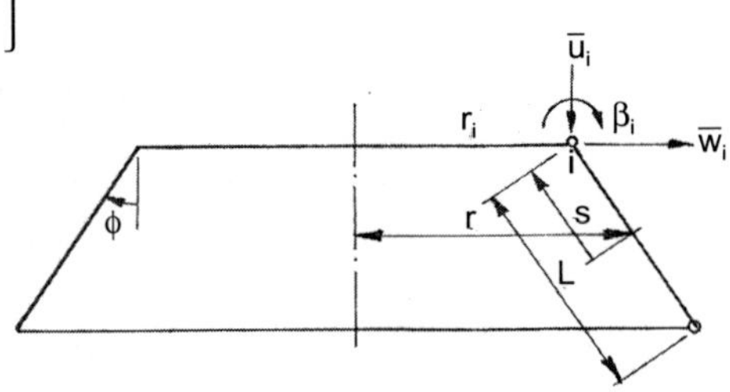

Fig. 12.4. Axi-symmetric shell element

It can be shown that the displacement function is

$$\{f\} = \begin{bmatrix} (1-s') & 0 & 0 & s' & 0 & 0 \\ 0 & (1-3s'^2+2s'^3) & L(s'-2s'^2+s'^3) & 0 & (3s'-2s'^3) & (-s'^2+s'^3)L \end{bmatrix} \begin{Bmatrix} u_i \\ w_i \\ \vdots \\ \left(\frac{dw}{ds}\right)_j \end{Bmatrix} \tag{12.19}$$

where $s' = \dfrac{s}{L}$

or, $$\{f\} = [N']\{X\} \tag{12.20}$$

Replacing the values of $\{X\}_i$ from eqn. (12.17) into eqn. (12.20), yields

$$\{f\} = [N'] \begin{bmatrix} [\lambda] & [0] \\ [0] & [\lambda] \end{bmatrix} \begin{Bmatrix} \{\overline{X}\}_i \\ \{\overline{X}\}_j \end{Bmatrix} \tag{12.21}$$

or, $$\{f\} = [N']\,[T]\,\{\overline{X}\} \tag{12.22}$$

where $[T]$ is the transformation matrix.

The strain matrix for the shell is given by

$$\{\varepsilon\} = \begin{Bmatrix} \varepsilon_s \\ \varepsilon_\theta \\ \kappa_s \\ \kappa_\theta \end{Bmatrix} = \begin{Bmatrix} du/ds \\ (w\cos\phi + u\sin\phi)/r \\ -d^2w/ds^2 \\ -\dfrac{\sin\phi}{r}\dfrac{dw}{ds} \end{Bmatrix} \tag{12.23}$$

Combining eqns. (12.19) and (12.23), yields

$$\{\varepsilon\} = \begin{bmatrix} -1/L & 0 & 0 & 1/L & 0 & 0 \\ \frac{1}{r}(1-s')\sin\phi & (1-3s'^2+2s'^3)\frac{\cos\phi}{r} & L(s'-2s'^2+s'^3)\frac{\cos\phi}{r} & \frac{s'}{r}\sin\phi & (3s'^2-2s'^3)\frac{\cos\phi}{r} & L(-s'^2+s'^3)\frac{\cos\phi}{r} \\ 0 & \frac{1}{L^2}(6-12s') & \frac{1}{L}(4-6s') & 0 & \frac{1}{L^2}(-6+12s') & \frac{1}{L}(2-6s') \\ 0 & 6(s'-s'^2)\frac{\sin\phi}{rL} & (-1+4s'-3s'^2)\frac{\sin\phi}{r} & 0 & -6(s'-s'^2)\frac{\sin\phi}{rL} & (2s'-3s'^2)\frac{\sin\phi}{r} \end{bmatrix} \begin{Bmatrix} u_i \\ w_i \\ \vdots \\ \left(\frac{dw}{ds}\right)_j \end{Bmatrix} \tag{12.24}$$

Equation (12.24) can be written as

$$\{\varepsilon\} = \left[[B_i']\,|\,[B_j']\right]\begin{Bmatrix}\{X\}_i\\ \hline \{X\}_j\end{Bmatrix} \tag{12.25}$$

Using the relation given by eqn. (12.17), eqn. (12.25) becomes

$$\{\varepsilon\} = \left[[B_i'][\lambda]\,|\,[B_j'][\lambda]\right]\begin{Bmatrix}\{\bar{X}\}_i\\ \hline \{\bar{X}\}_j\end{Bmatrix} \tag{12.26}$$

or

$$\{\varepsilon\} = [B]\{\bar{X}\} \tag{12.27}$$

The internal stress resultants for the shell element has been shown in Fig. 12.3. The 'stress' vector is given by

$$\{\sigma\} = \begin{Bmatrix}N_s\\ N_\theta\\ M_s\\ M_\theta\end{Bmatrix} \tag{12.28}$$

The stress strain relationship for the shell elements is given by

$$\{\sigma\} = [D]\,\{e\} \tag{12.29}$$

where

$$[D] = \frac{Et}{1-\nu^2}\begin{bmatrix}1 & \nu & 0 & 0\\ \nu & 1 & 0 & 0\\ 0 & 0 & t^2/12 & \nu t^2/12\\ 0 & 0 & \nu t^2/12 & t^2/12\end{bmatrix} \tag{12.30}$$

The stiffness matrix of the shell element is given by

$$[k]_e = \int_A [B]^T[D][B]\,dA \tag{12.31}$$

Now,

$$dA = 2\pi r\,ds = 2\pi rL\,ds' \tag{12.32}$$

Therefore, replacing the value of dA from eqn. (12.32) into eqn. (12.31), yields,

$$[k]_e = \int_0^1 [B]^T[D][B]\,2\pi rL\,ds' \tag{12.33}$$

Based on the geometry of the Fig. 12.4, it can be seen that

$$r = r_i + s\sin\phi \tag{12.34}$$

This value of r from eqn. (12.34) is to be substituted into eqn. (12.33) before the integration is carried out.

12.4 GENERAL SHELL FINITE ELEMENT OF TRIANGULAR SHAPE

We have already seen that the disadvantage of the flat shell element lies in the non-coupling between bending and stretching within each element, which results in the requirement of a large number of elements of the whole structure to achieve satisfactory accuracy. A shell element formulation is presented in the following, which can take care of the general shape [12.4].

The geometry of an arbitrary shallow shell element is shown in Fig. 12.5. The shell is defined by the height $\zeta(\xi, \eta)$ above the base plane in which ξ, η are taken as local coordinates and x, y as global coordinates. The base triangle has been indicated as 1′2′ 3′. Their base dimensions are a, b and c, and the rotation can be derived without much difficulty.

Special displacement functions have been used for ensuring conformity and high accuracy.

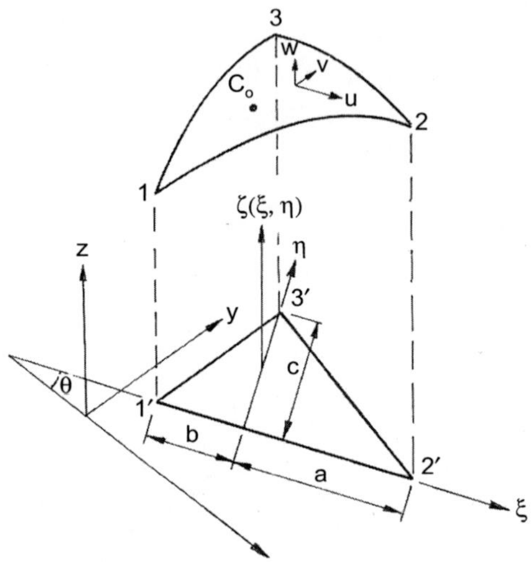

Fig. 12.5. Shell element geometry and coordinate system

12.4.1 Derivation of the Stiffness Matrix

The generalized displacements corresponding to the membrane action are u, v and their first derivative at each corner node plus u and v at the element's centroid. This gives a total of twenty degrees of freedom corresponding to u and v. The generalized displacements for w are w and its first and the second derivatives at the corner nodes. This results in a total of 18 d.o.f. corresponding to w. In all, the element then has 38 degrees of freedom. The displacement vector for the element is given by

$$\{X\}_e^T = \left\{ u_1 \left(\frac{\partial u}{\partial \xi}\right)_1 \left(\frac{\partial u}{\partial \eta}\right)_1 v_1 \left(\frac{\partial v}{\partial \xi}\right)_1 \left(\frac{\partial v}{\partial \eta}\right)_1 w_1 \left(\frac{\partial w}{\partial \xi}\right)_1 \left(\frac{\partial w}{\partial \eta}\right)_1 \left(\frac{\partial^2 w}{\partial \xi^2}\right)_1 \right.$$

$$\left. \left(\frac{\partial^2 w}{\partial \eta^2}\right)_1 \left(\frac{\partial^2 w}{\partial \xi \partial \eta}\right)_1 u_2 \cdots\cdots\cdots u_3 \cdots\cdots\cdots u_c \, v_c \right\} \tag{12.35}$$

The subscripts 1, 2, 3 denote the corners and 'c' denotes the centroid of the element.

The displacement *u*, *v* and *w* are expressed in polynomials as follows:

$$u = \alpha_1 + \alpha_2\xi + \alpha_3\eta + \alpha_4\xi^2 + \alpha_5\xi\eta + \alpha_6\eta^2 + \alpha_7\xi^3 + \alpha_8\xi^2\eta + \alpha_9\xi\eta^2 + \alpha_{10}\eta^3 \quad (12.36a)$$

$$v = \alpha_{11} + \alpha_{12}\xi + \alpha_{13}\eta + \alpha_{14}\xi^2 + \alpha_{15}\xi\eta + \alpha_{16}\eta^2 + \alpha_{17}\xi^3 + \alpha_{18}\xi^2\eta + \alpha_{19}\xi\eta^2 + \alpha_{20}\eta^3 \quad (12.36b)$$

$$w = \alpha_{21} + \alpha_{22}\xi + \alpha_{23}\eta + \alpha_{24}\xi^2 + \alpha_{25}\xi\eta + \alpha_{26}\eta^2 + \alpha_{27}\xi^3 + \alpha_{28}\xi^2\eta + \alpha_{29}\xi\eta^2 + \alpha_{30}\eta^3 + \alpha_{31}\xi^4 + \alpha_{32}\xi^3\eta + \alpha_{33}\xi^2\eta^2 + \alpha_{34}\xi\eta^3 + \alpha_{35}\eta^4 + \alpha_{36}\xi^5 + \alpha_{37}\xi^3\eta^2 + \alpha_{38}\xi^2\eta^3 + \alpha_{39}\xi\eta^4 + \alpha_{40}\eta^5 \quad (12.36c)$$

It may be noted that the displacement function *w* contains 20 constants whereas the degrees of freedom associated with *w* is only 18. Further, for the first time, we are encountering a situation where the displacement function is not evenly balanced between ξ and η. There is a special purpose for writing it in this manner. As *w* in eqn. (12.36c) does not contain $\xi^4\eta$, the requirement that the slope normal to the edge 12 is cubic automatically satisfied.

40 constants associated with eqns. (12.36a) to (12.36c) are to be evaluated. 38 equations can be obtained on the basis of substituting ξ, η values corresponding to each element of the displacement vector in eqn. (12.36a) to (12.36c). Two additional equations are to be sought. They are obtained on the basis of imposition of the conditions for cubic variation of normal slope along the remaining two edges which are however somewhat complicated.

In investigating the above conditions, only the fifth degree terms in eqn. (12.36c) are to be considered, since the lower order terms satisfy the conditions automatically. Consider the derivative normal to the line OP in Fig. 12.6. The derivation is given by

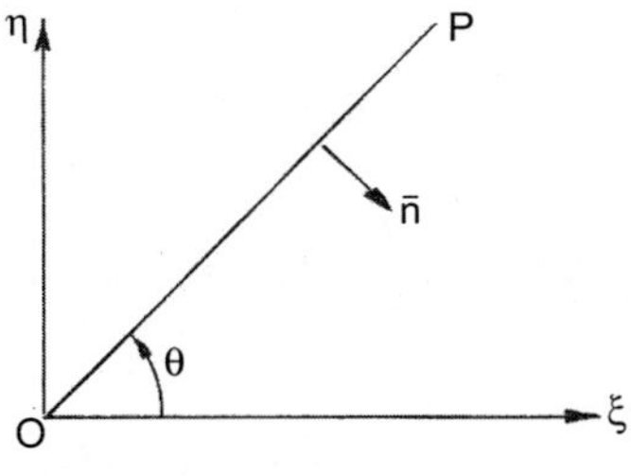

Fig. 12.6. Oblique boundary

$$\frac{\partial w}{\partial \eta} = \frac{\partial w}{\partial \xi}\sin\theta - \frac{\partial w}{\partial \eta}\cos\theta$$

$$= (5\alpha_{36}\,\xi^4 + 3\,\alpha_{37}\,\xi^2\,\eta^2 + 2\,\alpha_{38}\,\xi\eta^3 + \alpha_{39}\eta^4)\sin\theta - (2\,\alpha_{37}\,\xi^3\,\eta + 3\,\alpha_{38}\,\xi^2\,\eta^2 + 4\,\alpha_{39}\,\xi\eta^3 + 5\,\alpha_{40}\,\eta^4)\cos\theta \quad (12.37)$$

But on OP

$$x = s\cos\theta, \quad h = s\sin\theta \quad (12.38)$$

where *s* is the distance along OP. Substitution of eqn. (12.38) into eqn. (12.37) yields

$$\frac{\partial w}{\partial \eta} = s^4\{5\alpha_{36}\cos^4\theta + \alpha_{37}(3\cos^2\theta\sin^3\theta - 2\cos^4\theta\sin\theta) + \alpha_{38}(2\cos\theta\sin^4\theta - 3\cos^3\theta\sin^2\theta) + \alpha_{39}(\sin^5\theta - 4\cos^2\theta\sin^3\theta) - 5\alpha_{20}\cos\theta\sin^4\theta\} \quad (12.39)$$

Hence the condition that the normal derivative varies cubically along OP or any parallel line is that the expression in the parenthesis of eqn. (12.39) must vanish. To apply this condition to the edge 2′3′ of the triangular element of Fig. 12.4, the appropriate values of cos θ and sin θ must be inserted, that is

$$\cos\theta = \frac{a}{\sqrt{a^2+c^2}}, \quad \sin\theta = \frac{c}{\sqrt{a^2+c^2}} \tag{12.40}$$

Substitution of eqn. (12.40) into eqn. (12.39), yields

$$5a^4c\alpha_{36} + (3a^2c^3 - 2a^4c)\,\alpha_{37} + (-2ac^4 + 3a^3c^2)\,\alpha_{38} + (c^5 - 4a^2c^3)\,\alpha_{39} + 5ac^4\,\alpha_{40} = 0 \tag{12.41}$$

Similarly, the condition for the cubic variation of the normal slope along the edge 13 is

$$5b^4c\alpha_{36} + (3b^2c^3 - 2b^4c)\,\alpha_{37} + (-2bc^4 + 3b^3c^2)\,\alpha_{38} + (c^5 - 4b^2c^3)\,\alpha_{39} + 5bc^4\,\alpha_{40} = 0 \tag{12.42}$$

Therefore, with appropriate substitution the following equations result:

$$\underset{40\times 1}{\begin{Bmatrix} \{X\}_e \\ 0 \\ 0 \end{Bmatrix}} = \underset{40\times 40}{[T]} \quad \underset{40\times 1}{\{\alpha\}} \tag{12.43}$$

where $[T]$, the transformation matrix, is given in Table 12.1.

The determinant of the transformation matrix $[T]$ has the value

$$-64c^{34}(a+b)^{31}(a^2+c^2)(b^2+c^2)/729$$

which is non-zero for all practical purposes. Hence from eqn. (12.43), we get

$$\{\alpha\} = [T]^{-1}\begin{Bmatrix} \{X\}_e \\ 0 \\ 0 \end{Bmatrix} = [T_1]\{X\}_e \tag{12.44}$$

where $[T_1]$ is of the size 40 × 38 and contains first 38 columns of $[T]^{-1}$.

Following the shallow shell theory of Novozhilov (12.1), the membrane strains of the shell are given by

$$\left.\begin{aligned} \varepsilon_{\xi\xi} &= \frac{\partial u}{\partial \xi} - \frac{\partial^2 \zeta}{\partial \xi^2} w \\ \varepsilon_{\eta\eta} &= \frac{\partial v}{\partial \eta} - \frac{\partial^2 \zeta}{\partial \eta^2} w \\ \varepsilon_{\xi\eta} &= \frac{\partial u}{\partial \eta} + \frac{\partial v}{\partial \xi} - 2\frac{\partial^2 \zeta}{\partial \xi \partial \eta} w \end{aligned}\right\} \tag{12.45}$$

$\varepsilon_{\xi\xi}$, $\varepsilon_{\eta\eta}$ are axial strains in ξ and η directions and $\varepsilon_{\xi\eta}$ is the shear strain.

Table 12.1. Transformation Matrix

$$[T] = \begin{bmatrix} [S_1] & [0] & [0] \\ [0] & [S_1] & [0] \\ [0] & [0] & [S_2] \\ [S] & [0] & [0] \\ [0] & [S_3] & [0] \\ [0] & [0] & [S_4] \\ [S_5] & [0] & [0] \\ [0] & [S_5] & [0] \\ [0] & [0] & [S_6] \\ [S_7] & [0] & [0] \\ [0] & [S_7] & [0] \\ [0] & 0 & [S_8] \end{bmatrix}$$

where,

$$[S_1] = \begin{bmatrix} 1 & -b & 0 & b^2 & 0 & 0 & -b^3 & 0 & 0 & 0 \\ 0 & 1 & 0 & -2b & 0 & 0 & 3b^2 & 0 & 0 & 0 \\ 0 & 0 & 1 & 0 & -b & 0 & 0 & b^2 & 0 & 0 \end{bmatrix}$$

$$[S_2] = \begin{bmatrix} 1 & -b & 0 & b^2 & 0 & 0 & -b^3 & 0 & 0 & 0 & b^4 & 0 & 0 & 0 & 0 & -b^5 & 0 & 0 & 0 & 0 \\ 0 & 1 & 0 & -2b & 0 & 0 & 3b^2 & 0 & 0 & 0 & -4b^3 & 0 & 0 & 0 & 0 & 5b^4 & 0 & 0 & 0 & 0 \\ 0 & 0 & 1 & 0 & -b & 0 & 0 & b^2 & 0 & 0 & 0 & -b^3 & 0 & 0 & 0 & 0 & 0 & 0 & 0 & 0 \\ 0 & 0 & 0 & 2 & 0 & 0 & -6b & 0 & 0 & 0 & 12b^2 & 0 & 0 & 0 & 0 & -20b^3 & 0 & 0 & 0 & 0 \\ 0 & 0 & 0 & 0 & 1 & 0 & 0 & -2b & 0 & 0 & 0 & 3b^2 & 0 & 0 & 0 & 0 & 0 & 0 & 0 & 0 \\ 0 & 0 & 0 & 0 & 0 & 2 & 0 & 0 & -2b & 0 & 0 & 0 & 2b^2 & 0 & 0 & 0 & -2b^3 & 0 & 0 & 0 \end{bmatrix}$$

$$[S_3] = \begin{bmatrix} 1 & a & 0 & a^2 & 0 & 0 & a^3 & 0 & 0 & 0 \\ 0 & 1 & 0 & 2a & 0 & 0 & 3a^2 & 0 & 0 & 0 \\ 0 & 0 & 1 & 0 & a & 0 & 0 & a^2 & 0 & 0 \end{bmatrix}$$

$$[S_4] = \begin{bmatrix} 1 & a & 0 & a^2 & 0 & 0 & a^3 & 0 & 0 & 0 & a^4 & 0 & 0 & 0 & 0 & a^5 & 0 & 0 & 0 & 0 \\ 0 & 1 & 0 & 2a & 0 & 0 & 3a^2 & 0 & 0 & 0 & 4a^3 & 0 & 0 & 0 & 0 & 5a^4 & 0 & 0 & 0 & 0 \\ 0 & 0 & 1 & 0 & a & 0 & 0 & a^2 & 0 & 0 & 0 & a^3 & 0 & 0 & 0 & 0 & 0 & 0 & 0 & 0 \\ 0 & 0 & 0 & 2 & 0 & 0 & 6a & 0 & 0 & 0 & 12a^2 & 0 & 0 & 0 & 0 & 20a^3 & 0 & 0 & 0 & 0 \\ 0 & 0 & 0 & 0 & 1 & 0 & 0 & 2a & 0 & 0 & 0 & 3a^2 & 0 & 0 & 0 & 0 & 0 & 0 & 0 & 0 \\ 0 & 0 & 0 & 0 & 0 & 2 & 0 & 0 & 2a & 0 & 0 & 0 & 2a^2 & 0 & 0 & 0 & 2a^3 & 0 & 0 & 0 \end{bmatrix}$$

$$[S_5]=\begin{bmatrix}1 & 0 & c & 0 & 0 & c^2 & 0 & 0 & 0 & c^3\\ 0 & 1 & 0 & 0 & c & 0 & 0 & 0 & c^2 & 0\\ 0 & 0 & 1 & 0 & 0 & 2c & 0 & 0 & 0 & 3c^2\end{bmatrix}$$

$$[S_6]=\begin{bmatrix}1 & 0 & c & 0 & 0 & c^2 & 0 & 0 & 0 & c^3 & 0 & 0 & 0 & 0 & c^4 & 0 & 0 & 0 & 0 & c^5\\ 0 & 1 & 0 & 0 & c & 0 & 0 & 0 & c^2 & 0 & 0 & 0 & 0 & c^3 & 0 & 0 & 0 & 0 & c^4 & 0\\ 0 & 0 & 1 & 0 & 0 & 2c & 0 & 0 & 0 & 3c^2 & 0 & 0 & 0 & 0 & 4c^3 & 0 & 0 & 0 & 0 & 5c^4\\ 0 & 0 & 0 & 2 & 0 & 0 & 0 & 2c & 0 & 0 & 0 & 0 & 2c^2 & 0 & 0 & 0 & 0 & 2c^3 & 0 & 0\\ 0 & 0 & 0 & 0 & 1 & 0 & 0 & 0 & 2c & 0 & 0 & 0 & 0 & 3c^2 & 0 & 0 & 0 & 0 & 4c^3 & 0\\ 0 & 0 & 0 & 0 & 0 & 2 & 0 & 0 & 0 & 6c & 0 & 0 & 0 & 0 & 12c^2 & 0 & 0 & 0 & 0 & 20c^3\end{bmatrix}$$

$$[S_7]=\begin{bmatrix}1 & \frac{a-b}{3} & \frac{c}{3} & \frac{(a-b)^2}{9} & \frac{(a-b)c}{9} & \frac{c^2}{9} & \frac{(a-b)^3}{27} & \frac{(a-b)^2c}{27} & \frac{(a-b)c^2}{27} & \frac{c^3}{27}\end{bmatrix}$$

$$[S_8]=\begin{bmatrix}0 & 0 & 0 & 0 & 0 & 0 & 0 & 0 & 0 & 0 & 0 & 0 & 0 & 0 & 0 & 5a^4c & (3a^2c^3-2a^4c) & (-2ac^4-3a^3c^2) & (c^5-4a^2c^3) & 5ac^4\\ 0 & 0 & 0 & 0 & 0 & 0 & 0 & 0 & 0 & 0 & 0 & 0 & 0 & 0 & 0 & 5b^4c & (3b^2c^3-2b^4c) & (-2bc^4-3b^3c^2) & (c^5-4b^2c^3) & 5bc^4\end{bmatrix}$$

The bending strains are given by

$$\kappa_{\xi\xi}=-\frac{\partial^2 w}{\partial\xi^2},\ \kappa_{\eta\eta}=-\frac{\partial^2 w}{\partial\eta^2},\ \kappa_{\xi\eta}=-\frac{\partial^2 w}{\partial\xi\partial\eta} \tag{12.46}$$

For shallow shells, the area of the shell surface is approximately equal to its projected area. Therefore, the strain energy expression can be integrated over the base plane. The strain energy expression for shell is

$$\begin{aligned}U=\frac{Et}{2(1-\nu^2)}\iint\Bigg\{&\left[\left(\frac{\partial u}{\partial\xi}\right)^2+\left(\frac{\partial v}{\partial\eta}\right)^2+2\nu\left(\frac{\partial u}{\partial\xi}\right)\left(\frac{\partial v}{\partial\eta}\right)+\frac{1}{2}(1-\nu)\left(\frac{\partial u}{\partial\eta}+\frac{\partial v}{\partial\xi}\right)^2\right]\\ &-2\left[\left(\frac{\partial^2\zeta}{\partial\xi^2}+\nu\frac{\partial^2\zeta}{\partial\eta^2}\right)\frac{\partial u}{\partial\xi}+\left(\frac{\partial^2\zeta}{\partial\eta^2}+\nu\frac{\partial^2\zeta}{\partial\xi^2}\right)\frac{\partial v}{\partial\eta}+(1-\nu)\left(\frac{\partial^2\zeta}{\partial\xi\partial\eta}\right)\left(\frac{\partial u}{\partial\eta}+\frac{\partial v}{\partial\xi}\right)\right]w\\ &+\left[\left(\frac{\partial^2\zeta}{\partial\xi^2}\right)^2+\left(\frac{\partial^2\zeta}{\partial\eta^2}\right)^2+2\nu\frac{\partial^2\zeta}{\partial\xi^2}\frac{\partial^2\zeta}{\partial\eta^2}+2(1-\nu)\left(\frac{\partial^2\zeta}{\partial\xi\partial\eta}\right)^2\right]w^2\\ &+\frac{t^2}{12}\left[\left(\frac{\partial^2 w}{\partial\xi^2}\right)^2+\left(\frac{\partial^2 w}{\partial\eta^2}\right)^2+2\nu\frac{\partial^2 w}{\partial\xi^2}\frac{\partial^2 w}{\partial\eta^2}+2(1-\nu)\left(\frac{\partial^2 w}{\partial\xi\partial\eta}\right)^2\right]\Bigg\}\partial\xi\partial\eta\end{aligned} \tag{12.47}$$

where the integral is over the base triangle 1′ 2′ 3′ in Fig. 12.4. The function $\zeta(\xi,\eta)$ has been assumed to be of quadratic form given below:

$$\zeta(\xi,\eta)=C_1+C_2\xi+C_3\eta+C_4\xi^2+C_5\xi\eta+C_6\eta^2 \tag{12.48}$$

This implies that the shell element has constant curvatures and this is consistent with the approximation of the shallow shell theory. Only shape quantities required in calculating strain energy are the shell curvatures. They may be calculated from eqn. (12.48) or may be calculated directly. In order to use eqn. (12.48), ξ and η-values at 6 points (at 1′, 2′, 3′ and at mid-edges) are to be specified and then the constants are to be evaluated.

For obtaining the stiffness matrix of the element, eqns. (12.47), (12.36a) to (12.36c) and (12.44) are combined and the necessary integration is carried out and the strain-energy expression reduces to

$$U = \frac{Et}{2\left(1-\nu^2\right)}\{\alpha\}^T[k]_e\{\alpha\} \tag{12.49}$$

$[k]_e$ can be explicitly evaluated. In order to obtain this, eqns. (12.36a) to eqn. (12.36c) are rewritten as

$$\left.\begin{aligned} u &= \sum_{i=1}^{10}\alpha_i\xi^{m_i}\eta^{n_i} \\ v &= \sum_{i=11}^{20}\alpha_i\xi^{p_i}\eta^{q_i} \\ w &= \sum_{i=21}^{40}\alpha_i\xi^{r_i}\eta^{s_i} \end{aligned}\right\} \tag{12.50}$$

Based on the notations of eqn. (12.50), it can be shown that

$$\begin{aligned} k_{ij} = {} & m_i m_j\, F(m_i+m_j-2,\, n_i+n_j) + q_i q_j\, F(p_i+p_j,\, q_i+q_j-2) + \\ & \frac{1}{2}(1-\nu)[n_i n_j\, F(m_i+m_j,\, n_i+n_j-2) + p_i p_j\, F(p_i+p_j-2,\, q_i+q_j)] + \\ & \left[\frac{1}{2}(1-\nu)n_j p_i + \nu m_j q_i\right] F(m_j+p_i-1,\, n_j+q_i-1) + \\ & \left[\frac{1}{2}(1-\nu)n_i p_j + \nu m_i q_j\right] F(m_i+p_j-1,\, n_i+q_j-1) - \\ & \left(\frac{\partial^2\zeta}{\partial\xi^2}+\nu\frac{\partial^2\zeta}{\partial\eta^2}\right)[m_i\, F(m_i+r_j-1,\, n_i+s_j) + m_j\, F(m_j+r_i-1,\, n_j+s_i)] - \\ & \left(\frac{\partial^2\zeta}{\partial\eta^2}+\nu\frac{\partial^2\zeta}{\partial\xi^2}\right)[q_i\, F(p_i+r_j,\, q_i+s_j-1) + q_j\, F(p_j+r_i,\, q_j+s_i-1)] - \\ & (1-\nu)\frac{\partial^2\zeta}{\partial\xi\partial\eta}[n_i\, F(m_i+r_j,\, n_i+s_j-1) + n_j\, F(m_j+r_i,\, n_j+s_i-1) + \\ & p_i\, F(p_i+r_j-1,\, q_i+s_j) + p_j\, F(p_j+r_i-1,\, q_j+s_i) + \end{aligned}$$

$$\left[\left(\frac{\partial^2\zeta}{\partial\xi^2}\right)^2+\left(\frac{\partial^2\zeta}{\partial\eta^2}\right)^2+2\nu\frac{\partial^2\zeta}{\partial\xi^2}\frac{\partial^2\zeta}{\partial\eta^2}+2(1-\nu)\left(\frac{\partial^2\zeta}{\partial\xi\partial\eta}\right)^2\right]F(r_i+r_j,\ s_i+s_j)+$$

$$\frac{t^2}{12}\{r_ir_j(r_i-1)(r_j-1)F(r_i+r_j-4,\ s_i+s_j)+s_is_j(s_i-1)(s_j-1)F(r_i+r_j,\ s_i+s_j-4)+$$

$$[2(1-\nu)r_ir_js_is_j+\nu r_i\,s_j(r_i-1)(s_j-1)+\nu r_js_i(r_j-1)(s_i-1)]\times F(r_i+r_j-2,\ s_i+s_j-2)\} \quad (12.51)$$

where

$$F(m,n) = c^{n+1}[a^{m+1}-(-b)^{m+1}]\frac{m!n!}{(m+n+2)!} \quad (12.52)$$

It may be noted that i and j run from 1 to 40. Therefore, m_i and n_i are defined to be zero for $i > 10$, p_i and q_i are zero for $i < 11$ and $i > 20$ and r_i and s_i are zero for $i < 21$.

Combining eqns. (12.44) and (12.49) yields

$$U = \frac{Et}{2(1-\nu^2)}\{X\}_e^T[K]_1\{X\}_e \quad (12.53)$$

where

$$[K]_1 = [T_1]^T[k]_e[T_1] \quad (12.54)$$

$[k]_e$ is the stiffness matrix of size 38 × 38 in terms of generalized displacements in the local axis system ξ, η.

The displacement vectors in the local global axis system are related as follows:

$$\{X\}_e = [R]\{X_g\}_e \quad (12.55)$$

where $\{X_g\}_e$ is the displacement vector in the global degrees of freedom.

$$\{X_g\}_e = \left\{u_1\left(\frac{\partial u}{\partial x}\right)_1\left(\frac{\partial u}{\partial y}\right)_1 v_1\left(\frac{\partial v}{\partial x}\right)_1\left(\frac{\partial v}{\partial y}\right)_1 w_1\left(\frac{\partial w}{\partial x}\right)_1\left(\frac{\partial w}{\partial y}\right)_1\right.$$

$$\left.\left(\frac{\partial^2 w}{\partial x^2}\right)_1\left(\frac{\partial^2 w}{\partial y^2}\right)_1\left(\frac{\partial^2 w}{\partial x\partial y}\right)_1 u_2\,\dots\dots\dots,u_c\ v_c\right\} \quad (12.56)$$

Combining eqns. (12.53) to (12.55) yields

$$U = \frac{Et}{2(1-\nu^2)}\{X_g\}_e^T[K_g]_e\{X_g\}_e \quad (12.57)$$

where, $$[k_g]_e = [R]^T[T_1]^T[K]_1[T_1][R] \quad (12.58)$$

$[k_g]_e$ is the stiffness matrix in global coordinates of size 38 × 38.

12.4.2 Consistent Load Vector

From the principle of virtual work, the consistent load vector can be calculated. The applied loads in u, v and w directions are Q_u (ξ, η), Q_v (ξ, η), Q_w (ξ, η) respectively. The transformed load vector becomes

$$\{P\}_e = [R]^T [T_1]^T \{Q\} \tag{12.59}$$

where the entries in the column vector $\{Q\}$ are

$$Q_i = \left.\begin{matrix} \iint Q_u \xi^{m_i} \eta^{n_i} \, d\xi d\eta & i<11 \\ \iint Q_v \xi^{p_i} \eta^{q_i} \, d\xi d\eta & 10<i<20 \\ \iint Q_w \xi^{r_i} \eta^{s_i} \, d\xi d\eta & i>20 \end{matrix}\right\} \tag{12.60}$$

Table 12.2. Rotation Matrix [R]

$$[R] = \begin{bmatrix} [R_1] & & & & & & \\ & [R_2] & & & & & \\ & & [R_1] & & & & \\ & & & [R_2] & & & \\ & & & & [R_1] & & \\ & & & & & [R_2] & \\ & & & & & & [R_3] \end{bmatrix}$$

(submatrices not shown in the above matrix are null matrices)

where,

$$[R_1] = \begin{bmatrix} \cos\theta & 0 & 0 & \sin\theta & 0 & 0 \\ 0 & \cos^2\theta & \sin\theta\cos\theta & 0 & \sin\theta\cos\theta & \sin^2\theta \\ 0 & -\sin\theta\cos\theta & \cos^2\theta & 0 & -\sin^2\theta & \sin\theta\cos\theta \\ -\sin\theta & 0 & 0 & \cos\theta & 0 & 0 \\ 0 & -\sin\theta\cos\theta & -\sin^2\theta & 0 & \cos^2\theta & \sin\theta\cos\theta \\ 0 & \sin^2\theta & -\sin\theta\cos\theta & 0 & -\sin\theta\cos\theta & \cos^2\theta \end{bmatrix}$$

$$[R_2] = \begin{bmatrix} 1 & 0 & 0 & 0 & 0 & 0 \\ 0 & \cos\theta & \sin\theta & 0 & 0 & 0 \\ 0 & -\sin\theta & \cos\theta & 0 & 0 & 0 \\ 0 & 0 & 0 & \cos^2\theta & 2\sin\theta\cos\theta & \sin^2\theta \\ 0 & 0 & 0 & -\sin\theta\cos\theta & \cos^2\theta - \sin^2\theta & \sin\theta\cos\theta \\ 0 & 0 & 0 & \sin^2\theta & -2\sin\theta\cos\theta & \cos^2\theta \end{bmatrix}$$

$$[R_3] = \begin{bmatrix} \cos\theta & \sin\theta \\ -\sin\theta & \cos\theta \end{bmatrix}$$

The above integration is to be carried out over the base triangle 1′ 2′ 3′ of Fig. 12.5.

12.4.3 Condensation of Stiffness Matrix

It is convenient for this element to condense out the centroidal displacements u_c and v_c. Since these displacements are inside the element, they will remain unaffected when the elements are joined together. The matrix equation is partitioned as follows:

$$[K_g]_e \{X_g\}_e = \left[\begin{array}{c|c} [K]_o & [K]_{oc} \\ \hline [K]_{oc} & [K]_c \end{array}\right] \left\{\begin{array}{c} \{X_o\}_e \\ \hline \{X_c\}_e \end{array}\right\} = \left\{\begin{array}{c} \{P_o\}_e \\ \hline \{P_c\}_e \end{array}\right\} \tag{12.61}$$

As given in Sec. 5.4, the final equation of the condensed stiffness matrix becomes

$$\underset{36\times36}{[K]^e} \quad \underset{36\times1}{\{X_o\}_e} = \underset{36\times1}{\{P_o\}_e} \tag{12.62}$$

where

$$[K]_e = [K]_o - [K]_{oc}\,[K]_c^{-1}[K]_{oc}^T \tag{12.63}$$

$$[P]_c = [P_o]_e - [K]_{oc}\,[K]_c^{-1}[P_c]_e \tag{12.64}$$

12.5 ISOPARAMETRIC GENERAL SHELL ELEMENT

A shell element shape can be modeled by a three dimensional solid element that has thickness dimension considerably smaller than its other dimensions. Consider a typical shell element of Fig. 12.7. The external faces of the element are curved, while the sections across the thickness are generated by straight lines. Pair of points, *i*-top and *i*-bottom, each with given Cartesian prescribes the shape of the element.

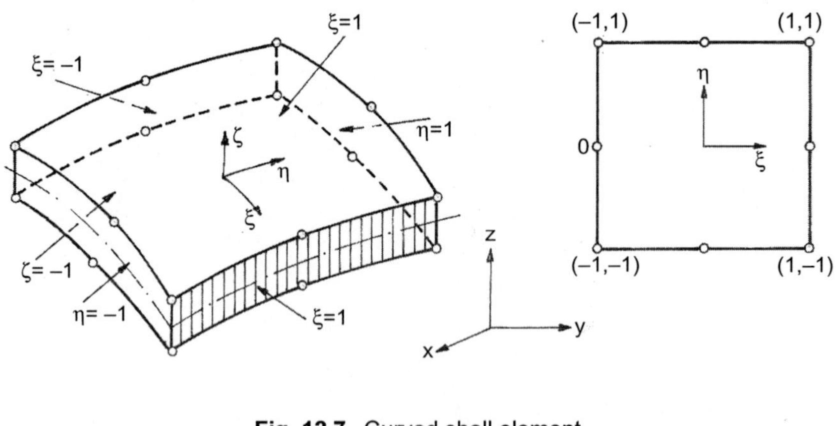

Fig. 12.7. Curved shell element

12.5.1 Geometry of the Shell Element

Let ξ, η be the curvilinear coordinates in the middle plane of the shell and ζ a linear coordinate in the thickness direction. If we further assume that ξ, η, ζ vary between – 1 and + 1 on the respective faces of the element, we can then write a relationship between the cartesian coordinates of any point of the shell and the curvilinear coordinates of the form

$$\begin{Bmatrix} x \\ y \\ z \end{Bmatrix} = \sum N_i(\xi,\eta)\frac{1+\zeta}{2}\begin{Bmatrix} x_i \\ y_i \\ z_i \end{Bmatrix}_{top} + \sum N_i(\xi,\eta)\frac{1-\zeta}{2}\begin{Bmatrix} x_i \\ y_i \\ z_i \end{Bmatrix}_{bottom} \tag{12.65}$$

where $N_i(\xi, \eta)$ are the shape functions of two-dimensional element.

It is convenient to write the relationship of eqn. (12.65) in a form specified by the 'vector' containing the upper and the lower points (*i.e.*, a vector of length equal to the shell thickness t_i) and the midsurface coordinates.

We can rewrite eqn. (12.65) as

$$\begin{Bmatrix} x \\ y \\ z \end{Bmatrix} = \sum N_i \begin{Bmatrix} x_i \\ y_i \\ z_i \end{Bmatrix}_{mid} + \sum N_i \frac{\zeta}{2}\overline{V}_{3i} \tag{12.66}$$

with

$$\overline{V}_{3i} = \begin{Bmatrix} x_i \\ y_i \\ z_i \end{Bmatrix}_{top} - \begin{Bmatrix} x_i \\ y_i \\ z_i \end{Bmatrix}_{bottom} = \begin{Bmatrix} x_j - x_k \\ y_j - y_k \\ z_j - z_k \end{Bmatrix} = t_i \begin{Bmatrix} l_{3i} \\ m_{3i} \\ n_{3i} \end{Bmatrix} \tag{12.67}$$

where

$$\begin{Bmatrix} l_{3i} \\ m_{3i} \\ n_{3i} \end{Bmatrix} = \frac{1}{t_i}\begin{Bmatrix} x_j - x_k \\ y_j - y_k \\ z_j - z_k \end{Bmatrix} \tag{12.68}$$

in which l_{3i}, m_{3i} and n_{3i} are the direction cosines of the line *jk* (Fig. 12.8). Therefore

$$\overline{V}_{3i} = t_i \begin{Bmatrix} l_{3i} \\ m_{3i} \\ n_{3i} \end{Bmatrix} \tag{12.69}$$

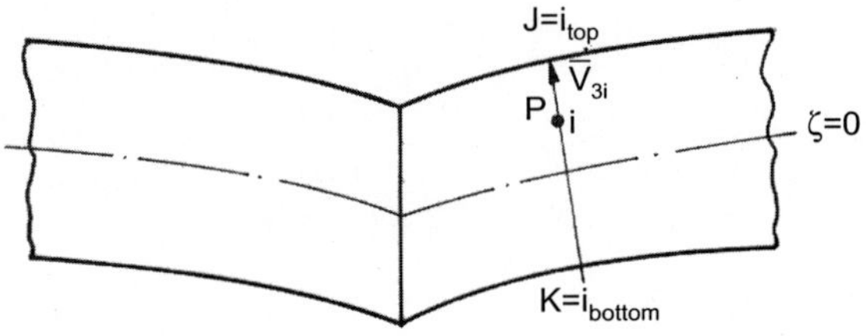

(a)

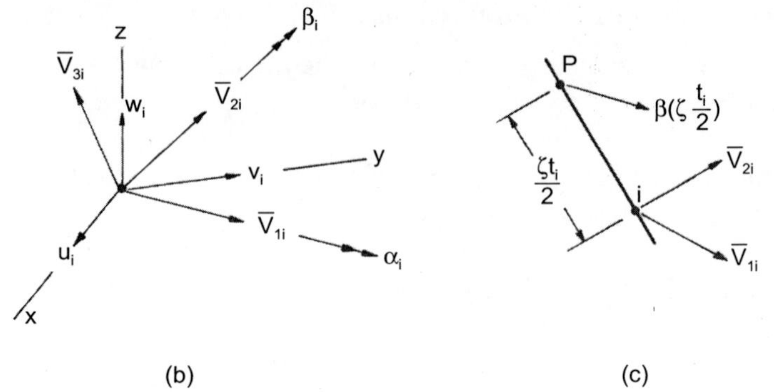

Fig. 12.8. (a) Typical node i with $\overline{V}_{3i}$, (b) Orthogonal vectors at node i. Directions of rotational d.o.f a_i and b_i translational d.o.f u_i, v_i, w_i (c) displacements at arbitrary point P on $\overline{V}_{3i}$

Equation (12.66) combined with eqns. (12.67) and (12.68), yields

$$\begin{Bmatrix} x \\ y \\ z \end{Bmatrix} = \sum N_i \begin{Bmatrix} x_i \\ y_i \\ z_i \end{Bmatrix} + \sum N_i \zeta \frac{t_i}{2} \begin{Bmatrix} l_{3i} \\ m_{3i} \\ n_{3i} \end{Bmatrix} \tag{12.70}$$

It may be noted that unlike eqn. (12.66), the suffix mid is eliminated in equation (12.70) and $x_i = (x_j + x_k)/2$and so on.

12.5.2 Displacement Field

The displacement field has now to be specified for the element. As the strains in the direction normal to the midsurface assumed to be negligible, the displacement throughout the element will be taken to be uniquely defined by the three cartesian components of the mid-surface node i displacement plus the displacement relative to the node i created by the rotation of $\overline{V}_{3i}$. The relative displacement components, shown in Fig. 12.8c, must be resolved into x, y, and z components before being added to the displacement of point P

$$u_p = u_i - \alpha_i \left(\zeta \frac{t_i}{2} \right) l_{2i} + \beta_i \left(\zeta \frac{t_i}{2} \right) l_{1i} \tag{12.71}$$

in which nodal rotations α and β are presumed to be small. Displacements of an arbitrary point in the element are

$$\begin{Bmatrix} u \\ v \\ w \end{Bmatrix} = \sum N_i \begin{Bmatrix} u_i \\ v_i \\ w_i \end{Bmatrix} + \sum N_i \zeta \frac{t_i}{2} \left[\overline{\mu_i} \right] \begin{Bmatrix} \alpha_i \\ \beta_i \end{Bmatrix} \tag{12.72}$$

u, v, w are the displacements in the directions of the global x, y, z axes. The matrix $[\overline{\mu}_i]$ which involves quantities such as l_{1i}, l_{2i} *etc*. are to be determined. As an infinity of vector dimensions normal to the given direction can be generated, a particular scheme has to be devised to ensure a unique definition. One such scheme is discussed in the following.

Vectors $\overline{V}_{1i}$ and $\overline{V}_{2i}$ in Fig. 11.8 are perpendicular to each other and to $\overline{V}_{3i}$. Thus, $\overline{V}_{1i}$ and $\overline{V}_{2i}$ are tangent to the midsurface, but they are not required to have a particular relation to cartesian coordinate directions. $\overline{V}_{1i}$ and $\overline{V}_{2i}$ are used to define the directions of nodal rotational d.o.f α_i and β_i, which are shared by all elements that share node i. Set of α_i and β_i directions may differ from node to node.

The Jacobian matrix is defined as

$$[j] = \begin{bmatrix} \dfrac{\partial x}{\partial \xi} & \dfrac{\partial y}{\partial \xi} & \dfrac{\partial z}{\partial \xi} \\ \dfrac{\partial x}{\partial \eta} & \dfrac{\partial y}{\partial \eta} & \dfrac{\partial z}{\partial \eta} \\ \dfrac{\partial x}{\partial \zeta} & \dfrac{\partial y}{\partial \zeta} & \dfrac{\partial z}{\partial \zeta} \end{bmatrix} \tag{12.73}$$

The basis of it is given later.

Equation (12.73) represents the midsurface tangent vector. First row contains the components of the vector $\overline{V}_{1i}$ in the global x, y and z directions

$$\overline{V}_{1i} = \left(\frac{\partial x}{\partial \xi}\right)_i i + \left(\frac{\partial y}{\partial \xi}\right)_i j + \left(\frac{\partial z}{\partial \xi}\right)_i k \tag{12.74}$$

Similarly,

$$\overline{V}_{2i} = \left(\frac{\partial x}{\partial \eta}\right)_i i + \left(\frac{\partial y}{\partial \eta}\right)_i j + \left(\frac{\partial z}{\partial \eta}\right)_i k \tag{12.75}$$

$$\overline{V}_{3i} = \overline{V}_{3i} = (\overline{V}_{1i} \times \overline{V}_{2i})\, t_i \tag{12.76}$$

Finally,

$$\overline{V}_{2i} = \overline{V}_{3i} = \overline{V}_{1i} \tag{12.77}$$

Direction cosines of $\overline{V}_{1i}$ and $\overline{V}_{2i}$ are given by dividing each vector by its magnitude. Therefore, $[\overline{\mu}_i]$ of eqn. (12.72) is given by

$$[\overline{\mu}_i] = \begin{bmatrix} -\dfrac{\overline{V}_{2i}}{\left|\overline{V}_{2i}\right|} & \dfrac{\overline{V}_{1i}}{\left|\overline{V}_{1i}\right|} \end{bmatrix} = \begin{bmatrix} -l_{2i} & l_{1i} \\ -m_{2i} & m_{1i} \\ -n_{2i} & n_{1i} \end{bmatrix} \tag{12.78}$$

where $|\overline{V}_{2i}|$ and $|\overline{V}_{1i}|$ are the magnitudes of vectors $\overline{V}_{2i}$ and $\overline{V}_{1i}$ respectively. Combining eqns. (12.70) and (12.73), we get

$$[J] = \begin{bmatrix} \sum\left(x_i + \frac{\zeta t_i}{2} l_{3i}\right)\frac{\partial N_i}{\partial \xi} & \sum\left(y_i + \frac{\zeta t_i}{2} m_{3i}\right)\frac{\partial N_i}{\partial \xi} & \sum\left(z_i + \frac{\zeta t_i}{2} n_{3i}\right)\frac{\partial N_i}{\partial \xi} \\ \sum\left(x_i + \frac{\zeta t_i}{2} l_{3i}\right)\frac{\partial N_i}{\partial \eta} & \sum\left(y_i + \frac{\zeta t_i}{2} m_{3i}\right)\frac{\partial N_i}{\partial \eta} & \sum\left(z_i + \frac{\zeta t_i}{2} n_{3i}\right)\frac{\partial N_i}{\partial \eta} \\ \sum \frac{t_i}{2} l_{3i} N_i & \sum \frac{t_i}{2} m_{3i} N_i & \sum \frac{t_i}{2} n_{3i} N_i \end{bmatrix} \quad (12.79)$$

Displacements of any arbitrary point in the element are given by eqn. (12.72).

12.5.3 Strains Inside the Element

The strains in terms of displacement derivatives are expressed as

$$\begin{Bmatrix} \varepsilon_x \\ \varepsilon_y \\ \varepsilon_z \\ \gamma_{xy} \\ \gamma_{yz} \\ \gamma_{zx} \end{Bmatrix} = \begin{bmatrix} 1 & 0 & 0 & 0 & 0 & 0 & 0 & 0 & 0 \\ 0 & 0 & 0 & 0 & 1 & 0 & 0 & 0 & 0 \\ 0 & 0 & 0 & 0 & 0 & 0 & 0 & 0 & 1 \\ 0 & 1 & 0 & 1 & 0 & 0 & 0 & 0 & 0 \\ 0 & 0 & 0 & 0 & 0 & 1 & 0 & 1 & 0 \\ 0 & 0 & 1 & 0 & 0 & 0 & 1 & 0 & 0 \end{bmatrix} \begin{Bmatrix} u_{,x} \\ u_{,y} \\ u_{,z} \\ v_{,x} \\ v_{,y} \\ v_{,z} \\ w_{,x} \\ w_{,y} \\ w_{,z} \end{Bmatrix} \quad (12.80)$$

$$\{\overline{\varepsilon}\}_{xyz} = [H]\,[\overline{u}]_{xyz} \quad (12.81)$$

where $[H]$ is the left matrix on the right hand side.

All six strain terms have been included in eqn. (12.80) because the shell mid-surface has no particular orientation with respect to the cartesian coordinates *xyz*. Again,

$$\frac{\partial u}{\partial \xi} = \frac{\partial u}{\partial x}\frac{\partial x}{\partial \xi} + \frac{\partial u}{\partial y}\frac{\partial y}{\partial \xi} + \frac{\partial u}{\partial z}\frac{\partial z}{\partial \xi} \quad (12.82)$$

Equation (12.82) can be written as

$$\frac{\partial u}{\partial \xi} = \begin{bmatrix} \frac{\partial x}{\partial \xi} & \frac{\partial y}{\partial \xi} & \frac{\partial z}{\partial \xi} \end{bmatrix} \begin{Bmatrix} \frac{\partial u}{\partial x} \\ \frac{\partial u}{\partial y} \\ \frac{\partial u}{\partial \xi} \end{Bmatrix} \quad (12.83)$$

Similarly,

$$\begin{Bmatrix} \dfrac{\partial u}{\partial \xi} \\ \dfrac{\partial u}{\partial \eta} \\ \dfrac{\partial u}{\partial \zeta} \end{Bmatrix} = \begin{bmatrix} \dfrac{\partial x}{\partial \xi} & \dfrac{\partial y}{\partial \xi} & \dfrac{\partial z}{\partial \xi} \\ \dfrac{\partial x}{\partial \eta} & \dfrac{\partial y}{\partial \eta} & \dfrac{\partial z}{\partial \eta} \\ \dfrac{\partial x}{\partial \zeta} & \dfrac{\partial y}{\partial \zeta} & \dfrac{\partial z}{\partial \zeta} \end{bmatrix} \begin{Bmatrix} \dfrac{\partial u}{\partial x} \\ \dfrac{\partial u}{\partial y} \\ \dfrac{\partial u}{\partial z} \end{Bmatrix} = [J] \begin{Bmatrix} \dfrac{\partial u}{\partial x} \\ \dfrac{\partial u}{\partial y} \\ \dfrac{\partial u}{\partial z} \end{Bmatrix} \tag{12.84}$$

In the above equation, [*J*] is the Jacobian matrix. This is already introduced in eqn. (12.73), which is defined in terms of nodal coordinates in eqn. (12.79)

Therefore,

$$\begin{Bmatrix} \dfrac{\partial u}{\partial x} \\ \dfrac{\partial u}{\partial y} \\ \dfrac{\partial u}{\partial z} \end{Bmatrix} = [J]^{-1} \begin{Bmatrix} \dfrac{\partial u}{\partial \xi} \\ \dfrac{\partial u}{\partial \eta} \\ \dfrac{\partial u}{\partial \zeta} \end{Bmatrix} \tag{12.85}$$

The derivatives of displacements in the global x, y, z coordinates are related to the local ξ, η, ζ (curvilinear) coordinates as

$$\begin{Bmatrix} u_{,x} \\ u_{,y} \\ u_{,z} \\ \cdot \\ \cdot \\ \cdot \\ w_{,z} \end{Bmatrix} = \begin{bmatrix} [J]^{-1} & 0 & 0 \\ 0 & [J]^{-1} & 0 \\ 0 & 0 & [J]^{-1} \end{bmatrix} \begin{Bmatrix} u_{,\xi} \\ u_{,\eta} \\ u_{,\zeta} \\ \cdot \\ \cdot \\ \cdot \\ w_{,\zeta} \end{Bmatrix} \tag{12.86}$$

or,

$$[\bar{u}]_{xyz} = [\bar{J}]^{-1} [\bar{u}]_{\xi\psi\zeta} \tag{12.87}$$

From eqn. (11.72) we get

$$\begin{Bmatrix} u_{,\xi} \\ u_{,\eta} \\ u_{,\zeta} \\ v_{,\xi} \\ \cdot \\ \cdot \\ w_{,\zeta} \end{Bmatrix} = \sum \begin{bmatrix} N_{i,\xi} & 0 & 0 & -\zeta t_i N_{i,\xi} l_{2i}/2 & \zeta t_i N_{i,\xi} l_{1i}/2 \\ N_{i,\eta} & 0 & 0 & -\zeta t_i N_{i,\eta} l_{2i}/2 & \zeta t_i N_{i,\eta} l_{1i}/2 \\ 0 & 0 & 0 & -t_i N_i l_{2i}/2 & -t_i N_i l_{1i}/2 \\ 0 & N_{i,\xi} & 0 & -\zeta t_i N_{i,\xi} m_{2i}/2 & -\zeta t_i N_{i,\xi} m_{1i}/2 \\ \cdot & \cdot & \cdot & \cdot & \cdot \\ \cdot & \cdot & \cdot & \cdot & \cdot \\ 0 & 0 & 0 & -t_i N_i n_{2i}/2 & t_i N_i n_{1i}/2 \end{bmatrix} \begin{Bmatrix} u_i \\ v_i \\ w_i \\ \alpha_i \\ \beta_i \end{Bmatrix} \tag{12.88}$$

or,

$$\{\bar{u}\}_{\xi\psi\zeta} = \sum [\,G_i\,]\{\,X_i\,\}_e^l \tag{12.89}$$

Combining eqns. (12.80), (12.87) and (12.89), yields

$$\begin{Bmatrix} \varepsilon_x \\ \varepsilon_y \\ \varepsilon_z \\ \cdot \\ \cdot \\ \gamma_{zx} \end{Bmatrix} = \sum [B_i] \begin{Bmatrix} u_i \\ v_i \\ w_i \\ \alpha_i \\ \beta_i \end{Bmatrix} \tag{12.90}$$

where $$[B_i] = \underset{6\times 9}{\sum [H]} \underset{9\times 9}{[J]^{-1}} \underset{9\times 5}{[G_i]} \tag{12.91a}$$

and $$[B] = \sum [B_i] \tag{12.91b}$$

The complete strain – displacement matrix $[B]$ is built of as many as 6 × 5 blocks of $[B_i]$ as there are nodes in the element.

12.5.4 Stress - Strain Relationship

The stress - strain relationship in cartesian coordinates is given by

$$\{\bar{\sigma}\}_{xyz} = [D]\{\bar{\varepsilon}\}_{xyz} \tag{12.92}$$

The stress - strain relationship in local coordinates is given by

$$\{\bar{\sigma}\}_{\xi\eta\zeta} = [D]\{\bar{\varepsilon}\}_{\xi\eta\zeta} \tag{12.93}$$

or $$\begin{Bmatrix} \sigma_1 \\ \sigma_2 \\ \sigma_3 \\ \tau_{12} \\ \tau_{23} \\ \tau_{31} \end{Bmatrix}_{\xi\eta\zeta} = \begin{bmatrix} D_{11} & D_{12} & 0 & 0 & 0 & 0 \\ D_{12} & D_{22} & 0 & 0 & 0 & 0 \\ 0 & 0 & 0 & 0 & 0 & 0 \\ 0 & 0 & 0 & G_{12} & 0 & 0 \\ 0 & 0 & 0 & 0 & 5G_{23}/6 & 0 \\ 0 & 0 & 0 & 0 & 0 & 5G_{31}/6 \end{bmatrix} \begin{Bmatrix} \varepsilon_1 \\ \varepsilon_2 \\ \varepsilon_3 \\ \gamma_{12} \\ \gamma_{23} \\ \gamma_{31} \end{Bmatrix}_{\xi\eta\zeta} \tag{12.94}$$

where directions 1 and 2 are tangent to the midsurface and direction 3 is normal to it. The factor of 5/6 accounts for the parabolic variation of the transverse shear strain through the thickness. Equation (12.94) has been contrived to make $s_3 = 0$.

The relationship of nodal degrees of freedom between the local and the global axes with the help of eqn. (12.78) is as follows:

$$\begin{Bmatrix} u_i \\ v_i \\ w_i \\ \alpha_i \\ \beta_i \end{Bmatrix} = \begin{bmatrix} 1 & 0 & 0 & 0 & 0 & 0 \\ 0 & 1 & 0 & 0 & 0 & 0 \\ 0 & 0 & 1 & 0 & 0 & 0 \\ 0 & 0 & 0 & -l_{2i} & -m_{2i} & -n_{2i} \\ 0 & 0 & 0 & l_{1i} & m_{1i} & n_{1i} \end{bmatrix} \begin{Bmatrix} u_i \\ v_i \\ w_i \\ \theta_{xi} \\ \theta_{yi} \\ \theta_{zi} \end{Bmatrix} \tag{12.95}$$

or, $$\{X_i\} = [T]\{X_i\}_e^g \tag{12.96}$$

12.5.5 Stiffness Matrix of the Shell Element

The strain energy of the shell element is given by

$$U = \int \{\varepsilon\}_{xyz}^T \, \{\bar{\sigma}\}_{xyz} \, dV \tag{12.97}$$

Using eqns. (12.90), (12.92) and (12.96), the stiffness matrix of the element in terms of global displacements is given by

$$[K]_e = \int_{-1}^{+1}\int_{-1}^{+1}\int_{-1}^{+1} [\bar{T}]^T [B]^T [D][B][\bar{T}] \det[j] \, d\xi d\eta d\zeta \tag{12.98}$$

where $[\bar{T}]$ is of the size $5n \times 6n$, $[B]$ $6 \times 5n$, $[D]$ 6×6 and n is the number of nodes per element. If the material properties are independent of ζ and if small errors are acceptable, then the thickness – direction integration can be done explicitly. In so doing one discards terms in $[J]$ that depend on ζ under the assumption that these terms are negligible if the element is not sharply curved.

For orthotropic material $[\bar{D}]$ is generated in the principal material directions and $[D]$ is to be obtained from $[\bar{D}]$ by coordinate transformation as

$$[D] = [T_\varepsilon]^T [\bar{D}][T_\varepsilon] \tag{12.99}$$

where

$$[T_\varepsilon] = \begin{bmatrix} l_1^2 & m_1^2 & n_1^2 & l_1 m_1 & m_1 n_1 & n_1 l_1 \\ l_2^2 & m_2^2 & n_2^2 & l_2 m_2 & m_2 n_2 & n_2 l_2 \\ l_3^2 & m_3^2 & n_3^2 & l_3 m_3 & m_3 n_3 & n_3 l_3 \\ 2l_1 l_2 & 2m_1 m_2 & 2n_1 n_2 & l_1 m_2 + l_2 m_1 & m_1 n_2 + m_2 n_1 & n_1 l_2 + n_2 l_1 \\ 2l_2 l_3 & 2m_2 m_3 & 2n_2 n_3 & l_2 m_3 + l_3 m_2 & m_2 n_3 + m_3 n_2 & n_2 l_3 + n_3 l_2 \\ 2l_3 l_1 & 2m_3 m_1 & 2n_3 n_1 & l_3 m_1 + l_3 m_1 & m_1 n_3 + m_3 n_1 & n_1 l_3 + n_3 l_1 \end{bmatrix} \tag{12.100}$$

where 1_1, l_2 *etc.* are direction cosines between the two axes system.

This transformation must be carried out at each Gauss point used in generating $[K]_e$ by numerical integration. Direction cosines needed in $[T_\varepsilon]$ are the direction cosines of vectors $\bar{V}_1$, $\bar{V}_2$ and $\bar{V}_3$ at the Gauss point. In turn these vectors can be found by shape function interpolation from nodal values,

$$\bar{V}_1 = \sum N_i \bar{V}_{1i}, \bar{V}_2 = \sum N_i \bar{V}_{2i}, \bar{V}_3 = \sum N_i \bar{V}_{3i} \tag{12.101}$$

where N_i are evaluated at the Gauss point in question.

12.6 VERTICALLY CURVED BEAM ELEMENT

A vertically curved circular beam is shown in Fig. 12.9. E, s, I, L are the material and geometric properties of the beam element. The beam subtends an angle ϕ at the centre. The element has two corner nodes. At each node, the unknowns are u, $\frac{\partial u}{\partial s}$, w and θ.

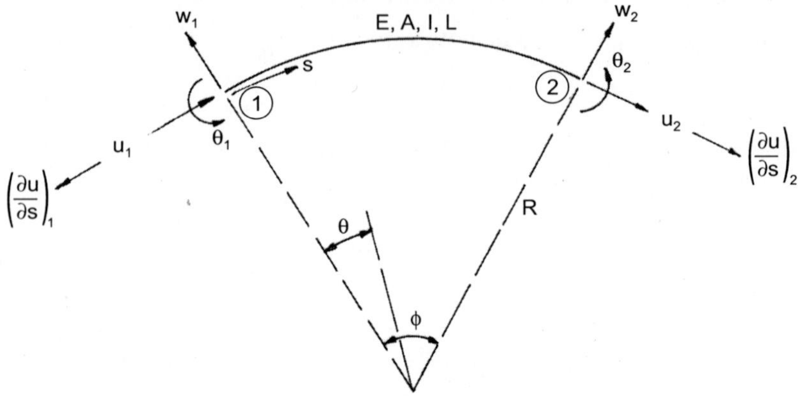

Fig. 12.9. Vertically curved circular beam element

The displacement functions are

$$\left.\begin{aligned} u &= N_1 u_1 + N_2 u_1' + N_3 u_2 + N_4 u_2' \\ w &= N_1 w_1 + N_2 \theta_1 + N_3 w_2 + N_4 \theta_2 \end{aligned}\right\} \tag{12.102}$$

where

$$\left.\begin{aligned} N_1 &= 1 - 3\xi^2 + 2\xi^3 \\ N_2 &= L\left(\xi - 2\xi^2 + \xi^3\right) \\ N_3 &= 3\xi^2 - 2\xi^3 \\ N_4 &= L\left(-\xi + \xi^3\right) \end{aligned}\right\} \tag{12.103}$$

when $\xi = \dfrac{s}{L} = \dfrac{R\theta}{R\phi} = \dfrac{\theta}{\phi}$

The strain energy of the beam is given by

$$U = \frac{EA}{2}\int_s \varepsilon^2 ds + \frac{EI}{L}\int_s \kappa^2 ds \tag{12.104}$$

where *e* and *k* are the axial strain and curvature of the middle surface.

$$\left.\begin{aligned} \varepsilon &= \frac{\partial u}{\partial s} + \frac{w}{R} = u' + \frac{w}{R} \\ \kappa &= \frac{s}{R}\frac{\partial u}{\partial s} - \frac{\partial^2 w}{\partial s^2} = \frac{s}{R}u' - w'' \end{aligned}\right\} \tag{12.105}$$

Combining eqns. (12.104) and (12.105)

$$U = \frac{EA}{2}\int_0^L (u')^2\, ds + \frac{EI}{2R^2}\int_0^L (u')^2\, ds + \frac{EA}{R}\int_0^L u'\, w\, ds - \frac{EI}{R}\int_0^L u'\, w''\, ds$$

$$+ \frac{EA}{2R^2}\int_0^L w^2\, ds + \frac{EI}{2}\int_0^L (w'')^2\, ds \tag{12.106}$$

The strain energy expression consists of axial, axial-flexural coupling and flexural behaviours.

Applying Castigliano's theorem, the element stiffness matrix is given by Yang and Kim (12.10)

$$[K]_e = [K]_{uu} + [K]_{uw} + [K]_{ww} \tag{12.107}$$

$$[k]_{uu} = \frac{EA}{5}\left(1+\frac{\alpha}{R^2}\right)\begin{bmatrix} \frac{6}{L} & & & \text{Symmetrical} \\ \frac{1}{2} & \frac{2L}{3} & & \\ -\frac{6}{L} & -\frac{1}{2} & \frac{6}{L} & \\ \frac{1}{2} & -\frac{L}{6} & -\frac{1}{2} & \frac{2L}{3} \end{bmatrix} \tag{12.108a}$$

$$[k]_{uw} = \frac{EA}{R}\begin{bmatrix} -\frac{1}{2} & & & \text{Symmetrical} \\ \frac{L}{10}+\frac{\alpha}{L} & \frac{\alpha}{2} & & \\ \frac{1}{2} & \frac{L}{10}+\frac{\alpha}{L} & \frac{1}{2} & \\ -\frac{L}{60}-\frac{\alpha}{L} & -\frac{L^2}{60}-\frac{\alpha}{2} & \frac{L}{10}+\frac{\alpha}{L} & -\frac{\alpha}{2} \end{bmatrix} \tag{12.108b}$$

$$[k]_{ww} = \frac{EA}{L}\begin{bmatrix} \frac{13\phi^2}{35}+\frac{12\alpha}{L^2} & & & \text{Symmetrical} \\ \frac{11L\phi^2}{210}+\frac{6\alpha}{2} & \frac{L^2\phi^2}{105}+4\alpha & & \\ \frac{9\phi^2}{70}+\frac{12\alpha}{L^2} & \frac{13L\phi^2}{420}-\frac{6\alpha}{L} & \frac{13\phi^2}{35}+\frac{12\alpha}{L^2} & \\ -\frac{13L\phi^2}{420}+\frac{6\alpha}{L} & -\frac{L^2\phi^2}{140}+2\alpha & -\frac{11L\phi^2}{210}-\frac{6\alpha}{L} & \frac{L^2\phi^2}{105}+4\alpha \end{bmatrix} \tag{12.108c}$$

$$\alpha = \frac{EI}{EA}$$

For vertically curved beams where thickness/radius ratio is small, $\alpha/R^2 \leq 1$ and α/R^2 may be neglected for such cases.

12.7 COMPUTER PROGRAM FOR THE FINITE ELEMENT ANALYSIS OF SHALLOW SHELLS OF GENERAL SHAPE USING TRIANGULAR ELEMENT

The computer program in 'C' as well as FORTRAN to solve a shallow shell of general shape by using the triangular element of Sec. 12.4 are given in the attached CD. It has already been shown that the element is quite sophisticated. The element stiffness is explicitly evaluated. However, the element is valid for shallow shell formulation only. The computer program has the provision of catering normal pressure on the surface as well as the vertical load.

A barrel vault cylindrical as shown in Fig. 12.10 subjected to vertical loading has been analysed on the basis of a 4 × 4 mesh division of the shell. The input data is given in Table 12.3. The nodal connectivity and the coordinates are generated by the computer program. The computer output primarily consists of nodal displacements and stress resultants, which are given in Table 12.4.

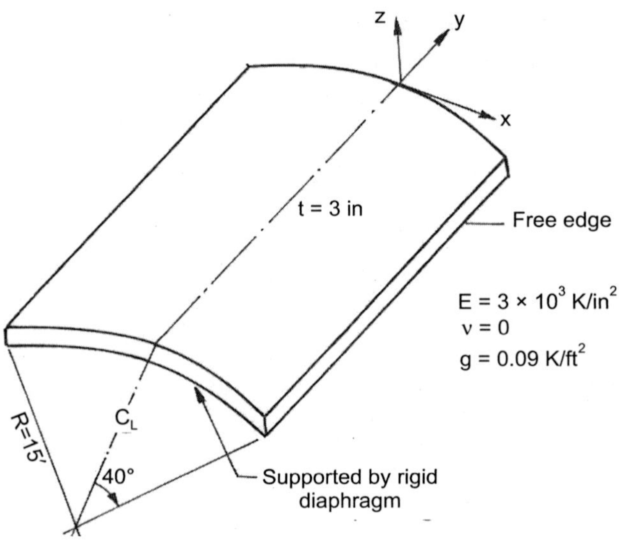

Fig. 12.10. Barrel vault

Table 12.3. Input Data for Shell Problem

```
SOLN. OF A DIAPHRAGM SUPPORTED CYLINDRICAL SHELL
CONSIDERING QUARTER SHELL (MESH 2 X 2)
43.2E4 0. 0.25 25. 1.E30 1.E30
2 2 7 1 0
16.06969 25.0
2 1
-0.09
1 1 1 1 0 0 0 1 1 0 1 1 0
2 1 1 0 0 0 0 1 1 0 1 0 0
3 1 1 0 0 0 0 1 1 0 1 0 0
4 1 0 1 0 0 0 0 1 0 0 1 0
7 1 0 1 1 1 0 0 1 1 0 1 0
8 0 0 0 1 1 0 0 0 1 0 1 0
9 0 0 0 1 1 0 0 0 1 0 1 0
```

Table 12.4. Computer Output for the Shell Problem

```
Shallow Shell FEM using Cowper's Element
emod=432000    pr=0.000000   th=0.250000  rx1=25  ry1=1e+030  rxy1=1e+030
nx=2     ny=2    nbp=7   nlc=1   nl=0
xlen=16.069690  ylen=25.000000
nnod = 9
x Co-ordinates
  0.00000e+000  8.03485e+000  1.60697e+001  0.00000e+000  8.03485e+000
  1.60697e+001  0.00000e+000  8.03485e+000  1.60697e+001
y co-ordinates
  0.00000e+000  0.00000e+000  0.00000e+000  1.25000e+001  1.25000e+001
  1.25000e+001  2.50000e+001  2.50000e+001  2.50000e+001
nelm = 8
Nodal connectivity
   5   4   1   1   2   5   6   5   2   2   3   6
   8   7   4   4   5   8   9   8   5   5   6   9

Iload = 2,       Icurve = 1
qo=  -9.00000e-002
Node number boundary condition code numbers
   1   1   1   1   0   0   0   1   1   0   1   1   0
   2   1   1   0   0   0   0   1   1   0   1   0   0
   3   1   1   0   0   0   0   1   1   0   1   0   0
   4   1   0   1   0   0   0   0   1   0   0   1   0
   7   1   0   1   1   1   0   0   1   1   0   1   0
   8   0   0   0   1   1   0   0   0   1   0   1   0
   9   0   0   0   1   1   0   0   0   1   0   1   0
rx= -4.00000e-002 ry= -1.00000e-030 rxy=  1.00000e-030
nbh = 60
       Node Global U, V, W Displacements
   1  -2.52771e-025   8.45449e-004   3.76523e-024
   2   1.64401e-024   2.31428e-003  -1.03472e-024
   3   2.73820e-024  -1.11845e-002  -1.30357e-023
   4  -2.01431e-024   7.60297e-004   2.14358e-002
   5   1.81474e-003   1.48234e-003  -5.76753e-002
   6   4.58028e-002  -7.30827e-003  -2.19612e-001
   7  -1.72833e-024   3.84346e-024   2.75567e-002
   8   2.62600e-003   3.96076e-024  -7.92146e-002
   9   6.33728e-002  -1.14202e-023  -3.06600e-001
Strain Energy =    1.22271e+003
node      nx           ny           nxy          mx           my          mxy
 1 -3.692e-020 -6.444e+000 -5.202e-001 -1.129e-022  1.301e-001 -5.839e-022
 2 -2.090e-020 -2.590e+000 -5.929e+000 -3.952e-022  1.408e-001 -5.235e-001
 3  1.286e-019 -8.413e+000 -8.452e+000 -1.250e-021  7.404e-002 -5.017e-001
 4  1.959e+002  1.090e+000  4.296e-001 -1.673e+000 -1.449e-001 -3.743e-022
 5 -4.909e+002  9.655e+000 -2.548e+000 -8.242e-001  1.615e-002 -3.450e-001
 6 -1.907e+003 -4.774e+001 -4.040e+000  1.385e-001  6.094e-001 -3.551e-001
 7  2.429e+002  5.846e+000 -1.511e-019 -1.962e+000 -4.803e-002  3.953e-021
 8 -6.901e+002  1.490e+001  2.125e+000 -9.085e-001  2.928e-001 -1.027e-022
 9 -2.670e+003 -5.449e+001 -8.700e+000  1.082e-001  9.076e-001  6.127e-023
```

REFERENCES AND SUGGESTED READINGS

12.1. V. V. Novozhilov, Theory of Thin Shells, P. Noordhoff, 1959.

12.2. D. G. Ashwell and R. H. Gallagher (Editors), Finite Elements for Thin Shells and Curved Members, John Wiley & Sons, 1976.

12.3. G. Wempner, Mechanics and Finite Elements of Shells, Applied Mechanics Review, v. 42, no. 5, 1989.

12.4. G. R. Cowper, G. M. Lindberg, and M. B. Olson, A Shallow Finite Element of Triangular Shape, International Journal of Solids and Structure, v. 6, 1970, pp. 1133–1156.

12.5. W. Flugge, Stresses In Shells, 2nd Edition, Springer-Verlag, 1973.

12.6. H. Kraus, Thin Elastic Shells, Wiley, New York, 1967.

12.7. S. Dasgupta and D. Sengupta, Horizontally Curved Isoparametric Beam Element with or without Elastic Foundation Including the Effect of Shear Deformation, Computers and Structures, v. 29, no. 6, 1988, pp. 967–973.

12.8. G. Sinha, A. H. Sheikh and M. Mukhopadhyay, A New Finite Element Model for the Analysis of Arbitrary Stiffened Shells, Finite Elements in Analysis and Design, v. 12, 1992, pp. 241–275.

12.9. P. L. Gould, Analysis of Shells and Plates, Springer-Verlag, 1988.

12.10. T. Y. Yang and H. W. Kim, Vibration and Buckling of Shells Under Initial Stress, AIAA J., v.11, no. 11, 1973, pp. 1525–1531.

EXERCISE 12

12.1. Develop a shell element by superimposing the constant strain triangle with non-compatible bending element having vertex degrees of freedom w, $\dfrac{\partial w}{\partial x}$ and $\dfrac{\partial w}{\partial y}$.

12.2. A doubly curved shell is modeled by flat triangular elements such as those of eqn. (12.2). Is the inter-elemental compatibility maintained of the edge displacements?

12.3. An isoparametric curved element is considered for the axi-symmetric shell as shown in Fig. 12.3. Derive the stiffness matrix of the element.

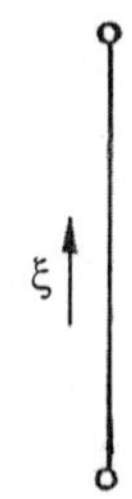

(a) parent element

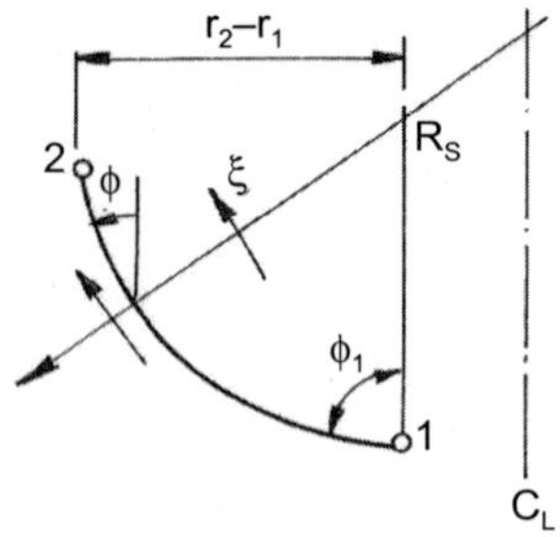

(b) Curvilinear element

Prob. 12.3.

12.4. Consider a thin-walled cylindrical shell which is symmetrically loaded, but without axial loads, *u*-displacement consideration not necessary. Using cubic polynomial for *w*, generate the 4×4 element stiffness matrix.

12.5. What difficulties are experienced when flat shell elements meeting at a node become coplanar? How are the difficulties obviated? Discuss all possibilities.

12.6. If the flat shell element is so constructed as to require only the continuity of displacements *u*, *v* and *w* at the corner nodes with continuity of the normal shape being imposed along the element sides, many of the difficulties encountered with the nodal assembly in global coordinates disappear. How?

12.7. How are you going to extend the analysis of Art. 12.5 to cater axi-symmetric curved thick shell?

12.8. What modifications will you have to perform on the analysis presented in Art. 12.5 to deal with thick plates?

12.9. Write a computer program in FORTRAN to analyse a shell of general shape for any arbitrary loading.

12.10. An end of an isoparametric bar element whose geometry is defined by the position of nodes along its centerline and vectors V_{2i} and V_{3i} that span its rectangular cross-section is shown below.

(*a*) Write an equation of geometry similar to eqn. (12.69)

(*b*) Write an equation of displacement analogous to eqn. (12.69)

12.11. (*a*) How can the transformation $[D]'$ to $[D]$ be made more computationally efficient?

(*b*) Rewrite eqn. (12.86) so that $[D]'$ instead of $[D]$ remains in the equation.

❑❑❑

13

SEMI-ANALYTICAL AND SPLINE FINITE STRIP METHOD OF ANALYSES OF PLATE BENDING

13.1 INTRODUCTION

Many practical structures have regular boundaries, *e.g.,* rectangular and circular plates. For them, the solution by the finite element analysis may entail a substantial computer time. The finite element analysis of plates requires a two-dimensional domain to be divided in both directions. As a consequence, a large number of simultaneous equations result. For the vibration and stability analyses of plates when the problem reduces to an eigenvalue problem, this becomes more uneconomical. For rectangular and circular plates, it was thus felt to develop a method, which is simpler than the conventional finite element method and which also takes less time for the computer solution. The finite strip method is a step forward in that direction.

The finite strip method is a simpler version of the finite element method presented so far. The plate is divided in one direction only, that is, the elements are in the form of strips. The displacement function is so chosen that it satisfies the boundary conditions in one direction. The unknowns in the problem are those at the nodal lines. Thus, a two-dimensional problem is reduced to a one-dimensional one. The rest of the procedure is same as the finite element method. It will be seen that the number of unknowns are significantly less than those needed for the analysis of the same plate by the finite element method.

The displacement function satisfies the boundary conditions along one direction of the plate, which converts the method to one that resembles series solution. Further, this method is similar to the finite element method. Due to these reasons the finite strip method is also termed as a semi-analytical finite element method.

This method has been applied to the analysis of various structures such as plates with inplane forces, plate bending, folded plates, shells and so on [13.1]. But this chapter is limited to the discussion of plate bending problems only.

13.2 BEAM FUNCTION

1. The characteristics function of a beam satisfying different boundary conditions can be expressed in a general form as

$$Y = C_1 \sin(\mu y/b) + C_2 \cos(\mu y/b) + C_3 \sinh(\mu y/b) + C_4 \cosh(\mu y/b) \qquad (13.1)$$

where b is the length of the beam.

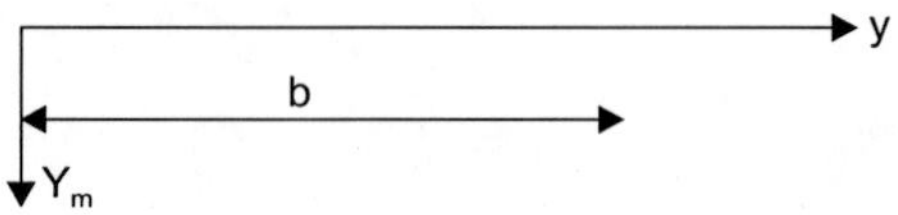

Fig. 13.13 A typical beam

The coordinate system is shown in Fig. 13.1. Equation (13.1) has been obtained from the vibration analysis of the beam treating it as a continuous system. Y represents to some scale deflected shape of the beam corresponding to different nodes. For mth node, eqn. (13.1) can be written as

$$Y_m = C_1 \sin(\mu_m y/b) + C_2 \cos(\mu_m y/b) + C_3 \sinh(\mu_m y/b) + C_4 \cosh(\mu_m y/b) \quad (13.2)$$

Therefore, the different deflected shapes are controlled by the parameter μ.

If a beam is considered to have both ends simply supported, then the boundary conditions are as follows:

At $\quad y = 0,\ Y = 0;\ d^2Y/dy^2 = 0 \qquad (13.3)$

At $\quad y = b,\ Y = 0;\ d^2Y/dy^2 = 0 \qquad (13.4)$

Putting the boundary conditions corresponding to eqn. (13.3) into eqn. (13.1), the following conditions are obtained.

$$C_2 + C_4 = 0$$
$$C_2 - C_4 = 0 \qquad (13.5)$$

Solving eqn. (13.5), the following values are obtained

$$C_2 = C_4 = 0 \qquad (13.6)$$

Substituting the conditions corresponding to eqn. (13.4) into eqn.(13.1), yields

$$\left.\begin{array}{l} C_1 \sin\mu + C_3 \sinh\mu = 0 \\ C_1 \sin\mu + C_3 \sinh\mu = 0 \end{array}\right\} \qquad (13.7)$$

Equation (13.7) gives

$$\left.\begin{array}{l} C_1 \sin\mu = 0 \\ C_3 \sinh\mu = 0 \end{array}\right\} \qquad (13.8)$$

Second of eqn. (13.8)

$$C_3 = 0 \qquad (13.9)$$

Therefore,

$$\sin\mu = 0$$

or, $\quad \mu = r\pi \qquad (13.10)$

$$r = 0, 1, 2, 3, \ldots\ldots., m, \ldots\ldots\ldots.$$

Therefore, the beam function for a beam simply supported at both ends is

$$Y_m = \sin\frac{\mu_m y}{b} \tag{13.11}$$

where $\mu_m = m\pi$

$m = 1, 2, 3, \ldots\ldots\ldots$

This is diagrammatically shown in Fig. 13.2

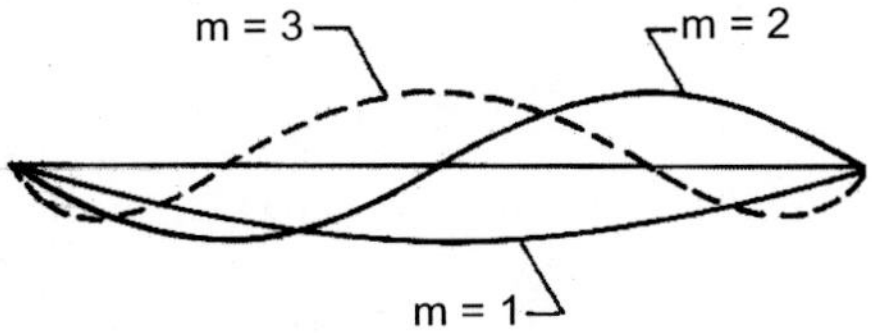

Fig. 13.2. Y_m for a simply supported beam

In this way, beam functions for beams having different boundary conditions can be derived. Other beam functions are given below.

2. One end simply supported, and the other end clamped

$$Y_m = \sin\frac{\mu_m y}{b} - \alpha_m \sinh\frac{\mu_m y}{b} \tag{13.12}$$

where $\mu_m = \dfrac{4m+1}{4}\pi$, $m = 1, 2, 3, \ldots\ldots\ldots$

$$\alpha_m = \frac{\sin\mu_m}{\sinh\mu_m}$$

3. Both ends clamped

$$Y_m = \frac{\sin\mu_m y}{b} - \sinh\frac{\mu_m y}{b} - \alpha_m\left(\cos\frac{\mu_m y}{b} - \cosh\frac{\mu_m y}{b}\right) \tag{13.13}$$

$$\mu_m = \frac{2m+1}{2}\pi,\ m = 1, 2, 3, \ldots\ldots.$$

$$\alpha_m = \frac{\sin\mu_m - \sinh\mu_m}{\cos\mu_m - \cosh\mu_m}$$

4. Both ends free

$$Y_m = \sin\frac{\mu_m y}{b} + \sinh\frac{\mu_m y}{b} - \alpha_m\left(\cos\frac{\mu_m y}{b} + \cosh\frac{\mu_m y}{b}\right) \tag{13.14}$$

$$\mu_m = \frac{2m+1}{2}\pi,\ m = 1, 2, 3, \ldots\ldots\ldots\ldots$$

$$\alpha_m = \frac{\sin\mu_m - \sinh\mu_m}{\cos\mu_m - \cosh\mu_m}$$

5. One end clamped, other end free

$$Y_m = \sin\frac{\mu_m y}{b} - \sinh\frac{\mu_m y}{b} - \alpha_m\left(\cos\frac{\mu_m y}{b} - \cosh\frac{\mu_m y}{b}\right) \quad (13.15)$$

$$\mu_1 = 1.875,\ \mu_2 = 4.694,\ \mu_3 = 7.855 \qquad m \geq 4,\ \mu_m = \frac{2m-1}{2}\pi$$

$$\alpha_m = \frac{\sin\mu_m + \sinh\mu_m}{\cos\mu_m + \cosh\mu_m}$$

6. One end simply supported and the other end free

$$Y_m = \sin\frac{\mu_m y}{b} + \alpha_m \sinh\frac{\mu_m y}{b} \quad (13.16)$$

$$\mu_m = \frac{4m-3}{4}\pi,\ m = 2, 3, \ldots\ldots\ldots.$$

$$\alpha_m = \frac{\sin\mu_m}{\sinh\mu_m}$$

7. One end clamped and the other end elastically restrained against rotation

$$Y_m = \sin\frac{\mu_m y}{b} - \sinh\frac{\mu_m y}{b} + \alpha_m\left(\cos\frac{\mu_m y}{b} - \cosh\frac{\mu_m y}{b}\right) \quad (13.17)$$

where $\alpha_m = \dfrac{\sin\mu_m - \sinh\mu_m}{\cosh\mu_m - \cos\mu_m}$

Different values of μ_m for various values of $\beta = C_R D/b$ where C_R is the restraining coefficient, D is the flexural rigidity of the beam, are given in Table 13.1.

8. One end simply supported and the other end elastically restrained against rotation

$$Y_m = \sin\frac{\mu_m y}{b} - \alpha_m \sinh\frac{\mu_m y}{b} \quad (13.18)$$

where $\alpha_m = \dfrac{\sin\mu_m}{\sinh\mu_m}$

Different values of μ_m for different values of $\beta = C_R D/b$ are given in Table 13.2.

Table 13.1. Values of μ_m for One Edge Clamped and the other Edge Elastically Estrained Against Rotation

β	m					
	1	2	3	4	5	6
0.01	4.686	7.782	10.898	14.015	17.134	20.254
0.05	4.547	7.585	10.661	13.751	16.850	19.958
0.10	4.431	7.450	10.522	13.615	16.720	19.834
1.00	4.042	7.134	10.257	13.388	16.523	19.660
1000.00	3.927	7.069	10.211	13.352	16.494	19.635

TABLE 13.2. Values of μ_m for one edge simply supported and the other edge elastically restrained against rotation

β	m					
	1	2	3	4	5	6
0.01	3.890	7.004	10.119	13.256	16.354	19.474
0.05	3.770	6.820	9.891	12.977	16.075	19.180
0.10	3.665	6.688	9.752	12.840	15.943	19.055
0.50	3.367	6.418	9.520	12.640	15.768	18.300
1.00	3.274	6.356	9.475	12.605	15.739	18.876
1000.00	3.142	6.284	9.425	12.567	15.708	18.850

9. Both edges possessing same degree of elastic restraint against rotation

$$Y_m = \sin\frac{\mu_m y}{b} - \sinh\frac{\mu_m y}{b} + \alpha_m\left(\cos\frac{\mu_m y}{b} - 2\mu_m\beta\sinh\frac{\mu_m y}{b} - \cosh\frac{\mu_m y}{b}\right) \quad (13.19)$$

where $\alpha_m = \dfrac{\sin\mu_m - \sinh\mu_m}{\cosh\mu_m - \cos\mu_m + 2\beta\mu_m\sinh\mu_m}$

Different values of μ_m for different degrees of coefficient of restraint are given in Table 13.3.

Table 13.3. Values of μ_m for Both Edges Elastically Restrained against Rotation (The Coefficient of Restraint Being Same at Both Edges)

β	m					
	1	2	3	4	5	6
0.01	4.642	7.711	10.802	13.895	16.991	20.089
0.05	4.374	7.329	10.339	13.375	16.431	19.592
0.10	4.156	7.069	10.066	13.106	16.172	19.257
0.50	3.577	6.547	9.613	12.712	15.827	18.950
1.00	3.399	6.428	9.525	12.643	15.770	18.902
10.00	3.173	6.399	9.436	12.575	15.715	18.855

The beam functions for the first 6 cases have been obtained from Ref. (13.1) and the next 3 from Ref. (13.2). The beam functions are orthogonal and possess the following properties:

$$\left.\begin{aligned} &\int_0^b Y_m\, Y_n\, dy = 0 \\ &\int_0^b Y_m^{iv}\, Y_n\, dy = 0 \end{aligned}\right\} \qquad (13.20)$$

where $m \neq n$.

However, $\int_0^b Y_m''\, Y_n\, dy = 0$ for opposite edges simply supported

13.3 MODEL OF THE PLATE

The plate is divided into a number of strips in either x or y direction. Figure 13.3 shows the plate divided in the x-direction. Each strip may have its individual elastic properties. A typical strip is shown in Fig. 13.4. At each nodal line i and j, there are two unknowns corresponding to each harmonic number (*i.e.*, corresponding to each value of m).

13.4 THE DISPLACEMENT FUNCTION

A displacement function of the form given below is chosen

$$w = \sum_{m=1}^{\gamma} f_m(x)\, Y_m \qquad (13.21)$$

where $f_m(x)$ contains the terms of that of the beam in x direction obtained from the finite element analysis of beam [eqn. (11.10)].

$$f_m(x) = [N]\, \{X_m\}_e \qquad (13.22)$$

Therefore, eqn. (13.21) becomes

$$w = \sum_{m=1}^{r} [N]_m\, \{X_m\}_e \qquad (13.23)$$

where

$$[N]_m = \left[\left(1-\frac{3x^2}{d^2}+\frac{2x^3}{d^3}\right)\left(x-\frac{2x^2}{d}+\frac{x^3}{d^2}\right)\left(\frac{3x^2}{d^2}-\frac{2x^3}{d^3}\right)\left(\frac{-x^2}{d}+\frac{x^3}{d^2}\right)\right] Y_m \qquad (13.24)$$

and

$$\{X_m\}_e = \begin{Bmatrix} W_{im} \\ \theta_{im} \\ W_{jm} \\ \theta_{jm} \end{Bmatrix} Y_m \qquad (13.25)$$

The total displacements along the sides *i* and *j* are

$$\begin{Bmatrix} W_i \\ \theta_i \\ W_j \\ \theta_j \end{Bmatrix} = \sum_{m=1}^{r} \begin{Bmatrix} W_{im} \\ \theta_{im} \\ W_{jm} \\ \theta_{jm} \end{Bmatrix} Y_m \tag{13.26}$$

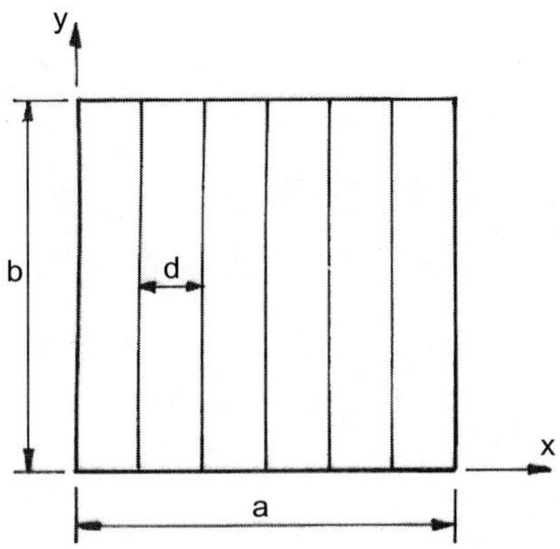

Fig. 13.3. Plate divided into strips

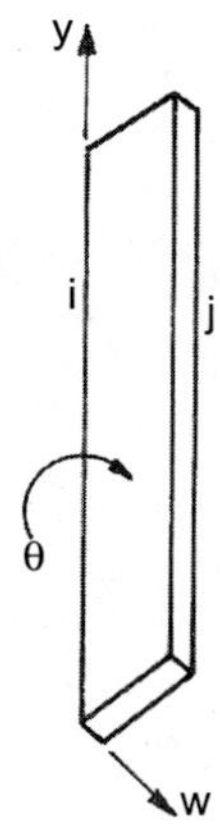

Fig. 13.4. A typical strip

13.5 CURVATURE-NODAL PARAMETER RELATIONSHIP

The curvatures of the plate strip are given by

$$\{\varepsilon\} = \begin{Bmatrix} -\dfrac{\partial^2 w}{\partial x^2} \\ -\dfrac{\partial^2 w}{\partial y^2} \\ 2\dfrac{\partial^2 w}{\partial x\, \partial y} \end{Bmatrix} \tag{13.27}$$

Performing the necessary differentiation of eqn. (13.23) as required in eqn. (13.27)

$$\{\varepsilon\} = \sum_{m=1}^{r} [B]_m \{X_m\}_e = [B]\{X\}_e \tag{13.28}$$

or,

$$\{\varepsilon\} = \left[[B]_1 [B]_2 \ldots\ldots [B]_r \right] \begin{Bmatrix} [X]_{1e} \\ [X]_{2e} \\ \vdots \\ [X]_{re} \end{Bmatrix} \tag{13.29}$$

where

$$[B]_m = \begin{bmatrix} \left(\frac{6}{d^2}-\frac{12x}{d^3}\right)Y_m & \left(\frac{4}{d}-\frac{6x}{d^2}\right)Y_m & \left(\frac{12x}{d^3}-\frac{6}{d^2}\right)Y_m & \left(\frac{2}{d}-\frac{6x}{d^2}\right)Y_m \\ -\left(1-\frac{3x^2}{d^2}-\frac{2x^3}{d^3}\right)Y''_m & -\left(x-\frac{2x^2}{d}+\frac{x^3}{d^2}\right)Y''_m & \left(-\frac{3x^2}{d^2}-\frac{2x^3}{d^3}\right)Y''_m & -\left(\frac{x^3}{d^2}-\frac{x^2}{d}\right)Y''_m \\ -\left(\frac{12x}{d^2}+\frac{12x^2}{d^3}\right)Y'_m & \left(2-\frac{8x}{d}-\frac{6x^2}{d^2}\right)Y'_m & -\left(\frac{12x^2}{d^2}-\frac{12x^3}{d^3}\right)Y'_m & \left(\frac{6x^2}{d^2}-\frac{4x}{d}\right)Y'_m \end{bmatrix} \tag{13.30}$$

13.6 MOMENT – CURVATURE RELATIONSHIP

For the rectangular orthotropic plate

$$\{\sigma\} = \begin{Bmatrix} M_x \\ M_y \\ M_{xy} \end{Bmatrix} = \begin{bmatrix} D_x & D_1 & 0 \\ D_1 & D_y & 0 \\ 0 & 0 & D_{xy} \end{bmatrix} \begin{Bmatrix} -\frac{\partial^2 w}{\partial x^2} \\ -\frac{\partial^2 w}{\partial y^2} \\ 2\frac{\partial^2 w}{\partial x \partial y} \end{Bmatrix}$$

or, $$\{\sigma\} = [D]\,\{\varepsilon\} \tag{13.31}$$

Combining eqns. (13.28) and (13.31), yields

$$\{\sigma\} = [D]\,[B]\,\{X\}_e \tag{13.32}$$

or, $$\{\sigma\} = \sum_{m=1}^{r} [D][B]_m\,\{X\}_{me} \tag{13.33}$$

13.7 STRIP STIFFNESS MATRIX

By the use of principle of virtual work or other variational principles, it can be shown that the strip stiffness matrix is given by

$$[K]_e = \int_0^b \int_0^d [B]^T [D] [B]\, dx\, dy \tag{13.34}$$

$$= \int_0^b \int_0^d \left[[B]_1 [B]_2 \cdots [B]_r\right]^T [D] \left[[B]_1 [B]_2 \cdots [B]_r\right] dx\, dy \tag{13.35}$$

$$= \int_0^b \int_0^d \begin{bmatrix} [B]_1^T [D] [B]_1 & [B]_1^T [D] [B]_2 & \cdots & [B]_1^T [D] [B]_r \\ [B]_2^T [D] [B]_1 & [B]_2^T [D] [B]_2 & \cdots & [B]_2^T [D] [B]_r \\ \vdots & \vdots & & \vdots \\ [B]_r^T [D] [B]_1 & [B]_r^T [D] [B]_2 & \cdots & [B]_r^T [D] [B]_r \end{bmatrix} dx\, dy$$

or,
$$[K]_e = \begin{bmatrix} [K]_{11e} & [K]_{12e} & \cdots & [K]_{1re} \\ [K]_{21e} & [K]_{22e} & \cdots & [K]_{2re} \\ \vdots & \vdots & & \vdots \\ [K]_{r1e} & [K]_{r2e} & \cdots & [K]_{rre} \end{bmatrix} \tag{13.36}$$

where
$$[K]_{mne} = \int_0^b \int_0^d [B]_{me}^T [D] [B]_{ne} \, dx \, dy$$

For the evaluation of $[K]_{11e}$ $[K]_{12e}$ *etc.* in eqn. (13.36), the following integral values are to be obtained

$$\int_0^b Y_m Y_n \, dy, \int_0^b Y_m'' Y_n'' dy, \int_0^b Y_m' Y_n' \, dy \text{ and } \int_0^b Y_n \, dy$$

The first two integrals are zero for all the beam functions for $m = n$. However, last two integrals simply supported ends. However, even in these cases these are assumed to be zero with the result that

$$[K]_{mn} = 0 \text{ when } m \neq n \tag{13.37}$$

Equation (13.36) can then be written as

$$[K]_e = \begin{bmatrix} [K]_{11e} & & & \\ & [K]_{22e} & & \\ & & \ddots & \\ & & & \ddots [K]_{re} \end{bmatrix} \tag{13.38}$$

The advantages and justification of the assumption of eqn. (13.37) will be discussed after the loading matrix.

13.8 LOADING MATRIX

All loads must be resolved into the beam function series in the y direction similar to the displacement function. A load of intensity 1/unit length in the z direction on a line parallel to y axis is expressed as

$$q = q_1 Y_1 + q_2 Y_2 + \ldots + q_r Y_r \tag{13.39}$$

or
$$q = \sum_{m=1}^{r} q_m Y_m \tag{13.40}$$

Multiplying both sides of eqn. (13.39) by Y_m and integrating with respect to y from 0 to b and noting the orthogonality property given by eqn. (13.20).

$$q_m = \frac{\int_0^b q \, Y_m \, dy}{\int_0^b q \, Y_m^2 \, dy} \tag{13.41}$$

For a uniformly distributed load from σ to b, eqn. (13.41) becomes

$$q_m = \frac{q \int_0^b Y_m \, dy}{\int_0^b Y_m^2 \, dy} \tag{13.42}$$

For a concentrated load P at $y = c$,

$$q_m = \frac{P\,Y_m(c)}{\int_0^b Y_m^2\,dy} \tag{13.43}$$

The loading matrix for the plate strip is given by (eqn. 11.56),

$$\{P\}_e = \int_0^b \int_0^b [N]^T q \; dx\,dy$$

$$\{P\}_e = \int_0^b \int_0^b \begin{Bmatrix} [N]_1^T \\ [N]_2^T \\ [N]_r^T \end{Bmatrix} (q_1 Y_1 + q_2 Y_2 + q_r Y_r)\; dx\,dy \tag{13.44}$$

Using the orthogonality relation $\int_0^b Y_m Y_n \; dy = 0$ for $m \neq n$, eqn. (13.44) can be simplified as

$$\{P\}_e = \int_0^b \int_0^b \begin{Bmatrix} [N]_1^T q_1 Y_1 \\ [N]_2^T q_2 Y_2 \\ \vdots \\ [N]_r^T q_r Y_r \end{Bmatrix} dx\,dy \tag{13.45}$$

$$\{P\}_e = \begin{Bmatrix} \{P\}_{1e} \\ \{P\}_{2e} \\ \vdots \\ \vdots \\ \{P\}_{re} \end{Bmatrix} \tag{13.46}$$

For simply supported strip subjected to uniform distributed load q, the following equation results

$$\{P\}_e = q\int_0^b Y_m \; dy \begin{Bmatrix} b/2 \\ b^2/12 \\ b/2 \\ -b^2/12 \end{Bmatrix} \tag{13.47}$$

13.9 FORCE DISPLACEMENT RELATIONSHIP

Using the well-known force-displacement relation, eqn (13.36) can be connected to eqn. (13.46) by the following relation.

$$\begin{bmatrix} [K]_{11e} & & & [0] \\ \vdots & [K]_{22e} & & \vdots \\ \vdots & & \ddots & \vdots \\ \vdots & & \ddots & \vdots \\ [0] & & & [K]_{rre} \end{bmatrix} \begin{Bmatrix} \{X\}_{1e} \\ \{X\}_{2e} \\ \vdots \\ \vdots \\ \{X\}_{re} \end{Bmatrix} = \begin{Bmatrix} \{P\}_{1e} \\ \{P\}_{2e} \\ \vdots \\ \vdots \\ \{P\}_{re} \end{Bmatrix} \tag{13.48}$$

Equation (13.48) is in uncoupled form. Therefore, solution can be separately sought for each harmonic term. Thus for harmonic term m, the corresponding eqn. is

$$[K]_{mme}\,\{X\}_{me} = \{P\}_{me} \qquad (13.49)$$

Equation (13.49) will give the values of displacements corresponding to mth harmonic term. Thus when all displacements for r harmonic terms are obtained, the final displacement will be the value evaluated after summing up as given by eqn. (13.26)

The advantage of assuming $\int_0^b Y_m Y_n\,dy$ and $\int_0^b Y_m'' Y_n\,dy$ to be zero for $m \neq n$ is now revealed. The greatest advantage lies in uncoupling the equation (13.48). Therefore, solution can be sought for individual harmonic term and this results in considerable saving in computation time.

These assumptions are however to be justified. The values of $\int_0^b Y_m' Y_n\,dy$ for $m \neq n$ for two functions given by eqns. (13.12) and (13.13) are given in Table (13.4). The most dominant terms are when $m = n$ and this may be the reason why results obtained by ignoring these integrals for $m \neq n$ have yielded exceedingly good results.

Table 13.4. Values of Integrals $C''_{mn} = \int_0^b Y_m'' Y_n\,dx(b=1)$

(*a*) One edge clamped, other edge simply supported, eqn. (13.12)

n	$m = 1$	2	3	4	5
1	– 5.7572	– 2.1424	1.9001	–1.6426	1.4292
2	– 2.1424	– 21.448	– 3.9096	3.8223	–3.583
3	1.9001	– 3.9096	–47.091	–5.5583	–5.6446
4	– 1.6424	3.8223	–5.5583	–82.459	–7.2169
5	1.4292	– 3.583	5.6446	–7.2169	–127.77

(*b*) Both edges clamped eqn. (13.13)

n	$m = 1$	2	3	4	5
1	– 12.699	0.23614	9.7041	– 0.1129	7.6117
2	0.23614	– 45.983	– 0.0243	17.133	– 0.0115
3	9.7041	– 0.0243	– 98.911	– 0.0019	– 24.349
4	0.1129	17.133	– 0.0019	– 171.59	– 0.0136
5	7.6117	– 0.0115	24.349	– 0.0136	– 264.37

13.10 SPLINE FINITE STRIP METHOD OF ANALYSIS OF PLATE BENDING

The semi-analytical finite strip method presented so far is indeed an elegant method for obtaining economic solution, specifically for regular shaped plates. It has been extensively used for various types of regular plate problems. But the method suffers from a number of drawbacks. It fails for plates having mixed boundary conditions, having internal cutouts, for continuous span, having discrete supports at strip ends. As the method uses the characteristic beam function in one direction which is continuously differentiable, it imposes limitations on the plate problems involving abrupt change of properties, concentrated and patch loads, internal supports *etc.*, since the second or the third derivatives should really be continuous. To overcome the difficulties mentioned above and to retain the advantages of the finite strip method, the spline finite strip method is developed [13.2–13.4].

13.10.1 The Spline Function

Spline is originally the name of a small flexible wooden strip employed by draughtsman as a tool for drawing a continuous smooth curve segment by segment. It has now been used as a mathematical tool for the solution of engineering problems.

A variety of spline functions are available. The spline function adopted here to represent displacement is the B_3-spline of equal section length. This is also known as cubic B_3-spline function. It is given by (*see* Fig. 13.5).

$$y = \sum_{i=-1}^{m+1} \alpha_i \phi_i \tag{13.50}$$

in which each local B_3 spline ϕ_i has non-zero values over four successive sections with section knot $x = x_i$ as the centre, is defined as

$$\phi_i = 1/6h^3 \begin{Bmatrix} 0, & x < x_{i-2} \\ (x - x_{i-2})^3, & x_{i-2} \le x \le x_{i-1} \\ h + 3h^2(x - x_{i-1}) + 3h(x - x_{i-1})^2 - 3(x - x_{i-1})^3, & x_{i-1} \le x \le x_i \\ h^3 + 3h^2(x_{i+1} - x) + 3h(x_{i-1} - x)^2 - 3(x_{i-1} - x)^3, & x_i \le x \le x_{i+1} \\ (x_{i+2} - x)^3, & x_{i+1} \le x \le x_{i+2} \\ 0, & x_{i+2} < x \end{Bmatrix} \tag{13.51}$$

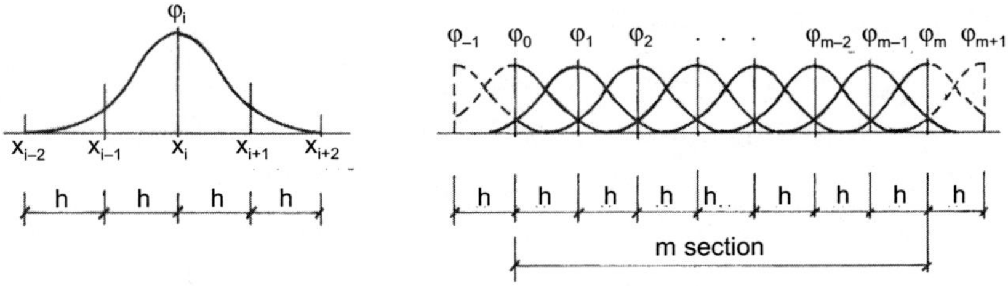

(a) (b)

Fig. 13.5. (a) Typical B_3 spline; (b) Basis of B_3-spline expression

The use of B_3-spline offers certain distinct advantages when compared with the conventional finite element method and the semi-analytical finite strip method.

1. It is computationally more efficient than the finite element method.

   When using B_3 splines as displacement functions, continuity is ensured up to second order (C_2-continuity). However, to achieve the same continuity conditions for the conventional finite element, it is necessary to have three times as many unknowns at the nodes.

2. It is more flexible than the semi-analytical finite strip method in the boundary conditions treatment.

   Only several local splines around the boundary point need to be amended to fit any specified boundary conditions.

The B_3-spline representations $[\Phi]$ can be easily managed to adapt to various prescribed boundary condition. Due to the localization of the B_3-splines, only three boundary local splines have to be amended, *i.e.*,

$$[\Phi] = \left[\bar{\phi}_{-1}, \bar{\phi}_0, \bar{\phi}_1, \phi_2, \phi_3, \phi_4 \ldots, \phi_{m-3}, \bar{\phi}_{m-1}, \bar{\phi}_m, \bar{\phi}_{m+1}\right] \quad (13.52)$$

in which the ϕ_i's are the standard local splines and the $\bar{\phi}_i$'s are the amended local splines. The amended scheme is given in Table 13.5. It may be mentioned that there are other possibilities of the amended scheme. The reader may check the amended spline by substituting ϕ_i's from equation (13.51).

Table 13.5. Amended Scheme for Boundary Local Splines

Boundary condition		Amended local splines $\bar{\phi}_{-1}$	$\bar{\phi}_{-0}$	$\bar{\phi}_1$
Free end	$y(x_0) \neq 0$, $y'(x_0) \neq 0$	ϕ_{-1}	$\phi_{-0} - 4\bar{\phi}_{-1}$	$\phi_1 - \frac{1}{2}\phi_0 + \phi_{-1}$
Simply supported end	$y(x_0) \neq 0$, $y'(x_0) \neq 0$	Eliminated	$\bar{\phi}_{-0} - 4\bar{\phi}_{-1}$	$\phi_1 - \frac{1}{2}\phi_0 + \phi_{-1}$
Clamped end	$y(x_0) = 0$	Eliminated	Eliminated	$\phi_1 - \frac{1}{2}\phi_0 + \phi_{-1}$
Sliding clamped end	$y(x_0) \neq c$	Eliminated	ϕ_0	$\phi_1 - \frac{1}{2}\phi_0 + \phi_{-1}$

13.10.2 Displacement Functions

Let us consider the formulation of rectangular plates only. A typical plate strip is shown in Fig. 13.6, which is rectangular. The width of the strip is b and length is a, which is divided into m sections, a knot is placed on each section.

Let $[\phi_1]$ and $[\phi_3]$ be B_3-spline representations for the displacements w of the nodal lines i and j respectively (m-sections) and $[\phi_2]$ and $[\phi_4]$ be B_3-spline representations for the rotations ϕ of nodal lines i and j, respectively (m sections), $\{w_i\}$ and $\{w_j\}$ be the displacement parameter vectors $[w_{i-1}, w_{i0}, w_{i1}, w_{i2}, \ldots.., w_{im-1}, w_{im}, w_{im+1}]^T$ and $[w_{j-1}, w_{j0}, w_{j1}, w_{j2}, \ldots\ldots.., w_{jm-1}, w_{jm}, w_{jm+1}]^T$ corresponding to ϕ_1 and ϕ_3, respectively, and $\{\theta_i\}$ and $\{\theta_j\}$ be the rotation parameter vectors $[\theta_{i-1}, \theta_{i0}, \theta_{i1}, \theta_{i2}, \ldots., \theta_{im-1}, \theta_{im}, \theta_{im+1}]^T$ and $[\theta_{j-1}, \theta_{j0}, \theta_{j1}, \theta_{j2}, \ldots\ldots.., \theta_{jm-1}, \theta_{jm}, \theta_{jm+1}]^T$ corresponding to ϕ_2 and ϕ_4 respectively. Then, with a cubic polynomial interpolation in the x-direction, the displacement function for the strip will be

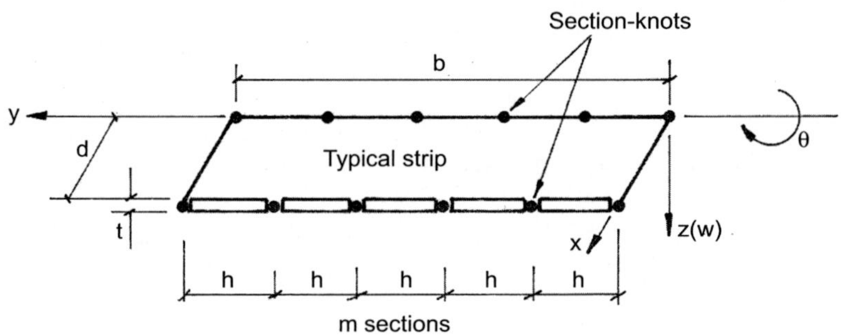

Fig. 13.6. Typical strip

$$\{f\} = w = [N_1, N_2, N_3, N_4]\begin{bmatrix} [\phi_1] & & & 0 \\ & [\phi_2] & & \\ & & [\phi_3] & \\ 0 & & & [\phi_4] \end{bmatrix}\begin{Bmatrix} \{w_i\} \\ \{\theta_i\} \\ \{w_j\} \\ \{\theta_j\} \end{Bmatrix} \quad (13.53)$$

where

$$N_1 = 1 - 3\bar{x}^2 + 2\bar{x}^3,\ N_2 = x(1 - 2\bar{x} + \bar{x}^2),\ N_3 = (3\bar{x}^2 - 2\bar{x}^3),\ N_4 = x(\bar{x}^2 - \bar{x}) \quad (13.54)$$

and $\bar{x} = x/d$. In concise matrix form,

$$\{f\} = [N]\,[\phi]\,\{\delta\} \quad (13.55)$$

13.10.3 Strain-Displacement Relationship

The strain-displacement relationship is

$$\{\varepsilon\} = \begin{Bmatrix} -\dfrac{\partial^2 w}{\partial x^2} \\ -\dfrac{\partial^2 w}{\partial y^2} \\ 2\dfrac{\partial^2 w}{\partial x\,\partial y} \end{Bmatrix} = \begin{bmatrix} -[N''] & 0 & 0 \\ 0 & -[N] & 0 \\ 0 & & 2[N'] \end{bmatrix} \begin{bmatrix} [\Phi] \\ [\Phi''] \\ \Phi' \end{bmatrix} [\delta] \quad (13.56)$$

where,

$$[N'] = \frac{\partial}{\partial x}[N] = \left[\frac{1}{d}(-6\bar{x}+6\bar{x}^2),(1-4\bar{x}+3\bar{x}^2),\frac{1}{d}(6\bar{x}-6\bar{x}^2),(3\bar{x}-2\bar{x})\right], \quad (13.57)$$

$$[N''] = \frac{\partial^2}{\partial x^2}[N] = \left[\frac{-6}{d^2}(1-2\bar{x}),\ \frac{-2}{d}(2-3\bar{x}),\ \frac{6}{d^2}(1-2\bar{x}),\ \frac{-2}{d}(1-3\bar{x})\right] \quad (13.58)$$

13.10.4 Stiffness Matrix

The stiffness matrix [*K*] for an orthotropic plate strip may be worked out by hand and expressed in an explicit form as follows:

$$[K] = \int [B]^T [D] [B]\, d(area) = \int \left[[\Phi]^T, [\Phi'']^T, [\Phi']^T \right]$$

$$\times \begin{bmatrix} -[N'']^T & 0 & 0 \\ 0 & -[N]^T & 0 \\ 0 & 0 & 2[N']^T \end{bmatrix} \begin{bmatrix} D_x & D_1 & 0 \\ D_1 & D_y & 0 \\ 0 & 0 & D_{xy} \end{bmatrix}$$

$$\times \begin{bmatrix} -[N''] & 0 & 0 \\ 0 & -[N] & 0 \\ 0 & 0 & 2[N'] \end{bmatrix} \begin{bmatrix} [\Phi] \\ [\Phi''] \\ [\Phi'] \end{bmatrix} d(area)$$

$$= \int_0^b \left[[\Phi]^T, [\Phi'']^T, [\Phi']^T \right] \begin{bmatrix} [C_1] & [C_3] & 0 \\ [C_2] & [C_4] & 0 \\ 0 & 0 & [C_5] \end{bmatrix} \begin{bmatrix} [\Phi] \\ [\Phi''] \\ [\Phi'] \end{bmatrix} dy$$

$$= \int_0^b \left\{ [\Phi]^T [C_1][\Phi] + [\Phi]^T [C_3][\Phi''] + [\Phi'']^T [C_2][\Phi] \right.$$

$$\left. + \left\{ [\Phi'']^T [C_4][\Phi''] + [\Phi']^T [C_5][\Phi'] \right\} dy \right. \quad (13.62)$$

where,

$$[C_1] = \int_0^d [N'']^T D_x [N''] dx, \qquad [C_2] = \int_0^d [N'']^T D_1 [N] dx$$

$$[C_3] = \int_0^d [N]^T D_1 [N''] dx \qquad [C_4] = \int_0^d [N]^T D_y [N] dx$$

$$[C_5] = \int_0^d [N']^T D_{xy} [N'] dx \tag{13.63}$$

where D_{xy}, D_y *etc.* are the usual orthotropic plate constants.

The integrations in equation (13.63) can be worked out explicitly since only simple polynomials are involved while the integrations for the coupling matrices $\int_0^b \phi_i^{(k)} \phi_j^{(l)} dy$ in equation (13.62) may be reduced to summation operations of the integrals $\int_0^b \phi_i^{(k)} \phi_j^{(l)} dy$ over a typical segment. Such segmental integrals have also been tabulated explicitly, which are given in Tables 13.6 – 13.10.

Table 13.6. Values of Cross-Segmental Integrations of Two Standard B_3-Splines, $\int_0^b \phi_i^{(k)} \phi_j^{(l)}$

$\phi_{(1)}$ / S_{ij} / $\phi_{(2)}$	Section L_1	Section L_2	Section L_3	Section L_4	
Section L_4	1	60	129	20	
Section L_3	60	933	1188	129	
Section L_2	129	1188	933	60	$\times \dfrac{1}{140} \times \dfrac{h}{36}$
Section L_1	20	129	60	1	

Table 13.7. Values of Cross-Segmental Integrations of Standard B_3-Spline and its First Derivative, $\int \phi_{(1)}\, \phi'_{(2)}\, dy$

$\phi_{(1)}$ S_{ij} $\phi_{(2)}$	Section L_1	Section L_2	Section L_3	Section L_4	
Section L_4	–1	–38	–71	–10	
Section L_3	–18	–183	–150	–9	
Section L_2	9	150	183	18	$\times \frac{1}{20} \times \frac{1}{36}$
Section L_1	10	71	38	1	

Table 13.8. Values of Cross-Segmental Integrations of First Derivatives of Standard B_3-Splines, $\int \phi'_{(1)}\, \phi'_{(2)}\, dy$

$\phi'_{(1)}$ S_{ij} $\phi'_{(2)}$	Section L_1	Section L_2	Section L_3	Section L_4	
Section L_4	–3	–36	21	18	
Section L_3	–36	–87	102	21	
Section L_2	21	102	–87	–36	$\times \frac{1}{10} \times \frac{1}{36h}$
Section L_1	18	21	–36	–3	

Table 13.9. Values of Cross-Segmental Integrations of Standard B_3-Spline and its Second Derivative, $\int \phi_{(1)} \phi''_{(2)}\, dy$

$\phi_{(1)}$ S_{ij} $\phi''_{(2)}$	Section L_1	Section L_2	Section L_3	Section L_4	
Section L_4	3	66	99	12	
Section L_3	6	–33	–132	–21	
Section L_2	–21	–132	–33	6	$\times \frac{1}{10} \times \frac{1}{36h}$
Section L_1	12	99	66	3	

Table 13.10. Values of Cross-Segmental Integrations of the Second Derivatives of Standard B_3-Spline, $\int \phi''_{(1)} \phi''_{(2)}\, dy$

$\phi''_{(1)}$ S_{ij} $\phi''_{(2)}$	Section L_1	Section L_2	Section L_3	Section L_4	
Section L_4	6	0	–18	12	
Section L_3	0	–18	36	–18	
Section L_2	–18	36	–18	0	$\times \frac{1}{36h^3}$
Section L_1	12	–18	0	6	

The explicit value of stiffness matrix is given in Table 13.11 [13.3].

13.10.5 The Loading Matrix

The loading matrix of the strip can be obtained by combining eqn. (11.56) with eqn. (13.55) and is given by

$$\{P\}_e = \int_0^b \int_0^a [\phi]^T [N]^T q \, dx \, dy \tag{13.64}$$

13.11 COMPUTER PROGRAM FOR THE SPLINE FINITE STRIP METHOD OF ANALYSIS OF PLATES IN BENDING

The computer programs in C and FORTRAN for the solution of plate bending problem by the spline finite strip method are given in the attached CD. The steps have been explained with comment statements. The stiffness matrix of a strip has been evaluated explicitly.

A simply supported square plate under uniform load has been solved with the help of the computer program. Only a quarter of a plate has been analyzed and symmetry conditions have been imposed at the symmetric edges. Input of the problem has been shown in Table 13.12 while the output has been presented in Table 13.13.

Table 13.11. Bending Stiffness Matrix of a Rectangular Strip

$$[K] = \frac{1}{420b^3}\left[\begin{array}{cccc|cccc|cccc|cccc}
5040 & & D_x & I_{1_{11}} & 2520 & b & D_x & I_{1_{12}} & -5040 & & D_x & I_{1_{13}} & 2520 & b & D_x & I_{1_{14}} \\
-504 & b^2 & D_1 & I_{2_{11}} & -462 & b^3 & D_1 & I_{2_{12}} & +504 & b^2 & D_1 & I_{2_{13}} & -42 & b^3 & D_1 & I_{2_{14}} \\
-504 & b^2 & D_1 & I_{3_{11}} & -42 & b^3 & D_1 & I_{3_{12}} & +504 & b^2 & D_1 & I_{3_{13}} & -42 & b^3 & D_1 & I_{3_{14}} \\
+156 & b^4 & D_y & I_{4_{11}} & +22 & b^5 & D_y & I_{4_{12}} & +54 & b^4 & D_y & I_{4_{13}} & -13 & b^5 & D_y & I_{4_{14}} \\
+2016 & b^2 & D_{xy} & I_{5_{11}} & +168 & b^3 & D_{xy} & I_{5_{12}} & -2016 & b^2 & D_{xy} & I_{5_{13}} & +168 & b^3 & D_{xy} & I_{5_{14}} \\
\hline
 & & & & 1680 & b^2 & D_x & I_{1_{22}} & -2520 & b & D_x & I_{1_{23}} & 840 & b^2 & D_x & I_{1_{24}} \\
 & & & & -56 & b^4 & D_1 & I_{2_{22}} & +42 & b^3 & D_1 & I_{2_{23}} & +14 & b^4 & D_1 & I_{2_{24}} \\
 & & & & -56 & b^4 & D_1 & I_{3_{22}} & +42 & b^3 & D_1 & I_{3_{23}} & +14 & b^4 & D_1 & I_{3_{24}} \\
 & & & & +4 & b^6 & D_y & I_{4_{22}} & +13 & b^5 & D_y & I_{4_{23}} & -3 & b^6 & D_y & I_{4_{24}} \\
 & & & & +224 & b^4 & D_{xy} & I_{5_{22}} & 168 & b^3 & D_{xy} & I_{5_{23}} & -56 & b^4 & D_{xy} & I_{5_{24}} \\
\hline
 & & & & & & & & 5040 & & D_x & I_{1_{33}} & -2520 & b & D_x & I_{1_{34}} \\
 & & & & & & & & -504 & b^2 & D_1 & I_{2_{33}} & +462 & b^3 & D_1 & I_{2_{34}} \\
 & & & & & & & & -504 & b^2 & D_1 & I_{3_{33}} & +42 & b^3 & D_1 & I_{3_{34}} \\
 & & & & & & & & +156 & b^4 & D_y & I_{4_{33}} & -22 & b^5 & D_y & I_{4_{34}} \\
 & & & & & & & & +2016 & b^2 & D_{xy} & I_{5_{33}} & -168 & b^3 & D_{xy} & I_{5_{34}} \\
\hline
 & & & & & & & & & & & & 1680 & b^2 & D_x & I_{1_{44}} \\
 & & & & & & & & & & & & -56 & b^4 & D_1 & I_{2_{44}} \\
 & & & & & & & & & & & & -56 & b^4 & D_1 & I_{3_{44}} \\
 & & & & & & & & & & & & +4 & b^6 & D_y & I_{4_{44}} \\
 & & & & & & & & & & & & +224 & b^4 & D_{xy} & I_{5_{44}}
\end{array}\right]$$

$$I_{1ij} = \int_0^a t\,\phi_i^T\,\phi_j\,dy,\ I_{2_{ij}} = \int_0^a t\,\phi_i''^T\,\phi_j\,dy,\ I_{3_{ij}} = \int_0^a t\,\phi_i^T\,\phi_j''dy,$$

$$I_{4ij} = \int_0^a t\,\phi_i''^T\,\phi_j''dy,\ I_{5_{ij}} = \int_0^a t\,\phi_i'^T\,\phi_j'\,dy$$

Table 13.12. Input Data for the Spline Problem

```
.5 .5 .01 10.92E6 .3 1.
8  8  2  3  2  3  0  1
0.5     0.5
0.0     0.5
0.5     0.0
```

Table 13.13. Computer Output for Plate-Bending by Spline Finite Strip Method

```
xl = 0.500        yl = 0.500      zl = 0.010      emod = 10920000.000
pr = 0.300        udl = 1.000
nelm =8 nint =8 nbdl =2 nbdr =3 nbdu =2 nbdd =3 npl =0  nsp =1
xsp =    5.00000e-001      ysp =    5.00000e-001
          OVERALL STIFFNESS MATRIX
          OVERALL LOADING MATRIX
          THE DISPLACEMENT MATRIX
 -1.76404e-034  9.27327e-022  3.73065e-021  6.29547e-021  8.12018e-021
  9.40646e-021  1.03057e-020  1.09018e-020  9.72081e-021  1.13534e-020
  3.13306e-035  1.36833e-019  1.45069e-003  2.89110e-003  5.56194e-003
  7.92384e-003  9.91410e-003  1.14934e-002  1.26364e-002  1.33277e-002
  2.02228e-002  4.19290e-019  1.25896e-020  8.92288e-005  1.78135e-004
  3.43359e-004  4.89642e-004  6.13011e-004  7.10950e-004  7.81850e-004
  8.24736e-004  1.25145e-003  4.42778e-020  9.52250e-020  1.38857e-003
  2.77567e-003  5.36569e-003  7.66402e-003  9.60543e-003  1.11480e-002
  1.22653e-002  1.29414e-002  1.96383e-002  4.16529e-019  2.05274e-020
  1.72203e-004  3.44033e-004  6.64902e-004  9.49829e-004  1.19060e-003
  1.38198e-003  1.52062e-003  1.60452e-003  2.43486e-003  8.48871e-020
  6.16039e-020  1.25823e-003  2.51605e-003  4.88794e-003  7.00952e-003
  8.81118e-003  1.02477e-002  1.12906e-002  1.19224e-002  1.80952e-002
  3.63007e-019  2.59240e-020  2.45684e-004  4.90979e-004  9.50828e-004
  1.36053e-003  1.70758e-003  1.98386e-003  2.18424e-003  2.30557e-003
  3.49897e-003  1.19746e-019  4.32815e-020  1.08798e-003  2.17578e-003
  4.23952e-003  6.09899e-003  7.68745e-003  8.95949e-003  9.88558e-003
  1.04477e-002  1.58599e-002  3.06195e-019  2.97446e-020  3.07675e-004
  6.14954e-004  1.19262e-003  1.70877e-003  2.14698e-003  2.49642e-003
  2.75013e-003  2.90387e-003  4.40728e-003  1.48558e-019  3.06854e-020
  8.92399e-004  1.78471e-003  3.48379e-003  5.02284e-003  6.34425e-003
  7.40672e-003  8.18251e-003  8.65424e-003  1.31396e-002  2.46962e-019
  3.24228e-020  3.56897e-004  7.13393e-004  1.38489e-003  1.98619e-003
  2.49764e-003  2.90607e-003  3.20290e-003  3.38288e-003  5.13461e-003
  1.71115e-019  2.10871e-020  6.80567e-004  1.36108e-003  2.65980e-003
  3.84042e-003  4.85790e-003  5.67868e-003  6.27946e-003  6.64537e-003
```

```
 1.00909e-002  1.86154e-019  3.41996e-020  3.92535e-004  7.84667e-004
 1.52422e-003  2.18745e-003  2.75236e-003  3.20395e-003  3.53241e-003
 3.73166e-003  5.66424e-003  1.87291e-019  1.32361e-020  4.58553e-004
 9.17080e-004  1.79333e-003  2.59170e-003  3.28149e-003  3.83920e-003
 4.24817e-003  4.49756e-003  6.83011e-003  1.24451e-019  3.52158e-020
 4.14094e-004  8.27783e-004  1.60855e-003  2.30935e-003  2.90674e-003
 3.38462e-003  3.73236e-003  3.94337e-003  5.98577e-003  1.97016e-019
 6.38704e-021  2.30630e-004  4.61247e-004  9.02281e-004  1.30461e-003
 1.65273e-003  1.93457e-003  2.14146e-003  2.26772e-003  3.44401e-003
 6.23087e-020  3.55466e-020  4.21308e-004  8.42211e-004  1.63677e-003
 2.35017e-003  2.95845e-003  3.44515e-003  3.79937e-003  4.01434e-003
 6.09354e-003  2.00260e-019 -6.21528e-034  2.14787e-019  4.01747e-019
 6.08060e-019  8.51322e-019  1.05241e-018  1.20987e-018  1.32280e-018
 1.43782e-018  1.41336e-018  4.39539e-035
Values of deflection w and moments Mx, My
  0.00406236      0.0479514      0.0479506
```

REFERENCES AND SUGGESTED READINGS

13.1. Y. K Cheung, Finite Strip Method in Structural Analysis, Pergamon Press, 1976.

13.2. Y. K Cheung, S.C. Fan and C. Q. Wu, Spline Finite Strip in Structural Analysis, Proc. The International Conference on Finite Element, Shaighai, China, 1982, pp. 704–709.

13.3. S.C. Fan and Y. K. Cheung, Flexural Vibration of Rectangular Plates with Complex Support Conditions, Journal of Sound and Vibration, v. 93, no. 1, 1984, pp. 181–194

13.4. W.Y. Li, Y. K Cheung and L. G. Tham, Spline Finite Strip Analysis of General Plates, Journal of Engineering Mechanics Division, Proc. ASCE, v. 112, 1986, pp. 43–54.

EXERCISE 13

13.1. Derive explicitly the stiffness matrix of a plate having all edges clamped, for harmonic number 1. Consider the plate as isotropic. The following are the values of different integrals:

$$a_m = \int_0^b Y_m^2\,dy,\quad c_m = \int_0^b Y_m''Y_m\,dy,\quad S_m = \int_0^b Y_m\,dy$$

$$e_m = \int_0^b Y_m'^2\,dy,\quad f_m = \int_0^b Y_m Y_m'\,dy,\quad d_m = \int_0^b Y_m' Y_m''dy$$

$$g_m = \int_0^b Y_m''^2\,dy$$

13.2. Determine the loading matrix corresponding to eqn. (13.41) for a load varying linearly at $y = 0$ to q_0 at $y = b$.

13.3. Based on the derivation of Prob. 13.1, determine the deflection for a square plate dividing it into two equal strips for a half plate and subjected to hydrostatic loading for the harmonic number 1.

13.4. Check, the orthogonality relationship for functions (13.14) and (13.18).

13.5. Deflection and bending moment at a point for a particular plate for different harmonic numbers are given below.

m	1	2	3	4	5	6	7
C_0	0.00523	0	0.00032	0	0.00012	0	0.00001
M_n	−0.11751	0	−0.0142	0	−0.00439	0	−0.0018

Find out the deflection and moment at that point

13.6. Derive the stiffness matrix of a skew strip.

❑❑❑

14

DYNAMIC AND INSTABILITY ANALYSES BY THE FINITE ELEMENT METHOD

14.1 INTRODUCTION

The finite element method which has been presented so far for the static analysis can be extended without much difficulty to the dynamic and the elastic instability analysis of structures. The procedure is similar. The mass matrix for dynamic analysis and the geometric matrix for instability analysis are to be determined. For more details than presented in this chapter, the reader can refer to relevant publications (14.1 – 14.10).

14.2 DYNAMIC ANALYSIS

By the application of the finite element technique, the continuous structure reduces to a system having multiple degrees of freedom. Dynamic analysis is of two types – free vibration analysis and response of the structure to forced vibrations; damping may or may not be considered depending on its importance. In forced vibrations, the external force that acts is a function of time. Depending on the nature of time variation of this force, the problem can be categorized into four types: harmonic, periodic, transient and random. The present discussion on vibration is limited to free vibration analysis only.

14.2.1 Torsional Vibration of Shafts

A shaft element of Fig. 14.1 has 2 nodes at its two ends. The unknown displacements at each end are the angles of twist ϕ_1 and ϕ_2. The displacement function is given by

$$f = \alpha_1 + \alpha_2 x \tag{14.1}$$

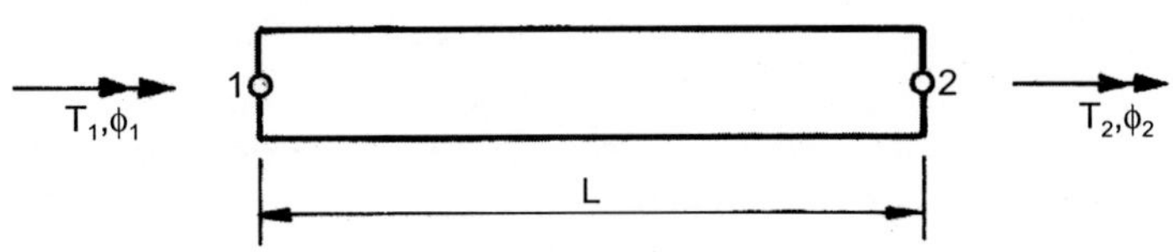

Fig. 14.1. Prismatic element

or, $$f = \phi = [1 \quad x]\begin{Bmatrix} \alpha_1 \\ \alpha_2 \end{Bmatrix} \tag{14.2}$$

The nodal values are

$$\begin{Bmatrix} \phi_1 \\ \phi_2 \end{Bmatrix} = \begin{bmatrix} 1 & 0 \\ 1 & L \end{bmatrix}\begin{Bmatrix} \alpha_1 \\ \alpha_2 \end{Bmatrix}$$

or, $$\begin{Bmatrix} \alpha_1 \\ \alpha_2 \end{Bmatrix} = \begin{bmatrix} 1 & 0 \\ 1 & L \end{bmatrix}^{-1}\begin{Bmatrix} \phi_1 \\ \phi_2 \end{Bmatrix} \tag{14.3}$$

Combining eqns. (14.2) and (14.3), yields

$$f = \phi = [1 \quad x]\begin{bmatrix} 1 & 0 \\ 1 & L \end{bmatrix}^{-1}\begin{Bmatrix} \phi_1 \\ \phi_2 \end{Bmatrix}$$

or, $$f = \phi = [N]\,\{X\}_e \tag{14.4}$$

where $$[N] = \left[\left(1 - \frac{x}{L}\right) \quad \frac{x}{L}\right] \tag{14.5}$$

General derivation for the static analysis still holds good. But two additional forces are created. The mass of the vibrating structure induces in an additional force. The inertia force is replaced by its static equivalent which is

$$-\bar{I}\,\ddot{\phi} \text{ in the present case}$$

where $\bar{I}$ is the mass moment of inertia per unit length. From eqn. (14.4), the following equation can be written as

$$\ddot{\phi} = [N]\{\ddot{X}\}_e \tag{14.6}$$

Equivalent nodal forces corresponding to the inertia force due to the mass of the structure can be written from eqn. (11.24) as

$$\{P\}_e^m = -\int_o^L [N]^T \bar{I}\,[N]\,dx\,\{\ddot{X}\}_e \tag{14.7}$$

or, $$\{P\}_e^m = -[M]_e\,\{\ddot{X}\}_e \tag{14.8}$$

where $$[M]_e = \int_o^L [N]^T \bar{I}\,[N]\,dx \tag{14.9}$$

Substituting the values of $[N]$ from eqn. (14.8) into eqn. (14.9), yields

$$[M]_e = \frac{\bar{I}\,L}{6}\begin{bmatrix} 2 & -1 \\ -1 & 2 \end{bmatrix} \tag{14.10}$$

$[M]_e$ is referred to as mass matrix. If $[M]_e$ is computed according to eqn. (14.9), it is referred to as consistent mass matrix.

In addition to the inertia of the mass, the other additional force that may be considered in the vibration analysis is the damping force. Similar to the derivation presented for the mass matrix, the damping matrix of the element can be shown to be

$$[C]_e = \int [N]^T \mu [N] dx \tag{14.11}$$

where μ is some numerical value of the viscous damping. Generation of the damping matrix creates difficulties, mainly because of the damping mechanism and the damping levels in the structures. As such it will not be discussed here.

Now, if all the elements are assembled, the final equations will be

$$[M]\{\ddot{X}\}+[C]\{\dot{X}\}+[K]\{X\} = \{P\} \tag{14.12}$$

If free undamped vibration is considered, then eqn. (14.12) becomes

$$[M]\{\ddot{X}\}+[K]\{X\} = \{0\} \tag{14.13}$$

Assuming, $\{X\} = a^{e^{ipt}} \{\psi\}$ (14.14)

So, $\{\ddot{X}\} = -ap^2 e^{ipt} \{\psi\}$ (14.15)

Combining eqns. (14.13) to (14.15), results

$$-p^2 [M] \{\psi\} + [K] \{\psi\} = \{0\}$$

or, $[K]^{-1} [M] \{\psi\} = \omega^2 [I] \{\psi\}$ (14.16)

where $\omega^2 = 1/p^2$ and $[I]$ is an identity matrix.

Equation (14.16) can be written as

$$[\overline{D}]\{\psi\} = \omega^2 [I]\{\psi\} \tag{14.17}$$

where $[\overline{D}] = [K]^{-1}[M]$ (14.18)

Equation (14.17) is a typical eigenvalue problem. There are various techniques available for their solution (14.1). A non-trivial solution of eqn. (14.17) will be

$$\det\left|[\overline{D}] - \omega^2 [I]\right| = 0 \tag{14.19}$$

14.2.1.1 An Example

If a uniform shaft having one end fixed, the other end free is divided into two equal elements and the node numbers are as shown in Fig. 14.2, then, the complete mass matrix becomes

$$[M] = \frac{\overline{I}L}{12}\begin{bmatrix} 2 & 1 \\ 1 & 4 \end{bmatrix}$$

and the complete stiffness matrix becomes

$$[K] = \frac{2GJ}{L}\begin{bmatrix} 1 & -1 \\ -1 & 2 \end{bmatrix}$$

Therefore, $$[\overline{D}] = \frac{L}{2GJ}\begin{bmatrix} 2 & 1 \\ 1 & 1 \end{bmatrix}\frac{\overline{I}L}{12}\begin{bmatrix} 2 & 1 \\ 1 & 4 \end{bmatrix} = \frac{\overline{I}L^2}{24GJ}\begin{bmatrix} 5 & 6 \\ 3 & 5 \end{bmatrix}$$

Equation (14.19) then becomes after putting $d = \dfrac{\bar{I}L^2}{24GJ}$

Let
$$\begin{vmatrix} 5d-\omega^2 & 6d \\ 3d & 5d-\omega^2 \end{vmatrix} = 0$$

or, $(5d-\omega^2) = 18d^2$

or, $5d-\omega^2 = \pm 4.2426d$

or, $\omega^2 = 0.7574d,\ 9.2426d$

Therefore, two values of the natural frequencies squared are

$$p^2 = 2.597\frac{GJ}{\bar{I}L^2} \text{ and } 31.687\frac{GJ}{\bar{I}L^2}$$

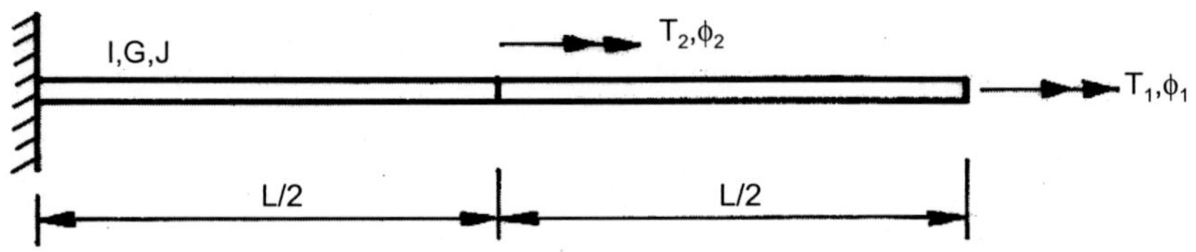

Fig. 14.2. Example of fixed-free shaft

14.2.2 Flexural Vibration of Beams

The consistent mass matrix is given by

$$[M] = \int [N]^T \rho A [N]\, dx \tag{14.20}$$

where ρ is the mass density of the material and A is the cross-sectional area of the uniform beam. Substituting $[N]$ from eqn. (14.20) and performing the necessary integration, eqn. (14.20) becomes

$$[M]_e = \frac{\rho AL}{420}\begin{bmatrix} 156 & & & \\ 22L & 4L^2 & \textit{symmetrical} & \\ 54 & 13L & 156 & \\ -13L & -3L^2 & -22L & 4L^2 \end{bmatrix} \tag{14.21}$$

Formation of the mass matrix in another form is also popular. The uniform beam having cross-sectional area A has been divided into four equal parts and the masses are lumped at the 5 stations – two end stations being supported, have not been numbered (Fig. 14.3). The masses between the stations have been lumped at the stations. Thus, $M_1 = \rho Aa$. The mass matrix in this case is

$$[M]_e = \begin{bmatrix} M_1 & 0 & 0 \\ 0 & M_1 & 0 \\ 0 & 0 & M_1 \end{bmatrix} \tag{14.22}$$

The elegance of eqn. (14.22) lies in the fact that $[M]_e$ is a diagonal matrix, and this leads to computational advantages.

For flexural vibrations of bar elements, it has been found that the consistent mass matrix formulation improves the results significantly. However, there is not much to gain by the consistent mass matrix formulation of torsional or longitudinal vibration problems.

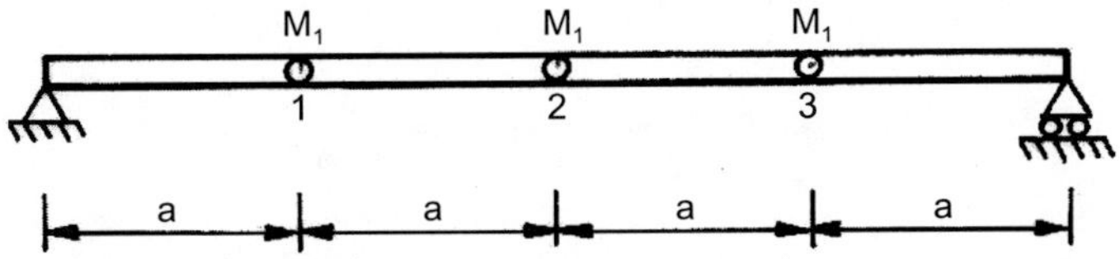

Fig. 14.3. Lumped mass of a beam

14.2.3 Inplane Vibration of Plates

The general expression for consistent mass matrix is

$$[M]_e = \int [N]^T \rho [N] dv \tag{14.23}$$

ρ is the mass density of the material of the structure.

For a plate of constant thickness t subjected to in-plane forces, eqn. (14.23) becomes

$$[M]_e = \rho t \int [N]^T [I] [N] dA \tag{14.24}$$

For a constant strain triangle, substituting $[N]$ from eqn. (14.10) into eqn. (14.24) performing necessary integration, yields

$$[M]_e = \rho \frac{t\Delta}{12} \begin{bmatrix} 2 & & & & & \\ 0 & 2 & & symmetrical & & \\ 1 & 0 & 2 & & & \\ 0 & 1 & 0 & 2 & & \cdots \\ 1 & 0 & 1 & 0 & 2 & \\ 0 & 1 & 0 & 1 & 0 & 2 \end{bmatrix} \tag{14.25}$$

14.2.4 Flexural Vibration of Plates

For the non-conforming plate bending element of Sec. 11.3, the shape function is related as follows

$$[N] = [C] [A]^{-1} \tag{14.26}$$

Therefore, the consistent mass matrix is

$$[M]_e = \rho t \int_A [N]^T [I] [N] dA \tag{14.27}$$

Substituting [*N*] from eqn. (14.26) into eqn. (14.27), yields

$$[M]_e = \rho t \int_A \left([A]^{-1}\right)^T [C]^T [C] [A]^{-1} \, dA \tag{14.28}$$

Equation (14.28) has been explicitly evaluated and the result is given below

$$[M]_e = \lambda \begin{bmatrix} 3454 & & & & & & & & & & & \\ -461 & 80 & & & & & & & \textit{symmetrical} & & & \\ 461 & -63 & 80 & & & & & & & & & \\ 1226 & -274 & 199 & 3454 & & & & & & & & \\ 274 & -60 & 42 & 461 & 80 & & \lambda = \dfrac{\rho abt}{6300} & & & & & \\ 199 & -42 & 40 & 461 & 63 & 80 & & & & & & \\ 1226 & -199 & 274 & 394 & 116 & 116 & 3454 & & & & & \\ -199 & 40 & -42 & -116 & -30 & -28 & -461 & 80 & & & & \\ -274 & 42 & -60 & -116 & -28 & -30 & -461 & 63 & 80 & & & \\ 394 & -116 & 116 & 1226 & 199 & 274 & 1226 & -274 & -199 & 3454 & & \\ 116 & -30 & 28 & 199 & 40 & 42 & 274 & -60 & -42 & 461 & 80 & \\ -116 & 28 & -30 & -274 & -42 & -60 & -199 & 42 & 40 & -461 & -63 & 80 \end{bmatrix} \tag{14.29}$$

14.2.4.1 Vibration of isoparametric bending element

The displacements of the plate are given by three quantities w, θ_x, θ_y. The nodal accelerations are

$$\{\ddot{f}_r\}^T = \{\ddot{w}_r \; \ddot{\theta}_{xr} \; \ddot{\theta}_{yr}\} \tag{14.30}$$

The inertia forces are produced due to the accelerations given by eqn. (14.30). The lateral inertia forces - $\rho t\ddot{w}$ result from transverse acceleration $\ddot{w}$ and ρ is the mass density of the plate. The rotational acceleration components $\ddot{\theta}_x$ and $\ddot{\theta}_y$ produce rotary inertia couples – $(\rho t^3/12)\,\ddot{\theta}_x$ and – $(\rho t^3/12)\,\ddot{\theta}_y$ respectively where t is the thickness of the plate.

From eqn. (14.24), the work done by nodal forces is

$$\{P\}_e = -\int [N]^T \{q\} \, dx \tag{14.31}$$

where $\{q\}$ are the inertia forces. They are given by

$$\{q\} = \begin{Bmatrix} \rho t\ddot{w} \\ \dfrac{\rho t^3}{12}\ddot{\theta}_x \\ \dfrac{\rho t^3}{12}\ddot{\theta}_y \end{Bmatrix} = \begin{bmatrix} \rho t & 0 & 0 \\ 0 & \dfrac{\rho t^3}{12} & 0 \\ 0 & 0 & \dfrac{\rho t^3}{12} \end{bmatrix} \begin{Bmatrix} \ddot{w} \\ \ddot{\theta}_x \\ \ddot{\theta}_y \end{Bmatrix}$$

or,

$$\{q\} = [a]\{\ddot{f}\} \tag{14.32}$$

Again from eqn. (11.99), the following relation is known

$$\{f\} = [N]\,\{X\}_e \tag{14.33}$$

Differentiating eqn. (14.33) twice with respect to time, yields

$$\{\ddot{f}\} = [N]\{\ddot{X}\}_e \tag{14.34}$$

Combining eqns. (14.32), (14.34) and (14.31)

$$\{P\}_e = \int_A [N]^T [a]\,[N]\,dA \tag{14.35}$$

From the vibration analysis of simply supported plate, it has been found that there was not much to gain by using consistent mass formulation of eqn. (14.35) over lumped mass matrix.

14.3 ELASTIC INSTABILITY ANALYSIS

The analysis presented here determines the intensity of the critical load of the elastic structure when bucking takes place. This load is important in design, as it indicates initiation of buckling. The treatment presented here is similar to that of the dynamic analysis. Just as the additional matrix required for the dynamic analysis over conventional static analysis is the mass matrix and that for the elastic instability analysis, it is the geometric stiffness matrix. The computer programs for static analysis require only some modifications to be applied to both dynamic analysis and elastic instability analysis of structures.

14.3.1 Column Instability Analysis

A uniform prismatic element is shown in Fig. 14.4. The member can deform by axial and flexural actions. Based on the assumptions made for flexural members, the strain is related to the displacement as follows

$$\varepsilon_x = \frac{du}{dx} - z\left(\frac{d^2w}{dx^2}\right) + \frac{1}{2}\left(\frac{dw}{dx}\right)^2 \tag{14.36}$$

where u and w are the displacements along x and z-axes. The first term of eqn. (14.36) is the axial strain component, the second term is the flexural strain component. The third term indicates the coupling of the flexural and axial components and the basis of this term is explained below.

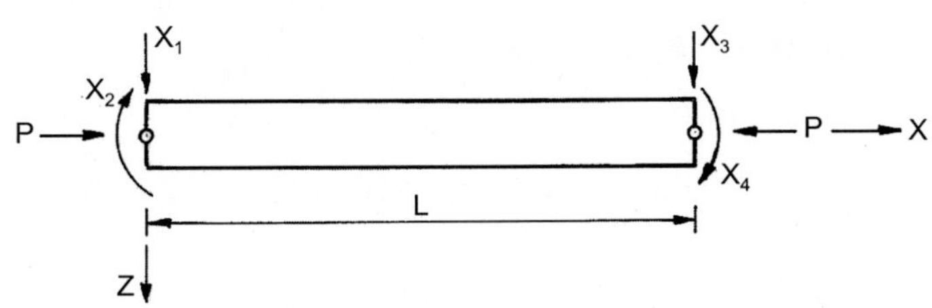

Fig. 14.4. Prismatic element

At the instant of buckling, two ends of the column move by, say, δx (Fig. 14.5 (c)). If the length of the curve between x and $x + dx$ in the displaced position is ds, then

$$ds = \sqrt{dx^2 + dw^2} = \left[1 + \left(\frac{dw}{dx}\right)^2\right]^{1/2} dx \tag{14.37}$$

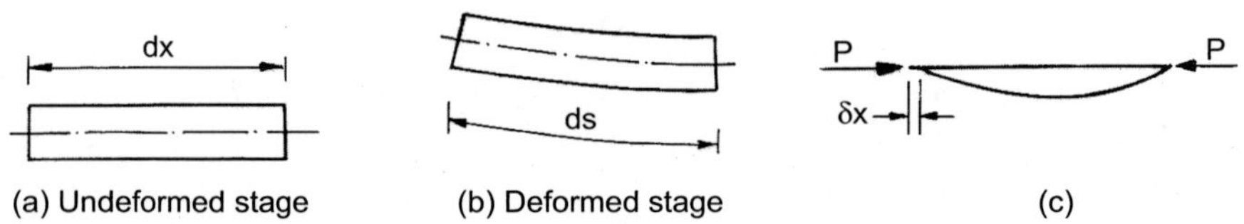

Fig. 14.5. Movement of the ends due to P

For small deflections and at the instant of buckling, $\frac{dw}{dx}$ is small compared to unity. Therefore, neglecting higher order terms on expansion of eqn. (14.37), yields

$$ds = \left[1 + \frac{1}{2}\left(\frac{dw}{dx}\right)^2\right] dx$$

or,

$$ds - dx = \frac{1}{2}\left(\frac{dw}{dx}\right)^2 dx \tag{14.38}$$

The total strain energy of the element is given by

$$U = \frac{1}{2}\int_v E\,\varepsilon_x^2\,dv \tag{14.39}$$

Substituting ε_x from eqn. (14.36) into eqn. (14.39), yields

$$U = \frac{E}{2}\int_L\int_A \left[\left(\frac{du}{dx}\right)^2 + z^2\left(\frac{d^2w}{dx^2}\right)^2 + \frac{1}{4}\left(\frac{dw}{dx}\right)^4 - 2z\left(\frac{du}{dx}\right)\left(\frac{d^2w}{dx^2}\right)\right.$$

$$\left. - z\left(\frac{d^2w}{dx^2}\right)^2\left(\frac{dw}{dx}\right)^2 + \left(\frac{du}{dx}\right)\left(\frac{dw}{dx}\right)^2\right] dA\,dx \tag{14.40}$$

Let us note the following geometric properties

$$\int_A dA = A,\ \int_A z\,d\,A = 0 \ \ and \ \int_A z^2 dA = I \tag{14.41}$$

On replacing the sectional properties into eqn. (14.40) results in

$$U = \frac{1}{2}\int_L \left[EA\left(\frac{du}{dx}\right)^2 + EI\left(\frac{d^2w}{dx^2}\right)^2 + EA\left(\frac{du}{dx}\right)\left(\frac{dw}{dx}\right)^2 + \frac{EA}{4}\left(\frac{dw}{dx}\right)^4\right] dx \tag{14.42}$$

Buckling represents flexural actions, therefore, the first term of eqn. (14.42) which indicates the strain energy due to pure axial deformation will not be considered. The last term of eqn. (14.42) is a high order term and it is also disregarded. Further, it may be noted that

$$P = EA\frac{du}{dx} \tag{14.43}$$

where tensile value of P is taken as positive. Equation (14.42) then becomes

$$U = \frac{1}{2}\int_L \left[EI\left(\frac{d^2w}{dx^2}\right)^2 + P\left(\frac{dw}{dx}\right)^2 \right] dx \tag{14.44}$$

Referring to beam element given in Section 11.2, eqn. (14.10) gives

$$w = [N]\,\{X\}_e \tag{14.45}$$

where $$[N] = \left[\left(1 - \frac{3x^2}{L^2} + \frac{2x^3}{L^3}\right) \left(x - \frac{2x^2}{L} + \frac{x^3}{L^2}\right) \left(\frac{3x^2}{L^2} - \frac{2x^3}{L^3}\right) \left(\frac{x^3}{L^2} - \frac{x^2}{L}\right) \right]$$

It has been already shown in Section 14.2.5, that the first part of the eqn. (14.44) is given by

$$\frac{1}{2}\int_L EI\left(\frac{d^2w}{dx^2}\right)^2 dx = \frac{1}{2}\{X\}_e^T \int_L [B]^T\,[D]\,[B]\,dx\{X\}_e \tag{14.46}$$

Differentiating w with respect to x in eqn. (14.45) gives

$$\frac{dw}{dx} = [H]\{X\}_e \tag{14.47}$$

where $$[H] = \left[\left(-\frac{6x}{L^2} + \frac{6x^2}{L^3}\right) \left(1 - \frac{4x}{L} + \frac{3x^2}{L^2}\right) \left(-\frac{6x^2}{L^3} + \frac{6x}{L^2}\right) \left(\frac{3x^2}{L^2} - \frac{2x}{L}\right) \right] \tag{14.48}$$

Therefore, $$\left(\frac{dw}{dx}\right)^2 = \{X\}_e^T\,[H]^T\,[H]\,\{X\}_e \tag{14.49}$$

Then, the second term of eqn. (14.44) becomes

$$\frac{1}{2}\int_L P\left(\frac{dw}{dx}\right)^2 dx = \frac{P}{2}\{X\}_e^T \int_L [H]^T\,[H]\,dx\,\{X\}_e \tag{14.50}$$

Equation (14.44) can now be written as (assuming P to be compressive)

$$U = \frac{1}{2}\{X\}_e^T \int_o^L [B]^T\,[D]\,[B]\,dx - P\int_o^L [H]^T\,[H]\,dx\{X\}_e$$

which on minimization yields (potential of the applied flexural load has been considered as zero)

$$[K]_e\,\{X\}_e - [G]_e\,\{X\}_e = 0 \tag{14.51}$$

where $[G]_e$ is called the geometric stiffness matrix of the element, $[K]_e$ is given by eqn. (11.23), $[G]_e$ is given by

$$[G]_e = \frac{P}{30}\begin{bmatrix} \frac{36}{L} & & & symmetrical \\ 3 & 4L & & \\ -\frac{36}{L} & -3 & \frac{36}{L} & \\ 3 & -L & -3 & 4L \end{bmatrix}$$

When the stiffness of all such elements are combined, the final equation becomes

$$[K]\{X\} = [G]\{X\} \qquad (14.52)$$

Equation (14.52) is an eigenvalue problem and can be compared with eqn. (14.17) for free vibration analysis. The solution of eqn. (14.52) will yield n values of P provided $[K]$ and $[G]$ are of size $(n \times n)$. The lowest value of P will correspond to the buckling load.

14.3.1.1 An illustrated example of a column

The critical load of a uniform column having both ends fixed is to be found by dividing the column into two equal elements (Fig. 14.6).

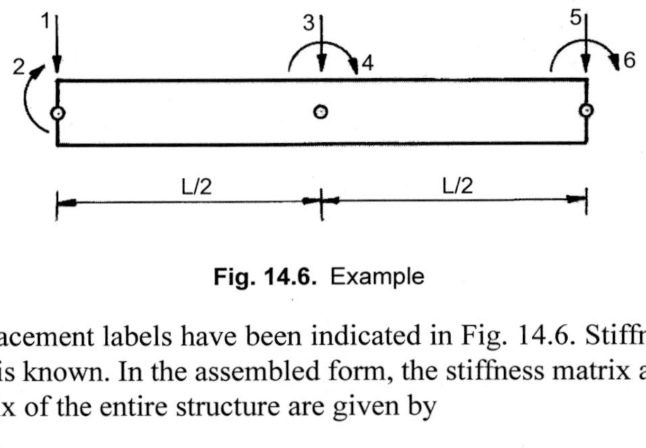

Fig. 14.6. Example

The displacement labels have been indicated in Fig. 14.6. Stiffness matrix of each element is known. In the assembled form, the stiffness matrix and geometric stiffness matrix of the entire structure are given by

$$[K]_e = \frac{8EI}{L^3}\begin{bmatrix} 12 & & & & & symmetrical \\ 3L & L^2 & & & & \\ -12 & -3 & 24 & & & \\ 3L & \frac{L^2}{2} & 0 & 2L^2 & & \\ 0 & 0 & -12 & -3 & 12 & \\ 0 & 0 & 3L & \frac{L^2}{2} & -3L & L^2 \end{bmatrix}$$

$$[G]_e = \frac{P}{15L}\begin{bmatrix} 36 & & & & & \\ \frac{3L}{2} & L^2 & & & symmetrical & \\ -36 & -\frac{3L}{2} & 72 & & & \\ \frac{3L}{2} & -\frac{L^2}{4} & -\frac{3L}{2} & 2L^2 & & \\ 0 & 0 & -36 & -\frac{3L}{2} & 36 & \\ 0 & 0 & \frac{3L}{2} & -\frac{L^2}{2} & -\frac{3L}{2} & 2L \end{bmatrix}$$

Displacements corresponding to 1, 2, 3 and 4 are restrained. Therefore, eqn. (14.52) becomes

$$\frac{8EI}{L^3}\begin{bmatrix} 24 & 0 \\ 0 & 2L^2 \end{bmatrix}\begin{Bmatrix} X_3 \\ X_4 \end{Bmatrix} = \frac{P}{15L}\begin{bmatrix} 72 & 0 \\ 0 & 2L^2 \end{bmatrix}\begin{Bmatrix} X_3 \\ X_4 \end{Bmatrix} \qquad \text{(a)}$$

Putting $\frac{PL^2}{15EI} = \lambda$, eqn. (a) can be written as

$$\begin{bmatrix} 192-72\lambda & 0 \\ 0 & 2L^2(8-\lambda) \end{bmatrix}\begin{Bmatrix} X_3 \\ X_4 \end{Bmatrix} = \begin{Bmatrix} 0 \\ 0 \end{Bmatrix} \qquad \text{(b)}$$

So, a non-trivial solution of eqn. (b) is

$$2L^2(192 - 72\lambda)(8 - \lambda) = 0$$

The least value of λ is given by

$$\lambda = \frac{8}{3}$$

or,

$$P_{cr} = \frac{8 \times 15EI}{3L^2} = \frac{40EI}{L^2}$$

The exact solution to this problem is $P_{cr} = 39.44\frac{EI}{L^2}$

14.3.2 Plate Instability Analysis

Strains can be shown to be related to the displacements as given by

$$\left.\begin{aligned} \varepsilon_x &= \frac{\partial u}{\partial x} - z\frac{\partial^2 w}{\partial x^2} + \frac{1}{2}\left(\frac{\partial w}{\partial x}\right)^2 \\ \varepsilon_y &= \frac{\partial v}{\partial y} - z\frac{\partial^2 w}{\partial y^2} + \frac{1}{2}\left(\frac{\partial w}{\partial y}\right)^2 \\ \gamma_{xy} &= \frac{\partial u}{\partial y} + \frac{\partial v}{\partial x} - 2z\frac{\partial^2 w}{\partial x \partial y} + \frac{\partial w}{\partial x}\frac{\partial w}{\partial y} \end{aligned}\right\} \qquad (14.53)$$

The strain energy of an isotropic plate is given by

$$U = \frac{D}{2}\int_A \left[\left(\frac{\partial^2 w}{\partial x^2}+\frac{\partial^2 w}{\partial y^2}\right)^2 + 2\frac{\partial^2 w}{\partial x^2}\cdot\frac{\partial^2 w}{\partial y^2} + 2(1-\nu)\left(\frac{\partial^2 w}{\partial x \partial y}\right)^2\right] dA$$

$$+\frac{1}{2}\int_A N_x\left(\frac{\partial w}{\partial x}\right)^2 dA + \int_A N_y\left(\frac{\partial w}{\partial y}\right)^2 dA + \int_A N_{xy}\left(\frac{\partial w}{\partial x}\right)\left(\frac{\partial w}{\partial y}\right) dA \quad (14.54)$$

where N_x and N_y are the uniform tensile forces per unit length along x and y-axes and N_{xy} the inplane shearing forces per unit length respectively. While deriving the expression for U, assumptions similar to that in the previous section have been made. The term within the third bracket is that due to flexural actions. This bending strain energy part is given

$$U_f = \frac{1}{2}\{X\}_e^T \int_A [B]^T [D] [B]\, dA \{X\}_e \quad (14.55)$$

with the usual notations adopted in the finite element plate bending chapter.

The strain energy part for the last three terms of eqn. (14.54) are derived as follows.

Introducing

$$[Q] = \begin{bmatrix} N_x & N_{xy} \\ N_{xy} & N_y \end{bmatrix} \quad (14.56)$$

and

$$[L] = \begin{Bmatrix} \frac{\partial w}{\partial x} \\ \frac{\partial w}{\partial y} \end{Bmatrix} \quad (14.57)$$

The last three terms of the strain energy expression can be written as

$$U = \frac{1}{2}\int_A [L]^T [Q] [L]\, dx \quad (14.58)$$

The displacement function of a plate element is given by

$$w = [N]\{X\}_e \quad (14.59)$$

Differentiating eqn. (14.57) according to eqn. (14.59), yields

$$\{L\} = [H]\{X\}_e \quad (14.60)$$

Substituting $\{L\}$ from eqn. (14.60) into eqn. (14.59), yields

$$U_i = \frac{1}{2}\{X\}_e^T \int_A [H]^T [Q] [H] dA \{X\}_e \quad (14.61)$$

Therefore, substituting (U_f and U_i) from eqns. (14.55) and (14.61) into eqn. (14.54), yields

$$U = \{X\}_e^T \frac{1}{2}\int_A [B]^T [D][B]\, dA\{X\}_e + \frac{1}{2}\{X\}_e^T \int_A [H]^T [Q][H]\, dA\{X\}_e \quad (14.62)$$

which on minimization (potential of the applied flexural load has been considered as zero), yields

$$[K]_e \{X\}_e + [G]_e \{X\}_e = \{0\} \quad (14.63)$$

where $[K]_e$ = plate stiffness matrix for flexure

and $[G]_e$ = geometric stiffness matrix

$$[G]_e = \int_A [H]^T [Q][H]\, dA \quad (14.64)$$

Equation (14.62) is an eigenvalue problem. On solution, the critical value of axial loads may be determined.

For rectangular element where a 12 term polynomial is assumed as the displacement function given by eqn. (14.29), the explicit value of $[G]_e$ is given below for the case when N_x only is present and other inplane forces are zero.

$$[G]_e = \frac{N_x}{630}\begin{bmatrix} 224 & & & & & & & & & & & \\ -84 & 1104 & & & & & & & & & & \\ -56 & 0 & 448 & & & & & & & & & \\ 0 & 0 & 0 & 96 & & & & & & & & \\ 0 & 78 & 0 & -36 & 48 & & & & & & & \\ 56 & -21 & -14 & 0 & 0 & 112 & & & & & & \\ 0 & -39 & 0 & 18 & -24 & 0 & 24 & & & & & \\ -21 & 204 & 0 & 78 & 0 & -42 & 0 & 552 & & & & \\ -14 & 0 & 112 & 0 & 0 & -28 & 0 & 0 & 224 & & & \\ 0 & -552 & -84 & 0 & -39 & 0 & 39 & -102 & -21 & 552 & & \\ 0 & 0 & 0 & -48 & 18 & 0 & -18 & -39 & 0 & 0 & 48 & \\ 0 & -102 & -21 & -39 & 0 & 0 & 0 & -276 & -42 & 102 & 39 & 276 \end{bmatrix} \quad (14.65)$$

REFERENCES AND SUGGESTED READINGS

14.1. K. J. Bathe, Finite Element Procedures, Prentice Hall, 1996.

14.2. G. Bergan et al., (Editors), Finite Elements in Non-linear Mechanics, Tapir Publishers, Norway, 1976.

14.3. A. Chajes, Principles of Structural Stability, Prentice Hall, 1974.

14.4. R. W. Clough and J. Penzien, Dynamics of Structures, McGraw-Hill, 1975.

14.5. R. H. Gallagher, Finite Element Analysis Fundamentals, Prentice Hall, 1975.

14.6. R. Glowinsky, E. Y. Rodin and O. C. Zienkiewicz, Energy Methods in Finite Element Analysis, John Wiley and Sons, 1979.

14.7. S. Levy and J. P. D. Wilkinson, The Component Element Method in Dynamics, McGraw-Hill, 1976.

14.8. M. Mukhopadhyay, Vibrations, Dynamics and Structural Systems, Oxford and IBH Publishing Co., 2nd Edition, 2000.

14.9. R. Szilard, Theory and Analysis of Plates: Classical and Numerical Methods, Prentice Hall, 1974.

14.10. G. B. Warburton, The Dynamical Behaviour of Structures, Pergamon Press, 1976.

EXERCISE 14

14.1. Consider an element of a shaft, subjected to torsion. The element has 3 nodes including one midside node. Find the consistent mass matrix and the stiffness matrix for the element.

14.2. Dividing the shaft of Prob. 14.1 into two elements, find the natural frequencies of the shaft. Consider both ends fixed.

14.3. Determine the natural frequencies of a bar under axial vibration. The bar, having one end fixed, other end free, is to be divided into two elements. The element will have 2 nodes, one at each of two ends.

14.4. Formulate the consistent mass matrix of an isoparametric linear beam element.

14.5. Determine the natural frequencies of a uniform beam, one end fixed and the other end free treating the entire beam as one element and using eqn. (14.21).

14.6. Repeat Problem 14.4 using an isoparametric quadratic beam element.

14.7. Calculate explicitly the consistent mass matrix of a plate vibrating in its own plane using isoparametric linear element.

14.8. Formulate consistent mass matrix of a plate vibrating in its own plane using a triangular element having 3 corner nodes and 3 mid-side nodes.

14.9. For a uniform column having one end fixed and the other end pinned, determine the critical load by dividing it into two equal elements.

14.10. Determine the critical load P for the frame. The frame is to be considered as three elements formed by three members.

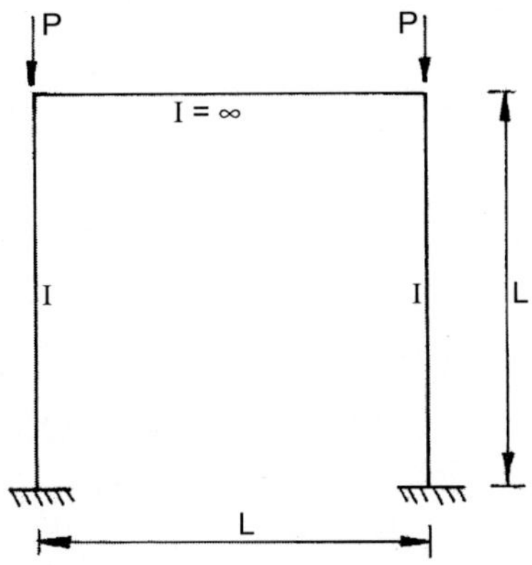

Prob. 14.10.

14.11. Determine the critical load P of the frame.

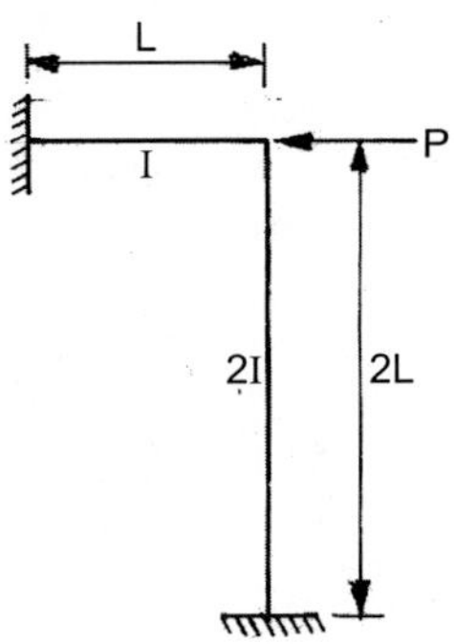

Prob. 14.11.

14.12. Determine the critical load by finite element analysis of torsional buckling of thin sections. The member is to be considered not free to warp given E, G, J and β where β is the warping constant.

14.13. Extend the analysis of Prob. 14.12 to the case of flexural-torsional buckling.

14.14. Using 12 term polynomial expression of eqn. (14.29) for rectangular plate bending. Determine the critical value of N_x for a square plate having all edges clamped. Divide the plate into four equal parts.

❑❑❑

15

THE FINITE DIFFERENCE METHOD FOR THE ANALYSIS OF BEAMS AND PLATES

15.1 INTRODUCTION

Another versatile numerical method that exists for the analysis of structural mechanics problems is the finite difference technique. The method is most general and can be applied to the solution of the problem of any field for which a differential equation is available. The method is suitable for the computer and can be easily programmed.

The differential equation is the starting point of the method. The continuum is divided in the form of a mesh and the unknowns in the problems are those at the nodes. The derivatives of the equation are expressed in the finite difference form. The differential equation, split in this discrete form, is applied at each node. This results in a set of simultaneous equations (15.1–15.3).

In the finite element method, the final differential equation is surpassed, whereas it is the differential equation to start with in the finite difference method. The approximation involved in the finite element method is physical in nature, as the actual continuum is replaced by finite elements. The element formulation is mathematically exact. The finite difference technique involves the exact representation of the continuum in terms of the differential equation and on this actual physical system, the mathematical model is approximated.

However, the finite difference method has several disadvantages. The derivatives of the approximated solution introduce inaccuracies. The imposition of boundary conditions has always been a problem with finite difference method, especially when the boundaries are irregular and complex. The method cannot be formulated with non-uniform and non-rectangular meshes.

15.2 FINITE DIFFERENCE REPRESENTATION OF DERIVATIVES

A variable w as a function of x is shown in Fig. 15.1. The function is divided into equally spaced interval of h. Let i be any station. Stations to its right are indicated as $i + 1$, $i+2$,.....and the corresponding values of the function are w_{i+1}, w_{i+2}.....

Stations to the left of '*i*' are designated as *i*–1, *i* – 2......and the corresponding values of the function are w_{i-1}, w_{i-2}.....

15.2.1 First Derivative

The first derivative is given by

$$\left(\frac{dw}{dx}\right)_i = \frac{1}{2h}\left(w_{i+1} - w_{i-1}\right) \tag{15.1}$$

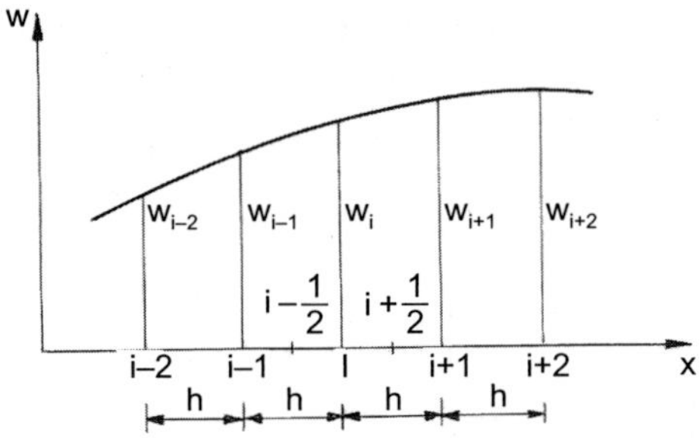

Fig. 15.1. A function

Equation (15.1) suggests that the slope of the curve between stations *i*–1 and *i*+1 has been considered to be constant.

15.2.2 Second Derivative

The second derivative can be approximated by

$$\left(\frac{d^2w}{dx^2}\right)_i = \frac{1}{h}\left[\left(\frac{dw}{dx}\right)_{i+1/2} - \left(\frac{dw}{dx}\right)_{i-1/2}\right] \tag{15.2}$$

It can be shown that

$$\left.\begin{aligned}\left(\frac{dw}{dx}\right)_{i+1/2} &= \frac{1}{h}\left(w_{i+1} - w_i\right)\\ \left(\frac{dw}{dx}\right)_{i-1/2} &= \frac{1}{h}\left(w_i - w_{i-1}\right)\end{aligned}\right\} \tag{15.3}$$

Combining eqns. (15.2) and (15.3) yields

$$\left(\frac{d^2w}{dx^2}\right)_i = \frac{1}{h^2}\left(w_{i+1} - 2w_i + w_{i-1}\right) \tag{15.4}$$

Bending moment is proportional to the second derivative.

15.2.3 Third Derivative

The shear force is a function of third derivative and it can be approximated as

$$\left(\frac{d^3w}{dx^3}\right)_i = \frac{1}{2h}\left[\left(\frac{d^2w}{dx^2}\right)_{i+1} - \left(\frac{d^2w}{dx^2}\right)_{i-1}\right] \tag{15.5}$$

Substituting the values of second derivatives from eqn. (15.4) into eqn. (15.5), it can be shown that

$$\left(\frac{d^3w}{dx^3}\right)_i = \frac{1}{2h^3}\left[w_{i+2} - 2w_{i+1} + 2w_{i-1} - w_{i-2}\right] \tag{15.6}$$

15.2.4 Fourth Derivative

Likewise, the fourth derivative can be shown to be

$$\left(\frac{d^4w}{dx^4}\right) = \frac{1}{h^4}\left[w_{i+2} - 4w_{i+1} + 6w_i - 4w_{i-1} + w_{i-2}\right] \tag{15.7}$$

In all the above finite difference expressions, central difference scheme has been adopted where it may be noted that the points are located symmetrically about the point at which the derivatives are of interest to us. Though there are two other schemes *viz.* the backward differences and the forward differences, it is the central difference scheme that is adopted for the solution of structural mechanics problems. The above four derivatives are shown schematically in Fig. 15.2

	$i-2$	$i-1$	i	$i+1$	$i+2$
$\left(\frac{dw}{dx}\right)_i = \frac{1}{2h}[\ \ ]$		-1		1	
$\left(\frac{d^2w}{dx^2}\right)_i = \frac{1}{h^2}[\ \ ]$		1	-2	1	
$\left(\frac{d^3w}{dx^3}\right)_i = \frac{1}{2h^3}[\ \ ]$	-1	$+2$		-2	1
$\left(\frac{d^4w}{dx^4}\right)_i = \frac{1}{h^4}[\ \ ]$	1	-4	6	-4	1

Fig. 15.2. Schematic representation of derivatives

15.3 ERRORS IN THE FINITE DIFFERENCE EXPRESSIONS

By applying Taylor's expansion of the functions of w about x, gives

$$w(x_{i+1}) = w(x_i) + \frac{h}{1!}w'(x_i) + \frac{h^2}{2!}w''(x_i) + \frac{h^3}{3!}w'''(x_i) + \ldots \tag{15.8}$$

$$w(x_{i-1}) = w(x_i) - \frac{h}{1!}w'(x_i) + \frac{h^2}{2!}w''(x_i) - \frac{h^3}{3!}w'''(x_i) + \ldots \tag{15.9}$$

Substituting for $w(x_i)$ in eqn. (15.9) from eqn. (15.8), yields

$$w(x_{i+1}) - w(x_{i-1}) = 2hw'(x_i) + \frac{2h^3}{3!}w'''(x_i) + \ldots \tag{15.10}$$

Arranging different terms of eqn. (15.10), gives

$$w'(x_i) = \frac{1}{2h}\left[w(x_{i+1}) - w(x_{i-1})\right] - \frac{h^2}{3!}w'''(x_i) - \frac{h^4}{5!}w^{v}(x_i) + \ldots \tag{15.11}$$

Therefore, the error in the first derivative *I* given by

$$\phi_1 = \phi_1 = -\frac{h^2}{3!}w'''(x_i) - \frac{h^4}{5!}w^v(x_i) - \ldots \tag{15.12}$$

Similarly, it can be shown that error in second, third and fourth derivatives is given by

$$\phi_2 = -\frac{2h^2}{4!}w^{iv}(x_i) - \frac{2h^4}{6!}w^{vi}(x_i) - \ldots \tag{15.13}$$

$$\phi_3 = -\frac{h^2}{4}w^v(x_i) - \ldots \tag{15.14}$$

$$\phi_4 = -\frac{h^2}{6}w^{vi}(x_i) - \ldots \tag{15.15}$$

15.4 EQUIVALENT CONCENTRATED LOAD

For the analysis of beams by the finite difference method, the concentrated loadings at the nodes are required for the analysis.

Three successive nodes are assumed to be simply supported and the support reactions are determined as shown in Fig. 15.3. They are then reversed to indicate the equivalent concentrated loading. For trapezoidal loading and two-degree parabolic loading, the equivalent concentrated loading has been shown in Fig. 15.4.

15.5 BOUNDARY CONDITIONS FOR BEAM BENDING

The boundary conditions for three cases – simple support, fixed support and free end have been considered here. The node '*i*' of Fig. 15.5 has been considered as the boundary of the member. The nodes *i* + 1 and *i* + 2 in Fig 15.6 are fictitious nodes outside the member.

15.5.1 Simple Support

The boundary conditions are

$$\left.\begin{aligned} w_i &= 0 \\ M_i &= 0 \end{aligned}\right\} \tag{15.16}$$

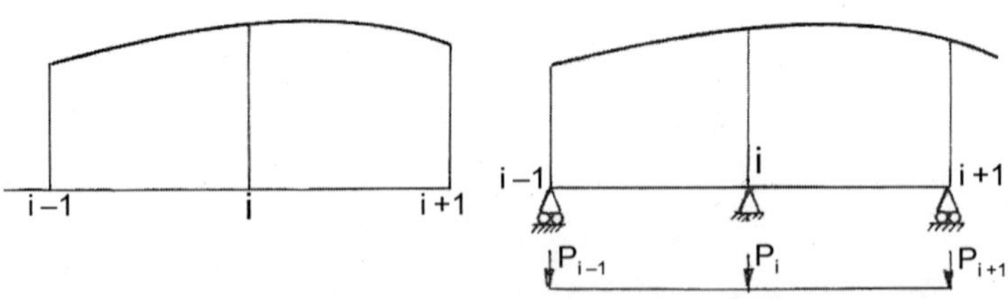

(a) Distributed load (b) Equivalent concentrated load

Fig. 15.3. Equivalent Concentrated Load

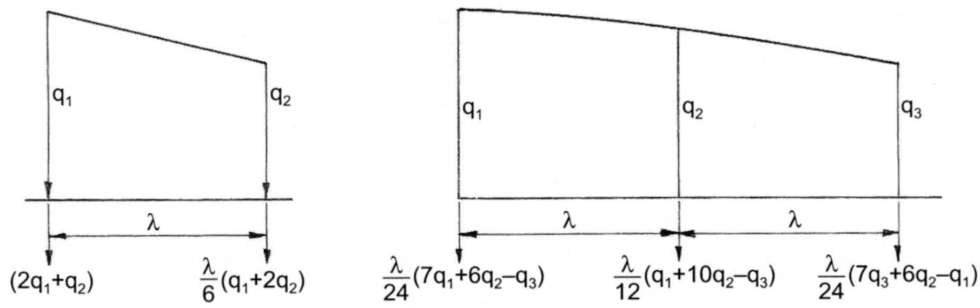

(a) Trapezoidal loading (b) Two-degree parabola

Fig. 15.4. Equivalent concentrated load for some loading cases

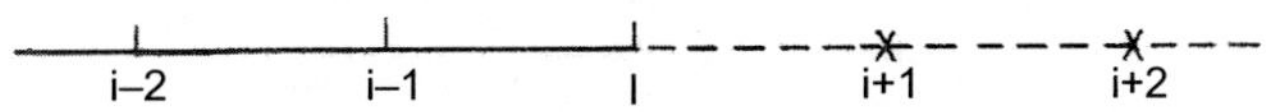

Fig. 15.5. Boundary node *i*

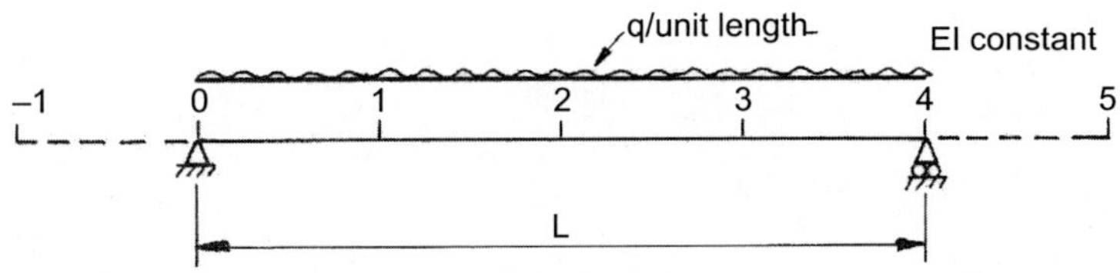

Fig. 15.6. A simply supported beam

Second of eqn. (15.16) can be expressed in terms of displacements as

$$w_{i-1} - 2w_i + w_{i+1} = 0 \tag{15.17}$$

15.5.2 Fixed End

The boundary conditions are

$$\left.\begin{aligned} w_i &= 0 \\ \left(\frac{dw}{dx}\right)_i &= 0 \end{aligned}\right\} \tag{15.18}$$

The second of eqn. (15.18) when expressed in the finite difference form becomes

$$w_{i+1} - w_{i-1} = 0 \tag{15.19}$$

15.5.3 Free End

The boundary conditions at the free end of a uniform beam are

$$\left.\begin{aligned} M_i &= 0 \\ V_i &= 0 \end{aligned}\right\} \tag{15.20}$$

When expressed in the finite difference form, eqn. (15.20) becomes

$$\left.\begin{aligned} w_{i-1} - 2w_i + w_{i+1} &= 0 \\ -w_{i-2} + 2w_{i-1} - 2w_{i+1} + w_{i+2} &= 0 \end{aligned}\right\} \qquad (15.21)$$

For non-uniform beam, second condition corresponding to eqn. (15.20) leads to considerable difficulty.

15.6 A STATICALLY DETERMINATE STATIC PROBLEM

That 'universal' problem of a simply supported prismatic beam under uniform load is solved here by the finite difference technique. The beam is divided into four equal parts each of length $l = L/4$. The nodes are numbered from 0 to 4 as shown in Fig 15.6. The followings are the differential equations associated with a beam problem.

$$EI\frac{d^2w}{dx^2} = -M \qquad (15.22)$$

$$\frac{d^2M}{dx^2} = -q \qquad (15.23)$$

and,

$$\frac{d^2}{dx^2}\left(EI\frac{d^2w}{dx^2}\right) = q \qquad (15.24)$$

For a uniform beam, eqn. (15.24) becomes

$$EI\frac{d^4w}{dx^4} = q \qquad (15.25)$$

Though it is always preferable to use the more general equation (eqn. (15.22)), but for this statically determinate problem, bending moments can be easily determined. Hence, the displacements at the nodes will be obtained first by the use of eqn. (15.22).

Equation (15.22) when expressed in the finite difference form becomes

$$\frac{1}{h^2}(w_{i-1} - 2w_i + w_{i+1}) = -\frac{M_i}{EI} \qquad (15.26)$$

The bending moments at nodes 1, 2, and 3 are given by

$$M_1 = M_3 = 3qL^2/32 \text{ and } M_2 = qL^2/8$$

Applying eqn. (15.26), first at node 1, *i.e.*, $i = 1$ it becomes

$$w_0 - 2w_1 + w_2 = -\frac{M_1h^2}{EI} \qquad (15.27)$$

Substituting the value M_1 and h into the eqn. (15.27) and noting that $w_0 = 0$, eqn. (15.27) becomes

$$-2w_1 + w_2 = -\frac{3}{512}\frac{qL^4}{EI} \qquad (15.28)$$

Applying eqn. (15.26) to nodes 2 and 3, yields

$$w_1 - 2w_2 + w_3 = -\frac{1}{128}\frac{qL^4}{EI} \tag{15.29}$$

$$w_2 - 2w_3 = -\frac{3}{512}\frac{qL^4}{EI} \tag{15.30}$$

Equations (15.28), (15.29) and (15.30) when written in matrix form becomes

$$\begin{bmatrix} -2 & 1 & 0 \\ 1 & -2 & 1 \\ 0 & 1 & -2 \end{bmatrix} \begin{Bmatrix} w_1 \\ w_2 \\ w_3 \end{Bmatrix} = -\frac{qL^4}{512EI} \begin{Bmatrix} 3 \\ 4 \\ 3 \end{Bmatrix} \tag{15.31}$$

On solution of eqn. (15.31), yields

$$\begin{Bmatrix} w_1 \\ w_2 \\ w_3 \end{Bmatrix} = \frac{qL^4}{2048EI} \begin{Bmatrix} 20 \\ 28 \\ 20 \end{Bmatrix} \tag{15.32}$$

Mid-span deflection is given by $w_2 = \dfrac{7qL^4}{512EI}$ which is about 5% in error when compared to the exact value.

The same problem is solved on the basis of eqn. (15.25). Eqn. (15.25) when expressed in the finite difference form becomes

$$\frac{EI}{h^4}\left(w_{i+2} - 4w_{i+1} + 6w_i - 4w_{i-1} + w_{i-2}\right) = p_i \tag{15.33}$$

It may reveal from eqn. (15.33) that the right hand side depends only on the intensity of loading at the node and is independent of the manner in which it is applied. Therefore, in order to substitute the nodal loading, eqn. (15.33) is modified as

$$\frac{EI}{h^3}\left(w_{i+2} - 4w_{i+1} + 6w_i - 4w_{i-1} + w_{i-2}\right) = p_i h = P_i \tag{15.34}$$

where P_i is the equivalent load at the node.

Applying boundary conditions of eqn. (15.16) and (15.17) gives

$$\left.\begin{aligned} w_0 &= 0 \\ w_{-1} + w_1 &= 0 \\ w_4 &= 0 \\ w_3 + w_5 &= 0 \end{aligned}\right] \tag{15.35}$$

Applying eqn. (15.34) to nodes 1, 2 and 3 and substituting the appropriate boundary conditions of (15.35), the following equations result.

$$\begin{bmatrix} 5 & -4 & 0 \\ -4 & 6 & -4 \\ 1 & -4 & 5 \end{bmatrix} \begin{Bmatrix} w_1 \\ w_2 \\ w_3 \end{Bmatrix} = -\frac{qL^4}{256EI} \begin{Bmatrix} 1 \\ 1 \\ 1 \end{Bmatrix} \tag{15.36}$$

On solution of eqn. (15.36), yields

$$\begin{Bmatrix} w_1 \\ w_2 \\ w_3 \end{Bmatrix} = \frac{qL^4}{512EI}\begin{Bmatrix} 5 \\ 7 \\ 5 \end{Bmatrix} \tag{15.37}$$

Once the deflections have been obtained, both the shearing force and the bending moment can now be determined.

Applying eqn. (15.26) to node 2 yields

$$\begin{aligned} M_2 &= -\frac{EI}{h^2}(w_1 - 2w_2 + w_3) \\ &= -\frac{16EI}{L^2}\times\frac{qL^4}{512EI}\times(-4) = \frac{qL^2}{8} \end{aligned}$$

Similarly, $$M_1 = \frac{16EI}{L^2}(0-2\times5+7)\frac{qL^4}{512EI} = \frac{3}{32}qL^2$$

The shear force equation for a uniform beam is

$$V = -EI\frac{d^3y}{dx^3} \tag{15.38}$$

When expressed in the finite difference form, eqn. (15.38) becomes

$$V_i = \frac{EI}{2h^3}(w_{i+2} - 2w_{i+1} + 2w_{i-1} - w_i\,) \tag{15.39}$$

Applying eqn. (15.39) to node 1 gives

$$V_1 = -\frac{64EI}{L^3}\frac{qL^4}{512EI}(5-2\times7+5+0) = \frac{qL}{4}$$

Similarly, $$V_2 = -\frac{64EI}{L^3}\frac{qL^4}{512EI}(2\times5-2\times5) = 0$$

15.7 A STATICALLY INDETERMINATE NSTATIC PROBLEM

The analysis of an indeterminate structural problem by the finite difference technique is not different from that of a determinate problem.

For the beam of Fig. 15.7 which is divided into 4 equal parts and the node numbers indicated, the boundary conditions at node 0 are

$$w_0 = 0 \text{ and } w_{-1} + w_1 = 0 \tag{15.40}$$

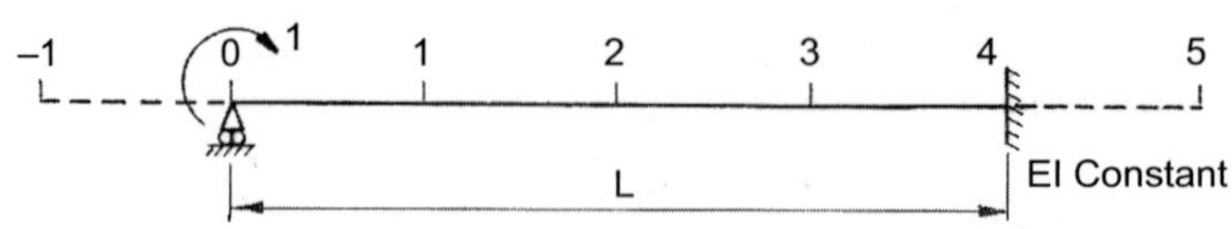

(a)

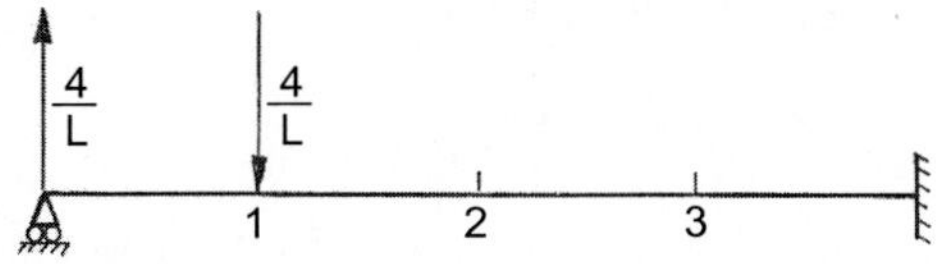

(b)

Fig. 15.7. A statically indeterminate problem

Boundary conditions at node 4 are

$$w_4 = 0 \text{ and } w_3 - w_5 = 0 \tag{15.41}$$

The couple applied at 0 is to be replaced by two equal and opposite forces as shown in Fig 15.7 (b).

Applying eqn. (15.34) at nodes 1, 2 and 3 and applying appropriate boundary conditions of (15.40) and (15.41), results in the following equations:

$$\frac{EI}{h^3}\begin{bmatrix} 5 & -4 & 1 \\ -4 & 6 & -4 \\ 1 & -4 & 7 \end{bmatrix}\begin{Bmatrix} w_1 \\ w_2 \\ w_3 \end{Bmatrix} = \frac{4}{L}\begin{Bmatrix} 1 \\ 0 \\ 0 \end{Bmatrix} \tag{15.42}$$

On solution of eqn. (15.42) yields

$$\begin{Bmatrix} w_1 \\ w_2 \\ w_3 \end{Bmatrix} = \frac{L^2}{704EI}\begin{Bmatrix} 26 \\ 24 \\ -10 \end{Bmatrix} \tag{15.43}$$

Slope at 0 is given by

$$\left(\frac{dw}{dx}\right)_0 = \frac{1}{2h}\left(w_1 - w_{-1}\right) \tag{15.44}$$

Combining eqn. (15.40) and (15.41) yields

$$\left(\frac{dw}{dx}\right)_0 = \frac{4}{2L} \times 2 \times \frac{26L^2}{704EI} = \frac{L}{6.77EI} \tag{15.45}$$

The exact value of the rotation at the end is $L/(4EI)$. The error is significant in this and this is primarily due to replacing the couple by two equal and opposite forces. With more divisions, the error will reduce.

15.8 FREE VIBRATION OF BEAMS

The beam of Fig. 15.7 is to be analysed for free vibration, *i.e.*, the natural frequencies and mode shapes are to be determined.

The differential equation of free vibration is given by

$$\frac{d^2}{dx^2}\left(EI\frac{d^2w}{dx^2}\right) = p^2 mw \tag{15.46}$$

where m is the mass per unit length.

For a uniform beam, eqn. (15.46) becomes,

$$EI\frac{d^4w}{dx^4} = p^2mw \tag{15.47}$$

When expressed in finite difference form, eqn. (15.47) becomes

$$\frac{EI}{h^4}\left(w_{i+2} - 4w_{i+1} + 6w_i - 4w_{i-1} + w_{i-2}\right) = p^2mw_i \tag{15.48}$$

Introducing $\lambda = p^2mh^4/EI$, eqn. (15.48) becomes

$$\left(w_{i+2} - 4w_{i+1} + 6w_i - 4w_{i-1} + w_{i-2}\right) = \lambda w_i \tag{15.49}$$

The boundary condition at nodes 0 and 4 are given be eqns. (15.40) and (15.41)

The beam is divided into 4 equal parts so that $h = L/4$. Equation (15.49) as is applied to nodes 1, 2 and 3 respectively and the appropriate boundary conditions are incorporated and the followings are the resulting equations:

$$\begin{bmatrix} 5 & -4 & 1 \\ -4 & 6 & -4 \\ 1 & -4 & 7 \end{bmatrix} \begin{Bmatrix} w_1 \\ w_2 \\ w_3 \end{Bmatrix} = \lambda \begin{bmatrix} 1 & 0 & 0 \\ 0 & 1 & 0 \\ 0 & 0 & 1 \end{bmatrix} \begin{Bmatrix} w_1 \\ w_2 \\ w_3 \end{Bmatrix} \tag{15.50}$$

Equation (15.50) is an eigenvalue problem. For non-trivial solution,

$$\begin{vmatrix} 5-\lambda & -4 & 1 \\ -4 & 6-\lambda & -4 \\ 1 & -4 & 7-\lambda \end{vmatrix} = 0 \tag{15.51}$$

The determinant of eqn. (15.51) is

$$(5-\lambda)\left[(6-\lambda)(7-\lambda)-16\right]+4\left[-4(7-\lambda)+4\right]+1\left[16-6+\lambda\right] = 0 \tag{15.52}$$

Equation (15.52) is a cubic equation where the lowest root is

$$\lambda_1 = 0.72$$

Therefore, $\dfrac{p^2h^4m}{EI} = 0.72$

or, $p_1 = 13.6\sqrt{\dfrac{EI}{mL^4}}$

The exact value of $p_1 = 15.9\sqrt{\dfrac{EI}{mL^4}}$

The other roots will give other natural frequencies.

Assume $w_1 = 1$ and substituting the value of $\lambda_1 = 0.72$ into second and third equations of eqn. (15.50), yields

$$\begin{aligned} 5.28w_2 - 4w_3 &= 4 \\ -4w_2 + 6.28w_3 &= -1 \end{aligned} \tag{15.53}$$

On solution, $w_2 = 1.231$ and $w_3 = 0.625$

Therefore, the mode shape in first mode is as Fig. 15.8.

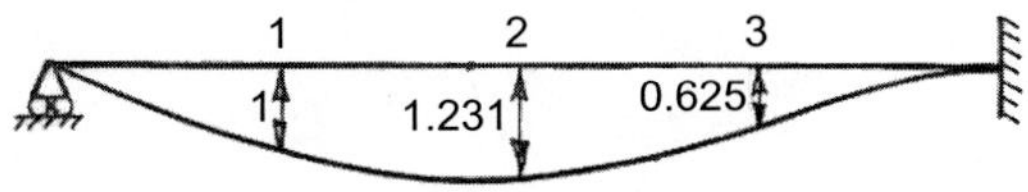

Fig. 15.8. Mode shape in the first mode

15.9 BUCKLING OF COLUMNS

The differential equation of axially loaded column is given by

$$EI\frac{d^2w}{dx^2} + pw = 0 \tag{15.54}$$

When expressed in the finite difference form, eqn. (15.54) becomes

$$\frac{EI}{h^2}\left(w_{i+1} - 2w_i + w_{i-1}\right) + pw_i = 0$$

or, $$w_{i+1} - (2-\lambda)w_i + w_{i-1} = 0 \tag{15.55}$$

where $$\lambda = \frac{ph^2}{EI}$$

Consider a column having both ends hinged. The column is uniform and is divided into four equal parts (Fig. 15.9). Applying eqn. (15.55) to nodes 1, 2 and 3 and substituting the appropriate boundary conditions, results in

$$\begin{bmatrix} -2+\lambda & 1 & 0 \\ 1 & -2+\lambda & 1 \\ 0 & 1 & -2+\lambda \end{bmatrix} \begin{Bmatrix} w_1 \\ w_2 \\ w_3 \end{Bmatrix} = \begin{Bmatrix} 0 \\ 0 \\ 0 \end{Bmatrix} \tag{15.56}$$

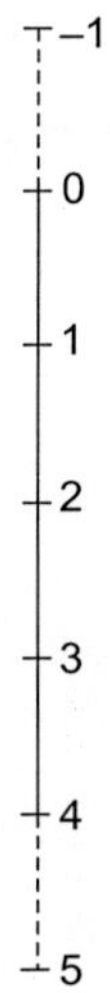

Fig. 15.9 Hinged – hinged column

Non-trivial solution of eqn. (15.56) is

$$(-2 + \lambda)\,[(-2 + \lambda)^2 - 1] - 1(-2 + \lambda) = 0$$

or, $\lambda^2 - 4\lambda + 2 = 0$

or, $\lambda^2 = -2 \pm \sqrt{2}$

Minimum value of $\lambda = 0.586$

$$\frac{ph^2}{EI} = 0.586 \text{ or } P_{cr} = 9.376\frac{EI}{L^2}$$

The answer differs from exact value by 5%.

15.10 FINITE DIFFERENCE REPRESENTATION OF THE PLATE EQUATION

The plate equation is given by

$$\frac{\partial^4 w}{\partial x^4}+2\frac{\partial^4 w}{\partial x^2 \partial y^2}+\frac{\partial^4 w}{\partial y^4}=\frac{p}{D} \tag{15.57}$$

The plate is divided into equally paced square mesh having $\Delta x = \Delta y = h$ (Fig. 15.10). The finite difference expressions are written with respect to the node *(i,j)*. '*i*' are numbered in the *x*-direction and '*j*'s are in the *y*-direction. Therefore

$$\left(\frac{\partial^4 w}{\partial x^4}\right)_{i,j} = \frac{1}{h^4}\left|w_{i+2,j}-4w_{i+1,j}+6w_{i,j}-4w_{i-1,j}+w_{i-2,j}\right| \tag{15.58}$$

$$\left(\frac{\partial^4 w}{\partial y^4}\right)_{i,j} = \frac{1}{h^4}\left|w_{i,j+2}-4w_{i,j+1}+6w_{i,j}-4w_{i,j-1}+w_{i,j-2}\right| \tag{15.59}$$

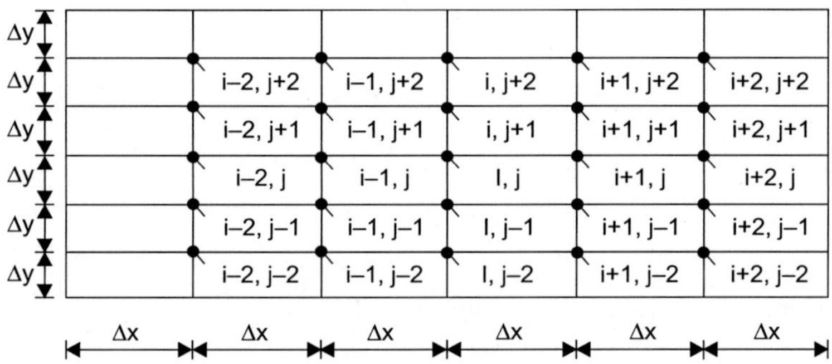

Fig. 15.10. Plate with mesh division

$$\left(\frac{\partial^4 w}{\partial x^2 \partial y^2}\right)_{i,j} = \left|\frac{\partial^2}{\partial y^2}\left(\frac{\partial^2 w}{\partial x^2}\right)\right|_{i,j}$$

$$= = \frac{1}{h^2}\left|\left(\frac{\partial^2 w}{\partial x^2}\right)_{i,j+1} - 2\left(\frac{\partial^2 w}{\partial x^2}\right)_{i,j} + \left(\frac{\partial^2 w}{\partial x^2}\right)_{i,j-1}\right|$$

$$= \frac{1}{h^4}\ [(w_{i+1,j+1}-2w_{i,j+1}+w_{i-1,j+1}) - 2\,(w_{i+1,j}-2w_{i,j}+w_{i-1,j})$$
$$+ (w_{i+1,j-1}-2w_{i,j-1}+w_{i-1,j-1})]$$

or,
$$\left(\frac{\partial^4 w}{\partial x^2 \partial y^2}\right)_{i,j} = \frac{1}{h^4}\ [w_{i+1,j+1}-2w_{i,j+1}+w_{i-1,j+1}-2w_{i+1,j}+4w_{i,j}-2w_{i-1,j}$$
$$+ w_{i+1,j-1}-2w_{i,j-1}+w_{i-1,j-1}] \tag{15.60}$$

The finite difference expressions for the bending moments are given by

$$(M_x)_{i,j} = -D\left[\frac{\partial^2 w}{\partial x^2} + \nu\frac{\partial^2 w}{\partial y^2}\right]_{i,j}$$

$$= -\frac{D}{h^2}[(w_{i+1,j} - 2w_{i,j} + w_{i-1,j}) + \nu(w_{i,j+1} - 2w_{i,j} + w_{i,j-1})] \quad (15.61)$$

$$= -\frac{D}{h^2}[(w_{i,j+1} - 2w_{i,j} + w_{i,j-1}) + \nu(w_{i+1,j} - 2w_{i,j} + w_{i-1,j})] \quad (15.62)$$

and
$$(M_x)_{i,j} = (1-\nu)D\left[\frac{\partial^2 w}{\partial x \partial y}\right]_{i,j}$$

$$= \frac{(1-\nu)D}{4h^2}[(w_{i+1,j+1} - w_{i+1,j-1} - w_{i-1,j+1} + w_{i-1,j-1})] \quad (15.63)$$

The finite difference expressions for shearing forces are

$$(V_x)_{i,j} = -D\left(\frac{\partial^3 w}{\partial x^3} + \frac{\partial^3 w}{\partial x \partial y^2}\right)_{i,j}$$

$$= -D(w_{i+2,j} - 2w_{i+1,j} + 2_{wi-1,j} - w_{i-2,j} + w_{i+1,j+1} - 2w_{i+1,j} + w_{i+1,j-1}$$
$$- w_{i-1,j+1} + 2w_{i-1,j} - w_{i-1,j-1}) \quad (15.64)$$

and
$$(V_y)_{i,j} = -D\left(\frac{\partial^3 w}{\partial y^3} + \frac{\partial^3 w}{\partial y \partial x^2}\right)_{i,j}$$

$$= -D(w_{i,j+2} - 2w_{i,j+1} + 2_{wi,j-1} - w_{i,j-2} + w_{i+1,j+1} - 2w_{i,j+1} + w_{i-1,j+1}$$
$$- w_{i+1,j-1} + 2w_{i,j-1} - w_{i-1,j-1}) \quad (15.65)$$

By substituting the finite difference expressions of the different terms of eqn. (15.57), the final equation for a node (*i, j*) is shown in Fig. 15.11 and is denoted as eqn. (15.66). For the sake of illustrating the method, an example is presented.

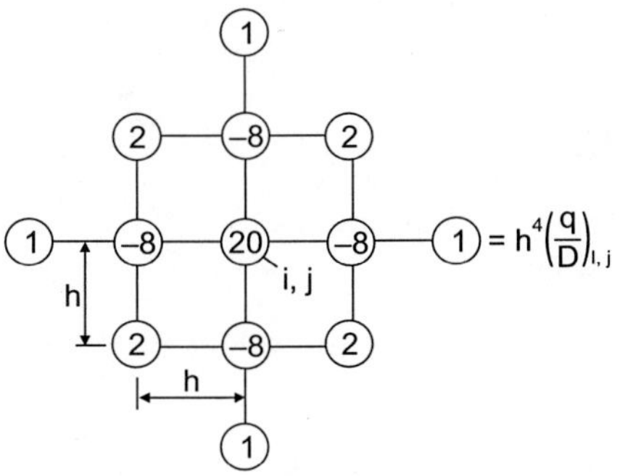

Fig. 15.11. Finite difference equation at node (*i, j*)

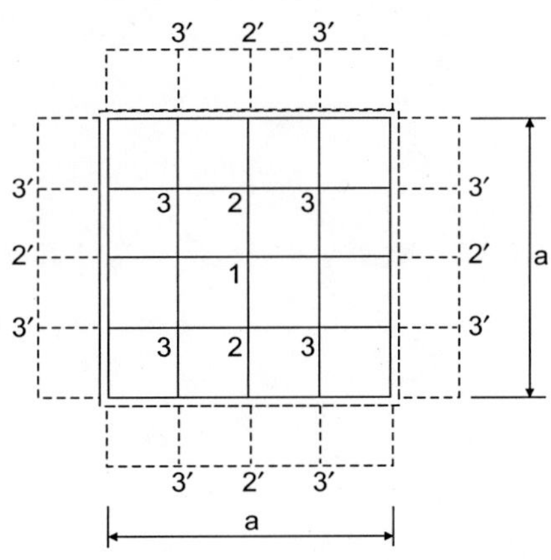

Fig. 15.12. The plate example

15.10.1 A Plate Example

A simply supported uniform square plate of Fig. 15.12 is divided into a 4 × 4 mesh size. It is uniformly loaded with a load having intensity q_0. The different nodes are designated in Fig. 15.12 taking into account the symmetry of the problem. The simply supported end conditions gives

$$w_2 = -w_{-2} \quad \text{and} \quad w_3 = -w_{-3} \tag{15.67}$$

If the plate dimension is a, then applying eqn. (15.66) at points 1, 2 and 3, gives

$$\begin{bmatrix} 20 & -32 & 8 \\ -8 & 24 & -16 \\ 2 & -16 & 20 \end{bmatrix} \begin{Bmatrix} w_1 \\ w_2 \\ w_3 \end{Bmatrix} = \frac{q_0 a^4}{256D} \begin{Bmatrix} 1 \\ 1 \\ 1 \end{Bmatrix} \tag{15.68}$$

Solution of the eqn. (15.68) gives

$$\begin{Bmatrix} w_1 \\ w_2 \\ w_3 \end{Bmatrix} = \frac{q_0 a^4}{D} \begin{Bmatrix} 0.00403 \\ 0.00293 \\ 0.00214 \end{Bmatrix} \tag{15.69}$$

Once the deflections at the nodal points are obtained, the bending moment and the shear force at any nodal point can be determined. For example, if M_x is required at node 1 of Fig. (15.12), then applying eqn. (15.61) at node 1 and assuming $v = 0.3$,

$$\begin{aligned} M_{x1} &= -\frac{16D}{a^2}(w_2 - 2w_1 + w_2) + 0.3(w_2 - 2w_1 + w_2) \\ &= -\frac{16D}{a^2} \times 2 \times 1.3 \times (w_2 - w_1) \\ &= 0.0458 q_0 a^2 \end{aligned}$$

REFERENCES AND SUGGESTED READINGS

15.1. D. D. McCracken and W. S. Dorn, Numerical Methods and FORTRAN Programming, John Wiley & Sons, 1966.

15.2. F. Schied, Numerical Analysis, Schaum Series, McGraw-Hill Book Co., 1968.

15.3. P. C. Wang, Numerical Methods in Structural Mechanics, John Wiley & Sons, 1980.

EXERCISE 15

15.1. For a uniform square plate under bending and as per mesh division shown, determine the deflection at nodal points.

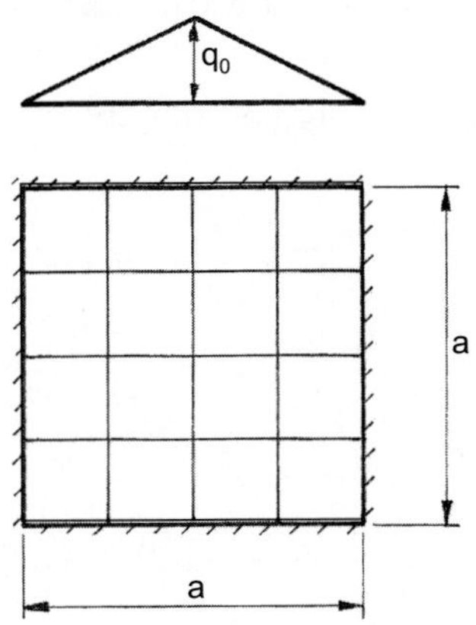

Prob. 15.1

15.2. Determine the buckling load for the columns.

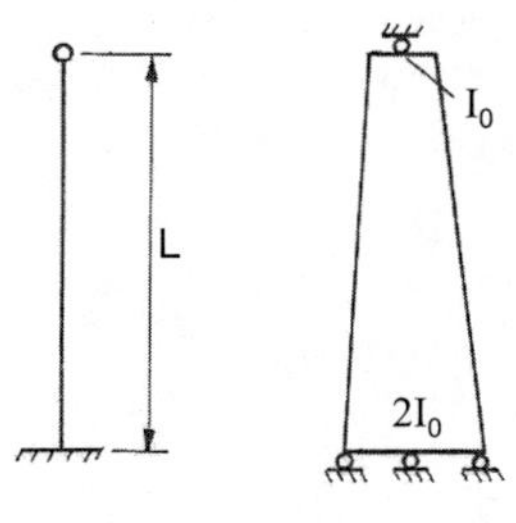

Prob. 15.2

15.3. Determine the natural frequencies of a uniform beam having both ends free. Divide the beam into 4 equal divisions.

15.4. Determine the buckling load of a square uniform plate dividing it into a mesh 3 x 3 size. The plate is simply supported and advantage may be taken of symmetry of the mode of buckling.

15.5. For a uniform rectangular plate having rectangular mesh division where $\Delta x = \Delta y$. Write finite difference expression for the plate equation.

15.6. For a uniform square plate having two opposite edges simply supported and the other two opposite edges free, determine the fundamental frequency on the basis of dividing the plate into 4×4 grid.

15.7. For a uniform rectangular plate of sides a and b having $a = 2b$, determine the nodal deflection by dividing the plate into 4×4 mesh divisions for a concentrated load P applied at mid-span.

15.8. Determine the critical load of a column having one end fixed and the other end free. The column is of uniform cross-section. The critical load is to be determined by applying finite difference method to higher order equation given below

$$w^{iv} + \frac{P}{EI} w'' = 0$$

15.9. For the uniform plate shown, determine the maximum bending moment at mid-span.

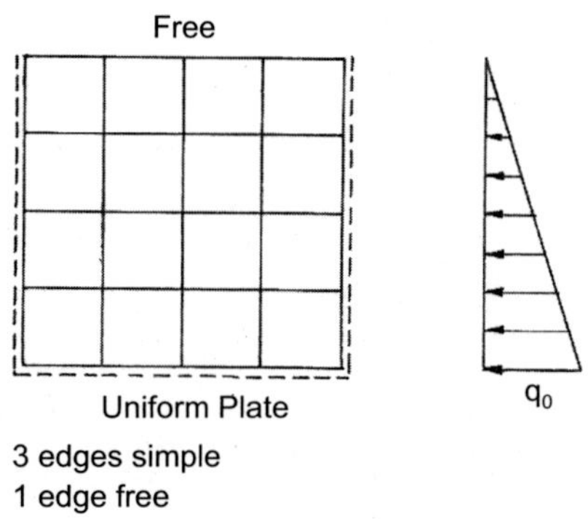

Prob. 15.9

15.10. Determine the critical load for the column.

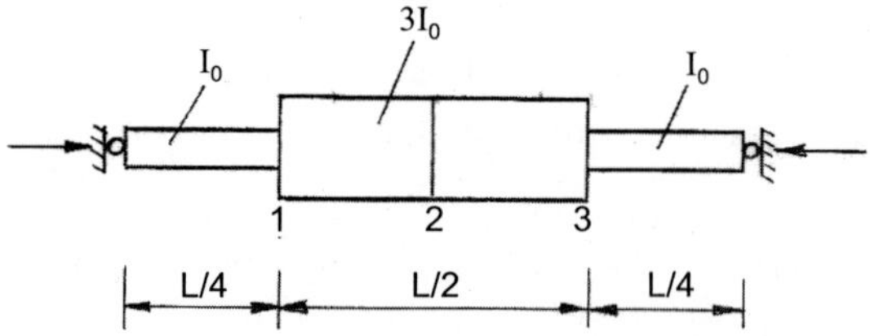

Prob. 15.10

❑❑❑

16

ADAPTIVE FINITE ELEMENT ANALYSIS

16.1 INTRODUCTION

The choice of optimum mesh size to obtain results within desired accuracy has been haunting the engineers since the inception of the finite element method. In order to achieve this, one of the popular approaches has been to study the convergence of results by making the mesh finer and finer. In such cases experience of the analysts based on previous computation rules of previous elements' shapes and sizes play a very significant role. But it necessitates many runs of the computer program. The approach is more time consuming and expensive as such. It is highly desirable to evolve an approach by which the refinement on mesh size becomes automatic, that is, accuracy to be achieved without any user intervention. In order to reach this goal, the error estimation of the solution is necessary. Attempts have been continuing to develop simple and reliable error estimates.

Thus, the approaches of arriving at accurate results obtained from the finite element method are as follows:

1. Experience-based discretization
2. Adaptive finite element technique

16.2 THE ADAPTIVE FINITE ELEMENT TECHNIQUE

The earlier practice based on the experience of the analysts is indeed an expensive procedure. An automatic mesh generator has thus become very much desirable. In the following the adaptive finite element method is discussed. Various topics involved in this are:

(*a*) superconvergent patch recovery technique, (*b*) error estimation, (*c*) error estimator, (*d*) refinement strategy of meshes and (*e*) adaptive mesh generator.

16.3 SUPERCONVERGENT PATCH RECOVERY TECHNIQUE

The obvious choice of calculating the derivatives (stresses) from the finite element approximation is to directly differentiate the finite element solution at salient points. For example, stresses or stress resultants are obtained by differentiating the displacement solution and using the appropriate constitutive relations. Direct differentiation will result in a lower order derivative which will be discontinuous. Points where accurate stresses are to be determined thus have inferior values, especially at the boundary of the elements and at the interelemental nodes. The superconvergent patch recovery technique (SPR) has been developed to evaluate the stresses accurately [16.1, 16.2].

The directly computed consistent stresses are

$$\{\sigma^h\} = [D][H]\{u^h\} \tag{16.1}$$

where

$[D]$ is the elasticity matrix,

$[H]$ is the differential operator which defines strain as

$$\{\varepsilon\} = [H]\{u^h\} \tag{16.2}$$

$\{u^h\}$ the finite element approximation is given by

$$\{u^h\} = [N]\ \{\overline{u}\} \tag{16.3}$$

where

$[N]$ is the shape function matrix

$\{\overline{u}\}$ is the nodal displacement matrix

It has already been indicated that stresses $\{\sigma^h\}$ obtained from eqn. (16.1) are less accurate as they do not possess interelemental continuity and have a low accuracy at the nodes and at the element boundaries. The objective of the recovery of finite element derivatives is to find nodal parameters $\{\overline{\sigma}_*\}$ such that smoothed continuous stress field $\{\sigma_*\}$ is defined by the basis function $[N]$ and nodal parameters $\{\overline{\sigma}_*\}$ as

$$\{\sigma_*\} = [N]\{\overline{\sigma}_*\} \tag{16.4}$$

In the SPR technique the domain is divided into finite element meshes which will result into many overlapping node patches. A node is connected to a number of elements which is referred to as a patch [Figs. 16.1 and 16.2]. In this recovery process, the nodal superconvergent stresses $\{\overline{\sigma}_*\}$ belong to a polynomial expansion σ_*^p of the same complete polynomial order p as that present in basis functions which are assumed valid over a node patch. Thus, σ_*^p within a node patch is given by

$$\sigma_*^p = \{P\}\{a\} \tag{16.5}$$

where

$\{P\}$ contains appropriate polynomial terms

and $\{a\}$ is a set of unknown parameters

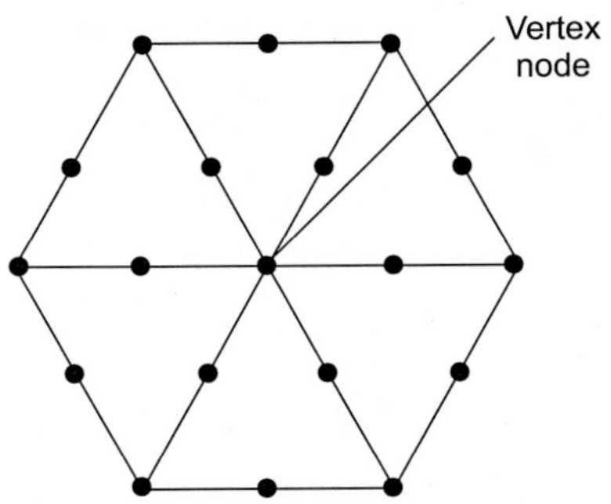

Fig. 16.1. Superconvergent points in a triangular element

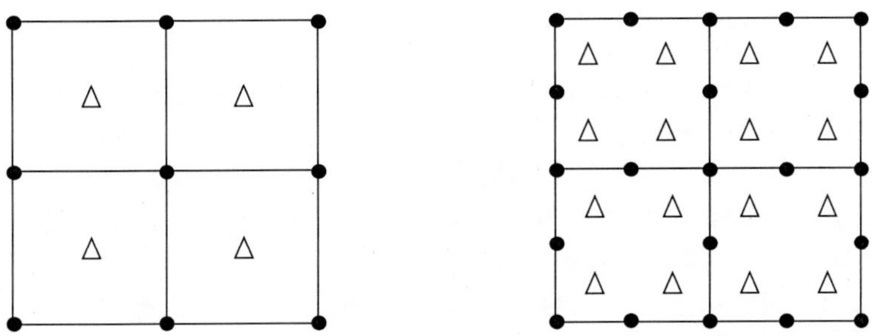

Fig. 16.2. Superconvergent points in a rectangular element

For one-dimensional problem of order p, $\{P\}$ and $\{a\}$ are expressed as

$$\{P\} = \{1 \quad x \quad x^2 \quad x^3 \quad \dots \quad x^p\} \tag{16.6}$$

$$\{a\} = \{a_1 \quad a_2 \quad a_3 \quad \dots \quad a_{p+1}\} \tag{16.7}$$

For two or three-dimensional problems, complete polynomial terms for the appropriate element order can be written. Therefore, two dimensions and linear expansions we have

$$\{P\} = \{1 \quad x \quad y\} \tag{16.8}$$

and for quadratic expansion

$$\{P\} = \{1 \quad x \quad y \quad x^2 \quad xy \quad y^2\} \tag{16.9}$$

In case of quadrilaterals in which terms additional to the complete polynomial occur, results indicate small improvement by using identical terms to those occurring in the shape function matrix $[N]$. Thus, for a bilinear quadrilateral it is

$$\{P\} = \{1 \quad x \quad y \quad xy\} \tag{16.10}$$

Thus, σ_*^p in a node patch for a quadratic triangular element can be written as

$$\sigma_*^p = \{1 \quad x \quad y \quad x^2 \quad xy \quad y^2\} \begin{Bmatrix} a_1 \\ a_2 \\ a_3 \\ a_4 \\ a_5 \\ a_6 \end{Bmatrix} \tag{16.11}$$

Now, the unknown coefficients $\{a\}$ are to be determined. For this a discrete least square fit is to be employed. The fit is made over a set of superconvergent sampling points existing in the chosen patch in order to minimize the function

$$F(a) = \sum_{i=1}^{n} [\sigma^h(x_i, y_i) - \{P(x_i, y_i)\}\{a\}]^2 \tag{16.12}$$

where (x_i, y_i) are the coordinates of a group of total number of sampling points n. However, it is important to recognize that such a suggestion is based on numerical experiments [16.1, 16.2] rather than vigorous mathematical analysis. The midside nodes of a triangular element and Gauss points of a quadratic element are superconvergent points for superconvergent stress extraction.

Total number of sampling points in each element are

$$n = mk \tag{16.13}$$

The implication of the minimization condition of $F(a)$ is

$$\sum_{i=1}^{n} \{P(x_i, y_i)\}^T \{P(x_i, y_i)\}\{a\} = \sum_{i=1}^{n} \{P(x_i, y_i)\}^T \{\sigma^h(x_i, y_i)\} \tag{16.14}$$

From eqn. (16.14), $\{a\}$ can be determined as

$$\{a\} = [A]^{-1}\{b\} \tag{16.15}$$

where

$$[A] = \sum_{i=1}^{h} \{P(x_i, y_i)\}^T \{P(x_i, y_i)\} \tag{16.16}$$

$$\{b\} = \sum_{i=1}^{n} \{P(x_i, y_i)\}^T \{\sigma^h(x_i, y_i)\} \tag{16.17}$$

Number of equations in each patch is modest and recovery is performed only for each vertex node. The procedure therefore is quite inexpensive. We may note that precisely the same matrix $[A]$ in the solution of each component of σ_*^p and a single inversion of $[A]$ is necessary.

Depending on the values of the parameter $\{a\}$, the recovered nodal values of $\{\bar{\sigma}_*\}$ are calculated by inserting appropriate coordinates into the expression for σ_*^p. The entire procedure can be repeated for all components of the stress or the formulation can be adjusted so that the unknown coefficients $\{a\}$ for each component of stress can be found in a single run. For accurate evaluation of nodal superconvergent stress $\bar{\sigma}_*$, following procedure is recommended [16.1, 16.2].

- In each patch only the nodes inside the patch are to be considered for interpolation using patch equation (16.11). Boundary nodes of a patch should not be interpolated from eqn. (16.11).
- The patches will frequently overlap for internal midside nodes. This suggests that such nodal values are frequently evaluated from more than one patch. For such nodes, average values of the recovered stresses are to be considered.
- In case where the vertex node is at the boundary of the domain, boundary patches should not be considered. Instead, the recovery of the boundary nodes should be determined from exterior patches that include them.

The matrix $[A]$ turns out to be singular in certain category of problems or numerically is very close to be singular. As such some special techniques such as singular value decomposition technique (SVA) may be applied for the inversion of $[A]$ matrix [16.3].

16.4 EXAMPLE OF VERIFICATION OF SPR

The problem to be modeled here is a portion of an infinite plate with a central circular hole subjected to unidirectional tensile load. The geometry of the problem is shown in Fig.16.3(a). Plane strain conditions have been assumed along with a Poisson's ratio of 0.3 and Young's modulus of 1000. The portion of the plate is analysed shown in Fig. 16.3(b). The convergence of the stress components has been observed at point 1 [Fig. 16.3 (b)]. The boundary conditions prescribed are such that on edges AB and ED, symmetry conditions are imposed and on edges BC and DC, the plate is loaded by tractions given by the analytical solution

$$\left.\begin{aligned}
\sigma_x &= 1-\frac{a^2}{r^2}\left(\frac{3}{2}\cos 2\theta+\cos 4\theta\right)+\frac{3}{2}\frac{a^4}{r^4}\cos 4\theta \\
\sigma_y &= -\frac{a^2}{r^2}\left(\frac{1}{2}\cos 2\theta-\cos 4\theta\right)-\frac{3}{2}\frac{a^4}{r^4}\cos 4\theta \\
\sigma_{xy} &= -\frac{a^2}{r^2}\left(\frac{1}{2}\sin 2\theta+\sin 4\theta\right)+\frac{3}{2}\frac{a^4}{r^4}\sin \theta
\end{aligned}\right\} \tag{16.18}$$

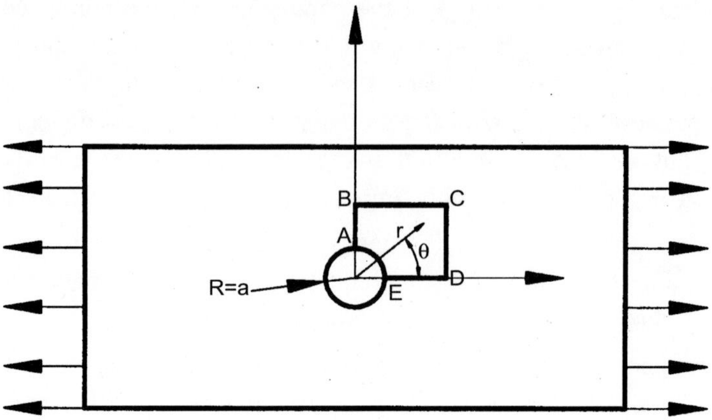

(a)

E = 1000
ν = 0.3
Thickness = 1.0
a = 1.0
Plane strain conditions

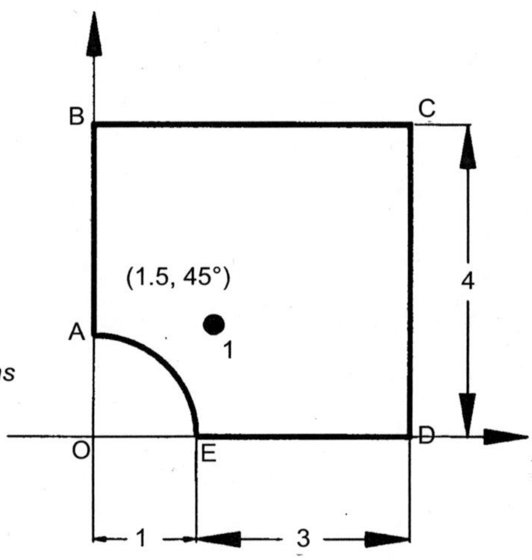

(b)

Fig. 16.3. Infinite plate with a central circular hole

Figure 16.4 shows semiuniform meshes that are considered for convergent study. Figure 16.5 shows the convergence of stress components at point 1. It may be observed that the stresses determined by the SPR technique are more accurate than the finite element solutions.

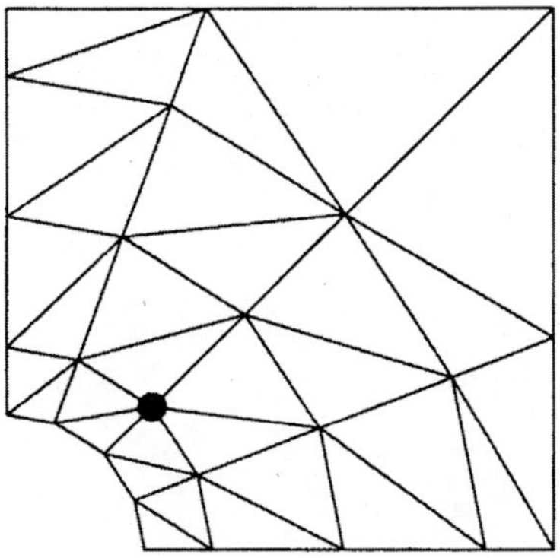

(a)
Mesh 1
Elements = 32; Nodes = 81

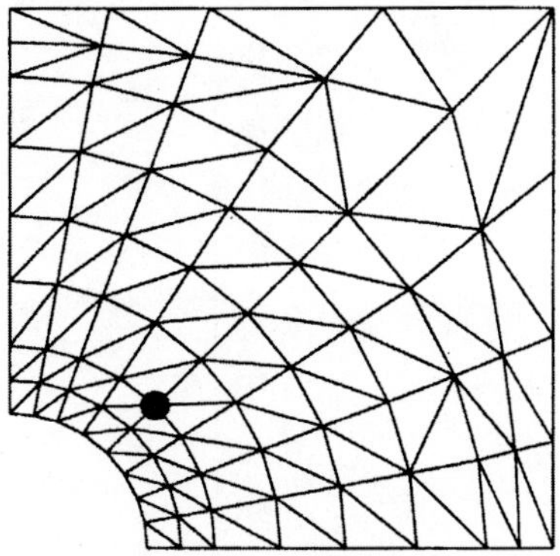

(b)
Mesh 2
Elements = 128; Nodes = 289

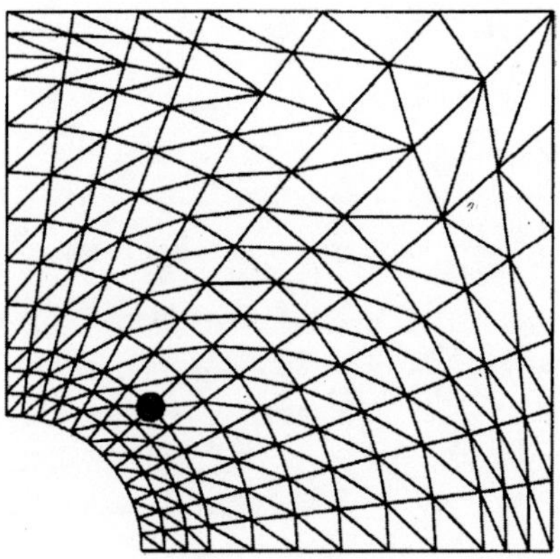

(c)
Mesh 3
Elements = 288; Nodes = 625

Fig. 16.4. Semi-uniform meshes used for accuracy observations

16.5 ERROR ESTIMATION

The finite element method is a numerical method. As such the solution obtained from it is an approximate one. The differences between the exact and approximate solutions are termed as discretization error or simply error. Thus, the error in displacement is given by

$$\{e_u\}=\{u\} - \{u^h\} \tag{16.19}$$

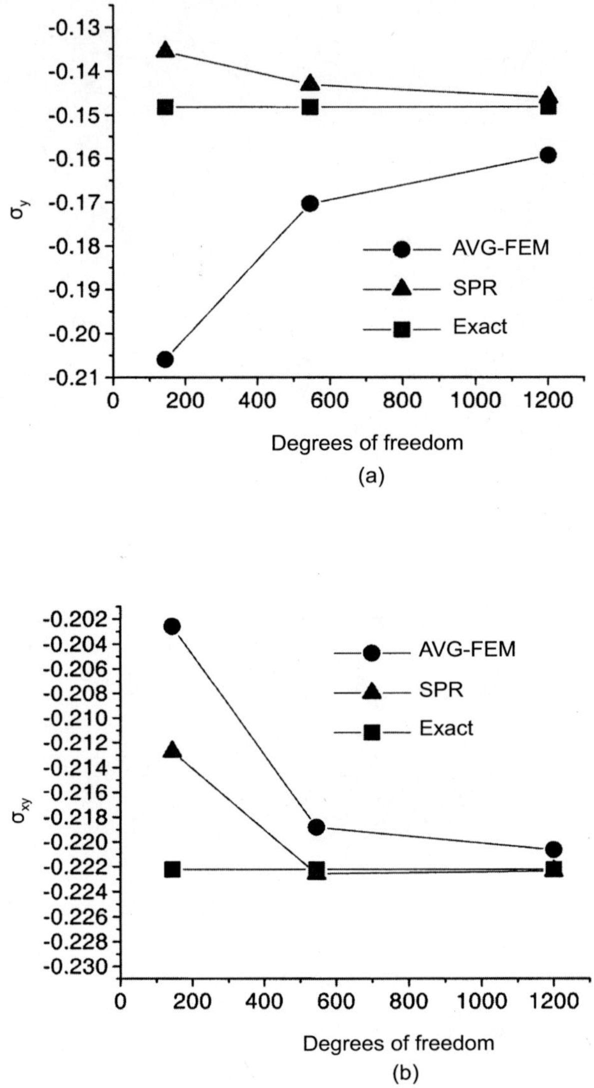

Fig. 16.5. Convergence of stress components point 1

where $\{u\}$ is the exact solution and $\{u^h\}$ is the finite element approximation. Similarly, the error in stresses is given by

$$\{e_\sigma\} = \{\sigma\} - \{\sigma^h\} \tag{16.20}$$

An analyst faces two important queries:

1. What is the procedure to determine reliability of the FEM solution?
2. How to determine the desired level of accuracy by efficiently controlling the error?

In order to seek an answer to both the above queries, estimation of the error is made which aids in adaptive computations. It is nothing but an iterative process in which refinement of approximations are continually made so that the numerical solution tends to near the 'exact' solution.

The error estimates are divided into two categories: (1) a priori error estimator and (2) posteriori error estimator. A priori error estimator provides only a qualitative description on the rate of convergence of the finite element solution. This error estimator is rather complex to use and difficult to implement.

A posteriori error estimator is used only after a particular FEM solution is obtained. This estimation can be divided into two categories. The computation residuals from the equilibrium equations and subsequent determination of interelemental traction jumps along element boundaries fall in the first category. In the second category stress smoothing technique is employed [16.4]. This error estimator is effective in many classes of application [16.5–16.9].

The procedures for adaptive finite element analysis are indicated as follows:

1. *h* - adaptivity - meshes are refined and/or coarsened to improve the finite element solutions.
2. *p* - adaptivity - by keeping mesh size unchanged, the order of the polynomial is varied for all elements to improve the finite element solution.
3. *h-p* adaptivity - both *h* and *p* types of refinement are adopted simultaneously for improving the finite element solution.

In order to use *p*-adaptivity for the problem, a new code is required to be developed. But the procedure is rather costly, but it gives a very accurate solution. However, *h-p* adaptivity will prove to be not only costly but complicated as well. *h*-adaptivity is most widely used in practice. We shall limit ourselves to *h*-adaptivity only.

In general, an automatic adaptive finite element modeling environment consists of the following:

1. An automatic adaptive mesh generator capable of producing the desired mesh configuration.
2. The finite element solver for class of problems under consideration.
3. A posteriori error estimator used for determining the discretization error in the requested norms.
4. Mesh indicators specify where and how to improve the mesh to obtain the desired level of accuracy.

Figure 16.6 shows a typical self-explanatory *h*-adaptive finite element scheme.

16.6 ZZ-ERROR ESTIMATOR

The ZZ-error estimator which is quantified on the basis of energy norm is an integral scalar quantity and is defined for elasticity problems as [16.1–16.10]

$$\|e_{es}\|_e^2 = \int_A (\{\sigma_*\} - \{\sigma^h\})^T [D]^{-1} (\{\sigma_*) - \{\sigma^h\})\, dA \tag{16.21}$$

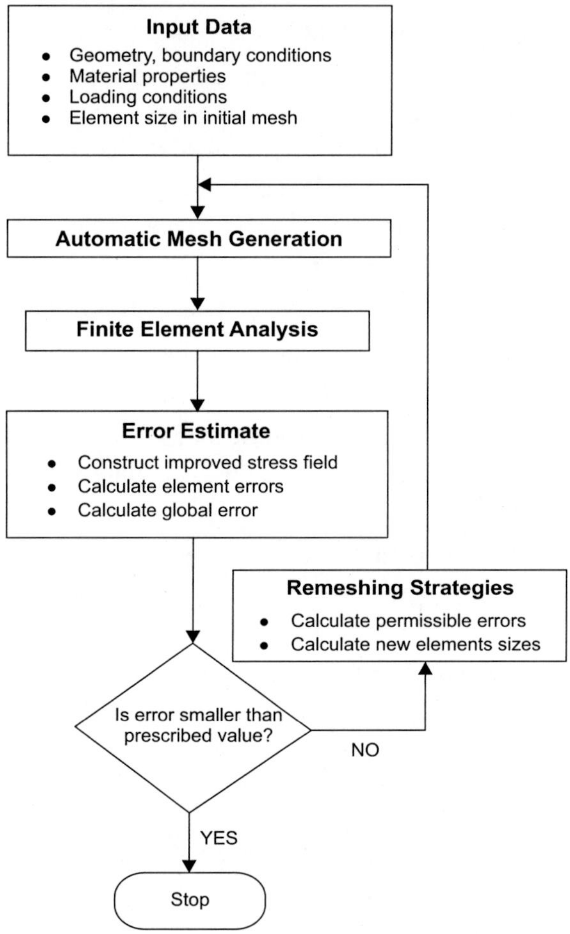

Fig. 16.6. A *h* - adaptive finite element scheme

where the integration is to be carried out over an elemental area. $\{\sigma_*\}$ is the smoothed stress field from a recovery procedure (for example, nodal averaging or superconvergent patch recovery). In eqn.(16.21), $\|e_{es}\|_e$ is the estimated discretization error of an element in energy norm.

The underlying principle of ZZ-error estimator is to use improved stress field which may be derived from post-processing of the finite element stress field rather than the exact field which is unknown. In the same manner, the square of the energy of an element is defined as

$$\| U \|_e^2 = \int_A \{\sigma^h\}^T [D]^{-1} \{\sigma^h\} dA \tag{16.22}$$

Summation of the squared values of $\|e_{es}\|_e$ and $\| U \|_e^2$ over all elements gives rise to squared energy norms for the whole domain

$$\| U \|^2 = \sum_{e=1}^{n} \|U\|_e^2 \tag{16.23}$$

where n is the total number of elements and the square of the energy norm is

$$\|e_{es}\|_e = \sum_{e=1}^{n} \|e_{es}\|_e^2 \tag{16.24}$$

The estimated relative percentage of error is defined as

$$\mu_{es} = \frac{\|e_{es}\|}{\sqrt{\|U\|^2 + \|e_{es}\|^2}} \times 100 \tag{16.25}$$

μ_{es} is widely accepted as the measure of accuracy of the computed solution. Thus, the adaptively refined mesh should satisfy the condition

$$\mu_{es} < \mu \tag{16.25}$$

where μ is the maximum permissible relative percentage of error or the target error set by the user.

The quantities $\| e_{es}\|_e^2$ and $\| U\|^2$ can be calculated by numerical integration such as Gaussian quadrature. Equations (16.21) and (16.22) require 3-point Gaussian integration scheme evaluation.

16.7 ZZ-REFINEMENT FRAMEWORK

It is basically a h-refinement strategy and is dependent on the h-convergent characteristics of the finite element method [16.9]. With trial functions in the displacement formulation of degree p, the error in stresses of the order of $O(h)^p$ or $O(NDF)^{-p/2}$(where NDF is the number of degrees of freedom) and is assumed to be equally applicable to the energy norm of the error. However, this is true if singularity exists in the domain. It can be shown that in general the rate of convergence of the estimated error for problems with singularity is

$$O(NDF)–[\min(\lambda, p)]/2 \tag{16.27}$$

where λ is a number associated with the intensity of singularity. For elasticity problems, it ranges from 0.5 for a nearly closed crack to 0.7 for 90° corner [16.11]

We defined optimised mesh as one in which the error in each element (in the limit) is constant. It is assumed in the optimal mesh that the energy norm error $\| e_{es} \|$ is equal in all elements. Thus, if the permissible error in the whole domain is

$$\mu \sqrt{\|U\|^2 + \|e_{es}\|^2} \tag{16.28}$$

then, in accordance with the definition of the optimum mesh, the requirement that the permissible error in any element should be

$$\bar{e}_m = \mu \sqrt{\frac{\|U\|^2 + \|e_{es}\|^2}{n}} \tag{16.29}$$

As mentioned above, the following two equations can be written as

$$\| e_{es} \|_e \cong c \cdot h^p_{old} \tag{16.30}$$

$$\bar{e}_m = c \cdot h^p_{new} \tag{16.31}$$

Dividing eqn. (16.30) by eqn. (16.31), we get

$$\frac{\bar{e}_m}{\|e_{es}\|_e} = \frac{h^p_{new}}{h^p_{old}} = \left(\frac{h_{new}}{h_{old}} \right)^p \tag{16.32}$$

Equation (16.32) can be written as

$$\frac{h_{new}}{h_{old}} = \left(\frac{\bar{e}_m}{\|e_{es}\|_e} \right)^{1/p} \tag{16.33}$$

which can be written in the final form of the introducing $\psi_e = \dfrac{\|e_{cs}\|_e}{\bar{e}_m}$ and $\beta_e = \dfrac{1}{p}$

$$h_{new} = \frac{h_{old}}{\psi_e^{\beta_e}} \tag{16.34}$$

Three conditions are possible depending on the ψ_e of the element

- $\psi_e > 1$, the element should be refined
- $\psi_e = 1$, no change in the element size is required
- $\psi_e < 1$, the element size may be derefined, if desired for computational efficiency.

In eqn. (16.34), h_{old} is the size of an element in current mesh and h_{new} is the new element size.

16.8 ADPTIVE MESH GENERATION

Considerable input data is to be prepared for carrying out the finite element analysis; out of which the data required as an input for mesh generation is quite large. The manual preparation of this data is tedious, time consuming and error prone. As such it is highly desirable to have a computerised or automatic generation of finite element mesh. This will require a minimum amount of data to be prepared for mesh generation. The mesh generation in successive cycle is then to be carried out without any user intervention. Attempts are being made to develop a highly effective and robust mesh generator which will be versatile and computationally efficient. Thus, the labour of creating a mesh manually is avoided. Very broadly automatic mesh generation methods are grouped into two categories [16.4].

(*a*) Methods that are based on mapping the domain or various subdomains into which it is divided into simple forms in which direct structured mesh forms can be avoided [Fig. 16.7, 16.8] and (*b*) methods that are based on joining a set of predetermined nodes to construct a mesh or to satisfy prescribed element shape and size requirements.

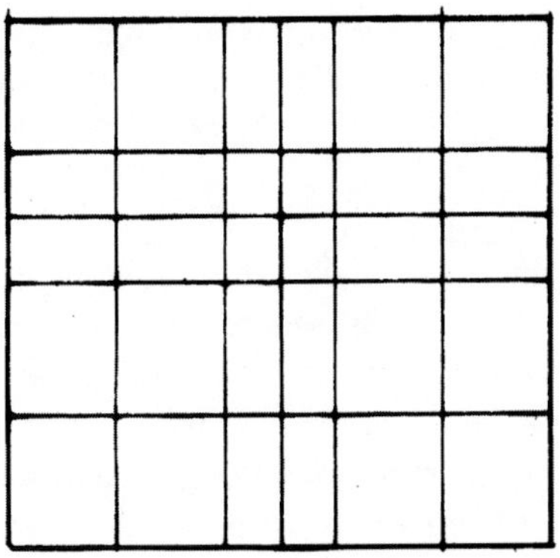

Fig. 16.7. Structured mesh

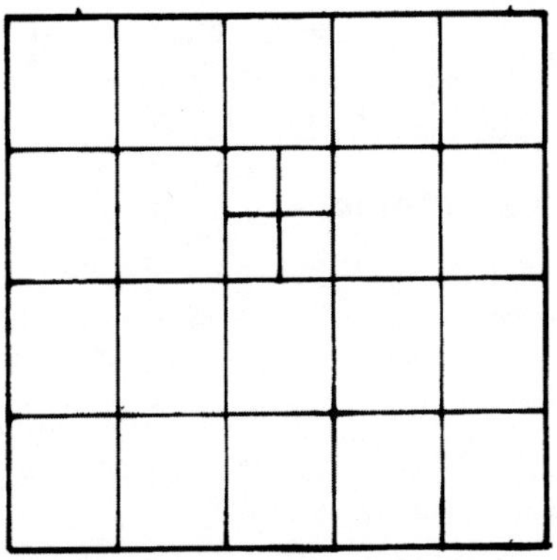

Fig. 16.8. Unstructured mesh

16.8.1 Mesh Generation Based on Mapping

In the first category are included methods such as isoparametric mapping concept [16.12] and those based on transfinite interpolation [16.13]. When mapping techniques are used in mesh generation, the nodal domain is to be subdivided into regular grids to induce a mesh. However, the manual process in subdivision of a domain is tedious and sometimes it is really difficult in case of a domain with complex geometry or simple geometry with complex mesh density. As the element sizes are controlled by the subdivision of simple subregions, more subregions are needed to generate the mesh that can accommodate changes in the desired element sizes from region to region. This has further restricted their use by the requirement of optimal mesh in adaptive analysis. For this reason we shall not discuss this any further.

In the second category many variants are possible. But as the mesh is constructed, it can in general achieve a prescribed density. Most of the techniques in the second category currently in use may be classified into following categories:

- Delaunay triangulation method
- Domain decomposition (Quadtree) method
- Advancing front technique

16.8.2 Delaunay Triangulation Method

The meshing techniques which utilize Delaunay's criterion are quite popular. The criterion states that the circumcircle of the element contains no points [16.14]. It generates triangular mesh a given set of nodes.

The boundary of the geometry is first meshed so as to generate a given set of nodes. The boundary nodes are triangulated according to Delaumay criterion. Nodes are then inserted incrementally into the mesh, redefining the triangle as a new node is inserted to maintain the Delaunay criterion. The method that is chosen for defining the locations of the interior nodes, distinguishes one Delauny algorithm from the other. An important property of the Delaunay triangulation is that it generates as close as possible to the equilateral triangle from a given set of nodes. Large number of researchers have utilized Delaunay criterion for triangulation [16.15–16.17].

16.8.3 Domain Decomposition Method (Quadtree)

Given a geometrical shape for the finite element analysis, there are several methods that may be used to produce the necessary nodal points and subsequent element mesh. Domain Decomposition Method is one such that is useful for generating necessary nodal points. In two-dimensional cases, domain decomposition method takes the form of a quadtree structure in which the corner point or the central point of each cell may be chosen as nodal points. Quadtree method is developed and implemented by Shephard and his coworkers [16.18–16.19].

Domain decomposition technique has the advantage of being easily created and accessed due to tree-like structure. However, the spacing of the resulting meshes does not transit smoothly from one location to the other which is an essential drawback. Thus, the shape of the mesh is not essentially good. Another disadvantage of the scheme is that it is difficult to fit the tree structure to the boundary of the arbitrary domain.

16.8.4 Advancing Front Technique

Advancing front technique (AFT) or moving front method is a technique in which the adaptive refinement of the grid is accomplished by means of a remeshing procedure which is based on the information provided by the computer solution [16.7, 16.20, 16.21]. This technique starts and completes the meshing in a systematic, ordered and highly programmable manner. Unlike Delaunay's triangulation, AFT does not depend on any geometrical criterion, but on logical steps to mesh a given complicated domain. logical steps to mesh a given complicated domain. The AFT was developed by Peraire *et al.* [16.19]. The AFT is composed of four distinct phases of meshing a domain in a desired way. They are:

- Generation of the background mesh
- Geometrical representation of the domain
- Generation of boundary nodes
- Generation of elements

16.8.4.1 Structured and unstructured mesh

Structured and unstructured quadrilateral meshes are shown in Figs. 16.7 and 16.8. From the figure it can be seen that the bisection is made to regular mesh lines running to the boundary to retain the four-neighbour structure. This strategy is simple to implement. However, it clearly refines many more elements than necessary. The customary way of avoiding the excess of refinement is to introduce irregular nodes when the edges of a refined element meet the middle of a coarser one (Fig. 16.8). The mesh is no longer structured.

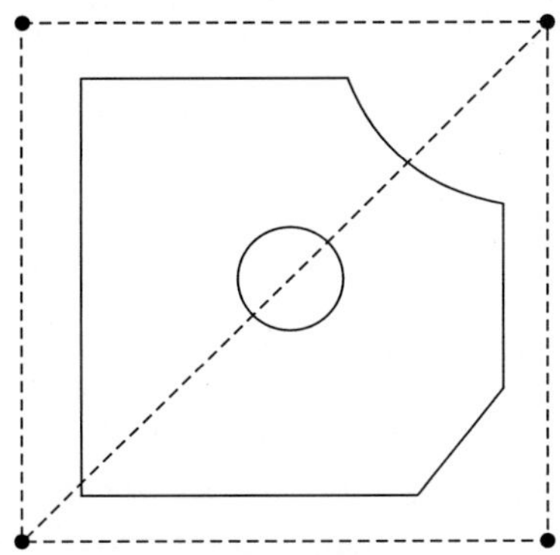

Fig. 16.9. A simple background mesh

16.8.4.2 The background mesh

Our discussion is limited to triangular elements only. The background mesh is simply a structured or unstructured mesh [Figs. 16.9, 16.10]. The first step in mesh generation is to construct by hand coarse background mesh of triangular elements which completely covers the solution domain of interest as shown in Fig. 16.9. The background mesh may have any arbitrary orientation and need not align with the boundaries of the domain. For the generation of triangular elements it is convenient to define a node spacing function δ. It represents the side of an equilateral triangle that is yet to be generated. At each node on the background mesh the nodal value δ is required to be specified at the beginning of the mesh generation. The background mesh is used to provide a piecewise linear spatial distribution of the parameter δ over the mesh to be generated.

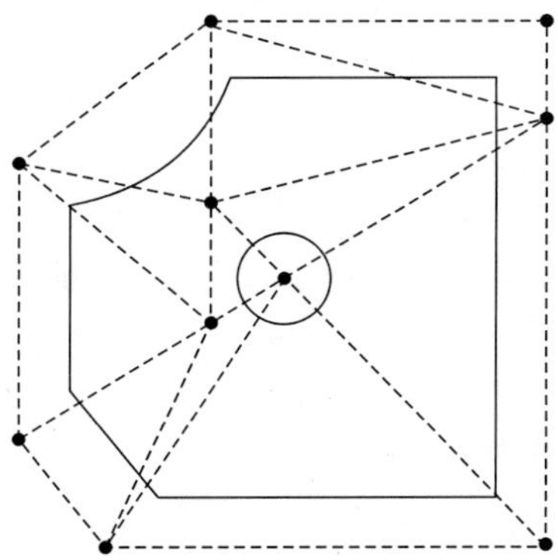

Fig. 16.10. A complex background mesh

For uniformly distributed elements in a domain, very simple shaped background grids are sufficient [Fig. 16.9]. Relatively complicated background grids are required for possible variation of the sizes of the elements in the domain [Fig. 16.10]. During the mesh generation process, local values of δ will be obtained by linear interpolation of the triangles of the background grid between specified values. In course of the adaptive analysis, the current mesh becomes background mesh for the next mesh generation, thus allowing for a more flexible definition of spatial variation of the mesh generation parameter δ. Referring to eqn. (16.34),

h_{new} is defined for an element. During adaptive mesh refinement, the δ-value of a node can be computed as

$$\delta = \frac{\sum_{e=1}^{n} h_{new}^{e}}{m} \tag{16.35}$$

where m is a number of elements connected to a vertex node or simply a node.

16.8.4.3 Geometrical representation of the domain

In advancing front technique, the boundary of the solution domain is represented by the union of closed loops of curved segments. There is only one closed loop for simple connected regions whereas for multi-connected regions, there will be as many internal loops as the number of openings inside the domain. Each loop is composed of convenient number of segments and each segment will have definite starting and ending points. End-points are always the intersection points of a segment with its adjacent segments.

Thus, as shown in Fig. 16.11, the domain has two closed loops: one, the outer boundary and the other, the inner hole. The outer loop has segments and the inner loop has two curvilinear segments 7 and 8. However, the maximum number of segments to define a loop is arbitrary. The segments of the interior are defined in a clockwise fashion. This means that as the boundary curve is traversed, the region to be triangulated lies always to the left as shown [Fig. 16.11].

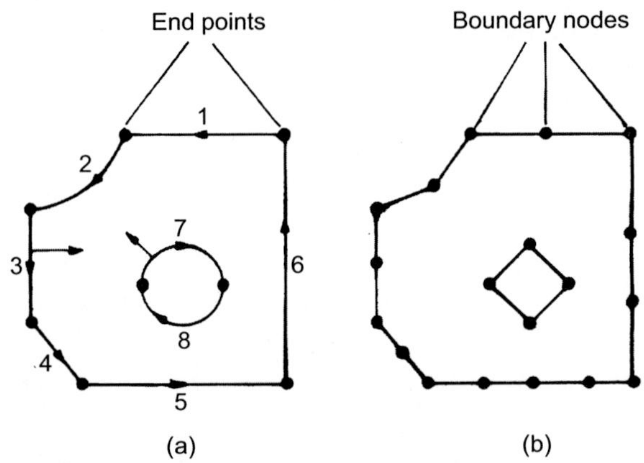

Fig. 16.11. (a) Boundary segments, end points, loops and meshing
(b) Conversion of domain into polygon object upon joining the boundary nodes

The curvilinear boundary segments can be approximated by spline functions. The spline function is represented by control points which belong to boundary segments. The number and position of control points selected by the user in each segment should be done in such a way that the spline curves can accurately describe the boundary segment of the loop and consequently the entire boundary of the domain. The control points should include the end points of a segment [Fig. 16.12].

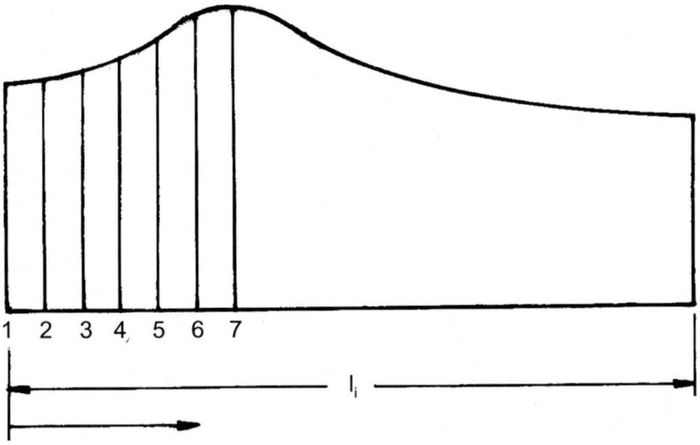

Fig. 16.12. Curve with spline interpolation

16.8.4.4 Generation of boundary nodes

After defining each segment of the boundary with the help of spline functions, next step is to generate the nodes on the boundary of the domain of interest. The first step towards the generation of boundary nodes is to place the points of intersection of the segments. Additional nodes of the boundary segments of the entire domain have to be placed before beginning the process of generation of triangles. Each boundary segment is then considered in turn and nodal points are generated in boundary segments with spacing of points being determined by interpolated values of δ over the background mesh. Thus, the boundary domain is discretized into a polygon with sides connected to each other at a boundary node [Fig. 16.11] by generating nodes on the spline curves which represent boundary segments.

Suppose that the boundary of the domain consists of M segments. For a segment of typical length $l_i (i = 1,2,3, \ldots, M)$, uniformly distributed sample points is first placed along the curve with spline interpolation [Fig. 16.12]. A piecewise linear interpolation of nodal spacing values on the midpoint are lengths (obtained by joining two successive sample points) for all arc lengths are to be obtained [Fig. 16.13]. Then the trapezoidal rule may be applied for evaluating the integral,

$A_i = \int_0^{l_i} \frac{1}{\delta(s)} dl$ where $\delta(s)$ is termed as node spacing function, s being the arc length of the segment i.

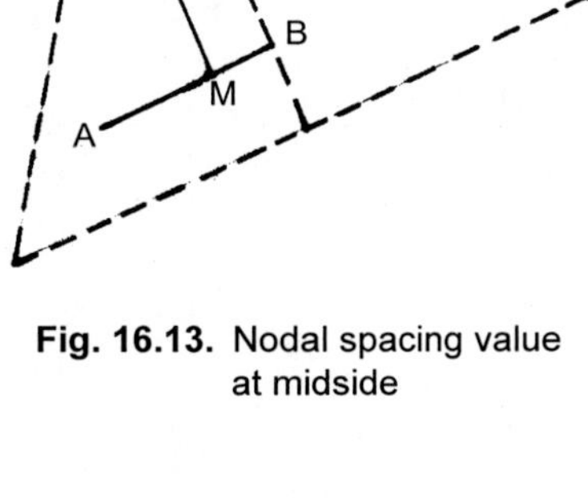

Fig. 16.13. Nodal spacing value at midside

Let H_i be the number of sides and H_{i+1} is the number of boundary nodes of the chosen segment, then H_i is chosen to be the nearest integer to

$$A_i = \int_0^{l_i} \frac{1}{\delta(s)} dl \tag{16.36}$$

note that H_i is an integer.

Once the number of nodes are obtained, their positions are to be determined (Fig. 16.14). The position of the node $n_k(k = 0,1,2,\ldots H_i)$ to be generated by the boundary curve is defined by S_K and t is defined by the equation given below.

$$K = \frac{H_i}{A_i} \int_0^{S_K} \frac{1}{\delta(s)} dl \tag{16.37}$$

Similar procedure is adopted for other segments.

Equation (16.37) can be rewritten as

$$\frac{KA_i}{H_i} = \int_0^{S_K} \frac{1}{\delta(s)} dl \tag{16.38}$$

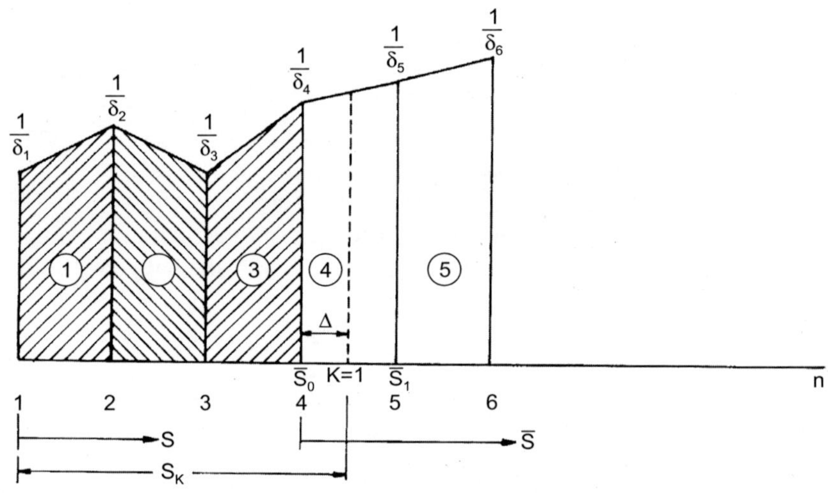

Fig. 16.14. Curves approximated as straight lines

With $K = 1$, the distance of the first node from the 0th node is to be determined. By solving eqn. (16.36), the area of each trapezoid is to be stored. Adding these areas from $s = 0$, it will indicate the location of the node at a particular trapezoid. The distance of the node for that trapezoid is to be determined. Let the node fall in 4th trapezoid [Fig. 16.14]. The equation of the edge of the trapezoid is given by [Fig. 16.14].

$$\frac{1}{\delta(\bar{s})} = m\bar{S} + \bar{c} \tag{16.39}$$

assuming a straight line variation considered at the 4th station. The boundary conditions are:

at $$\overline{S} = 0, \quad \frac{1}{\delta(\overline{s})} = \frac{1}{\delta(4)} \tag{16.40a}$$

and $$\overline{S} = s \quad \frac{1}{\delta(3)} = \frac{1}{\delta(s)} \tag{16.40b}$$

Substituting the conditions of eqns. (16.40a) and (16.40b) into eqn. (16.39), m and $\overline{c}$ can be evaluated.

Therefore,

$$K\frac{A_i}{H_i} = \text{Area of the trapezoid } [(1) + (2) + (3)] + \left[\frac{1}{\delta(4)} + \frac{1}{\delta(\Delta)}\right] / 2 \times \Delta \tag{16.41}$$

where $\Delta = S_K - S_4$

Likewise, values of other node positions in a segment are obtained.

16.8.4.5 Generation of triangular elements

Before the generation starts, a generation front has to be set up. At the start of the process, the front consists of a sequence of straight line segments which connect consecutive boundary nodes. During the generation process, any straight line segment which is available to form an element side is termed active whereas any segment which is not active is removed from the front. Thus, while the domain boundary will always remain the same, the generation front will change continuously and has to be updated whenever a new element is found. The updating process is illustrated in Fig. 16.15. The element generation steps are as follows [16.20, 16.22].

1. An active side (with nodes A and B) in the front is chosen as a base.
2. Determine the node spacing value δ_M at the midpoint M of AB by interpolating about the background mesh [Fig. 16.13].

Determine δ_1 according to

$$\delta_1 = \begin{cases} 0.55AB & \text{if} \quad \delta_M < 0.55AB \\ \delta_M & \text{if} \quad 0.55\,AB < \delta_M < 2AB \\ 2AB & \text{if} \quad \delta_M > 2AB \end{cases}$$

The above inequalities have been suggested by Peraire *et al.* [16.19] to ensure geometrical compatibility and also to ensure that element with excessive distortion is not generated.

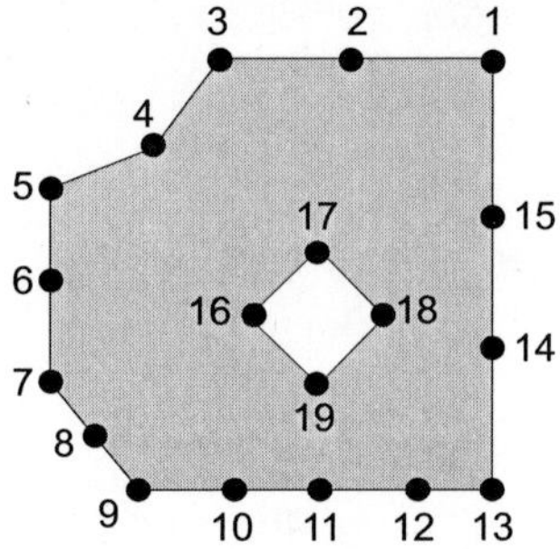

1	2	3	4	5	6	7	8	9	10	11	12	13	14	15	16	17	18	⊗	19	20
2	3	4	5	6	7	8	9	10	11	12	13	14	15	1	17	18	19	⊗	20	16

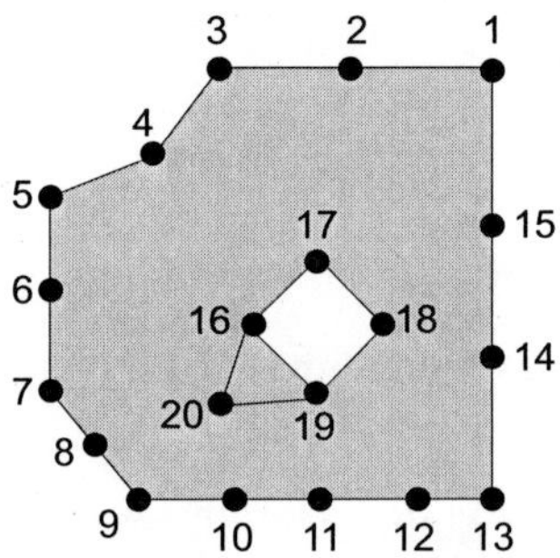

Fig. 16.15. Updatation of front during triangle generation process

3. Construct a point c, δ_1 from M (Fig. 16.13)
4. Determine all the active nodes (nodes still available to be a part of the new element), which lie within the circle with centre at C and radius n_{IAB}. However, there is no unique choice for n_I (n_I=5 has been chosen in Ref. 16.19). These nodes are ordered according to their distance from C with the first node in the list closest to C and are denoted by $N_1, N_2 \ldots, N_e$ (N_1 being the first node).
5. If $AN_1 > 1.5\delta_1$ and $BN_1 > 1.5\delta_1$, then place C at the head of the list.
6. The required connecting point N_j is then taken to be the first node in the list which is such that the interior of the triangular ABN_j does not contain any other node N_i in the list (excluding C) and such that the line MN_j does not intersect any existing side of the front. The coordinates are transferred back to the original space and the front is updated as indicated previously.
7. Triangle generation process ceases when the number of active sides of the front reduces to zero.

For comparison purpose, the effectivity index of error estimates is presented. This index is defined as

$$\theta = \frac{\|e_{es}\|}{\|e\|_{exact}} \tag{16.42}$$

16.8.4.6 Example of h-adaptive finite element analysis

The test problem that is considered concerns the plane stress analysis of a square plate with a square cutout loaded by uniform tension on opposite outer edges as shown in Fig 16.16. Derefinement schemes have not been implemented. Due to the symmetry, one quarter of the plate is modeled as shown in Fig. 16.16 (b). This example has been used extensively in the literature as a benchmark problem for adaptive refinement technique. The exact value of $\|U\|^2$ is 0.311329399.

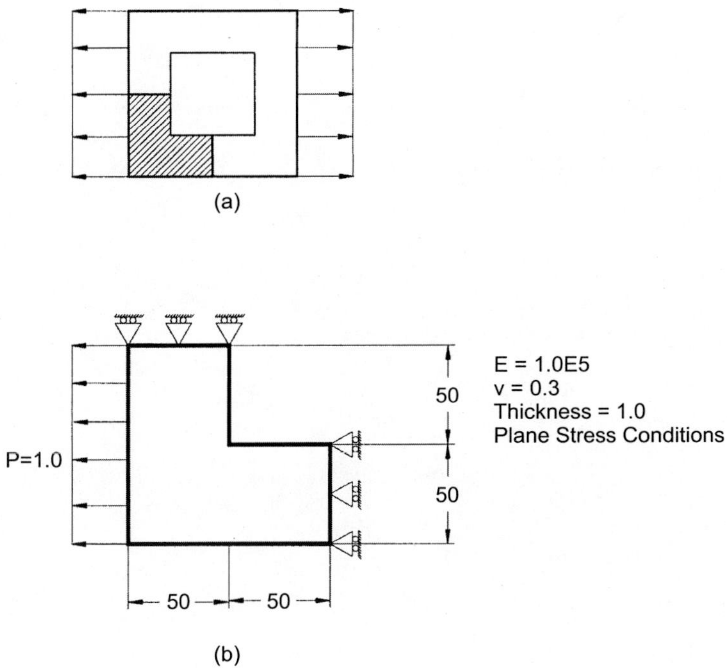

Fig. 16.16. L-shaped domain

Table 16.1 summarizes the *h*-adaptive finite element analysis of the domain for a target error of 3% Fig. 16.17 shown *h*-adaptive meshing sequence.

Table 16.1. Summary of *h*-Adaptive Finite Element Analysis

Target error 3.0%					
Mesh	Nelm	Node	$\|\|e_{es}\|\|^2$	$\mu_{es}\%$	θ
1	18	49	9.87365e-3	17.8	0.895
2	138	307	1.68943e-3	4.48	1.64
3	285	620	1.76370e-4	2.04	1.168

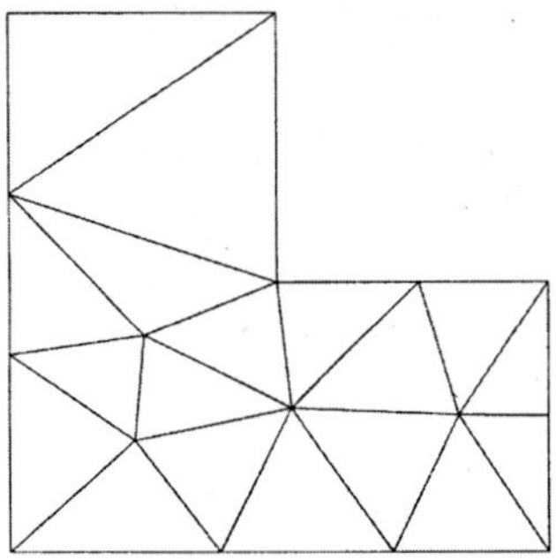

(a)
Mesh 1: Starting Mesh

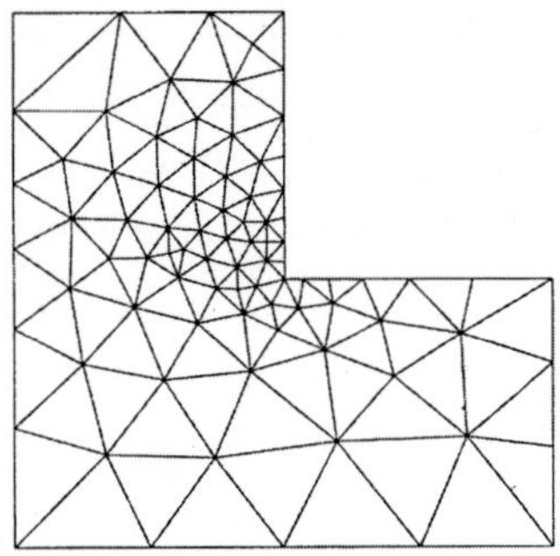

(b)
Mesh 2

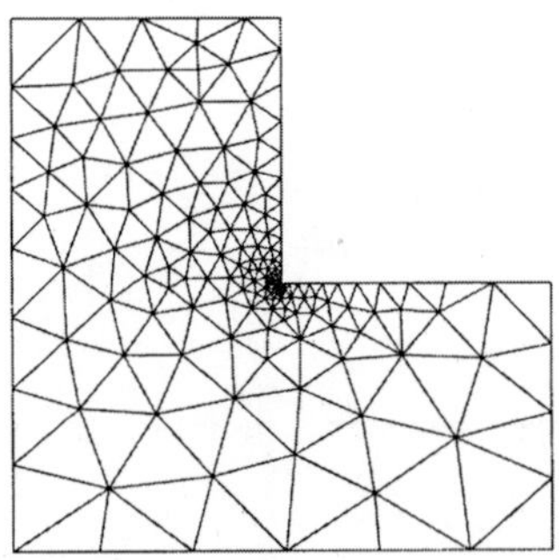

(c)
Mesh 3

Fig. 16.17. *h*-adaptive match sequence

REFERENCES

16.1. O.C. Zienkiewicz and J.Z. Zhu, The Superconvergent Patch Recovery and a Posteriori Error Estimate, Part I: The Recovery Technique, IJNME, v.33, 1992, pp. 1331–1364.

16.2. O.C. Zienkiewicz and J.Z. Zhu, The Superconvergent Patch Recovery and a Posteriori Error Estimate, Part II: Error Estimates and Adaptivity, IJNME, v.33, 1992, pp. 1365–1382.

16.3. W.H. Press, S.A. Teukdsky, W.I. Velteshink and B.P. Flannery, Numerical Recipies in FORTRAN : The Art of Scientific Computing, Cambridge University Press, New Delhi, 1992.

16.4. O.C. Zienkiewicz and J.Z. Zhu, Adaptivity and Mesh Generation, IJNME, v.32, 1991, pp. 183–210.

16.5. O.C. Zienkiewicz, Y.C. Liu and G.C. Huang, Error Estimation and Adaptivity in Flow Formulation of Forming Problems, IJNME, v.25, 1988, pp. 23–42.

16.6. M. Ainsworth, J.Z. Zhu, A.W. Craig and O.C. Zienkiewicz, Analysis in Zienkiewicz - Zhu Error Estimator in the Finite Element Method, IJNME, v.28, 1989, pp. 2161–2174.

16.7. O.C. Zienkiewicz, G.C. Huang and Y.C. Liu, Adaptive FEM Computation of Forming Process - Application to Porous and Non-porous Materials, IJNME, v.30, 1990, pp. 1527–1553.

16.8. O.C. Zienkiewicz, J.Z. Zhu and N.G. Gong, Effective and Practical Adaptive Analysis Proceedings for the Finite Element Method, IJNME, v.28, 1989, pp. 879–891.

16.9. O.C. Zienkiewicz and J.Z. Zhu, Error Esimations and Adaptive Refinement for Plate Bending Problem, IJNME, v.28, 1989, pp. 2839–2883.

16.10. O.C. Zienkiewicz and J.Z. Zhu, A Simple Error Estimator and Adaptive Procedures for Practical Engineering Analysis, IJNME, v.24, 1987, pp. 337–351.

16.11. O.C. Zienkiewicz and R.L Taylor, The Finite Element Method, v.1 and v.2, McGraw-Hill International Edition, 1988.

16.12. O.C. Zienkiewicz and D.V. Phillips, An Automatic Mesh Generation Scheme for the Plane and Curved Surfaces by Isoparametric Coordinates, IJNME, v.3, 1971, pp. 519–528.

16.13. W.J. Gordon and C.A. Hall, Construction of Curvilinear Coordinate System and Application to Mesh Generation, IJNME, v.7, 1973, pp. 461–477.

16.14. B.N. Delaunay, Sur la sphere *vide* Izvesia Academia Nauk, SSSR, VII Seria, v.72, 1934, pp. 793–800.

16.15. S.H. Lo, Delaunay Traingulation of Non-convex Planar Domains, IJNME, v.28, 1989, pp. 2695–2707.

16.16. H. Borouchaki and P.I. George, Optimal Delaunay Point Insertion, IJNME, v.39, 1997, pp. 3407–3437.

16.17. R.W. Lewis, Y. Zheng and A.S. Usmani, Aspects of Adaptive Mesh Generation Based on Domain Decomposition and Delaunay Triangulation, Finite Element in Analysis and Design, v.20, 1995, pp. 47–70.

16.18. M.S. Shephard, Approaches to Automatic Generation and Control of Finite Element Meshes, Applied Mechanics Review, Trans ASME, v.41, 1988, pp. 169–188.

16.19. M.S. Shephard, Automatic Generation of Finite Element Models Solving Large Size Problems in Mechanics, (Edited by M. Papradrakis) John Wiley, New York, 1993, pp. 431–460.

16.20. J. Peraire, M. Vabdati, K. Morgan and O.C. Zienkiewicz, Adaptive Meshing for Possible Flow Computations, Journal of Computational Physics, v.72, 1987, pp. 449–466.

16.21. S.H. Lo, A New Mesh Generation Scheme for Arbitrary Planar Domains, IJNME, v.21, 1985, pp. 1403–1425.

16.22. K.S.R.K. Murthy, An Investigation on *h*-adaptive Finite Element Analysis of Planar Mixed mode Crack Problems, Doctoral Dissertation, Department of Ocean Engineering and Naval Architecture, Indian Institute of Technology, Kharagpur, 2003.

❑❑❑

17

GEOMETRICAL NON-LINEAR FINITE ELEMENT ANALYSIS

17.1 INTRODUCTION

All the chapters of the book have so far dealt with problems which are linear. Non-linearity in structural mechanics may be classified under the following two heads:

(*a*) Material non-linearity

(*b*) Geometrical non-linearity

Material non-linearity arises in problems where the stress is not linearly proportional to strain. This results due to the presence of plasticity, creep or other complex relations. For this category of problems, small strains and small displacements are considered, which signifies that the geometry of the elements remains basically unchanged during the loading process.

Geometric non-linearity arises from the consideration of the non-linear strain-displacement relationship. It encompasses large strains and large displacements. For example, in the linear analysis of a plate, the middle surface is assumed to remain unstretched. If the membrane action is considered in the plate flexure, it will result in the non-linear strain-displacement relationship. In many cases, large displacements may occur without causing large strains. The elastic postbuckling behaviour of the structure comes under this category.

Non-linear problems may be encountered which involve both the material non-linearity and geometric non-linearity. It will thus involve non-linear constitutive relationship as well as non-linear strain-displacement relation.

17.2 NON-LINEAR EQUATION SOLVING PROCEDURES

A set of non-linear equations in the finite element analysis can be written as

$$\{\psi\}\{X\} = [K_S\{X\}]\{X\} - \{\overline{P}\} = \{0\} \tag{17.1}$$

A few basic techniques exist for obtaining the solution of non-linear equations. Fundamentals of some of them are discussed here.

17.2.1 Direct Iteration Method

Figure 17.1 illustrates the direct iteration method of one variable. For a multivariable problem, the same concept can be extended. In this direct iteration procedure, one starts with an initial value $\{X\} = \{X\}^0$.

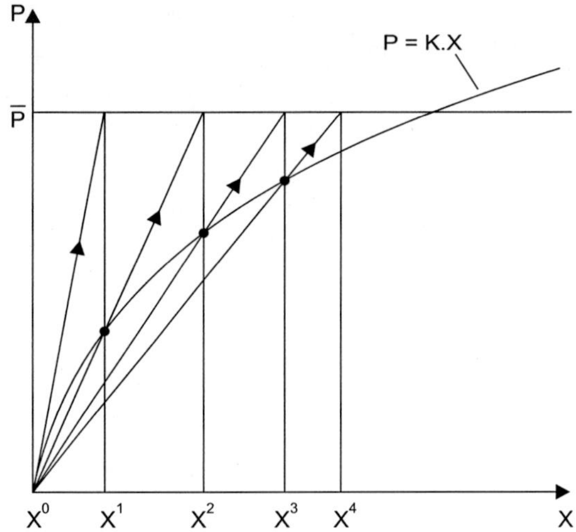

Fig. 17.1. Direct iteration technique

The first approximation of $\{X\}$ is thus

$$\{X\}^1 = [K_s^0]^{-1}\{\overline{P}\} \tag{17.2}$$

where $\quad [K_s^0] = [K\{X\}]$

Next, $[K_S\{X\}]$ is modified according to $\{X\}^1$. The second approximation of $\{X\}$ is

$$\{X\}^2 = [K_S^1]^{-1}\{\overline{P}\} \tag{17.3}$$

The process is repeated and after nth iteration, it is

$$\{X\}^n = ([K]_s^{n-1})^{-1}\ \{\overline{P}\} \tag{17.4}$$

The iteration is continued till the values of the two successive iterations are quite close. In order to list the closeness, various norms and convergence criteria are used. One such is given below.

$$\{e\} = \{X\}^n - \{X\}^{n-1} \tag{17.5}$$

$$|e| = \max e_i \text{ (maximum value)}$$

or

$$|e| = \sqrt{\{e\}^T\{e\}} \tag{17.6}$$

Convergence is attained when

$$|e| = \leq \alpha$$

where α is the tolerance of error.

While using direct iteration procedure, one may have to be careful on the possibility of the divergence of the results. One of the disadvantages of the method is that the direct iteration procedure is to be used for every load level and the results of a particular load level cannot be used for next load level.

17.2.2 Newton - Raphson Method

In order to solve eqn. (17.1), suppose an approximate solution of the following form has been reached.

$$\{X\} = \{X\}^n \tag{17.7}$$

The function $\{\psi\}(\{X\})$ is expressed in terms of Talyor's series and higher order terms are ignored to get the improved solution as

$$\{\psi\}(\{X\})^{n+1} = \{\psi\}(\{X\}^n) + \left(\frac{d\{\psi\}}{d\{X\}}\right)_n \{\Delta X\}^n = 0 \tag{17.8}$$

where $$\{X\}^{n+1} = \{X\}^n + \{\Delta X\}^n \tag{17.9}$$

In the above equation

$$\frac{d\{\psi\}}{d\{X\}} = [K(\{X\})]_T \tag{17.10}$$

$[K(\{X\})]_T$ represents the tangent stiffness matrix.

Improved values of $\{X\}^{n+1}$ is obtained from eqn. (17.9) by calculating

$$\{\Delta X\}^{n+1} = ([K]_T^n)^{-1} (\{P\} - [K]_S^n \{X\}^n) \tag{17.11}$$

Figure 17.2 illustrates the Newton - Raphson method for a single variable. For every incremental solution, a new set of linearised equation is required to be solved.

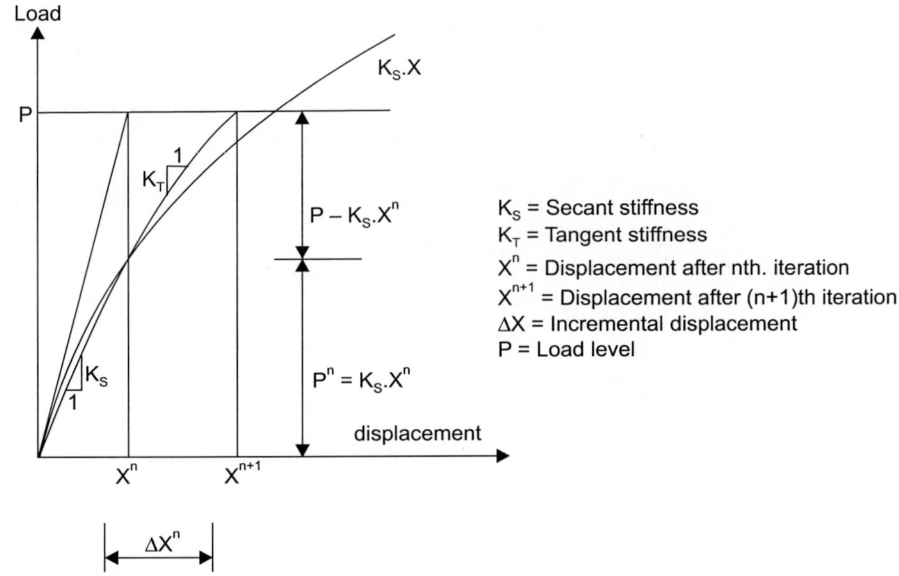

Fig. 17.2. Newton-Raphson method

17.2.3 Modified Newton - Raphson method

In Newton - Raphson method, eqn. (17.11) is to be solved at each iteration stage which is time consuming. So the method is modified to reduce the computational effort. An approximation given below is introduced

$$[K]_T^n = [K]^0 \quad (17.12)$$

The modified algorithm then is

$$([K]^0)^{-1} (\{P\} - [K]_S^n \{X\}^n) = \{\Delta X\}^{n+1} \quad (17.13)$$

The solution procedure thus obtained will save the cost of solution but the rate of convergence will now be slow. Figure 17.3 gives an illustration of the procedure. For certain problems, if the process is found to be slow, then $[K]_T$ may be updated after some iterations.

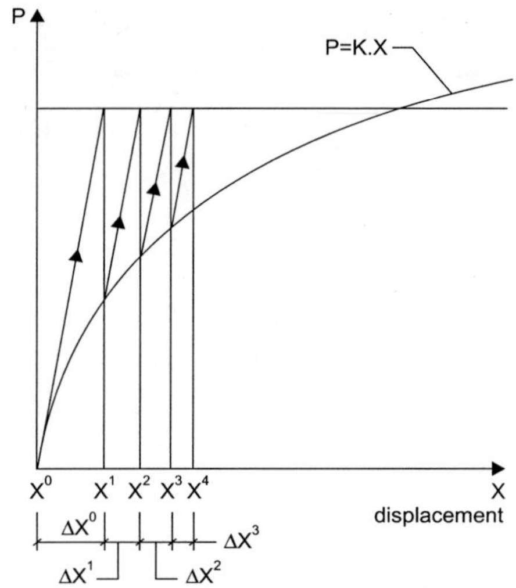

Fig. 17.3. Modified Newton-Raphson method

17.2.4 Incremental Techniques

The solution of non-linear equations can be carried out by treating the non-linearity as piecewise linear. In this technique, the load is divided into a number of small increments which may be equal or unequal. One proceeds from one load level to its next increment and the solution for this increment is assumed to be linear. Solution of displacements is thus obtained for each increment of load. In order to obtain the total displacement at any stage of loading, all these displacement increments are to be added.

Let the initial load and displacement given by $\{P\}^0$ and $\{X\}^0$ for an undeformed body, are null vectors. If the total load is divided into M increments, then

$$\{P\} = \{P\}^0 + \sum_{i=1}^{M} \{\Delta P\}^i \quad (17.14)$$

After applying n th increment, the load is

$$\{P\}^n = \{P\}^0 + \sum_{i=1}^{n} \{\Delta P\}^i \quad (17.15)$$

Similarly, the displacements after nth increment are

$$\{X\}^n = \{X\}^0 + \sum_{i=1}^{n} \{\Delta X\}^i \quad (17.16)$$

The increments of displacements are to be evaluated by considering the stiffness matrix at the end of the previous increment

$$[K(\{X\}^{i-1})]_T^{i-1} \{\Delta X\}^i = \{\Delta P\}^i \qquad \text{for } i = 1, 2, 3, \ldots, M \quad (17.17)$$

It is to be noted that the tangent stiffness matrix is used in the formulation of eqn. (17.17). The incremental technique is explained in Fig. 17.4

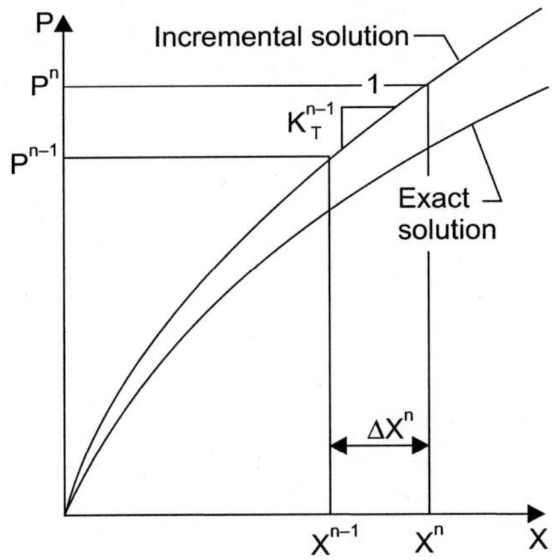

Fig. 17.4. Incremental technique

17.3 FORMULATION OF THE GEOMETRIC NON-LINEAR PROBLEM

There are two approaches used for the formulation of the geometric non-linear problem – the Eulerian approach and the Lagrangian approach. In the former, the formulation is done on the basis of changed geometry at each stage. In the Lagrangian method, the displacements are referred to the original configuration. The treatment here is limited to the Lagrangian approach only. In this context two different notations are quite common. Mallet and Marcal [17.1] [17.2] have formulated the non-linear problem on the basis of N-notation. The notations of Zienkiewicz and his coworkers [17.3] are referred as *B*-notation. It can be shown that final matrices obtained by both notations are identical.

17.3.1 Equilibrium Equations

The equilibrium equations may be obtained by application of virtual principles as

$$\{\psi\} = \int_v [B]^T \{\sigma\} dV - \{P\}_e = \{0\} \tag{17.18}$$

where $\{\psi\}$ denotes the sum of external and internal generalised forces and $[B]$ is obtained from the following relationship.

$$d\{\varepsilon\} = [B]\, d\{X\}_e \tag{17.19}$$

Again matrix $[B]$ may be expressed as

$$[B] = [B]_0 + [B(\{X\})]_L \tag{17.20}$$

where $[B]_0$ is the linear part and is same as that derived in the earlier chapters. $[B]_L$ is the non-linear part and is dependent on displacements.

If the strains are reasonably small, we can still write the elastic relationship

$$\{\sigma\} = \{D\}\{\varepsilon\} \tag{17.21}$$

where
$$\{\varepsilon\} = \left([B]_0 + \frac{1}{2}[B]_L\right)\{X\}_e \tag{17.22}$$

It will be explained later how this is obtained.

With the help of eqns. (17.20) – (17.22), eqn. (17.18) becomes

$$\int_v ([B]_0 + [B]_L)^T [D]\left([B]_0 + \frac{1}{2}[B]_L\right) dV\{X\}_e - \{P\}_e = \{0\}$$

or,
$$[K]_S \{X\}_e = \{P\}_e = \{0\} \tag{17.23}$$

where $[K]_S$ is the secant stiffness matrix and is given by

$$[K]_S = \int_V ([B]_0 + [B]_L)^T [D]\left([B]_0 + \frac{1}{2}[B]_L\right) dV \tag{17.24}$$

Equation (17.24) when expanded becomes

$$[K]_S = \int_V [B]_0^T [D][B]_0 dV + \frac{1}{2}\int_V [B]_0^T [D][B]_L dV$$

$$+\int_V [B]_L^T [D][B]_0 dV + \frac{1}{2}\int_V [B]_L^T [D][B]_L dV \tag{17.25}$$

Equation (17.25) can be written as

$$[K]_S = [K]_0 + [K]_{LS} \tag{17.26}$$

where, $[K]_0$ = linear part of the stiffness matrix

$$= \int_V [B]_0^T [D][B]_0 dV$$

and $[K]_{LS}$ = non-linear part of the secant stiffness matrix

$$= \frac{1}{2}\int_v [B]_0^T [D][B]_L dV + \int_V [B]_L^T [D][B]_0 dV + \frac{1}{2}\int_V [B]_L^T [D][B]_L dV \tag{17.27}$$

17.3.2 Incremental Equilibrium Equation

The tangent stiffness matrix given by eqn. (17.11) may be obtained from the incremental equilibrium equations. Taking appropriate variation of eqn. (17.18) with respect to $\{X\}$this may be obtained as

$$d\{\psi\} = \int_V d[B]^T \{\sigma\} dV + \int_V [B]^T d\{\sigma\} dV - d\{P\}_e = [K]_T d\{X\}_e - d\{P\}_e \tag{17.28}$$

Substituting eqn. (17.20) and eqn. (17.22) into eqn. (17.28), it yields to

$$\int_V d([B]_0 + [B]_L)^T \{\sigma\} dV + \int_V ([B]_0 + [B]_L)^T \{D\}([B]_0 + [B]_L)\, dV\, d\{X\}_e + d\{P\}_e = \{0\}$$

or, $$[K]_\sigma\, d\{X\}_e + ([K]_0 + [K]_{LT})\, d\{X\}_e = d\{P\}_e \tag{17.29}$$

where

$$[K]_0 = \int_v [B]_0^T [D][B]_0\, dV \tag{17.30a}$$

$$[K]_{LT} = \int_v [B]_0^T [D][B]_L\, dV + \int_v [B]_L^T [D][B]_0\, dV + \int_v [B]_L^T [D][B]_L\, dV \tag{17.30b}$$

and $$[K]_\sigma\, d\{X\}_e = \int_v d[B]_L^T \{\sigma\}\, dV \tag{17.30c}$$

It will be seen that $[K]_\sigma$ is a symmetric matrix dependent on the stress level. This matrix is referred to as the initial stress matrix or the geometric matrix.

Thus, the tangent stiffness matrix consists of three parts. They are

$$[K]_T = [K]_0 + [K]_{LT} + [K]_\sigma \tag{17.31}$$

17.4 LARGE DEFLECTION ANALYSIS OF PLATES IN B-NOTATION

Figure 17.5 indicates the general pattern of variation of deflection at a point in the plate with the load. It may be noted that the non-linear deflection in plate is always less than linear deflection. In such cases lateral displacements are coupled. Taking the middle plane of the plate as the reference plane for the analysis, strain matrix can be given by

$$\{\varepsilon\} = \begin{Bmatrix} \dfrac{\partial u}{\partial x} \\ \dfrac{\partial v}{\partial y} \\ \dfrac{\partial u}{\partial y} + \dfrac{\partial v}{\partial x} \\ --- \\ -\dfrac{\partial^2 w}{\partial x^2} \\ -\dfrac{\partial^2 w}{\partial y^2} \\ -\dfrac{2\partial^2 w}{\partial x \partial y} \end{Bmatrix} + \begin{Bmatrix} \dfrac{1}{2}\left(\dfrac{\partial w}{\partial x}\right)^2 \\ \dfrac{1}{2}\left(\dfrac{\partial w}{\partial y}\right)^2 \\ \left(\dfrac{\partial w}{\partial x}\right)\left(\dfrac{\partial w}{\partial y}\right) \\ --- \\ 0 \\ 0 \\ 0 \end{Bmatrix} \tag{17.32}$$

or, $$\{\varepsilon\} = \begin{Bmatrix} \{\varepsilon^p\} \\ \hline \{\varepsilon^b\} \end{Bmatrix} \tag{17.33}$$

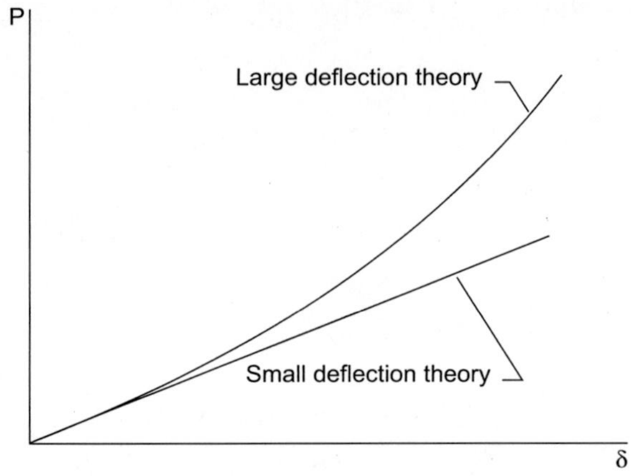

Fig. 17.5. Load-deflection curve of a plate

Similarly, $\{\varepsilon\}$, $\{\sigma\}$ and $[D]$ can be splitted into two parts: one is related to inplane action and other is related to the bending action.
Therefore, the stress – resultant vector is

$$\{\sigma\} = \begin{Bmatrix} \{\sigma^p\} \\ \{\sigma^b\} \end{Bmatrix} \tag{17.34}$$

where

$$\{\sigma^p\} = \begin{Bmatrix} N_x \\ N_y \\ N_{xy} \end{Bmatrix}, \quad \{\sigma^b\} = \begin{Bmatrix} M_x \\ M_y \\ M_{xy} \end{Bmatrix}$$

And the rigidity matrix is

$$[D] = \begin{bmatrix} [D^p] & [0] \\ [0] & [D^b] \end{bmatrix} \tag{17.36}$$

For the isotropic plate problem, the submatrices in the above equation are

$$[D^p] = \frac{Et}{1-\nu^2}\begin{bmatrix} 1 & \nu & 0 \\ \nu & 1 & 0 \\ 0 & 0 & \frac{1-\nu}{2} \end{bmatrix} \tag{17.37a}$$

$$[D^b] = \frac{Et^3}{12(1-\nu^2)}\begin{bmatrix} 1 & \nu & 0 \\ \nu & 1 & 0 \\ 0 & 0 & \frac{1-\nu}{2} \end{bmatrix} \tag{17.37b}$$

From eqns. (17.32) and (17.33), one obtains

$$[\varepsilon^p] = [\varepsilon^p]_0 + [\varepsilon^p]_L \tag{17.38}$$

where,

$$[\varepsilon^p]_0 = \begin{Bmatrix} \dfrac{\partial u}{\partial x} \\ \dfrac{\partial v}{\partial y} \\ \dfrac{\partial u}{\partial y} + \dfrac{\partial v}{\partial x} \end{Bmatrix} \tag{17.39a}$$

$$[\varepsilon^p]_L = \begin{Bmatrix} \dfrac{1}{2}\left(\dfrac{\partial w}{\partial x}\right)^2 \\ \dfrac{1}{2}\left(\dfrac{\partial w}{\partial y}\right)^2 \\ \dfrac{\partial w}{\partial x}\cdot\dfrac{\partial w}{\partial y} \end{Bmatrix} \tag{17.39b}$$

Equation. (17.30) can be written as

$$[\varepsilon^p]_L = \frac{1}{2}\begin{bmatrix} \dfrac{\partial w}{\partial x} & 0 \\ 0 & \dfrac{\partial w}{\partial y} \\ \dfrac{\partial w}{\partial y} & \dfrac{\partial w}{\partial x} \end{bmatrix} \begin{Bmatrix} \dfrac{\partial w}{\partial x} \\ \dfrac{\partial w}{\partial y} \end{Bmatrix} = \frac{1}{2}[A]\{\theta\} \tag{17.40}$$

where,

$$[A] = \begin{bmatrix} \dfrac{\partial w}{\partial x} & 0 \\ 0 & \dfrac{\partial w}{\partial y} \\ \dfrac{\partial w}{\partial y} & \dfrac{\partial w}{\partial x} \end{bmatrix} \tag{17.41}$$

$$\{\theta\} = \begin{Bmatrix} \dfrac{\partial w}{\partial x} \\ \dfrac{\partial w}{\partial y} \end{Bmatrix} \tag{17.42}$$

The deflection w of the plate can be expressed as

$$w = [N]^b \{X\}_e \tag{17.43}$$

where $[N]^b$ is the shape function matrix for w.

Equation (17.42) can thus be written with the help of eqn. (17.43) as:

$$\{\theta\} = \begin{Bmatrix} \dfrac{\partial}{\partial x}[N]^b\{X\}_e \\ \dfrac{\partial}{\partial y}[N]^b\{X\}_e \end{Bmatrix} = \begin{bmatrix} \dfrac{\partial}{\partial x}[N]^b \\ \dfrac{\partial}{\partial y}[N]^b \end{bmatrix}\{X\}_e = [G]\{X\}_e \tag{17.44}$$

$$[G] = \begin{bmatrix} \dfrac{\partial}{\partial x}[N]^b \\ \dfrac{\partial}{\partial y}[N]^b \end{bmatrix} \tag{17.45}$$

Combining eqns. (17.40) and (17.44), yields

$$\{\varepsilon^p\}_L = = \frac{1}{2}[A][G]\{X\}_e \tag{17.46}$$

From eqns. (17.33) and (17.38), the strain vector can be written as

$$\{\varepsilon\} = \begin{Bmatrix} \{\varepsilon^p\}_0 \\ \{\varepsilon^b\}_0 \end{Bmatrix} + \begin{Bmatrix} \{\varepsilon^p\}_L \\ \{0\} \end{Bmatrix} = \{\varepsilon\}_0 + \{\varepsilon\}_L \tag{17.47}$$

It has been shown a number of times in the previous chapters that

$$\{\varepsilon\}_0 = [B]_0\,\{X\}_e \tag{17.48}$$

Further,

$$\{\varepsilon\}_L = \begin{Bmatrix} \dfrac{1}{2}[A][G] \\ \{0\} \end{Bmatrix}\{X\}_e = \frac{1}{2}[B]_L\{X\}_e \tag{17.49}$$

Therefore,

$$\{\varepsilon\} = \left([B]_0 + \frac{1}{2}[B]_L\right)\{X\}_e \tag{17.50}$$

This is directly used in eqn. (17.22).

The variation of the strain can be written as

$$d\{\varepsilon\} = d\{\varepsilon_0\} + d\{\varepsilon\}_L = [B]_0 d\{X\}_e + \begin{Bmatrix} d\{\varepsilon^p\}_L \\ \{0\} \end{Bmatrix} \tag{17.51}$$

Taking the variation of eqn. (17.40), yields

$$d\{\varepsilon^p\}_L = d\left(\frac{1}{2}[A]\{\theta\}\right) = \frac{1}{2}d[A]\{\theta\} + \frac{1}{2}[A]d\{\theta\} \tag{17.52}$$

We may note that

$$d[A]\{Y\} = \begin{bmatrix} d\left(\dfrac{\partial w}{\partial x}\right) & 0 \\ 0 & d\left(\dfrac{\partial w}{\partial y}\right) \\ d\left(\dfrac{\partial w}{\partial y}\right) & d\left(\dfrac{\partial w}{\partial x}\right) \end{bmatrix} \begin{Bmatrix} y_1 \\ y_2 \end{Bmatrix}$$

$$= \begin{bmatrix} y_1 & 0 \\ 0 & y_2 \\ y_2 & y_1 \end{bmatrix} d \begin{Bmatrix} \dfrac{\partial w}{\partial x} \\ \dfrac{\partial w}{\partial y} \end{Bmatrix} \tag{17.53}$$

It gives

$$d[A]\{\theta\} = [A]\, d\{\theta\} \tag{17.54}$$

Combining eqns. (17.52) and (17.54) yields

$$d\{\varepsilon^p\}_L = [A]\, d\{\theta\} = [A][G]\, d\{X\}_e \tag{17.55}$$

Hence, (17.56)

$$[B]_L^p = [A]\,[G] \tag{17.56}$$

and

$$[B]_L = \begin{bmatrix} [A][G] \\ [0] \end{bmatrix} \tag{17.57}$$

Combining eqns. (17.51), (17.55), (17.56) and (17.57) yields

$$d\{\varepsilon\} = ([B]_0 + [B]_L)\, d\{X\}_e \tag{17.58}$$

With $[B]_0$ and $[B]_L$, the stiffness matrix for the plate element is given by eqn. (17.25). It should be noted that the volume integral will be replaced by area integral, as the contribution across the thickness direction is taken into account in the rigidity matrix (eqn. (17.37)).

Similarly, the stiffness matrix can be obtained from eqn. (17.29). The derivation of the geometric matrix $[K]_\sigma$ is presented below.

Equation (17.30), is reproduced.

$$[K]_\sigma\, d\{X\}_e = \int_v d[B]_L^T\,\{\sigma\} dv$$

Using eqn. (17.34), (17.35), (17.41) and (17.57), the above equation becomes

$$[K]_\sigma\, d\{X\}_e = \int_A [G]^T \begin{bmatrix} d\dfrac{\partial w}{\partial x} & 0 & d\dfrac{\partial w}{\partial y} \\ 0 & d\dfrac{\partial w}{\partial y} & d\dfrac{\partial w}{\partial x} \end{bmatrix} \begin{Bmatrix} N_x \\ N_y \\ N_{xy} \end{Bmatrix} dA$$

$$= \int_A [G]^T \begin{Bmatrix} d\dfrac{\partial w}{\partial x} N_x + d\dfrac{\partial w}{\partial y} N_{xy} \\ d\dfrac{\partial w}{\partial y} N_y + d\dfrac{\partial w}{\partial x} N_{xy} \end{Bmatrix} dA$$

$$= \int_A [G]^T \begin{bmatrix} N_x & N_{xy} \\ N_{xy} & N_y \end{bmatrix} \begin{Bmatrix} d\dfrac{\partial w}{\partial x} \\ d\dfrac{\partial w}{\partial y} \end{Bmatrix} dA$$

$$= \int_A [G]^T \begin{bmatrix} N_x & N_{xy} \\ N_{xy} & N_y \end{bmatrix} d \begin{Bmatrix} \dfrac{\partial w}{\partial x} \\ \dfrac{\partial w}{\partial y} \end{Bmatrix} dA \tag{17.59}$$

Using eqn. (17.42), eqn. (17.59) becomes

$$[K]_\sigma \, d\{X\}_e = \int_A [G]^T [S] \, d\{\theta\} \, dA \tag{17.60}$$

where

$$[S] = \begin{bmatrix} N_x & N_{xy} \\ N_{xy} & N_y \end{bmatrix} \tag{17.61}$$

Replacing $\{\theta\}$ from eqn. (17.44) with eqn. (17.60), yields

$$[K]_\sigma \, d\{X\}_e = \int_A [G]^T [S][G] dA \, d\{X\}_e \tag{17.62}$$

Therefore,

$$[K]_\sigma = \int_A [G]^T [S] [G] \, dA \tag{17.63}$$

17.5 LARGE DEFLECTION ANALYSIS OF PLATES IN N-NOTATION

It can be shown that [17.1, 17.2] the equilibrium equations can be written as

$$\left([N_0] + \frac{1}{2}[N_1] + \frac{1}{3}[N_2] \right) \{X\}_e = \{P\}_e \tag{17.64}$$

where,

$$[N_0] = \int_A [B]_0^T [D][B]_0 \, dA \tag{17.65a}$$

$$[N_1] = \int_A ([B]_0^T [D][B]_L + [B]_L^T [D][B]_0 + [G]^T [S]_0 [G]) \, dA \tag{17.65b}$$

$$[N_2] = \int_A ([B]_L^T [D][B]_L + [G]^T [S]_L [G])\, dA \tag{17.65c}$$

The matrices $[S]_0$ and $[S]_L$ are defined later.

In the above equations, $[N_0]$ is independent of displacement $\{X\}_e$, and $[N_1]$ is linearly dependent on $\{X\}_e$ and $[N_2]$ is quadratically dependent upon $\{X\}_e$. With respect to B-notation, the secant stiffness matrix eqn. (17.25) results in an unsymmetric matrix while the secant stiffness matrix given by eqn. (17.64) in N-notation results in a symmetric matrix. It can be shown that eqn. (17.25) and (17.64) are identical.

Substituting the value from eqns. (17.65a), (17.65b), (17.65c) into eqn. (17.64), yields

$$\begin{aligned}[K]_s = {} & \int_A [B]_0^T [D][B]_0\, dA + \frac{1}{2}\int_A [B]_0^T [D][B]_L\, dA \\ & + \frac{1}{2}\int_A [B]_L^T [D][B]_0\, dA + \frac{1}{3}\int_A [B]_L^T [D][B]_L\, dA \\ & + \frac{1}{2}\int_A [G]^T [S]_0 [G]\, dA + \frac{1}{3}\int_A [G]^T [S]_0 [G]\, dA \end{aligned} \tag{17.66}$$

Explicitly,

$$[S] = [S]_0 + [S]_L \tag{17.67}$$

where $[S]$ is given by eqn. (17.61). $[S]_0$ corresponds to the linear part of the strain $\{\varepsilon^p\}_0$ (*see* eqn. (17.39a)) and $[S]_L$ corresponds to non-linear part of strain $\{\varepsilon^p\}_L$ (eqn. (17.39b)).

Thus,

$$[S]_0 = \begin{bmatrix} N_x^0 & N_{xy}^0 \\ N_{xy}^0 & N_y^0 \end{bmatrix}, \qquad [S]_L = \begin{bmatrix} N_x^L & N_{xy}^L \\ N_{xy}^L & N_y^L \end{bmatrix} \tag{17.68}$$

where $$\begin{Bmatrix} N_x^o \\ N_y^0 \\ N_{xy}^0 \end{Bmatrix} = [D^p]\{\varepsilon^p\}_0, \qquad \begin{Bmatrix} N_x^L \\ N_y^L \\ N_{xy}^L \end{Bmatrix} = [D^p]\{\varepsilon^p\}_L \tag{17.69}$$

Now, taking the fifth integral of eqn. (17.66) it may be processed with the help of eqn. (17.44) as

$$\int_A [G]^T [S]_0\, [G]\, dA\, \{X\}_e = \int_A [G]^T [S]_0\, \{\theta\}\, dA \tag{17.70}$$

Using eqn. (17.42) and (17.68), the equation becomes

$$\int_A [G]^T [S]_0 [G]\, dA \{X\}_e = \int_A [G]^T \begin{bmatrix} N_x^0 & N_{xy}^0 \\ N_{xy}^0 & N_y^0 \end{bmatrix} \begin{Bmatrix} \dfrac{\partial w}{\partial x} \\ \dfrac{\partial w}{\partial y} \end{Bmatrix} dA$$

$$= \int_A [G]^T \begin{Bmatrix} N_x^0 \dfrac{\partial w}{\partial x} + N_{xy}^0 \dfrac{\partial w}{\partial y} \\ N_{xy}^0 \dfrac{\partial w}{\partial x} + N_y^0 \dfrac{\partial w}{\partial y} \end{Bmatrix} dA = \int_A [G]^T \begin{bmatrix} \dfrac{\partial w}{\partial x} & 0 & \dfrac{\partial w}{\partial y} \\ 0 & \dfrac{\partial w}{\partial y} & \dfrac{\partial w}{\partial x} \end{bmatrix} \begin{Bmatrix} N_x^o \\ N_y^0 \\ N_{xy}^0 \end{Bmatrix} dA \quad (17.71)$$

With the help of eqns. (17.35), (17.37a), (17.39) and (17.41), the above equation becomes

$$\int_A [G]^T [S]_0 [G] dA \{X\}_e = \int_V [G]^T [A]^T [D^p] \{\varepsilon^p\}_0 \, dV$$

$$= \int_A \left[([A][G])^T [0] \right] \begin{bmatrix} [D^p] & [0] \\ [0] & [D^b] \end{bmatrix} \begin{Bmatrix} \{\varepsilon^p\}_0 \\ \{\varepsilon^b\}_0 \end{Bmatrix} dA \quad (17.72)$$

Combining eqns. (17.31), (17.47), (17.48), (17.57) and (17.72) yields

$$\int_V [G]^T [S]_0 [G] \, dV \{X\}_e = \int_A [B]_L^T [D][B]_0 \, dA \{X\}_e$$

Following the steps as above, the last integral of eqn. (17.66) may be processed as

$$\int_A [G]^T [S]_L [G] dA \{X\}_e = \int_A [G]^T [A]^T [D^p] \{\varepsilon^p\}_L \, dA$$

$$= \int_A \left(([A][G])^T [0] \right) \begin{bmatrix} [D^p] & [0] \\ [0] & [D^b] \end{bmatrix} \begin{Bmatrix} \{\varepsilon^p\}_L \\ \{0\} \end{Bmatrix} dA$$

$$= \frac{1}{2} \int_A [B]_L^T [D][B]_L \, dA \{X\}_e \quad (17.73)$$

Substituting the appropriate terms from eqns. (17.72) and (17.73) into the secant stiffness matrix expressions given by eqn. (17.60), yields

$$[K]_S \{X\}_e = \int_V [B]_0^T [D][B]_0 \, dV \{X\}_e + \frac{1}{2} \int_v [B]_0^T [D][B]_L \, dV \{X\}_e$$

$$+ \frac{1}{2} \int_v [B]_L^T [D][B]_0 \, dV \{X\}_e + \frac{1}{3} \int_v [B]_L^T [D][B]_L \, dV \{X\}_e$$

$$+ \frac{1}{2} \int_v [B]_L^T [D][B]_0 \, dV \{X\}_e + \frac{1}{6} \int_v [B]_L^T [D][B]_L \, dV \{X\}_e$$

or

$$[K]_S \{X\}_e = \int_v [B]_0^T [D][B]_0 \, dV \{X\}_e + \frac{1}{2} \int_v [B]_0^T [D][B]_L \, dV \{X\}_e$$

$$+ \int_v [B]_L^T [D][B]_0 \, dV \{X\}_e + \frac{1}{2} \int_v [B]_L^T [D][B]_L \, dV \{X\}_e \quad (17.74)$$

Equation (17.74) is identical to eqn. (17.25), the secant stiffness matrix derived in B-notation.

17.6 EXAMPLE OF A PIN-JOINTED BAR

Figure 17.6 presents a pin-jointed bar. The derivation is presented for various matrices required for carrying out large deflection analysis of this bar [17.4]. The derivation has been carried out in the local *x-y* coordinate system. Nodal displacements are (u_1, v_1) and (u_2, v_2).

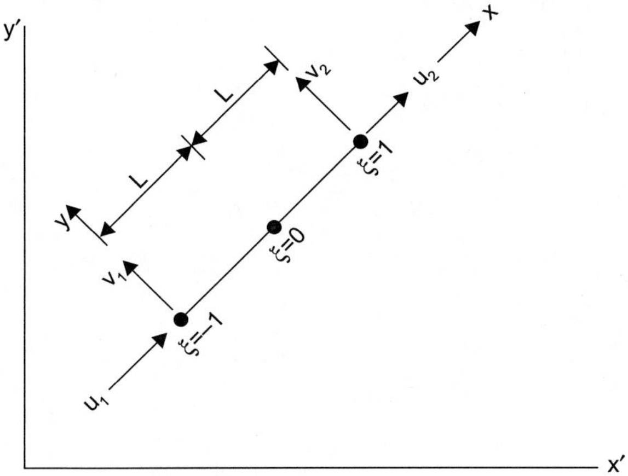

Fig. 17.6. A pin-jointed bar

The displacement within an element may be expressed in the terms of nodal displacements as

$$\begin{Bmatrix} u \\ v \end{Bmatrix} = \begin{bmatrix} N_1 & 0 & N_2 & 0 \\ 0 & N_1 & 0 & N_2 \end{bmatrix} \begin{Bmatrix} u_1 \\ v_1 \\ u_2 \\ v_2 \end{Bmatrix} \tag{17.75}$$

where

$$N_1 = \frac{1}{2}(1-\xi),\ N_2 = \frac{1}{2}(1+\xi) \text{ and } \xi = \frac{x}{L} \tag{17.76}$$

The nodal displacement matrix is

$$\{X\}_e^T = \{u_1 \quad v_1 \quad u_2 \quad v_2\} \tag{17.77}$$

From the above equation, we get

$$\frac{du}{dx} = \frac{1}{L}.\frac{du}{d\xi},\ \frac{dv}{dx} = \frac{1}{L}.\frac{dv}{d\xi},\ dv = a.L.d\xi \tag{17.78}$$

where *a* is the cross-sectional area of the bar.

For this problem, the strain is

$$\varepsilon_x = \frac{du}{dx} + \frac{1}{2}\left[\left(\frac{du}{dx}\right)^2 + \left(\frac{dv}{dx}\right)^2\right] \tag{17.79}$$

or, $$\{\varepsilon\} = \{\varepsilon\}_0 + \{\varepsilon\}_L \tag{17.80a}$$

where $$\{\varepsilon\}_0 = \frac{du}{dx}, \quad \{\varepsilon\}_L = \frac{1}{2}\left[\left(\frac{du}{dx}\right)^2 + \left(\frac{dv}{dx}\right)^2\right] \tag{17.80b}$$

The constitutive relation for the bar is

$$\{\sigma\} = [D]\{\varepsilon\} \tag{17.81a}$$

or,

$$\{\{S\}_0 \{S\}_L\} = [D]\{\{\varepsilon\}_0 \{\varepsilon\}_L\} \tag{17.81b}$$

where, $$[D] = E \tag{17.81c}$$

From eqns. (17.75), (17.76) and (17.78), we get

$$\frac{du}{dx} = \frac{dN_1}{dx}u_1 + \frac{dN_2}{dx}u_2 = \frac{1}{2L}(-u_1 + u_2) = \frac{u_2 - u_1}{2L} \tag{17.82}$$

Therefore, $$\{\varepsilon\}_0 = \frac{du}{dx} = \frac{\bar{u}}{l_0} \tag{17.83}$$

where $$\bar{u} = u_2 - u_1, \; l_0 = 2L$$

Similarly, $$\frac{dv}{dx} = \frac{\bar{v}}{l_0} \tag{17.84}$$

where $$\bar{v} = v_2 - v_1$$

Thus, the non-linear strain is

$$\{\varepsilon\}_L = \frac{1}{2}\left[\left(\frac{\bar{u}}{l_0}\right)^2 \left(\frac{\bar{v}}{l_0}\right)^2\right] \tag{17.85}$$

Combining eqns. (17.81), (17.83) and (17.85), yields

$$\{\{S\}_0 \{S\}_L\} = E\left\{\frac{\bar{u}}{l_0} \; \frac{1}{2}\left[\left(\frac{\bar{u}}{l_0}\right)^2 + \left(\frac{\bar{v}}{l_0}\right)^2\right]\right\} \tag{17.86}$$

where, $$\{S\}_0 = E\left(\frac{\bar{u}}{l_0}\right) \quad \text{and} \quad \{S\}_L = \frac{E}{2}\left[\left(\frac{\bar{u}}{l_0}\right)^2 + \left(\frac{\bar{v}}{l_0}\right)^2\right] \tag{17.87}$$

Now, the non-linear strain given by eqn. (17.80) may be expressed as

$$\{\varepsilon\}_L = \frac{1}{2}\begin{bmatrix}\frac{\partial u}{\partial x} & \frac{\partial v}{\partial x}\end{bmatrix}\begin{Bmatrix}\frac{\partial u}{\partial x} \\ \frac{\partial v}{\partial x}\end{Bmatrix} = [A]\{\theta\}$$

Thus,

$$[A] = \begin{bmatrix}\frac{\partial u}{\partial x} & \frac{\partial v}{\partial x}\end{bmatrix} \tag{17.88}$$

And $\{\theta\}$ may be expressed in terms of nodal displacement vector $\{X\}_e$ with the help of eqn. (17.75) as

$$[\theta] = \begin{Bmatrix} \dfrac{\partial u}{\partial x} \\ \dfrac{\partial v}{\partial x} \end{Bmatrix} = \begin{Bmatrix} \dfrac{\partial [N]}{\partial x} \\ \dfrac{\partial [N]}{\partial x} \end{Bmatrix} \{X\}_e = [G]\{X\}_e \tag{17.89}$$

where
$$[G] = \begin{bmatrix} \dfrac{dN_1}{dx} & 0 & \dfrac{dN_2}{dx} & 0 \\ 0 & \dfrac{dN_1}{dx} & 0 & \dfrac{dN_2}{dx} \end{bmatrix}$$

From the eqn. (17.82), the linear strain matrix may be written as

$$[B]_0 = \begin{bmatrix} \dfrac{\partial N_1}{\partial x} & 0 & \dfrac{\partial N_2}{\partial x} & 0 \end{bmatrix} = \begin{bmatrix} -\dfrac{1}{l_0} & 0 & \dfrac{1}{l_0} & 0 \end{bmatrix} \tag{17.90}$$

Now, the non-linear strain matrix may be obtained with the help of eqn. (17.76), (17.83), (17.84), (17.88) and (17.79) as

$$[B]_L = \begin{bmatrix} \dfrac{\partial u}{\partial x} & \dfrac{\partial v}{\partial x} \end{bmatrix} \begin{bmatrix} \dfrac{\partial N_1}{\partial x} & 0 & \dfrac{\partial N_2}{\partial x} & 0 \\ 0 & \dfrac{\partial N_1}{\partial x} & 0 & \dfrac{\partial N_2}{\partial x} \end{bmatrix}$$

$$= \begin{bmatrix} \dfrac{\bar{u}}{l_0} & \dfrac{\bar{v}}{l_0} \end{bmatrix} \begin{bmatrix} -\dfrac{1}{l_0} & 0 & \dfrac{1}{l_0} & 0 \\ 0 & -\dfrac{1}{l_0} & 0 & \dfrac{1}{l_0} \end{bmatrix}$$

$$= \frac{1}{l_0^2}\begin{bmatrix} -\bar{u} & -\bar{v} & \bar{u} & \bar{v} \end{bmatrix} \tag{17.91}$$

Substituting eqns. (17.81c) and (17.90) with eqn. (17.30a) gives

$$[K]_0 = \int_V [B]_0^T [D] [B]_o \, dV$$

$$= aL\int_{-1}^{1} \frac{1}{l_0} \begin{bmatrix} -1 \\ 0 \\ 1 \\ 0 \end{bmatrix} E\frac{1}{l_0}\{-1 \quad 0 \quad 1 \quad 0\} d\xi$$

$$= \frac{Ea}{l_0} \begin{bmatrix} 1 & 0 & -1 & 0 \\ 0 & 0 & 0 & 0 \\ -1 & 0 & 1 & 0 \\ 0 & 0 & 0 & 0 \end{bmatrix} \tag{17.92}$$

In a similar manner, the other matrices required for the formation of secant and tangent stiffness matrices can be derived.

17.7 COMPUTER PROGRAM FOR GEOMETRICALLY NON-LINEAR ANALYSIS OF PLATES

The computer program in C and FORTRAN for the solution of geometrically non-linear problem of bare plates is given in the attached CD. Isoparametric quadratic plate bending element presented in section 14.7 has been used for obtaining the solution.

Notations used in the computer program are explained at the beginning. Also indicated are various steps explained with 'comment' statements. Though the present program enables the analysis of rectangular or parallelogrammic plates, yet with a little more modification, the computer program can be used for the large deflection analysis of plates of general shapes.

The main features of the program are:

1. Automatic mesh generation.
2. Generation of linear element stiffness matrices.
3. Generation of overall linear matrix in single array according to skyline storage scheme.
4. Generation of element loading matrix.
5. Generation of overall loading matrix.
6. Fitting of boundary conditions of the structure.
7. Calculations of linear nodal displacements.
8. Generation of non-linear element stiffness matrices – both tangent stiffness matrix and secant stiffness matrix.
9. Formation of overall non-linear tangent stiffness matrix and total stiffness matrix (linear plus non-linear tangent stiffness matrix).
10. Newton-Raphson iteration technique for solving large deflection problem.
11. Calculation of stress resultants.

A major elegance in the computer program lies in its optimum utilisation of memory allocations, as it crunches the matrices by removing zeroes. It further has the provision of taking advantage of the symmetry of the plate. The non-linear secant stiffness matrix in the computer program is written in B-notations, so it is unsymmetric. As such it may be noted that the non-linear load calculation in the Newton-Raphson technique is performed at the element level and the overall load needed in the interaction is achieved from the summation of these element loads. A tolerance limit is specified for the interaction process.

All stiffness matrices are calculated by using Gauss's quadrature technique of numerical integration. Solutions of calculations are obtained by using Gauss's solution procedure.

The preparation of input data is very simple. The input data for the plate of Fig. 17.7 is given in Table 17.1. The computer output for that has been given in Table 17.2. The plate has all the edges clamped.

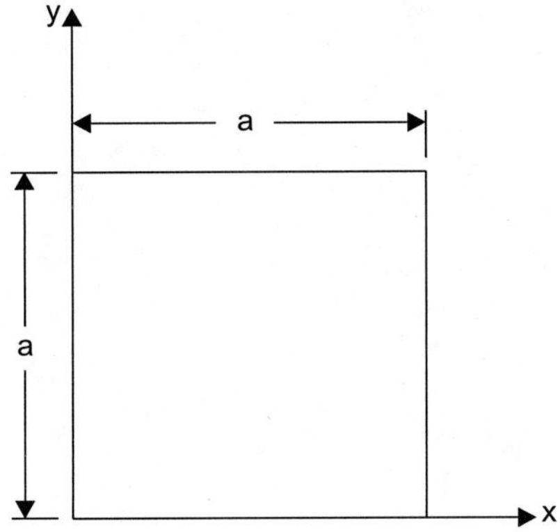

a = 300 in, t = 3 in
v = 0.316, E = 30 × 10^6 Psi,
q = 0.3 lb/in^2

Fig. 17.7. A square plate

Table 17.1. Input Data for the Non-linear Plate Problem

```
300.0 300.0 90.0 0.3
4 4 8 5 2 1 1 12
 2 2 2 2
3.0e08 0.316 3.0
2 1.0 1.0
3 -1.0 1.0
5 -1.0 -1.0
9 -1.0 1.0
6 1.0 -1.0
7 -1.0 -1.0
10 1.0 1.0
11 -1.0 1.0
8 1.0 -1.0
12 1.0 1.0
14 1.0 -1.0
15 -1.0 -1.0
163
3.0 3.0
1
1
1 0.09
38.3 63.4 95.0 134.9 184.0 245.0 318.0 402.0 17.79
```

Table 17.2. Computer Output for the Non-linear Plate Problem

```
 a              b              phai   ql
    300.00         300.00         90.000         .30000
     nx    ny    nn  nfree  nq  nlc  npt  nspt
      4     4     8     5    2    1    1   12
    nbon1    nbon2      nbon3    nbon4
      2     2     2     2
          emodp      prp    thick
    .30000E+09  .31600       3.0000
 exl eyl=        75.000000           75.000000
     nds(1)            163
 points of displacement are            163
 nt=           325nbh=             85ncen=           163ntot=         24055nfree4=
         10
  th1 th2 gr thick
    3.0000         3.0000         .00000         3.0000
 element stiffness matrix generation starts
     ind=              1
nonl=               1
 nll,   toll     =               1     9.000000E-02
 alf are          38.300000
  the present alf is        38.300000
err=        12.494700
err=   1.045809E-01
err=   9.173726E-02
err=   1.070604E-01
err=   8.132938E-02
    displacement are        3.930287E-02
    element no ...zi    eta,displaceemnt
          2        1.0000              1.0000              .20819E-14
      420.540600      1330.825000        -46.452770     -17904.030000
  -36708.720000     -2439.932000       -418.881700     266438.100000
    element no ...zi    eta,displaceemnt
          3       -1.0000              1.0000              .20819E-14
      420.540600      1330.825000        -46.452780      -8049.217000
  -33594.600000     -2439.932000        863.809900     264767.800000
    element no ...zi    eta,displaceemnt
          5       -1.0000             -1.0000              .82664E-15
     -71.389440       -22.559060         -3.273324      -6561.609000
   -2609.120000      -237.560300       1302.911000       1771.461000
    element no ...zi    eta,displaceemnt
          9       -1.0000              1.0000              .82664E-15
     -71.389440       -22.559060         -3.273324      -6570.386000
   -2636.894000      -237.560300       4903.771000        -17.545690
    element no ...zi    eta,displaceemnt
          6        1.0000             -1.0000              .39303E-01
     -43.610440      -199.119700        131.009800      17650.350000
   15256.610000      1340.942000      -1655.323000    -272952.800000
    element no ...zi    eta,displaceemnt
          7       -1.0000             -1.0000              .39303E-01
     177.987600      -129.094700        127.976000      -4758.072000
    8175.547000      1968.741000        271.321500    -264918.400000
    element no ...zi    eta,displaceemnt
         10        1.0000              1.0000              .39303E-01
       4.460321       -46.997020        -94.725400      10161.910000
   -8440.975000     -2513.179000        967.573800     240480.200000
```

```
  element no ...zi    eta,displaceemnt
      11      -1.0000            1.0000             .39303E-01
    226.058400        23.028000        -97.759250    -12246.510000
 -15522.040000     -1885.380000      -3357.863000    234889.800000
  element no ...zi    eta,displaceemnt
       8       1.0000           -1.0000             .27253E-14
   -147.375800       -46.570740        -10.109550    -12447.280000
  -8275.946000       689.889200       2081.333000    -12821.710000
  element no ...zi    eta,displaceemnt
      12       1.0000            1.0000             .27253E-14
   -147.375700       -46.570740        -10.109550    -10883.950000
  -3328.700000       689.889000      -2985.272000     18481.820000
  element no ...zi    eta,displaceemnt
      14       1.0000           -1.0000             .12262E-14
    218.778900       692.338400         32.382250    -13037.360000
 -26479.450000      2429.966000      -1044.802000    -87092.560000
  element no ...zi    eta,displaceemnt
      15      -1.0000           -1.0000             .12262E-14
    218.778900       692.338400         32.382250     -4290.706000
 -23715.510000      2429.966000       1849.126000    -89187.210000
```

Results obtained by using the computer program developed here are presented in Table 17.3 and compared them with those obtained by Levy [17.5]. Results have been presented in the following non-dimensional form.

Central deflection: $w = \dfrac{w}{t}$ Extreme fibre stress: $T_{cs} = \dfrac{\sigma a^2}{Et^2}$

where a is the plate dimension and t is the plate thickness.

Table 17.3 presents results for different load factors. A mesh of 8x8 for the whole plate has been chosen and a tolerance of 1 percent ($\alpha = 0.01$) has been considered in the analysis. Results show good agreement for displacement and stresses.

Table 17.3. Comparison of Displacements and Stresses for all Edges Clamped Plate

Sl. No.	Load factor	Central deflection		Central stresses		Edge stresses	
		Exact (17.5)	Present	Exact (17.5)	Present	Exact (17.5)	Present
1	17.79	0.2370	0.7346	2.60	2.63	5.48	5.34
2	38.30	0.4710	0.4660	5.20	5.48	11.52	11.26
3	63.40	0.6950	0.6862	8.00	7.28	18.03	15.29
4	95.00	0.9120	0.8987	11.10	11.13	25.32	25.07
5	134.90	1.1210	1.1025	13.30	13.87	33.50	33.19
6	184.00	1.3230	1.2979	15.90	16.85	42.40	42.05
7	245.00	1.5210	1.4890	19.20	19.27	52.80	51.88
8	318.00	1.7140	1.6750	21.90	22.04	63.90	62.45
9	402.00	1.9020	1.8526	25.10	24.82	75.80	73.44

REFERENCES AND SUGGESTED READINGS

17.1. R. Mallet and P. Marcal, Finite Element Analysis of Non-linear Structures, Journal of Structural Division, Proc. ASCE, v. 94, 1968, pp. 2081 – 2105.

17.2. S. Rajasekaran and D. W. Murray, On Incremental Finite Element Matrices, Journal of Structural Division, Proc. ASCE, v. 99, 1973, pp. 2423–2438.

17.3. O. C. Zienkiewicz and R. L. Taylor, The Finite Element Method, v. I and II, 4th Edition, McGraw-Hill, 1989.

17.4. P. C. Wang, Numerical Methods in Structural Mechanics, John Wiley and Sons, 1980

17.5. S. Levy, Square Plate with Clamped Edges under Normal Pressure Producing Large Deflections, NASA Technical Note, 847, 1942.

❑❑❑

18

FINITE ELEMENT METHOD OF ANALYSIS OF STIFFENED PLATES

18.1 INTRODUCTION

We have seen in the previous chapters that the finite element method has proved to be the most versatile amongst the different numerical methods. One of the important categories of structures is the stiffened plate, which has considerable application in civil, aerospace and ship and various other categories of structures. The stiffened plate is assumed to consist of two parts: the plate and the stiffener. If the stiffener centroid is coincident with the plate middle surface, then no inplane stresses will be developed due to stiffener bending (Fig. 18.1). If the centroid of the stiffener and the plate middle surface does not coincide, the inplane stress will be developed in the plate due to stiffener bending (Fig. 18.2). The term eccentric stiffened plates will be used to designate this structure. The finite element analysis for the first category of structures will involve only the bending displacements as the unknowns, while the latter category of structures will involve both inplane and bending displacements as unknowns.

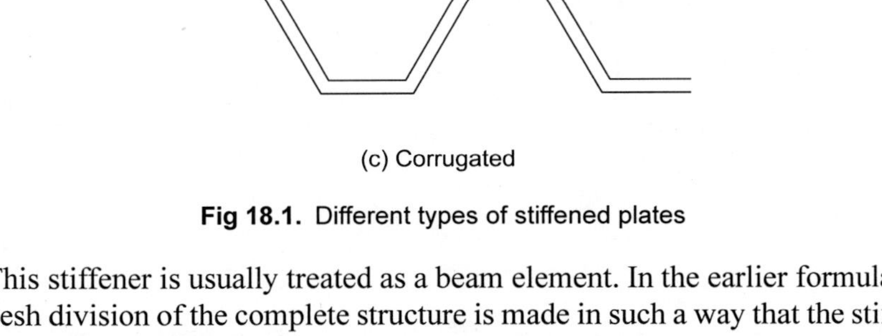

(a) Rib Stiffened

(b) Cellular

(c) Corrugated

Fig 18.1. Different types of stiffened plates

This stiffener is usually treated as a beam element. In the earlier formulations the mesh division of the complete structure is made in such a way that the stiffener is placed along the nodal line [18.1,18.2]. Later investigations have been conducted to model the stiffener in such a way that it can be placed anywhere within the plane element and not necessarily on the nodal lines [18.3,18.4]. It introduces considerably flexibility in the analysis. The mesh division can be made according to the requirements of the stress output and resolutions sought and not according to stiffener locations.

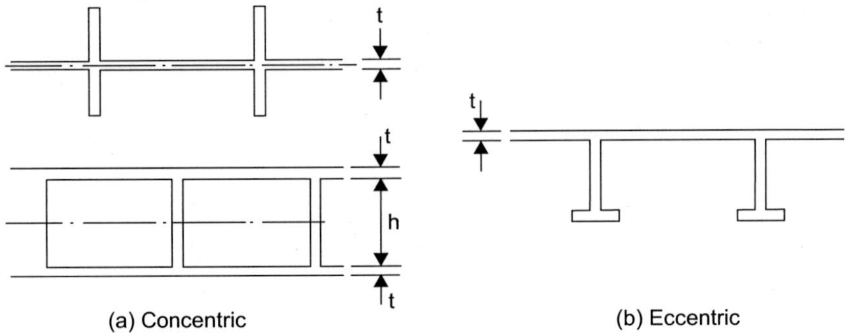

Fig 18.2. Concentrically and eccentrically stiffened plates

18.2 MODELING THE PLATE AND THE STIFFENER

In the case of concentric stiffened plates, both the plate and the stiffener undergo bending deformation only. The stiffened plate for such cases is to be analysed as a plate-bending problem. Only bending and torsional displacements are to be considered for the analysis of the beam. However, the analysis presented in the following is that of eccentrically stiffened plate from which the concentrically stiffened plate can be formulated by eliminating inplane stress part of the former.

18.3 RECTANGULAR STIFFENED PLATE BENDING ELEMENT

It is convenient to consider the plate middle surface as the reference axis. Though the load acts normal to the middle plate of the plate, the plate as such will be subjected to inplane and bending deformations when the stiffener is placed eccentric to it. In small deflection theory, the inplane and the plate-bending displacement functions can be assumed independently.

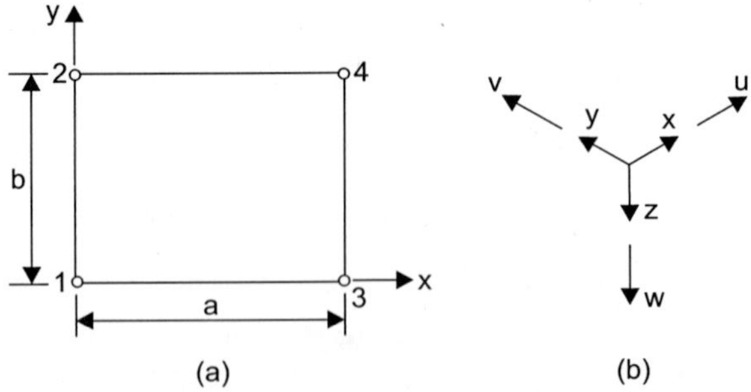

Fig. 18.3. A rectangular element

For structures having regular shaped boundaries, rectangular elements can be effectively employed in the analysis. A typical rectangular element having sides *a* and *b* is shown in Fig. 18.3. The thickness of the element is *t*. Therefore, there are 20 degrees of freedom for the element. The displacement vector is given by

$$\{X\}_e^T = \begin{matrix} \{u_1\ v_1\ u_2\ v_2\ u_3\ v_3\ u_4\ v_4\ w_1\ \theta_{x1}\ \theta_{y1} \\ w_2\ \theta_{x2}\ \theta_{y2}\ w_3\ \theta_{x3}\ \theta_{y3}\ w_4\ \theta_{x4}\ \theta_{y4}\} \end{matrix} \tag{18.1}$$

It may be noted that inplane displacements at the four nodes are numbered first and then the bending displacements with the usual notations. The plate stress part of the four nodal rectangular plate elements has been discussed in Sec. 9.7 and the rectangular plate-bending element in Sec. 11.3. As such let us delve straight into the stiffener formulation.

18.3.1 Stiffness Matrix of the Stiffener Element

The stiffener is considered as eccentric to and integral with the plate. As such it is assumed to be placed along the nodal line parallel to the *x* axis. Due to the requirement of the conformity of displacements between the plate and the stiffener, the following displacement functions for the stiffener are assumed.

$$\begin{Bmatrix} u \\ w \\ \phi \end{Bmatrix} = \begin{bmatrix} 1 & x & 0 & 0 & 0 & 0 & 0 & 0 \\ 0 & 0 & 1 & x & x^2 & x^3 & 0 & 0 \\ 0 & 0 & 0 & 0 & 0 & 0 & 1 & x \end{bmatrix} \begin{Bmatrix} \alpha'_1 \\ \alpha'_2 \\ \alpha'_3 \\ \alpha'_4 \\ \alpha'_5 \\ \alpha'_6 \\ \alpha'_7 \\ \alpha'_8 \end{Bmatrix} \tag{18.2}$$

or,

$$\{F_s\} = [C]\{\alpha\} \tag{18.3}$$

where ϕ is the angle twist .

The stiffener displacements are shown in Fig. 18.4. along with u and w, ϕ the angle of twist has been assumed as an independent displacement. Figure 18.4 indicates 4 generalised displacements u, w, θ_y $\left(= \dfrac{\partial w}{\partial x}\right)$ and θ_x $\left(= \dfrac{\partial w}{\partial y} = \phi\right)$. Substituting the nodal values in eqn. (18.3).

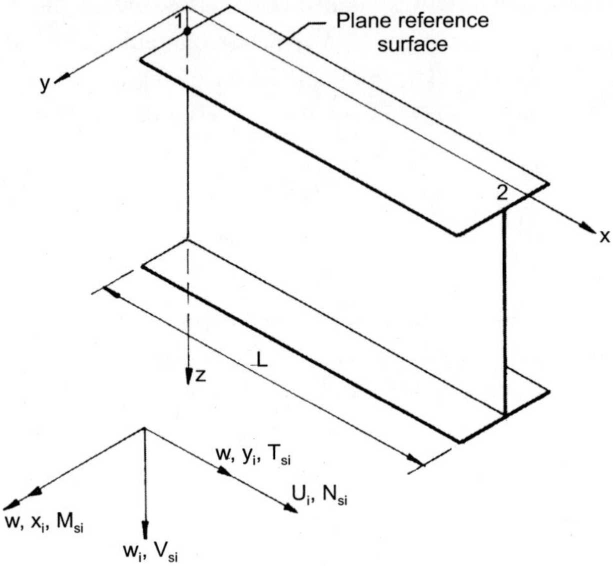

Fig. 18.4. A stiffener element

$$\begin{Bmatrix} \alpha'_1 \\ \alpha'_2 \\ \alpha'_3 \\ \alpha'_4 \\ \alpha'_5 \\ \alpha'_6 \\ \alpha'_7 \\ \alpha'_8 \end{Bmatrix} = \begin{bmatrix} 1 & 0 & 0 & 0 & 0 & 0 & 0 & 0 \\ -\frac{1}{L} & 0 & 0 & 0 & \frac{1}{L} & 0 & 0 & 0 \\ 0 & 1 & 0 & 0 & 0 & 0 & 0 & 0 \\ 0 & 0 & 1 & 0 & 0 & 0 & 0 & 0 \\ 0 & -\frac{3}{L^2} & -\frac{2}{L^2} & 0 & 0 & \frac{3}{L^2} & -\frac{1}{L} & 0 \\ 0 & \frac{2}{L^2} & \frac{1}{L^2} & 0 & 0 & -\frac{2}{L^2} & \frac{1}{L^2} & 0 \\ 0 & 0 & 0 & 1 & 0 & 0 & 0 & 0 \\ 0 & 0 & 0 & -\frac{1}{L} & 0 & 0 & 0 & \frac{1}{L} \end{bmatrix} = \begin{Bmatrix} u_1 \\ w_1 \\ w,_{x1} \\ w,_{y1} \\ u_2 \\ w_2 \\ w,_{x2} \\ w,_{y2} \end{Bmatrix} \tag{18.4}$$

or,

$$\{\alpha\} = [Q]\{X_s\}_e \tag{18.5}$$

where the subscript s refers to the stiffener.

Combining eqns. (18.3) and (18.5), we get

$$\{f_s\} = [N_s]\{X_s\}_e \tag{18.6}$$

where $[N_s] = [C]\ [Q]$ whose explicit values are given below by eqn. (18.7)

$$[N_s]^T = \begin{bmatrix} 1-\dfrac{x}{l} \\ 1-\dfrac{3x^2}{l^2}+\dfrac{2x^3}{L^3} \\ x-\dfrac{2x^2}{L}+\dfrac{x^3}{L^2} \\ 1-\dfrac{x}{L} \\ \dfrac{x}{L} \\ \dfrac{3x^2}{L^2}-\dfrac{2x^3}{L^3} \\ -\dfrac{x^2}{L}+\dfrac{x^3}{L^2} \\ \dfrac{x}{L} \end{bmatrix} \tag{18.7}$$

The strain components for the stiffener are

$$\{\varepsilon_s\} = \begin{Bmatrix} \dfrac{\partial u}{\partial x} \\ -\dfrac{\partial^2 w}{\partial x^2} \\ \dfrac{\partial \phi}{\partial x} \end{Bmatrix} \tag{18.8}$$

Using eqn. (18.3), eqn. (18.8) can be written as

$$\{\varepsilon_s\} = \begin{bmatrix} 0 & 1 & 0 & 0 & 0 & 0 & 0 & 0 \\ 0 & 0 & 0 & 0 & 0 & -2 & -6x & 0 \\ 0 & 0 & 0 & 0 & 0 & 0 & 0 & 1 \end{bmatrix} [Q]\{X_s\}_e \tag{18.9}$$

or,

$$\{\varepsilon_s\} = [B_s]\{X_s\}_e \tag{18.10}$$

The 'stress-strain' relation of the stiffener is given by

$$\begin{Bmatrix} N_s \\ M_s \\ T_s \end{Bmatrix} = \begin{bmatrix} EA_x & ES_x & 0 \\ ES_x & EI_x & 0 \\ 0 & 0 & GJ_x \end{bmatrix} \begin{Bmatrix} \dfrac{\partial u}{\partial x} \\ -\dfrac{\partial^2 w}{\partial x^2} \\ \dfrac{\partial \phi}{\partial x} \end{Bmatrix} \tag{18.11}$$

or, $$\{\sigma_s\} = [D_s]\{\varepsilon_s\} \tag{18.12}$$

Here A_x = the cross-sectional area of the x-directional stiffener

S_x = first moment of the area of the x-directional stiffener with respect to the middle surface of the plate

I_x = the moment of inertia of the x-directional stiffener with respect to the reference surface

J_x = the polar moment of inertia of the x-directional stiffener

Combining eqns. (18.10), and (18.12) we get

$$\{\sigma_s\} = [D_s][B_s]\{X_s\}_e \tag{18.13}$$

It can be shown that the stiffness matrix of the stiffener is given by

$$[K_s]_e = \int_0^L [B_s]^T [D_s] [B_s] \, dx \tag{18.14}$$

$[K_s]_e$ has been evaluated explicitly and is given by eqn. (20.15)

$$[K_s]_e = \begin{bmatrix} \frac{EA_x}{L} & 0 & -\frac{ES_x}{L} & 0 & -\frac{EA_x}{L} & 0 & \frac{ES_x}{L} & 0 \\ 0 & \frac{2EI_x}{L^3} & \frac{6EI_x}{L^2} & 0 & 0 & -\frac{12EI_x}{L^3} & \frac{6EI_x}{L^2} & 0 \\ -\frac{ES_x}{L} & \frac{6EI_x}{L^2} & \frac{4EI_x}{L} & 0 & \frac{ES_x}{L} & -\frac{6EI_x}{L^2} & \frac{2EI_x}{L} & 0 \\ 0 & 0 & 0 & \frac{GJ_x}{L} & 0 & 0 & 0 & -\frac{GJ_x}{L} \\ -\frac{EA_x}{L} & 0 & \frac{ES_x}{L} & 0 & \frac{EA_x}{L} & 0 & -\frac{ES_x}{L} & 0 \\ 0 & -\frac{12EI_x}{L^3} & -\frac{6EI_x}{L^2} & 0 & 0 & \frac{12EI_x}{L^3} & -\frac{6EI_x}{L^2} & 0 \\ \frac{ES_x}{L} & \frac{6EI_x}{L^2} & \frac{2EI_x}{L} & 0 & -\frac{ES_x}{L} & -\frac{6EI_x}{L^2} & \frac{4EI_x}{L} & 0 \\ 0 & 0 & 0 & -\frac{GJ_x}{L} & 0 & 0 & 0 & \frac{GJ_x}{L} \end{bmatrix} \begin{Bmatrix} u_1 \\ w_1 \\ w,_{x1} \\ w,_{y1} \\ u_2 \\ w_2 \\ w,_{x2} \\ w,_{y2} \end{Bmatrix} \tag{18.15}$$

The stiffness matrix of the y directional stiffener can be similarly obtained.

18.4 ISOPARAMETRIC STIFFENED PLATE BENDING ELEMENT

One of the most popular elements for the analysis of plate structures is the isoparametric element. Like the previous section, let us treat the plate part and stiffener part separately. We put forward the formulation of the stiffener in a more general manner than done previously. The restriction that the stiffener should be placed on the nodal line is removed and it may to now be placed anywhere within the plate element and need not be parallel to any edge of the plate.

18.4.1 Stiffness Matrix of Arbitrarily Oriented Eccentric Stiffener

An arbitrarily oriented stiffener placed within the plate element and its isoparametric mapping is shown in Fig. 18.5.

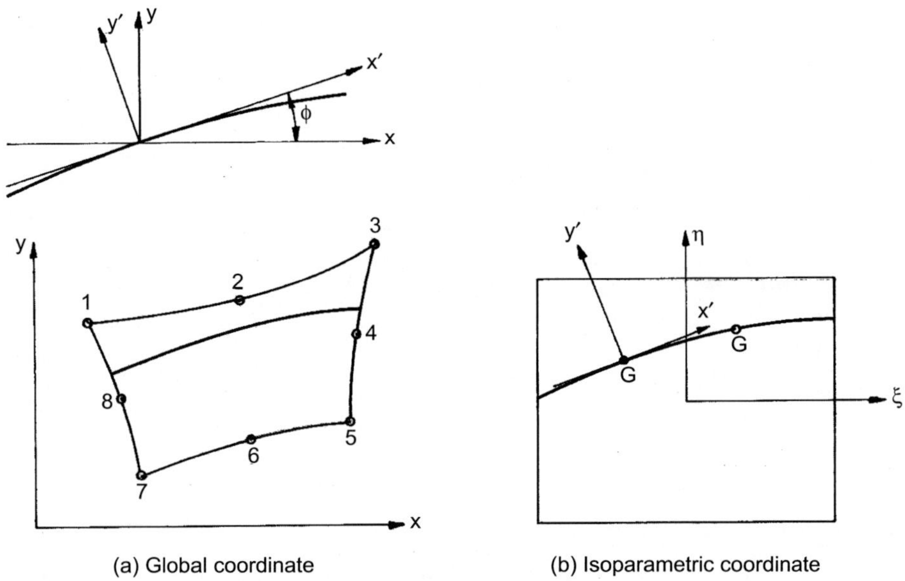

Fig. 18.5. Arbitrary oriented stiffener within the plate element

The displacement field of the stiffener is given by

$$\{f\} = \begin{Bmatrix} U \\ V \\ W \end{Bmatrix} = \begin{Bmatrix} u' - z\,\theta_{x'}(x') \\ v' - z\,\theta_{y'}(x') \\ w + c'\theta_{y'}(x') \end{Bmatrix} \tag{18.16}$$

where c' is the distance of a point on the stiffener cross-section measured from y axis (Fig. 18.6). x'- y' are the coordinate system tangential to the Gauss points. The strain components are given by

$$\{\varepsilon_{x'}\} = \begin{Bmatrix} \dfrac{\partial u'}{\partial x'} - z\dfrac{\partial \theta x'}{\partial x'} \\ \dfrac{\partial w}{\partial x'} + c'\dfrac{\partial \theta y'}{\partial x'} - \theta_{x'} \\ -z\dfrac{\partial \theta y'}{\partial x'} \end{Bmatrix} \tag{18.17}$$

The generalized strain components of the stiffener in the x'- y' coordinate system are given by

$$\{\bar{\varepsilon}_{x'}\} = \begin{Bmatrix} \dfrac{\partial u'}{\partial x'} \\ -\dfrac{\partial \theta_{x'}}{\partial x'} \\ \dfrac{\partial w_{x'}}{\partial x'} - \theta_{x'} \\ -\dfrac{\partial \theta_{y'}}{\partial x'} \end{Bmatrix} \tag{18.18}$$

The relationship between $\{\varepsilon_{x'}\}$ and $\{\bar{\varepsilon}_{x'}\}$ is given by

$$\{\varepsilon_{x'}\} = [H_s]\{\bar{\varepsilon}_{x'}\} \tag{18.19}$$

where

$$[H_s] = \begin{bmatrix} 1 & z & 0 & 0 \\ 0 & 0 & 1 & -c' \\ 0 & 0 & 0 & z \end{bmatrix} \tag{18.20}$$

The xy-coordinate axis system is related to $x'y'$- axis system is as follows:

$$\left.\begin{aligned} x &= x'\cos\phi - y'\sin\phi \\ y &= x'\sin\phi + y'\cos\phi \end{aligned}\right\} \tag{18.21}$$

The relationship between local displacement in the x'-y' axis system and global displacement in the x-y coordinate is

$$\begin{Bmatrix} u' \\ \theta_{x'} \\ \theta_{y'} \end{Bmatrix} = \begin{bmatrix} \cos\phi & \sin\phi & 0 & 0 \\ 0 & 0 & \cos\phi & \sin\phi \\ 0 & 0 & -\sin\phi & \cos\phi \end{bmatrix} \begin{Bmatrix} u \\ v \\ \theta_x \\ \theta_y \end{Bmatrix} \tag{18.22}$$

Using the relations given in eqns. (18.21) and (18.22), eqn. (18.18) can be expressed in terms of x-y coordinate system

$$\{\bar{\varepsilon}_{x'}\} = \begin{Bmatrix} \dfrac{\partial u}{\partial x}\cos^2\phi + \dfrac{1}{2}\left(\dfrac{\partial u}{\partial y} + \dfrac{\partial v}{\partial x}\right)\sin 2\phi + \dfrac{\partial v}{\partial y}\sin^2\phi \\ -\dfrac{\partial \theta_x}{\partial x}\cos^2\phi - \dfrac{1}{2}\left(\dfrac{\partial \theta_x}{\partial y} + \dfrac{\partial \theta_y}{\partial x}\right)\sin 2\phi - \dfrac{\partial \theta_y}{\partial y}\sin^2\phi \\ \left(\dfrac{\partial w}{\partial x} - \theta_x\right)\cos\phi + \left(\dfrac{\partial w}{\partial y} - \theta_y\right)\sin\phi \\ -\dfrac{1}{2}\left(-\dfrac{\partial \theta_x}{\partial x} + \dfrac{\partial \theta_y}{\partial y}\right)\sin 2\phi + \dfrac{\partial \theta_x}{\partial y}\sin^2\phi - \dfrac{\partial \theta_y}{\partial x}\cos^2\phi \end{Bmatrix} \tag{18.23}$$

or,
$$\{\bar{\varepsilon}_{x'}\} = [T]\{\bar{\varepsilon}_x\} \tag{18.24}$$

where $[T]$ is the transformation matrix and is given by

$$[T]^T = \begin{bmatrix} \cos^2\phi & 0 & 0 & 0 \\ \sin^2\phi & 0 & 0 & 0 \\ \frac{1}{2}\sin 2\phi & 0 & 0 & 0 \\ 0 & \cos^2\phi & 0 & -\frac{1}{2}\sin 2\phi \\ 0 & \sin^2\phi & 0 & \frac{1}{2}\sin 2\phi \\ 0 & \frac{1}{2}\sin 2\phi & 0 & \cos^2\phi \\ 0 & \frac{1}{2}\sin 2\phi & 0 & -\sin^2\phi \\ 0 & 0 & \cos\phi & 0 \\ 0 & 0 & \sin\phi & 0 \end{bmatrix} \tag{18.25}$$

and $\{\bar{\varepsilon}_x\}$ is the generalized strain vector in the x-y coordinates and is given by

$$\{\bar{\varepsilon}_x\}^T = \left\{\frac{\partial u}{\partial x}\ \frac{\partial v}{\partial y}\ \left(\frac{\partial u}{\partial y}+\frac{\partial v}{\partial x}\right)\ -\frac{\partial \theta_x}{\partial x}\ -\frac{\partial \theta_y}{\partial y}\ -\frac{\partial \theta_y}{\partial x}\ -\frac{\partial \theta_x}{\partial y}\ \left(\frac{\partial w}{\partial x}-\theta_x\right)\left(\frac{\partial w}{\partial y}-\theta_y\right)\right\}$$

(18.26)

Using the shape function relation of the type given by eqn. (14.98), eqn. (18.26) can be written as

$$\{\bar{\varepsilon}_x\}^T = \sum_{r=1}^{n} [B_{sr}]\{X_r\}_e = [B_s]\{X\}_e \tag{18.27}$$

where,

$$[B_{sr}] = \begin{bmatrix} \frac{\partial N_r}{\partial x} & 0 & 0 & 0 & 0 \\ 0 & 0 & 0 & -\frac{\partial N_r}{\partial x} & 0 \\ 0 & 0 & \frac{\partial N_r}{\partial x} & -N_r & 0 \\ 0 & 0 & 0 & 0 & \frac{\partial N_r}{\partial x} \end{bmatrix} \tag{18.28}$$

The stress-strain relationship for the stiffener is given by

$$\{\sigma_{x'}\} = [D_s]\{\varepsilon_{x'}\} \tag{18.29}$$

where,

$$\{\sigma_{x'}\}^T = \{\sigma_{x's} \quad \tau_{x'y's} \quad \tau_{zx's}\} \tag{18.30}$$

$$[D_s] = \begin{bmatrix} E & 0 & 0 \\ 0 & G & 0 \\ 0 & 0 & G \end{bmatrix} \tag{18.31}$$

Combing eqns. (18.19) and (18.24), eqn. (18.29) can be written as

$$\{\sigma_{x'}\} = [D_s]\,[H_s]\,[T]\,[B_s]\,\{X\}_e \tag{18.32}$$

Using the principle of virtual work, the element stiffness matrix of the stiffener is expressed as

$$[K_s]_e = \int [B_s]^T\,[T]^T\,[D_s]\,[T]\,[B_s]\,dx' \tag{18.33}$$

where

$$[\overline{D}_s] = \int_{-b/2}^{b/2}\int_{-d_1/2}^{d_2} \begin{bmatrix} E & & & \\ Ez & Ez^2 & sym & \\ 0 & 0 & G & \\ 0 & 0 & Gc' & G(c'^2+z^2) \end{bmatrix} dy'\,dz \tag{18.34}$$

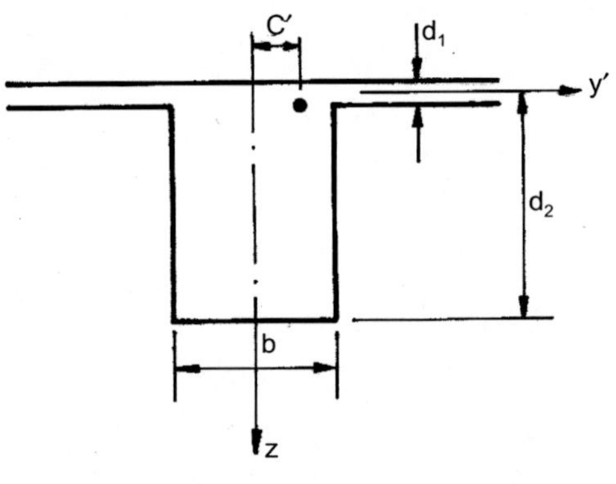

Fig. 18.6. Cross-section of the stiffener

On integration, eqn. (18.34) reduces to

$$[\overline{D}_s] = \begin{bmatrix} EA_s & & & \\ ES_{x's} & EI_{x's} & sym & \\ 0 & 0 & S_{is} & 0 \\ 0 & 0 & 0 & GJ_s \end{bmatrix} \tag{18.35}$$

where

A_s = cross-sectional area of the stiffener

$I_{x's}$ = second moment of the stiffener cross-sectional area about the reference axis

S_{is} = shear rigidity of the stiffener

$J_{x's}$ = polar moment of inertia of the stiffener cross-sectional area

$S_{x's}$ = first moment of the stiffener cross-sectional area about the reference axis

$\iint (y'^2 + z^2)\, dy'\, dz$ is interpreted as the polar moment of inertia of the cross-section. $\iint G\, y'\, dy'\, dz$ is zero for sections symmetrical about z axis and lying in the x-y plane. However, for sections, which are not symmetrical about z axis, this quantity has been taken to be zero. The element stiffness matrix of the stiffener given by eqn. (18.32) is evaluated numerically.

The main elegance of the formulation of the stiffener presented above reveals that the stiffener can be placed anywhere within the plate element. As such the mesh division is no longer dependent on the stiffener location. A careful study of the terms involved in the determination of $[K_s]_e$ for the stiffener will immediately reveal that the contribution of the stiffener is reflected at all eight nodes of the plate element in which it is placed and that there is a mutual coupling at all these nodes. No doubt the model gives a considerable advantage to the user.

REFERENCES AND SUGGESTED READINGS

18.1. T. S. Koko and M. D. Olson, Non-linear Analysis of Stiffened Plates Using Superelements, International Journal for Numerical Methods for Engineering, v. 31,1991, pp. 319–413.

18.2. C. S. Rossow and A. K. Ibrahimkhail, Constrained Method of Analysis of Stiffened Plates, Computers and Structures, v. 8,1978, pp. 51–60.

18.3. M. Mukhopadhyay and S. K. Satsangi, Isoparametric Stiffened Plate-bending Element for the Analysis of Ships' Structures, Transaction of Royal Institute of Naval Architects, v.126, 1984, pp. 144–151.

18.4. D. Venugopal Rao, A. H. Sheikh and M. Mukhopadhyay, Finite Element Large Deflection Analysis of Stiffened Plates, v. 47,no. 6, 1993, pp. 987–993.

❑❑❑

19

SELECTED TOPICS

A few selected topics of special interest have been put forward in this chapter.

19.1 RAYLEIGH-RITZ METHOD

In many problems of continuum mechanics, a functional can be formed that describes the problem. In Rayleigh-Ritz method, approximate solutions of the variables of the functional are assumed. The classical solution of the differential equation is thus avoided. Substitution of approximate solutions results in a set of algebraic equations. Incorporation of more terms in assumed solutions will yield more accurate result (19.1).

In Rayleigh-Ritz method, an approximate solution of variable of the following form is assumed

$$u = \sum_i \beta_i F_i \tag{19.1}$$

where β_i is the generalized coordinate whose value is to be determined.

One of the major requirement is that the function, $F_i = F_i(x)$ must be admissible, that is, it must satisfy compatibility conditions and essential boundary conditions. A functional of the following type is assumed

$$\pi_p = \int_{x_1}^{x_2} I(u, u_{,x}, x)\, dx \tag{19.2}$$

where $u_x = \dfrac{\partial u}{\partial x}$ with boundary conditions $u(x_1) = u(x_2) = 0$

The principle of stationary potential energy is one the many variational principles. It states that the functional π_p is also stationary with respect to β_1, β_2, …, β_8. This will result in equations containing β_i coefficients.

$$\frac{\partial \pi_p}{\partial \beta_i} = 0, \quad i = 1, 2, \ldots, n \tag{19.3}$$

Equations of the type of eqn. (19.3) are linear in terms of β_i where π_p is a quadratic function of u and $u_{,x}$.

19.1.1 An Example

Determine the central deflection of the beam of Fig. 19.1

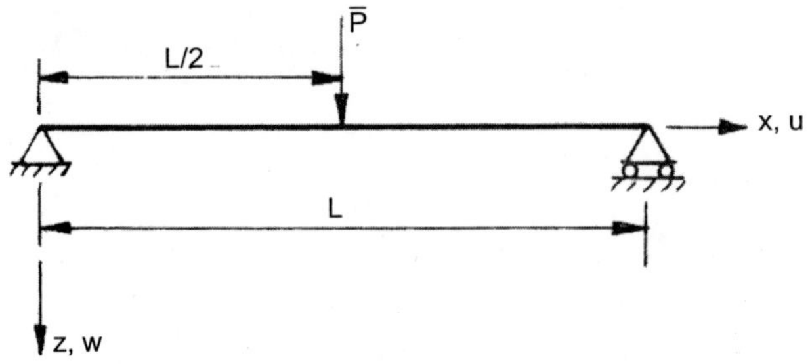

Fig. 19.1. Uniform beam under a central point load

An approximate solution for the deflection, that is, w-displacement is assumed

$$w = \beta_1 \sin\frac{\pi x}{L} + \beta_2 \sin\frac{3\pi x}{L} \tag{19.4}$$

Equation 19.4 satisfies the boundary conditions of the problem, that $w = 0$, and $M = 0$ at $x = 0$ and at $x = L$. Only two terms have been considered here. Incorporation of more terms would result in more number of equations. Equation 19.4 is a symmetric. The functional is given by

$$\pi_p = \frac{EI}{2}\int_o^L \left(\frac{d^2 w}{dx^2}\right)^2 dx - \bar{P}w_c \tag{19.5}$$

In eqn. (19.5), the first term corresponds to the internal energy of the beam and the second to the potential of loads. w_c is the central deflection.

Substituting w from eqn. (19.4) into eqn. (19.5), yields

$$\pi_p = \frac{EI}{2}\int_o^L \left\{\beta_1\left(\frac{\pi}{L}\right)^2 \sin\frac{\pi x}{L} + \beta_2\left(\frac{3\pi}{L}\right)^2 \sin\frac{3\pi x}{L}\right\}^2 dx$$

$$- \bar{P}\left\{\beta_1 \sin\left(\frac{\pi}{2}\right) + \beta_2 \sin\left(\frac{3\pi}{2}\right)\right\} \tag{19.6}$$

After carrying out the necessary operations and applying orthogonality relationship

$$\pi_p = \frac{EI}{2}\left\{\beta_1^2\left(\frac{\pi}{L}\right)^4 \frac{L}{2} + \beta_2^2\left(\frac{3\pi}{L}\right)^4 \frac{L}{2}\right\} - \bar{P}(\alpha_1 - \alpha_3) \tag{19.7}$$

Applying the principle of stationary potential energy given by eqn. (19.3), yields

$$\left.\begin{aligned}\frac{\partial \pi_p}{\partial \beta_1} &= \frac{EI}{2}\left\{2\beta_1\left(\frac{\pi}{L}\right)^4 \frac{L}{2}\right\} - \bar{P} = 0 \\ \frac{\partial \pi_p}{\partial \beta_2} &= \frac{EI}{2}\left\{2\beta_2\left(\frac{3\pi}{L}\right)^4 \frac{L}{2}\right\} + \bar{P} = 0\end{aligned}\right\} \tag{19.8}$$

Solution of eqn. (19.8) gives

$$\beta_1 = \frac{2\bar{P}}{EI}\frac{L^3}{\pi^4}, \quad \beta_2 = -\frac{2\bar{P}}{EI}\left(\frac{L^3}{\pi^4}\right)\left(\frac{1}{3}\right)^4 \tag{19.9}$$

Substituting β_1, β_2 from eqn. (19.9) into eqn. (19.4), yields

$$w = \frac{2\bar{P}L^3}{EI\pi^4}\left(\sin\frac{\pi x}{L} - \frac{1}{3^4}\sin\frac{3\pi x}{L}\right) \tag{19.10}$$

If n terms instead of 2 terms are considered, then for $n \to \infty$, this solution for central deflection is given by

$$w_c = \frac{2\bar{P}}{EI}\frac{L^3}{\pi^4}\left(1 + \frac{1}{3^4} + \frac{1}{5^4} + \ldots\right) \tag{19.11}$$

Equation (19.11) is an infinite series. On evaluation, the value of w_c coincides with the exact result

$$w_c = \frac{\bar{P}L^3}{48\,EI} \tag{19.12}$$

19.2 RAYLEIGH-RITZ FINITE ELEMENT METHOD

The finite element method is nothing but Rayleigh-Ritz method. Instead of using approximating function for the whole field, functions are assumed in the piecewise fashion from the nodal degrees of freedom of the field. As shown in the previous section, the functional for the structural problem will consist of the sum of the internal strain energy and the potential of the vector. The variational principle of the type of eqn. (19.3) is to be applied. The method is explained with the help of an example.

An axially loaded bar divided into three elements is shown in Fig. 19.2. A typical element of length L is shown. Axial strain is given by

$$\varepsilon_x = \frac{du}{dx} \tag{19.13}$$

Using interpolation functions, eqn. (19.13) becomes

$$\{\varepsilon_x\} = [B]\{X\}_e \tag{19.14}$$

where $$[B] = \frac{d}{dx}[N] = \left[-\frac{1}{L} \quad \frac{1}{L} \right] \tag{19.15}$$

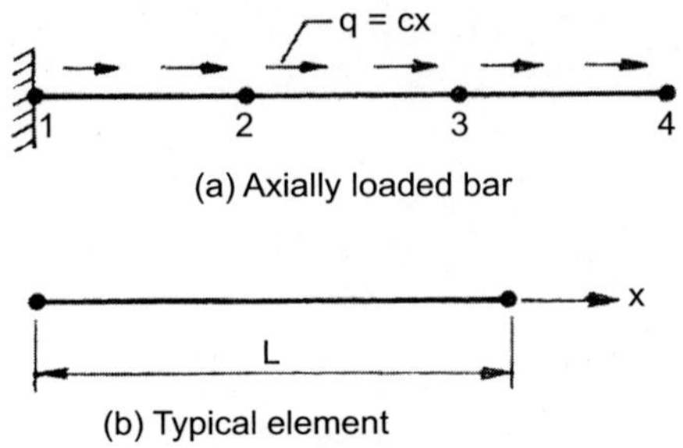

Fig. 19.2. Axially loaded bar

The strain energy of the element is

$$U = \int_0^L \frac{1}{2} E\, \varepsilon_x^2 A\, dx = \frac{1}{2}\int_0^l \varepsilon_x^T . EA . \varepsilon_x\, dx$$

or, $$U = \frac{1}{2}\{X\}_e^T\,[k]_e\{X\}_e \tag{19.16}$$

where $$[k]_e = \int_0^L [B]^T . AE . [B]\, dx \tag{19.17}$$

The potential of the load is given by

$$\Omega = -\int_0^L u\,\sigma_x\, A\, dx = -\int_0^L u^T . q\, dx \tag{19.18}$$

For the bar having axial force,

$$\sigma_x = \frac{q}{A} \tag{19.19}$$

Therefore,

$$\Omega = -\{X\}_e^T\{P\}_e \tag{19.20}$$

where $\{P\}_e$ is the consistent load vector as

$$\{P\}_e = \int_o^L [N]^T q\, dx \tag{19.21}$$

Applying eqn. (19.3), that is,

$$\frac{\partial \pi_p}{\partial \{X\}_e} = \{0\} \tag{19.22}$$

It will give

$$[k]_e\{X\}_e = \{P\}_e$$

19.3 WEIGHTED RESIDUAL METHODS

For certain physical problems, the functional may not be obtained. This problem arises when the differential equation of the problem contains certain derivatives of the odd order. Examples may be cited of fluid mechanics problems where for some types of flow, only the differential equations and boundary conditions are available. These types of problems can still be solved by the finite element method which is applied by means of a weighted residual method. Like the Rayleigh-Ritz method, the differential equations of physical problem form a part of the integral expressions which is used by the weighted residual method. Functional and residual formulations are both known as weak forms of stating governing differential equations ('Weak' form is because the conditions are enforced in an average or integral sense). The differential equations themselves are mentioned as strong form (it is 'strong' because the conditions are enforced at every point).

The governing differential equations and natural or non-essential or Neumann boundary conditions of a physical problem are written as

$$Du - f = 0 \text{ in domain } V \tag{19.23}$$

$$Bu - g = 0 \text{ on the boundary } S \text{ of } V \tag{19.24}$$

where, D, B = differential operators

u = dependent variable, for example, displacement

x = independent variable, for example, coordinate

f, g = functions of x or constants or zero

For beam bending, eqn. (19.23) becomes

$$EI\, w,_{xxxx} = q \tag{19.25}$$

Thus, $$D = EI\frac{d^4}{dx^4},\quad u = w \text{ and } f = q$$

Equation (19.24) when applied to beam bending results in the following two equations:

$$EIw,_{xx} - M_B = 0$$

$$EIw,_{xxx} - V_B = 0$$

where M_B, V_B are bending moment and shear force respectively at the end of the beam. If u is the exact solution of the differential (19.23), $\bar{u}$ is assumed to be an approximate solution. The difference between the exact and the approximate solution can be expressed in the form of residuals, that is,

$$R_D = D\bar{u} - f \text{ (interior residual)} \tag{19.27}$$

$$R_B = B\bar{u} - g \text{ (boundary residual)} \tag{19.28}$$

If $\bar{u}$ happens to be close to u, the residual will be small. A variety of methods known as weighted residual techniques exist for solving partial differential equations. The basic objectives of the weighted residual methods are to formulate a scheme by which small residuals will be obtained. This results in a set of algebraic equations. Equations of the following type making weighted average of the residuals as zero are obtained.

$$\int_v W_i R_D \, dv = 0 \tag{19.29}$$

where W_i represents a set of weighted functions.

The accuracy of the solution depends to a great extent on the choice of the approximating function $\bar{u}$ (also known as trial function) and the number of terms included in it.

19.4 GALERKIN METHOD

Galerkin method is one of the weighted residual methods. In this the weighting function is assumed same as the approximating function. Further, non-essential boundary conditions are introduced by using boundary integral R_B in combination with integration by parts. The method is explained with an example.

The equilibrium equation of a prismatic beam is given by

$$EI\frac{d^4w}{dx^4} = P \tag{19.30}$$

Galerkin equation then becomes

$$\int_o^l \left\{ EI\frac{d^4w}{dx^4} - P \right\} \delta w \, dx = 0 \tag{19.31}$$

Equation (19.31) is integrated by parts twice which yields

$$\int_0^L EI\left(\frac{d^2w}{dx^2}\right)\left(\frac{d^2\delta w}{dx^2}\right)dx - \int_o^L p\,\delta w\,dx - \left[EI\frac{d^2w}{dx^2}.\frac{d\delta w}{dx} - EI\frac{d^3w}{dx^3}\delta w\right]_{x=o}^{x=L} = 0 \tag{19.32}$$

The boundary conditions for the simply supported beam (Fig. 19.1) are

at $\quad x = 0, L \qquad M = EI\dfrac{d^2w}{dx^2} = 0$ (19.33a)

$$w = 0 \tag{19.33b}$$

The above boundary conditions when substituted in eqn. (19.32) indicate

$$\frac{d\delta w}{dx} = 0, \qquad \delta w = 0 \tag{19.34}$$

Thus, eqn. (19.32) reduces to

$$\int_0^L EI\left(\frac{d^2w}{dx^2}\right)\left(\frac{d^2\delta w}{dx^2}\right) dx = \int_0^L p\,\delta w\,dx \tag{19.35}$$

Boundary conditions in w, $\dfrac{dw}{dx}$ are called essential and are satisfied exactly.

Let us formulate the beam problem when both ends are considered free, which is a more general case. The boundary conditions then will be

$$EI\frac{d^2w}{dx^2} = \overline{M}, \qquad -EI\frac{d^3w}{dx^3} = \overline{V} \tag{19.36}$$

where $\overline{M}$ and $\overline{V}$ are the applied bending moment and shear force respectively.

The boundary residuals are given by

$$R'_B = \left\{EI\frac{d^2w}{dx^2} - \overline{M}\right\} \tag{19.37a}$$

$$R''_B = \left\{-EI\frac{d^3w}{dx^3} - \overline{V}\right\} \tag{19.37b}$$

Instead of eqn. (19.31), Galerkin's equation then becomes

$$\int_0^L \left\{EI\frac{d^4w}{dx^4} - p\right\}\delta v\,dx = \left[\left(EI\frac{d^3w}{dx^3} + \overline{V}\right)\delta w - \left(EI\frac{d^2w}{dx^2} - \overline{M}\right)\frac{d\,\delta x}{dx}\right]_{x=0}^{x=L} \tag{19.38}$$

It may be noted that boundary error in moment is weighted with rotations and shear with displacements.

Integrating eqn. (19.38) by parts twice results

$$\int_0^L EI\left(\frac{d^2w}{dx^2}\right)\left(\frac{d^2\delta w}{dx^2}\right)dx = \int_0^L p\,\delta w\,dx + \left[\overline{V}\delta v + \overline{M}\frac{d\,\delta v}{dx}\right]_{x=0}^{x=L} \tag{19.39}$$

It may further be noted that eqn. (19.39) is second order whereas the differential equation of equilibrium is fourth order.

19.4.1 An Example of Galerkin Method

An example is presented for the determination of critical load of a plate in in-plane shear by Galerkin method. A square plate having all edges simply supported is subjected to uniform shearing forces. N_{xy} applied along all the four edges as shown in Fig. 19.3.

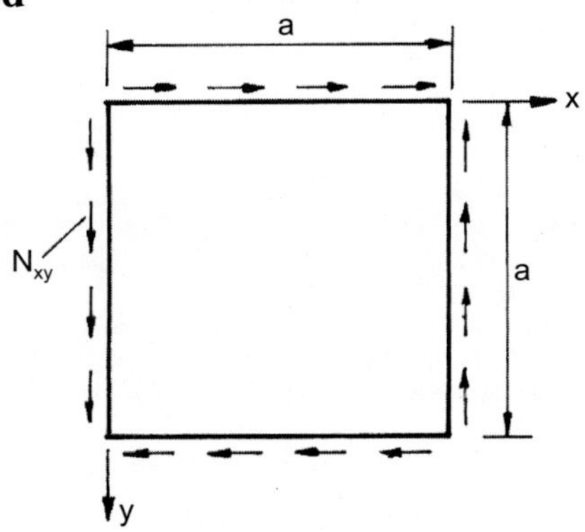

Fig. 19.3. Simply supported plate in pure shear

The deflection function of the plate satisfying the boundary conditions is expressed as

$$w = C_1\sin\frac{\pi x}{a}\sin\frac{\pi y}{a} + C_2\sin\frac{2\pi x}{a}\sin\frac{2\pi y}{a} \tag{19.40}$$

The equilibrium equation of the plate is

$$\frac{\partial^4 w}{\partial x^4} + 2\frac{\partial^4 w}{\partial x^2 \partial y^2} + \frac{\partial^4 w}{\partial y^2} + 2\frac{N_{xy}}{D}\frac{\partial^2 w}{\partial x \partial y} = 0 \tag{19.41}$$

The weighting functions are

$$W_1 = \frac{\partial w}{\partial C_1} = \sin\frac{\pi x}{a}\sin\frac{\pi y}{a} \tag{19.42a}$$

$$W_2 = \frac{\partial w}{\partial C_2} = \sin\frac{2\pi x}{a}\sin\frac{2\pi y}{a} \tag{19.42b}$$

The Galerkin residual equation is

$$\int_0^a\int_0^a W_i\left(\frac{\partial w^4}{\partial x^4} + 2\frac{\partial w^4}{\partial x^2 \partial y^2} + \frac{\partial w^4}{\partial y^4}\right) dx\,dy = 0 \tag{19.43}$$

Substituting appropriate values from eqns. (19.40), (19.42a) and (19.42b) into eqn. (19.43) yields

$$\int_0^a\int_0^a \left[\frac{4C_1\pi^4}{a^4}\sin^2\frac{\pi x}{a}\sin^2\frac{\pi y}{a} + \frac{64C_2\pi^4}{a^4}\sin\frac{2\pi x}{a}\sin\frac{2\pi y}{a}\sin\frac{\pi x}{a}\sin\frac{\pi y}{a}\right.$$

$$+\frac{2N_{xy}}{D}\left(\frac{C_1\pi^2}{a^2}\cos\frac{\pi x}{a}\sin\frac{\pi x}{a}\cos\frac{\pi y}{a}\sin\frac{\pi y}{a}\right.$$

$$\left.\left.+\frac{4C_2\pi^2}{a^2}\cos\frac{2\pi x}{a}\sin\frac{\pi x}{a}\cos\frac{2\pi y}{a}\sin\frac{\pi y}{a}\right)\right] dx\,dy = 0 \tag{19.44a}$$

$$\int_0^a\int_0^a \left[\frac{4C_1\pi^4}{a^4}\sin\frac{\pi x}{a}\sin\frac{2\pi x}{a}\sin\frac{\pi y}{a}\sin\frac{2\pi y}{a} + \frac{64C_2\pi^4}{a^4}\right.$$

$$\sin^2\frac{2\pi x}{a}\sin^2\frac{2\pi y}{a} + \frac{2N_{xy}}{D}\left(\frac{C_1\pi^2}{a^2}\cos\frac{\pi x}{a}\sin\frac{2\pi x}{a}\cos\frac{\pi y}{a}\sin\frac{2\pi y}{a}\right.$$

$$\left.\left.+\frac{4C_2\pi^2}{a^2}\cos\frac{2\pi x}{a}\sin\frac{2\pi x}{a}\cos\frac{2\pi y}{a}\sin\frac{2\pi y}{a}\right)\right] dx\,dy = 0 \tag{19.44b}$$

The following definite integrals are to be noted

$$\int_0^a \sin^2\frac{m\pi x}{a}dx = \frac{a}{2} \tag{19.45a}$$

$$\int_0^a \sin\frac{m\pi x}{a}\sin\frac{n\pi y}{a}dx = 0 \quad \text{for } m \neq n \tag{19.45b}$$

$$\int_0^a \cos\frac{m\pi x}{a}\sin\frac{n\pi x}{a}dx = 0 \quad \text{for } m \neq n \tag{19.45c}$$

$$\int_0^a \cos\frac{2\pi x}{a}\sin\frac{\pi x}{a}dx = -\frac{2a}{3\pi} \tag{19.45d}$$

$$\int_0^a \sin\frac{2\pi x}{a}\cos\frac{\pi x}{a}dx = \frac{4a}{3\pi} \tag{19.45e}$$

Using the above relations, eqns. (19.44a) and (19.44b) reduce to

$$\left.\begin{aligned} \frac{4C_1\pi^4}{a^4}\left(\frac{a}{2}\right)^2 + \frac{2N_{xy}}{D}\frac{4C_2\pi^2}{a^2}\left(-\frac{2a^2}{3\pi}\right) = 0 \\ \frac{64C_2\pi^4}{a^4}\left(\frac{a}{2}\right)^2 + \frac{2N_{xy}}{D}\frac{C_1\pi^2}{a^2}\left(\frac{4a^2}{3\pi}\right) = 0 \end{aligned}\right\} \tag{19.46}$$

or,

$$\left.\begin{aligned} \frac{\pi^4}{a^2}C_1 + \frac{32N_{xy}}{9D}C_2 = 0 \\ \frac{32N_{xy}}{9D}C_1 + \frac{16\pi^4}{a^2}C_2 = 0 \end{aligned}\right\} \tag{19.47}$$

Non-trivial solution of homogeneous equation (19.47) is given by

$$\begin{vmatrix} \dfrac{\pi^4}{a^2} & \dfrac{32N_{xy}}{9D} \\ \dfrac{32N_{xy}}{9D} & \dfrac{16\pi^4}{a^2} \end{vmatrix} = 0 \tag{19.48}$$

The determinant yields the critical load of inplane shear and is given by

$$\left(N_{xy}\right)_{cr} = 11.1\frac{\pi^2}{a^2}D \tag{19.49}$$

19.5 GALERKIN FINITE ELEMENT METHOD

The differential equation of a uniform bar under axial load is

$$EA\,\sigma_{,xx} + q = 0 \tag{19.50}$$

For an axially loaded bar

$$\sigma_x = E\,\varepsilon_x = E\,u_{,x} \tag{19.51}$$

Combining eqns. (19.50) and (19.51) yields

$$EA\,u_{,xx} + q = 0 \tag{19.52}$$

At the free end of the bar, $F = 0$. At the other end where an axial force is applied, the non-essential boundary condition is given by

$$EA\,u_{,x} - F = 0 \tag{19.53}$$

Essential boundary conditions deal with the prescribed values of u.

Let the bar be divided into 'n' elements with each element is of length L. For each element, the displacement function is given by

$$\bar{u} = [N]\{X\}_e \tag{19.54a}$$

$$[N] = \begin{bmatrix} \dfrac{L-x}{L} & \dfrac{x}{L} \end{bmatrix}$$

where

$$\{X\}_e^T = \{u_1 \quad u_2\}$$

Weights W_i in Galerkin method are

$$W_i = \frac{\partial \bar{u}}{\partial u_i} = N_i \tag{19.55}$$

$$\left.\begin{aligned} N_1 &= \frac{L-x}{L} \\ N_2 &= \frac{x}{L} \end{aligned}\right\} \tag{19.56}$$

Galerkin's residual equation becomes

$$\sum_{j=1}^{nl} \int_0^L N_i \left(EA\bar{u}_{,xx} + q \right) dx = 0 \tag{19.57}$$

Equation (19.54) is to be repeated for all shape functions.

As the total number of shape functions are equal to the total number of degrees of freedom, total number of Galerkin residual equations will be equal to the total d.o.f. Integration eqn. (19.47) for the ith element, yields

$$\int_0^L N_i\, EA\bar{u}_{,xx}\, dx = \left[N_i\, EA\bar{u}_{,x} \right]_o^L - \int_0^L N_{i,x}\, EA\bar{u}_{,x}\, dx \tag{19.58}$$

Using nonessential boundary condition of eqn. (19.43) in eqn. (19.48), eqn. (19.47) becomes

$$\sum_{j=1}^{nl} \int_0^L \left(-N_{i,x}\, EA\bar{u}_{,x} + N_i q \right) dx + \sum_{j=1}^{nl} \left[N_i\, F_i \right]_o^L = 0 \tag{19.59}$$

From eqn. (19.54a) and (19.54b) gives

$$\bar{u}_{,x} = \begin{bmatrix} -\dfrac{1}{L} & \dfrac{1}{L} \end{bmatrix} \begin{Bmatrix} u_1 \\ u_2 \end{Bmatrix} = [B]\{X\}_e \tag{19.60}$$

Combining eqn. (19.59) and (19.60) gives

$$\sum_{j=1}^{nl} \int_0^L [B]^T [EA][B]\, dx\, \{X\}_e = \sum_{j=1}^{nl} \int_0^L [N]^T q\, dx + \sum_{j=1}^{nl} \left[[N]^T F \right]_0^L \tag{19.61}$$

Except for the last term, eqn. (19.61) reduces to the standard formula

$$[K]\{X\} = \{P\} \tag{19.62}$$

The last term of eqn. (19.51) needs some explanation.

The shape function matrix for a typical element can be written at its ends $x = 0$ and $x = L$ as

$$[N]_0 = [1 \quad 0] \text{ and } [N]_L = [0 \quad 1]$$

For two adjacent elements (j–1) and j as shown in Fig. 19.4, last summation in eqn. (19.6) produces the terms

$$\begin{array}{llll} & -\begin{Bmatrix}1\\0\end{Bmatrix}F_{j-1} & & \\ \text{node } a & & & \\ \text{node } b & & +(\begin{Bmatrix}0\\1\end{Bmatrix}F_{j-1} & \\ & & -\begin{Bmatrix}1\\0\end{Bmatrix}F_j & \\ \text{node } c & & & +\begin{Bmatrix}0\\1\end{Bmatrix}F_j \end{array} \qquad (19.64)$$

On assembly, the resultant axial force at node b is $(F_{j-1} - F_j)$. This is identified with the externally applied load.

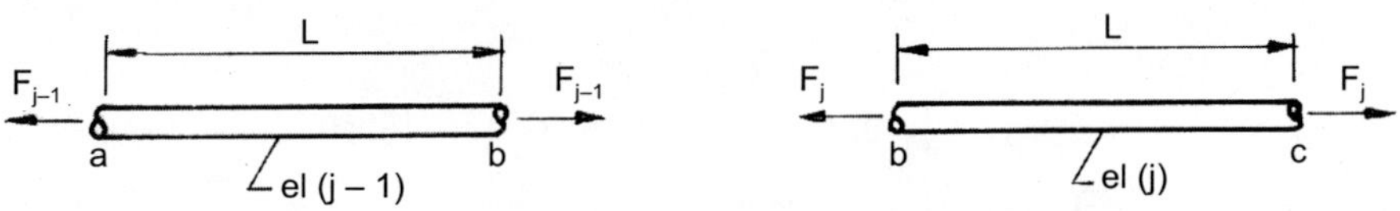

Fig. 19.4. Bar elements with nodal loads

19.6 TORSIONAL STIFFNESS OF PRISMATIC BEAM ELEMENT

Total torsion acting on a thin-walled beam is split into two parts.

$$T(x) = T_x(x) + T_\omega(x) \qquad (19.65)$$

where T_x is the St. Venant torsion and T_ω is the warping moment.

$$T_x = GJ\,\theta'(x) \qquad (19.66)$$

$$T_w = -EI_w\,\theta'''(x) \qquad (19.67)$$

E and G are Young's modulus and shear modulus of elasticity respectively. J is St. Venant's torsional constant, I_w is warping torsional stiffness for determination of which one may refer to Refs. 19.2 and 19.3.

Kawai [19.4] has developed a suitable beam element to take care of torsion. He represented angle of twist in the following form for a 2-noded beam element

$$\theta(x) = \alpha_1 + \alpha_2 x + \alpha_3 x^2 + \alpha_4 x^3 \qquad (19.68)$$

or, $$\theta(x) = \begin{bmatrix} 1 & x & x^2 & x^3 \end{bmatrix} \begin{Bmatrix} \alpha_1 \\ \alpha_2 \\ \alpha_3 \\ \alpha_4 \end{Bmatrix} \tag{19.69}$$

Equation (19.69) can be represented as

$$\theta = [Q]\{\alpha\} \tag{19.70}$$

where

$$[Q] = [1 \quad x \quad x^2 \quad x^3]$$
$$\{\alpha\}^T = \{\alpha_1 \quad \alpha_2 \quad \alpha_3 \quad \alpha_4\} \tag{19.71}$$

Each node of the beam element has two degrees of freedom θ and $\theta'\left(=\dfrac{d\theta}{dx}\right)$. The nodal displacement vector is

$$\{X\}_e^T = \{\ \theta_1 \quad \theta'_1 \quad \theta_2 \quad \theta'_2\} \tag{19.72}$$

The corresponding force vector is written as

$$\{P\}_e^T = \{T_{x1} \quad T_{\omega 1} \quad T_{x2} \quad T_{\omega 2}\} \tag{19.73}$$

The angle of twist can be expressed in terms of nodal displacements following the steps mentioned in previous chapters

$$\theta = [Q]\,[A]^{-1}\,\{X\}_e \tag{19.74}$$

where $$[A]^{-1} = \begin{bmatrix} 1 & 0 & 0 & 0 \\ 0 & 1 & 0 & 0 \\ -\dfrac{3}{L^2} & -\dfrac{2}{L} & \dfrac{3}{L^2} & -\dfrac{1}{L} \\ \dfrac{2}{L^3} & \dfrac{1}{L^2} & -\dfrac{2}{L^3} & \dfrac{1}{L^2} \end{bmatrix} \tag{19.75}$$

Based on eqn. (19.74),

$$\theta' = [B]_1\,\{X\}_e \tag{19.76}$$

where $$[B]_1 = \left[\frac{-6}{L}\left(\xi-\xi^2\right) \ \ (1-4\xi+3\xi^2) \ \ \frac{6}{L}\left(\xi-\xi^2\right) \ \ (-2\xi+3\xi^2)\right] \tag{19.77}$$

with $$\xi = \frac{x}{L}$$

Similarly,

$$\theta''' = [B]_2\,\{X\}_e \tag{19.78}$$

where $$[B]_2 = \frac{6}{L^2}\left[\frac{2}{L} \quad 1 \quad -\frac{2}{L} \quad 1\right] \tag{19.79}$$

From eqn. (19.66) and substituting θ' from eqn. (19.76) in it

$$T_x = J\,[B]_1\,\{X\}_e \tag{19.80}$$

Similarly, from eqn. (19.67) and (19.78), we get

$$T_\omega = -EI_\omega [B]_2 \{X\}_e \tag{19.81}$$

Equations (19.80) and (19.81) reveal that St. Venant's torsion T_x varies parabolically within the element and the warping torsion T_ω is constant along the element.

The element stiffness matrix consists of two parts

$$[k]_e = [k]_j + [k]_\omega \tag{19.82}$$

where $[k]_j$ is the St. Venant's torsional stiffness matrix and $[k]_\omega$ is the warping torsional stiffness matrix.

$$[k]_j = \int_0^L [B]_1^T GJ [B]_1 \, dx \tag{19.83}$$

$$[k]_\omega = \int_o^L [B]_1^T (-EI_\omega) [B]_2 \, dx \tag{19.84}$$

Explicit evaluation of eqn. (19.83) and (19.84), yields

$$[k]_j = \frac{GJ}{L} \begin{bmatrix} \frac{6}{5} & & \textit{symmetrical} & \\ \frac{L}{10} & \frac{2L^2}{15} & & \\ -\frac{6}{5} & \frac{L}{10} & \frac{6}{5} & \\ \frac{L}{10} & -\frac{L^2}{30} & \frac{-L}{10} & \frac{2L^2}{15} \end{bmatrix} \tag{19.85}$$

and

$$[k]_\omega = \frac{EI_\omega}{L^3} \begin{bmatrix} 12 & & \textit{symmetrical} & \\ 6L & 4L^2 & & \\ -12 & -6L & 12 & \\ 6L & 2L^2 & -6L & 4L^2 \end{bmatrix} \tag{19.86}$$

19.7 TORSION OF NON-CIRCULAR SECTIONS

The differential equation in terms of stresses for a bar subjected to torsion is [19.5, 19.6]

$$\frac{\partial_{T_{xz}}}{\partial y} - \frac{\partial_{T_{yz}}}{\partial x} = -2G\theta \tag{19.87}$$

where θ is the angle of twist per unit length.

The stresses are expressed in terms of stress functions as

$$\tau_{xz} = G\theta \frac{\partial \psi}{\partial y} \quad \text{and} \quad \tau_{yz} = -G\theta \frac{\partial \psi}{\partial x} \tag{19.88}$$

where ψ is the stress function.

Combining eqns. (19.87) and (19.88) results

$$\nabla^2\psi = -2 \tag{19.89}$$

$$\psi = 0 \text{ at the boundary}$$

Torque carried by the section is given by

$$T = 2G\theta \iint_A \psi \, dA \tag{19.90}$$

The functional of the boundary value problem is

$$\pi_p = \frac{G\theta^2}{2} \iint_A \left[\left(\frac{d\psi}{dx} \right)^2 + \left(\frac{d\psi}{dy} \right)^2 - 4\psi \right] dx\, dy \tag{19.91}$$

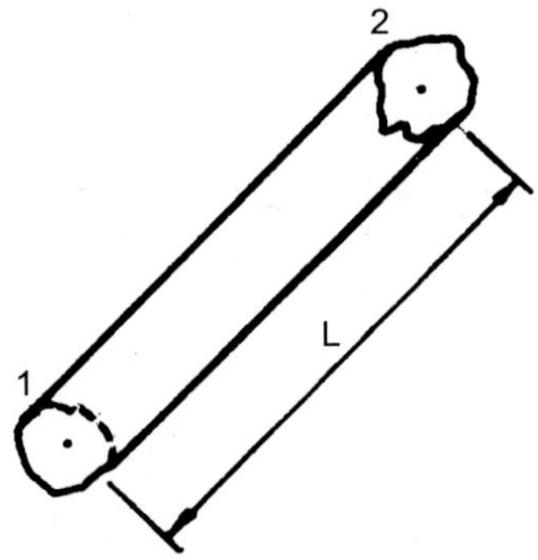

Fig. 19.5. Torsion element

Let us assume that the section consists of triangular elements. A typical element is shown in Fig. 19.5. ψ for the element is expressed as

$$\psi = N_1\psi_1 + N_2\psi_2 + N_3\psi_3 \tag{19.92}$$

Substituting from eqn. (19.92) into eqn. (19.91), yields

$$\pi_p = \frac{G\theta^2}{2} \iint_A \left[\left\{ \sum_{L=1}^{3} \frac{\partial N_i}{\partial x} \psi_i \right\}^2 + \left\{ \sum_{L=1}^{3} \frac{\partial N_i}{\partial y} \psi_i \right\}^2 - 4\left\{ \sum_{L=1}^{3} N_i \psi_i \right\} \right] dx\, dy \tag{19.93}$$

Applying the principle of stationary potential

$$\frac{\partial \pi_p}{\partial \psi_j} = \frac{G\theta^2}{2} \iint_A \left[2\left(\sum_{L=1}^{3} \frac{\partial N_i}{\partial x} \psi_i \right) \frac{\partial N_j}{\partial x} \right.$$

$$\left. +2\left(\sum_{i=1}^{3} \frac{\partial N_j}{\partial x} \psi_i \right) \frac{\partial N_j}{\partial y} - 4N_j \right] dx\, dy = 0 \tag{19.94}$$

$$j = 1, 2, 3$$

We can write eqn. (19.94) in matrix notations as

$$\iint_A \left(\left[\frac{\partial N_i}{\partial x} \cdot \frac{\partial N_j}{\partial x} + \frac{\partial N_i}{\partial y} \cdot \frac{\partial N_j}{\partial y} \right] \{ \psi_i \} - 2\{ N_j \} \right) dx\, dy = 0 \tag{19.95}$$

Equation (19.95) can be written as

$$[k]_e \{\psi\} = \{P\}_e \tag{19.96}$$

where
$$k_{ij} = \iint_A \left(\frac{\partial N_i}{\partial x} . \frac{\partial N_j}{\partial x} + \frac{\partial N_i}{\partial y} . \frac{\partial N_j}{\partial y} \right) dx\, dy \tag{19.97}$$

$$\{P\}_e = 2 \iint_A [\, N \,]\, dx\, dy \tag{19.98}$$

The explicit evaluation of element stiffness matrix is identical to that presented in Section 9.2.

The values of shape function mentioned in eqn. (19.92) are given in terms of nodal coordinates as

$$\left.\begin{aligned} N_1 &= \frac{a_1 + b_1 x + c_1 y}{2\Delta} \\ N_2 &= \frac{a_2 + b_2 x + c_2 y}{2\Delta} \\ N_3 &= \frac{a_3 + b_3 x + c_3 y}{2\Delta} \end{aligned}\right\} \tag{19.99}$$

where

$$\begin{aligned} a_1 &= x_2 y_3 - y_2 x_3 & b_1 &= y_2 - y_3 & c_1 &= x_3 - x_2 \\ a_2 &= x_3 y_1 - y_3 x_1 & b_2 &= y_3 - y_1 & c_1 &= x_1 - x_3 \\ a_3 &= x_1 y_2 - y_1 x_2 & b_3 &= y_1 - y_2 & c_3 &= x_2 - x_1 \end{aligned}$$

$$2\Delta = \det \begin{bmatrix} 1 & x_1 & y_1 \\ 1 & x_2 & y_2 \\ 1 & x_3 & y_3 \end{bmatrix}$$

$$= 2 \times \text{area of the triangle 123}$$

Therefore, k_{ij} of eqn. (19.97) is given by

$$k_{ij} = \frac{1}{4\Delta}(b_i b_j + c_i c_j) \tag{19.100}$$

The stiffness matrix becomes

$$[k]_e = \frac{1}{4\Delta} \begin{bmatrix} b_1 b_1 + c_1 c_1 & b_1 b_2 + c_1 c_2 & b_1 b_3 + c_1 c_3 \\ b_2 b_1 + c_2 c_1 & b_2 b_2 + c_2 c_2 & b_2 b_3 + c_2 c_3 \\ b_3 b_1 + c_3 c_1 & b_3 b_2 + c_3 c_2 & b_3 b_3 + c_3 c_3 \end{bmatrix} \tag{19.101}$$

Similarly, it can be shown that

$$\{P\}_e^T = \frac{2\Delta}{3} \{1 \quad 1 \quad 1\} \tag{19.102}$$

19.8 AXI-SYMMETRICAL ELEMENT

Many practical problems can be encountered involving axi-symmetric solids (such as solids of revolution) subjected to axially symmetric loading. Due to the symmetry of the problem, the stress components are independent of angular coordinate θ.

The mathematical problem is similar to that of plane stress and plane strain. By virtue of symmetry, the state of strain and therefore the complete state of stress is fully defined by two components of displacements in any plane along its axis of symmetry. For all stress and strain components all derivatives with respect to θ will vanish. As such v, $\gamma_{r\theta}$, $\gamma_{\theta z}$, $\tau_{r\theta}$ and $\tau_{\theta z}$ are zero [Fig. 19.6]. The radial and axial coordinates are denoted as r and z with u and w being the corresponding displacement. The nonzero stress components then are σ_r, σ_θ, σ_z and τ_{rz}.

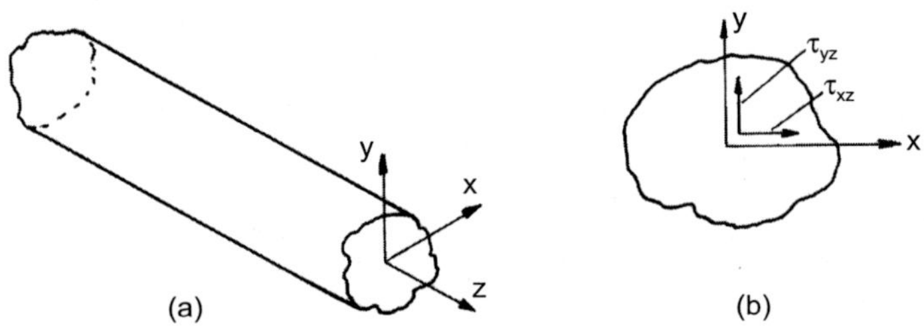

Fig. 19.6. Bar subjected to torsion

In plane stress or plain strain components, as either the stress or strain in the direction normal to the plane is zero, the strain energy of the plate is associated with three strain components. A major difference in the treatment creeps in the case of axi-symmetric solids. Any displacement in the radial direction will induce in strain in the circumferential direction. The circumferential stress in not zero, the work done by it in the strain energy expression should be considered.

The strain-displacement relations for the non-zero strain then are

$$\varepsilon_r = \frac{\partial u}{\partial r}, \quad \varepsilon_\theta = \frac{u}{r}, \quad \varepsilon_z = \frac{\partial w}{\partial z}, \quad \gamma_{rz} = \frac{\partial u}{\partial z} + \frac{\partial w}{\partial r} \tag{19.103}$$

The constitutive relation for the isotropic case is given by

$$\begin{Bmatrix} \sigma_r \\ \sigma_z \\ \sigma_\theta \\ \tau_{rz} \end{Bmatrix} = \frac{E}{(1+\nu)(1-2\nu)} \begin{bmatrix} 1-\nu & \nu & \nu & 0 \\ & 1-\nu & \nu & 0 \\ & & 1-\nu & 0 \\ & \textit{symmetrical} & & \dfrac{1-2\nu}{2} \end{bmatrix} \begin{Bmatrix} \varepsilon_r \\ \varepsilon_z \\ \varepsilon_\theta \\ \gamma_{rz} \end{Bmatrix} \tag{19.104}$$

The axi-symmetric ring element has triangular cross-section with nodes i, j and m (Fig. 19.7). Each node has two displacements u and w. The algebra involved in the determination of shape function is identical to the CST element.

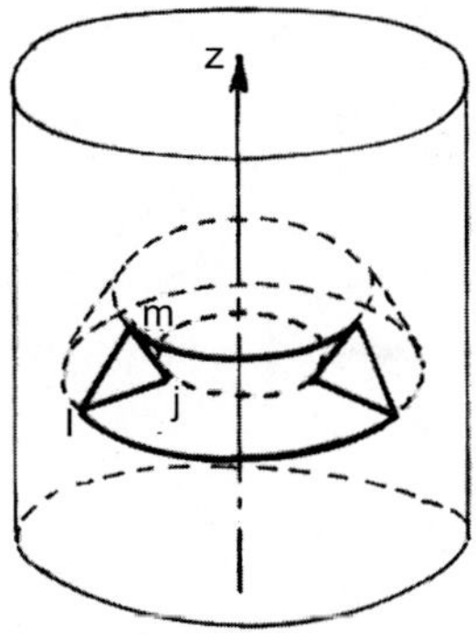

Fig. 19.7. Axi-symmetric solid

The displacement field is therefore given by

$$\{f\} = \begin{Bmatrix} u \\ w \end{Bmatrix} = \begin{bmatrix} IN_1 & IN_2 & IN_3 \end{bmatrix} \{X\}_e \tag{19.105}$$

where

$$N_i = \frac{a_i + b_i r + c_i z}{2\Delta} \tag{19.106}$$

$$\{X\}_e^T = \{u_1 \quad w_1 \quad u_2 \quad w_2 \quad u_3 \quad w_3\}$$

a_i, b_i *etc.* are explained in Chapter 9.

The strain-displacement relationship is given by

$$\{\varepsilon\} = \begin{Bmatrix} \varepsilon_r \\ \varepsilon_z \\ \varepsilon_\theta \\ \gamma_{rz} \end{Bmatrix} = \begin{Bmatrix} \dfrac{\partial u}{\partial r} \\ \dfrac{u}{r} \\ \dfrac{\partial w}{\partial z} \\ \dfrac{\partial u}{\partial z} + \dfrac{\partial w}{\partial r} \end{Bmatrix} = \frac{1}{2\Delta} \begin{bmatrix} b_1 & 0 & b_2 & 0 & b_3 & 0 \\ \dfrac{N_1}{r} & 0 & \dfrac{N_2}{r} & 0 & \dfrac{N_3}{r} & 0 \\ 0 & c_1 & 0 & c_2 & 0 & c_3 \\ c_1 & b_1 & c_2 & b_2 & c_3 & b_3 \end{bmatrix} \tag{19.107}$$

or,
$$\{\varepsilon\} = [B]\,\{X\}_e \tag{19.108}$$

It may be noted in eqn. (19.107) that strains are no longer constant within the element as in CST, but are functions of r and z.

Strain-strain relationship is given in eqn. (19.104)

$$\{\sigma\} = [D]\,\{\varepsilon\} \tag{19.109}$$

In the derivation of the stiffness matrix, the volume integral has to be taken over the whole ring of material.

$$dV = 2\pi r dA = 2\pi r\, dr\, dz \tag{19.110}$$

The stiffness matrix of axi-symmetric element then becomes

$$[k]_e = 2\pi \int\int [B]^T [D] [B] r \, dr \, dz \tag{19.110}$$

The evaluation of the element stiffness matrix requires the integration to be carried out. Numerical integration may be performed.

Zienkiewicz [19.7, 19.8] suggested a simple procedure to evaluate r and z at the centroidal point, that is,

$$\left.\begin{aligned} r &= \frac{r_i + r_j + r_m}{3} \\ z &= \frac{z_i + z_j + z_m}{3} \end{aligned}\right\} \tag{19.111}$$

The above values of r and z are substituted into $[B]$ matrix.

19.9 THREE-DIMENSIONAL ELEMENTS

Shape functions are given below for the solid elements of serendipity family.

19.9.1 Linear Element (8 Nodes) (Fig. 19.8a)

$$N_r = \frac{1}{8}(1+\xi_o)(1+\eta_o)(1+\zeta_o) \tag{19.112}$$

$$\xi_o = \xi\xi_r, \; \eta_o = \eta\eta_r, \; \zeta_o = \zeta\zeta_r$$

19.9.2 Quadratic Element (20 Nodes) (Fig. 19.8b)

Corner nodes

$$N_r = \frac{1}{8}(1+\xi_o)(1+\eta_o)(1+\zeta_o)(\xi_o + \eta_o + \zeta_o - 2) \tag{19.113a}$$

Typical mid-side node

$$\xi_r = \pm 1, \; \eta_r = 0, \; \zeta_r = \pm 1$$

$$N_r = \frac{1}{4}(1-\eta^2)(1+\xi_o)(1+\zeta_o) \tag{19.113b}$$

19.9.3 Cubic Element (32 Nodes) (Fig. 19.8c)

Corner node

$$N_r = \frac{1}{64}(1+\xi_o)(1+\eta_o)(1+\zeta_o)\left[9(\xi^2 + \eta^2 + \zeta^2) - 19\right] \tag{19.114a}$$

$$\xi_r = \pm 1, \; \eta_r = \pm\frac{1}{3}, \; \zeta_r = \pm 1$$

$$N_r = \frac{9}{64}(1-\eta^2)(1+9\eta_o)(1+\xi_o)(1+\zeta_o) \tag{19.114b}$$

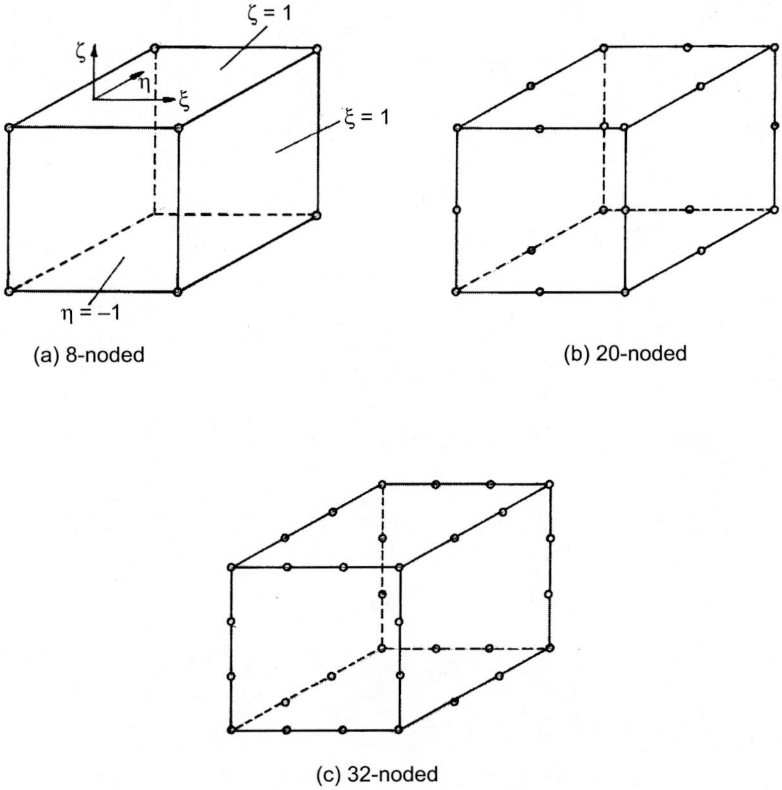

Fig. 19.8. Isoparametric brick element

19.9.4 16 Noded Solid (Fig. 19.9a)

Corner nodes

$$N_r = \frac{1}{8}(1+\xi_o)(1+\eta_o)(1+\zeta_o)(\xi_o+\eta_o-1) \tag{19.115a}$$

Typical mid-side nodes (2, 6, 10, 14)

$$N_r = \frac{1}{4}(1-\xi^2)(1+\eta_o)(1+\zeta_o) \tag{19.115b}$$

Typical mid-side node (4, 8, 12, 16)

$$N_r = \frac{1}{4}(1-\eta^2)(1+\xi_o)(1+\zeta_o) \tag{19.115c}$$

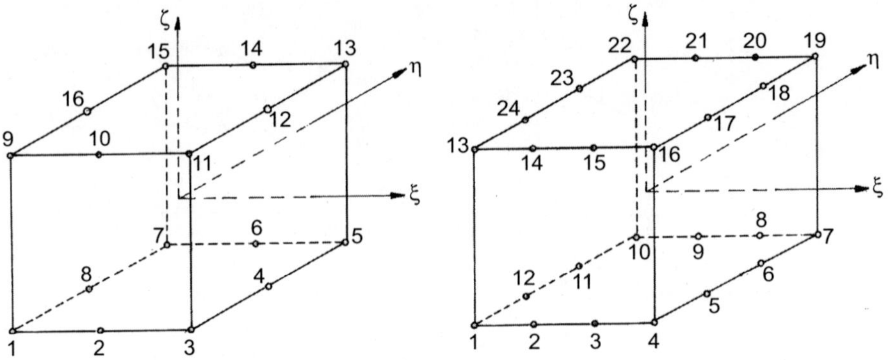

Fig. 19.9. Solid element

19.9.5 24 Noded Solid (Fig. 19.9b)

Corner nodes

$$N_r = \frac{1}{64}(1+\xi_o)(1+\eta_o)\left[-10+9\left(\xi^2+\eta^2\right)\right](1+\zeta_o) \quad (19.116a)$$

Mid-side nodes (5, 6, 11, 12, 17,18, 23, 24)

$$N_r = \frac{9}{64}(1+\xi_o)\left(1-\eta^2\right)(1+9\eta_o)(1+\zeta_o) \quad (19.116b)$$

Mid-side nodes (2, 3, 8, 9, 14, 15, 20, 21)

$$N_r = \frac{9}{64}(1+\eta_o)\left(1-\xi^2\right)(1+9\zeta_o)(1+\zeta_o) \quad (19.116c)$$

REFERENCES AND SUGGESTED READINGS

19.1. R. D. Cook, D. S. Malkus and M. E. Plesha, Concepts and Applications of Finite Element Analysis, 3rd Edition, John Wiley & Sons, 1989.

19.2. V. Fidyosev, Strength of Materials, Mir Publishers, Moscow, Russia, 1968.

19.3. L. R. Herrman, Elastic Torsional Analysis of Irregular Shapes, Journal of Engineering Mechanics Division, ASCE, EM6, December 1965, pp. 11–19.

19.4. T. Kawai, The Application to the Finite Element Method to Ship Structures, Computers & Structures, v. 3, 1973, pp. 1175–1194.

19.5. S. Timoshenko and J. N. Goodier, Theory of Elasticity, 2nd Edition, McGraw-Hill, 1951.

19.6. I. H. Shames and C. L. Dym, Energy and Finite Element Methods in Structural Mechanics, McGraw-Hill Book Co., 1985.

19.7. O. C. Zienkiewicz, R. Lewis and E. Stagg (Editors), Numerical Methods in Offshore Engineering, John Wiley & Sons, 1977.

19.8. O. C. Zienkiewicz, and R. L. Taylor, The Finite Element Method, v. 1, 4th Edition, McGraw-Hill, 1989.

❑❑❑

APPENDIX I

FIXED-END FORCES

Table A.1. Values of Fixed End Forces of Prismatic Beams for Various Types F Load Conditions

Loadings	Fixed-end Forces
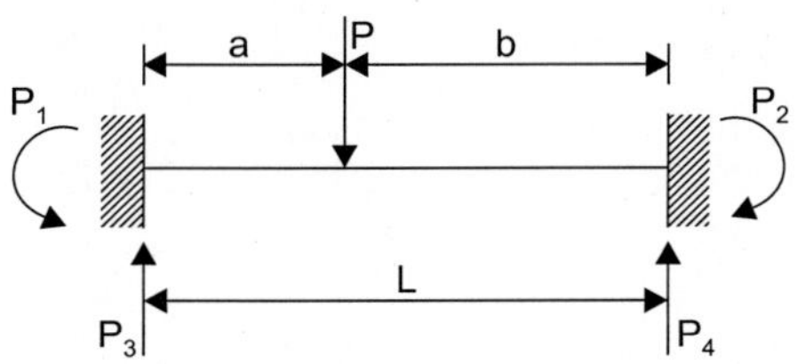	$P_1 = \frac{Pab^2}{L^2}, P_2 = \frac{Pa^2b}{L^2}$ $P_3 = P\left(\frac{b}{L} + \frac{ab^2}{L^3} - \frac{a^2b}{L^3}\right)$ $P_4 = P\left(\frac{a}{L} + \frac{a^2b}{L^3} - \frac{ab^2}{L^3}\right)$
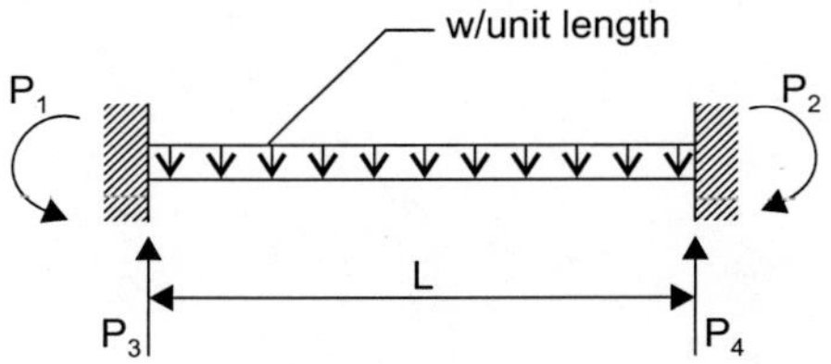	$P_1 = P_2 = \frac{wL^2}{12}$ $P_3 = P_4 = \frac{wL}{2}$
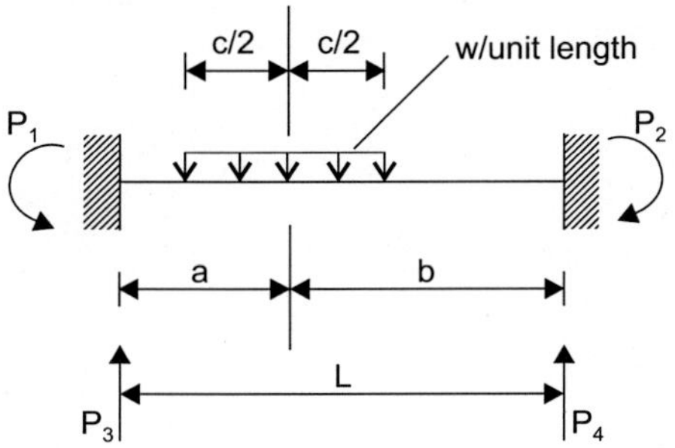	$P_1 = \frac{wC}{12L^2}\left[C^2(L-3a)+12a^2b\right]$ $P_2 = \frac{wC}{12L^2}\left[C^2(L-3b)+12ab^2\right]$ $P_3 = \frac{wCb}{12L^2} + \frac{P_1 - P_2}{L}$ $P_4 = \frac{wCa}{12L^2} + \frac{P_2 - P_1}{L}$

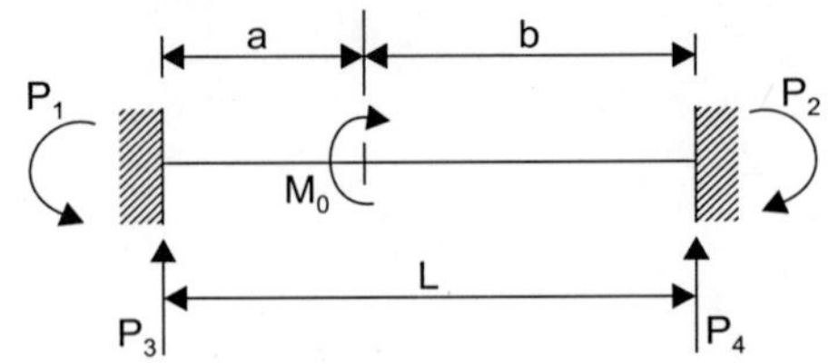

$$P_1 = -\frac{M_0 b}{L}\left(2 - \frac{3b}{L}\right)$$

$$P_2 = \frac{M_0 a}{L}\left(2 - \frac{3a}{L}\right)$$

$$P_3 = -P_4 = 6\frac{M_0 ab}{L^3}$$

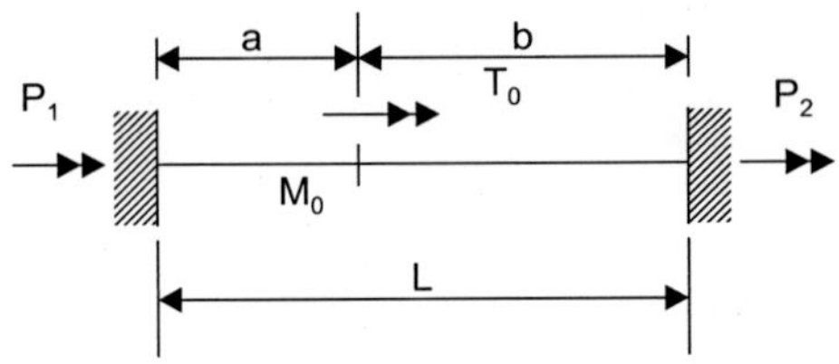

$$P_1 = -\frac{T_0 b}{L}$$

$$P_2 = -\frac{T_0 a}{L}$$

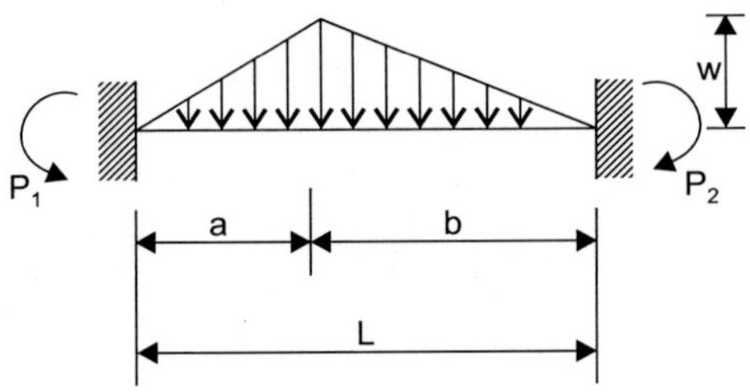

$$P_1 = \frac{wL^2}{12} - \frac{wa^2}{60}\left[10 - 10\frac{a}{L} + 3\left(\frac{a}{L}\right)^2\right] - \frac{wb^2}{60}\left[\frac{b}{L} + 3\left(\frac{b}{L}\right)^2\right]$$

$$P_2 = \frac{wb^2}{12} - \frac{wa^2}{60}\left[5\frac{a}{L} - 3\left(\frac{a}{L}\right)^2\right] - \frac{wa^2}{60}\left[10 - 10\frac{b}{L} + 3\left(\frac{b}{L}\right)^2\right]$$

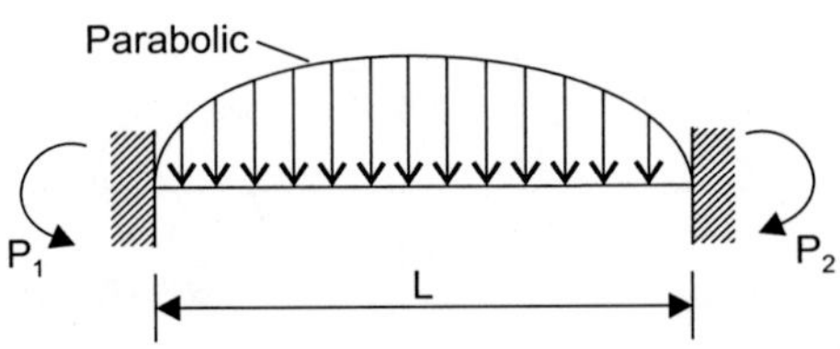

$$P_1 = P_2 = \frac{wL^2}{15}$$

❑❑❑

APPENDIX II

All the source files are given in the attached CD. The computer programs are arranged under two Folders. The folder FORTRAN contains all the computer programs written in FORTRAN and those under the folder C are in C. The followings are the list of FORTRAN files included:

1. DIR_TR.F: Direct stiffness method for the analysis of plane truss.
2. DIR_FR.F: Direct stiffness method for the analysis of plane frame.
3. DIR_GR.F: Direct stiffness method for the analysis of grid.
4. SUB_ST.F: Substructure analysis of plane truss.
5. CST_PS.F: Plane stress analysis using constant strain triangle.
6. ISO_PS.F: Plane stress analysis using isoparametric element.
7. ISO_PL_BEND.F: Plate bending analysis using isoparametric bending element.
8. SHELL.F: Analysis of shells using shallow shell triangular element.
9. SPLINE.F: Spline finite strip analysis of plate bending.
10. NONLN_FEA.F: Geometric non-linear finite element analysis of plates.

The following computer programs are given in C under the folder C:

1. dir_tr.c: Direct stiffness method for the analysis of plane truss.
2. dir_fr.c: Direct stiffness method for the analysis of plane frame.
3. dir_gr.c: Direct stiffness method for the analysis of grid.
4. sub_st.c: Substructure analysis of plane truss.
5. cst_ps.c: Plane stress analysis using constant strain triangle.
6. iso_ps.c: Plane stress analysis using isoparametric element.
7. iso_pl_bend.c: Plate bending analysis using isoparametric bending element.
8. shell.c: Analysis of shells using shallow shell triangular element.
9. spline.c: Spline finite strip analysis of plate bending.
10. nonln_fea.c: Geometric non-linear finite element analysis of plates.

❑❑❑

SUBJECT INDEX

A

Adaptive finite element 156, 345, 353
Adaptive mesh generation 368
Advancing front technique 357, 358, 360
Area coordinate 245
Assembly of element 178
Axi-symmetric element 420

B

Background mesh 358, 359, 361
Banded matrix 60
Beam function 292, 293, 299
Body forces 13, 139, 144
Boundary conditions 13, 16, 90
Buckling 3, 163, 164

C

Castigliano's theorems 16
Characteristics function 292
Compatibility conditions 4, 403
Complementary energy 15-17, 157
Concentric stiffened 392
Consistent load 275, 406
Consistent mass matrix 314, 316, 317
Constant strain triangle 165, 288, 317
Continuum 155, 156, 158
Convergence criteria 172, 175, 181

D

Damping matrix 315
Delaunay Triangulation 357, 358, 368
Determination of displacements 18
DKT element 240, 241, 261
Domain decomposition 357, 358, 368

E

Eccentric stiffener 397
Eigenvalue 291, 315, 322
Equilibrium conditions 4, 7, 15
Equilibrium matrix 127, 128
Error estimation 345, 351, 367
Errors in the finite difference 331

F

Finite difference method 160, 329, 332
Finite element 3, 13, 18
Finite strip 160, 161, 163
Flat shell 261, 262, 267
Free vibration 163, 313, 322
Frontal solution 107, 122

G

Galerkin finite element 411
Galerkin method 160, 408, 409
Gauss quadrature 192, 193
Geometric stiffness matrix 319, 322, 325

H

h-adaptivity 353
Hermitian polynomial 231, 233
Higher order element 158
Horizontally curved beam 251, 252
Hybrid element 157

I

Interpolation function 183, 216, 233
Isoparametric element 165, 183-185, 187
Isoparametric general shell element 276

J

Jacobian 188, 189, 191

K

Kinematic indeterminacy 9, 10, 60

L

Lack of fit 49, 50
Lagrangian Interpolation 195, 196
Loads 2-4, 13, 19
Lumped mass matrix 319

M

Mesh 155, 156, 161

N

Natural coordinates 195, 216
Newton-Cotes formula 191, 192
Non-linear problems 369
Numerical integration 159, 189, 191

O

Orthogonality relationship 312, 404
Orthotropic plates 148, 204

P

p-adaptivity 353
Pascal triangle 172
Patch test 247-250
Plane strain 142, 143, 165
Plane stress 141-143, 153, 157
Plate bending 153, 157, 158
Potential energy 15, 16, 157
Prestrain 49, 50
Principle of contragredience 126, 127
Principle of virtual work 13-15, 182, 220

R

Rayleigh-Ritz finite element 405
Rayleigh-Ritz method 160, 403, 405
Rectangular element 144, 165, 173
Reduced integration 237-239, 260
Rotation Matrix 73, 83, 275
Rotational restraint 153

S

Semi-analytical methods 160
Settlement of supports 49, 52
Shape functions 184-186, 195, 197

Shear deformation 56, 57, 101
Shells 1, 3, 139
Shells of revolution 265
Skew element 227, 232
Skew support 90
Smoothed stress 239, 256, 354
Space frame 8, 9, 81
Spline finite strip 302, 309-311, 425
Statical indeterminacy 6-7, 19, 28
Stiffened plate 165, 391, 392
Stiffener element 393, 394
Structural analysis 2-5, 10, 13
Substructure technique 107, 108, 110
Superconvergent patch recovery 345, 346, 354
Surface force 139, 157

T

Three-dimensional element 261, 420
Transfinite interpolation 357
Transformation matrix 74, 76, 78
Triangular element 162, 165, 172

V

Vertically curved beam 283
Vibration 157, 160, 163
Virtual work 13-15, 74, 75

W

Weighted residual method 407

Z

zz-error estimator 354, 355
zz-refinement 355

❑❑❑